Springer Series in

MATERIALS SCIENCE 80

Springer Series in

MATERIALS SCIENCE

Editors: R. Hull R. M. Osgood, Jr. J. Parisi H. Warlimont

The Springer Series in Materials Science covers the complete spectrum of materials physics, including fundamental principles, physical properties, materials theory and design. Recognizing the increasing importance of materials science in future device technologies, the book titles in this series reflect the state-of-the-art in understanding and controlling the structure and properties of all important classes of materials.

71 **Dissipative Phenomena in Condensed Matter**
Some Applications
By S. Dattagupta and S. Puri

72 **Predictive Simulation of Semiconductor Processing**
Status and Challenges
Editors: J. Dabrowski and E.R. Weber

73 **SiC Power Materials**
Devices and Applications
Editor: Z.C. Feng

74 **Plastic Deformation in Nanocrystalline Materials**
By M.Yu. Gutkin and I.A. Ovid'ko

75 **Wafer Bonding**
Applications and Technology
Editors: M. Alexe and U. Gösele

76 **Spirally Anisotropic Composites**
By G.E. Freger, V.N. Kestelman, and D.G. Freger

77 **Impurities Confined in Quantum Structures**
By P.O. Holtz and Q.X. Zhao

78 **Macromolecular Nanostructured Materials**
Editors: N. Ueyama and A. Harada

79 **Magnetism and Structure in Functional Materials**
Editors: A. Planes, L. Mañosa, and A. Saxena

80 **Micro- and Macro-Properties of Solids**
Thermal, Mechanical and Dielectric Properties
By D.B. Sirdeshmukh, L. Sirdeshmukh, and K.G. Subhadra

81 **Metallopolymer Nanocomposites**
By A.D. Pomogailo and V.N. Kestelman

82 **Plastics for Corrosion Inhibition**
By V.A. Goldade, L.S. Pinchuk, A.V. Makarevich and V.N. Kestelman

83 **Spectroscopic Properties of Rare Earths in Optical Materials**
Editors: G. Liu and B. Jacquier

84 **Hartree–Fock–Slater Method for Materials Science**
The DV–X Alpha Method for Design and Characterization of Materials
Editors: H. Adachi, T. Mukoyama, and J. Kawai

85 **Lifetime Spectroscopy**
A Method of Defect Characterization in Silicon for Photovoltaic Applications
By S. Rein

86 **Wide-Gap Chalcopyrites**
Editors: S. Siebentritt and U. Rau

87 **Micro- and Nanostructured Glasses**
By D. Hülsenberg and A. Harnisch

88 **Introduction to Wave Scattering, Localization and Mesoscopic Phenomena**
By P. Sheng

Volumes 20–70 are listed at the end of the book.

D.B. Sirdeshmukh
L. Sirdeshmukh
K.G. Subhadra

Micro- and Macro-Properties of Solids

Thermal, Mechanical and Dielectric Properties

With 209 Figures and 153 Tables

Springer

Professor D.B. Sirdeshmukh
Professor L. Sirdeshmukh
Professor K.G. Subhadra
Kakatiya University
Physics Department
Warangal 506 009, India
E-mail: dsirdeshmukh@yahoo.com

Series Editors:

Professor Robert Hull
University of Virginia
Dept. of Materials Science and Engineering
Thornton Hall
Charlottesville, VA 22903-2442, USA

Professor R. M. Osgood, Jr.
Microelectronics Science Laboratory
Department of Electrical Engineering
Columbia University
Seeley W. Mudd Building
New York, NY 10027, USA

Professor Jürgen Parisi
Universität Oldenburg, Fachbereich Physik
Abt. Energie- und Halbleiterforschung
Carl-von-Ossietzky-Strasse 9–11
26129 Oldenburg, Germany

Professor Hans Warlimont
Institut für Festkörper-
und Werkstofforschung,
Helmholtzstrasse 20
01069 Dresden, Germany

ISSN 0933-033X

ISBN-10 3-540-31785-6 Springer Berlin Heidelberg New York

ISBN-13 978-3-540-31785-2 Springer Berlin Heidelberg New York

Library of Congress Control Number: 2006928433

Springer is a part of Springer Science+Business Media.
springer.com

Typesetting: Data prepared by the Authors and by SPi using a Springer TEX macro package
Cover concept: eStudio Calamar Steinen
Cover production: *design & production* GmbH, Heidelberg

Printed on acid-free paper SPIN: 11554769 57/3100/SPi 5 4 3 2 1 0

To the memory of our mentors

Prof. R. Satyanarayan
Prof. V.T. Deshpande
Prof. P.G. Puranik
Prof. K. Venkata Ramaiah

Preface

This book deals with some micro and macroproperties of solids. Microproperties are perceived at the lattice level and are generally studied by diffraction or spectroscopic methods. The lattice constant, its temperature variation (measured through the property of thermal expansion), the amplitudes of atomic thermal vibrations (reflected in the Debye–Waller factor) and the colour centres are some examples of microproperties. In contrast, macroproperties are studied through measurements on material in bulk. Elastic properties, hardness, dielectric properties and melting temperature are examples of macro-properties.

The approach is to discuss in detail the physics of some select properties. Theoretical as well as experimental aspects are kept in view. The beginnings of studies of these properties can be traced to the earlier part of the last century. Due to the basic nature of these properties, there has been continuing research interest and constant refinement of the experimental methods. New levels of accuracy in measurement have made it possible to observe second-order changes like the effects of temperature, pressure, magnetic field, radiation, impurities and other defects; thin film and particle size effects have also been studied. Apart from bringing out the fundamental aspects, the book also provides considerable space for a discussion of current trends in research in the form of a comprehensive 'overview'. Typically, in each chapter, the earliest reference pertains to the period 1910–1930 and the latest to the period 1995–2005. The behaviour of a variety of materials like metals, alloys, ionic crystals, semiconductors, mixed valence compounds, optoelectronic materials and biomaterials is discussed vis-à-vis these properties.

Chapter 1 deals with the lattice constant which is a fundamental attribute of a crystal lattice. Various methods of accurate determination of lattice constants are discussed. The accuracy in lattice constant measurement has now reached almost the limit as it is of the same order as the accuracy in wavelength measurements. Effects of irradiation, impurities, deuteration of hydrogen-containing compounds and particle size are discussed. Chapter 2 is on thermal expansion of solids. A variety of experimental methods are described including some novel techniques which use holography and gamma ray absorption. The intimate relation between thermal expansion and anharmonicity of lattice vibrations is brought out and the important role of thermal expansion in throwing light on thermally generated defects is pointed out. Chapter 3 is devoted to the Debye–Waller factor which is related to the amplitudes of atomic thermal vibrations. Originally introduced as a correction for X-ray diffraction intensities, it has emerged as a powerful solid-state probe. It is related to the lattice dynamical models, the interatomic bond strength, the lattice strain and surface forces. The hardness of crystals forms the subject of Chap. 4. Starting with microhardness methods, recent techniques of ultra and nanohardness are discussed. A variety of aspects are included like temperature and pressure variation and effects of magnetic field, irradiation and chemical bonding. An interesting new observation is the study of phase transitions through micro-Raman spectroscopy of indentations obviating the use of a diamond anvil. Chapter 5 on the dielectric behaviour of materials starts with the basics of dielectrics. Experimental techniques for different frequency ranges are considered. It is shown that dielectric properties throw much light on such diverse aspects as anharmonicity, spectroscopic phenomena, defects and chemical bonding. Various conduction mechanisms are discussed including polaron conduction. The dielectric behaviour of organic compounds and biomaterials is considered along with inorganic compounds.

To complement the experimental approach, the theoretical approach to solid-state properties is developed in Chap. 6. The evaluation of thermal parameters like the Debye temperature and Gruneisen constants and also mechanical properties like elastic constants is discussed. A new method for the evaluation of the Raman mode Gruneisen parameter of fluorite type crystals from dielectric properties is included. A comprehensive treatment of mixed crystal physics is given in Chap. 7. The properties of mixed crystals are intermediate between those of the parent compounds. This creates the possibility of controlling the properties by controlling the mixed crystal composition. Several properties like hardness, dielectric constant, formation energy of defects, effective ionic charge, colour centre wavelengths, melting points, Debye–Waller factors and transition temperatures are discussed with reference to their composition dependence. Chapter 8 on the elastic properties of solids is mainly focused on the serious problem of discrepancies in elastic properties and some possible checks.

The choice of the properties included in this book was, to some extent, influenced by the research interests of the authors. Considerable amount of work was generated through the research programmes undertaken by the authors over several decades. Instead of including the results of our work in the 'overview', they are presented as a separate section entitled 'Some of our results' in each chapter. We had the privilege of having Prof. K.A. Gschneidner (Jr.) and Dr. B.J. Beaudry (Iowa State University), Prof. B.S. Shah (Saurashtra University) and Dr. B.R. Rao (Indian Institute of Chemical Technology) as collaborators. A large number of Ph.D. and M.Phil. students were also associated with our work. While reference has been made to them at appropriate places as co-authors, we would like to place on record our appreciation of their role in our research programmes.

We are thankful to Dr. K.S. Rajam and Dr. H.C. Barshilia (National Aerospace Laboratory, Bangalore) for sharing information on nanoindentation. Thanks are also due to Prof. K.G. Nickel (University of Tubingen) for providing material on micro-Raman spectroscopy of indentations. The chapter; Elastic Properties of Solids – 'A Critical Analysis' is a modified version of a recent Review Article: 'Consistency Checks on Elastic Properties of Solids' published by two of the authors (DBS, KGS) in the Journal of Materials Science. Grateful thanks are offered to colleagues at Kluwer Publishers and their successors, Springer, for kind permission to use substantial material from the review.

We would like to thank Prof. K.G. Bansigir (Jiwaji, University), Dr. Ch. V. Purushotham Reddy (Chaitanya Educational Institutes) and Mr. Vinod Kumar (South Asian Publishers) for much interest. Thanks are offered to colleagues at the Kakatiya University for their support and cooperation.

Finally, we are grateful to Dr. C. Ascheron of Springer-Verlag for timely suggestions and valuable advice during the various stages of the preparation of this book. The prompt response and patient guidance received from colleagues at the Hiedelberg office of Springer and at SPi, India are acknowledged.

Warangal, India
June 2006

Dinker Sirdeshmukh
Lalitha Sirdeshmukh
K.G. Subhadra

Acknowledgements

The following publishers are thanked for permission to reproduce illustrations and material from their publications:

1. American Institute of Physics
2. American Physical Society
3. Current Science Association
4. Elsevier Ltd.
5. Indian Academy of Sciences
6. Institute of Physics (UK)
7. International Union of Crystallography
8. Kluwer Publishers
9. Physical Society of Japan
10. Springer-Verlag
11. Wiley - VCH

Contents

1

Lattice Constant – A Solid State Probe

1.1 Introduction

The lattice constants are basic crystallographic parameters as they represent the dimensions of the unit cell. The number of lattice constants varies from 1 to 6 as we pass from the most symmetric (cubic) to the least symmetric (triclinic) crystal class. From the lattice constants, other crystal parameters and properties like molar volume, density and ionic radii can be estimated. The variations in lattice constants with temperature and pressure yield values of the thermal expansion coefficient and compressibility, respectively. The effects of thermally generated defects, doping and radiation induced defects can be followed through small, but significant, changes in the lattice constants. When determined with accuracy, the lattice constant provides rich information about the crystal and is a powerful solid state probe.

1.2 Experimental Methods

1.2.1 Principle

The determination of lattice constant is based on Bragg's law of X-ray diffraction illustrated in Fig. 1.1 and stated in (1.1).

$$n\lambda = 2d \sin\theta, \tag{1.1}$$

where, λ is the wavelength, d the interplanar spacing and θ the Bragg angle and n is the diffraction order parameter. A set of indices h, k, l, (called the Miller indices) is associated with each plane. Thus, (1.1) can be rewritten as

$$\sin^2\theta = \lambda^2/4d_{hkl}^2. \tag{1.2}$$

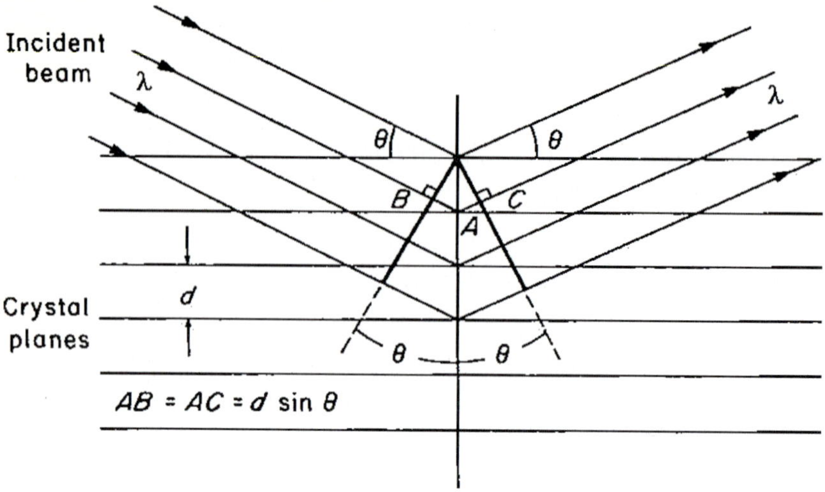

Fig. 1.1. Bragg's law

Expressing $d_{\rm hkl}$ in terms of the lattice constant 'a' and the Miller indices h, k, l, (1.2) takes the form

$$\sin^2\theta = (\lambda^2/4a^2)(h^2 + k^2 + l^2) \tag{1.3}$$

for a cubic crystal. The order parameter n is included in the Miller indices.

There are similar but more complicated equations for crystals of lower symmetry. Thus, in principle, the lattice constant can be determined with an experimental set-up that permits recording of X-ray diffraction reflections and the measurement of the associated Bragg angles.

1.2.2 Experimental Techniques

Several techniques are in use for recording X-ray diffraction patterns. For lattice constant determination, generally, powder methods are preferred. A comprehensive discussion of these methods is available in literature [1.1–1.4]. Some of the methods are discussed here.

The Debye–Scherrer Camera (DSC)

The geometry of this camera is shown in Fig. 1.2. Monochromatic X-rays enter through the collimator and fall on a needle-like powder specimen. The X-rays are diffracted both in the front and in the backward direction by different lattice planes. In the figure, lines with single arrow indicate incident and undiffracted rays and those with double arrows indicate diffracted rays. The undiffracted rays are absorbed in the exit port. The photographic film stays pressed against the inner surface of the cylindrical camera of radius R.

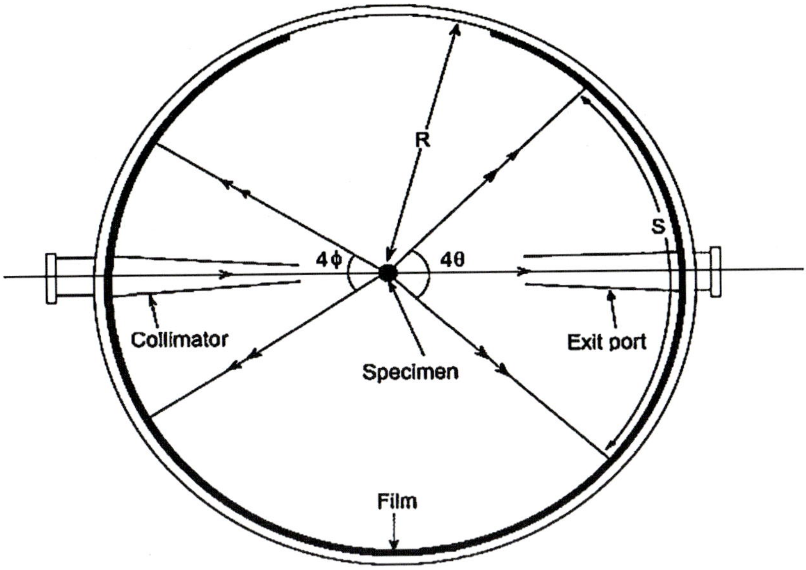

Fig. 1.2. Geometry of the Debye–Scherrer camera

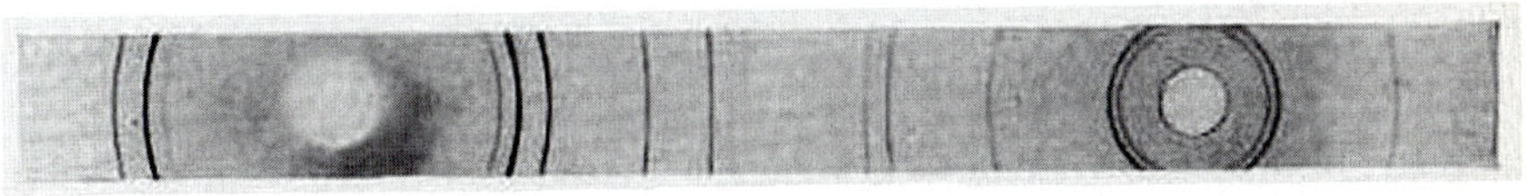

Fig. 1.3. Debye–Scherrer photograph of Al

A typical photograph taken with this camera is shown in Fig. 1.3. Denoting the diameter of a Bragg reflection by S, the Bragg angle is given by

$$\theta = (S/4R)(180/\pi)^\circ \quad (1.4)$$

for forward reflection and

$$\theta = (\pi/2) - \phi = [(\pi/2) - (S/4R)]\ (180/\pi)^\circ \quad (1.5)$$

for back reflection. Generally, the diameter of a DSC is 11.46 cm. The mounting of the film is called 'Straumanis mounting'. It facilitates recording of the ring systems in front reflection ($\theta < 90^\circ$) as well as back reflection ($\theta > 90^\circ$). Further, by measuring the distance between the centers of the two punched holes, the effective radius of the camera can be determined.

Unicam-Type Camera (UC)

A critical examination of the commercially available Unicam camera revealed that it has the following drawbacks:

1. It employs the Bradley–Jay mounting with two film strips. The center of the ring system is not included on either strip making the use of knife edges unavoidable.
2. The specimen centering device is not convenient to operate and is not easily accessible. In the original design, the specimen is suspended from above.
3. The platinum-wound heater is excellent but its spherical design makes any repair impossible and replacement is too expensive.

A Unicam-type cylindrical camera with 19 cm diameter has been designed and fabricated [1.5]. While the design is essentially similar to the Unicam model, some of the drawbacks in the original design have been overcome. As this camera was indigenously fabricated in the authors' laboratory, the design and operation are discussed in detail.

The principle of the cylindrical camera is shown in Fig. 1.4. This is similar to the Debye–Scherrer camera (Fig. 1.2) but with a difference in film mounting which will be discussed later. The film forms a cylinder at the axis of which the powder sample in the form of a thin cylindrical rod is located; for thoroughly random orientation of the crystallites, the specimen is rotated about its own axis. The incident rays enter through a collimator and the direct ray is stopped by a beam stopper. In a back reflection camera, the diffracted rays travel backward (with respect to the incident beam) along a cone of angle 4ϕ where $\phi = (\pi/2) - \theta$.

The essential parts of the camera are shown in Fig. 1.5. These are: the main base (A), the base plate (B), the specimen holder (C), the central brass

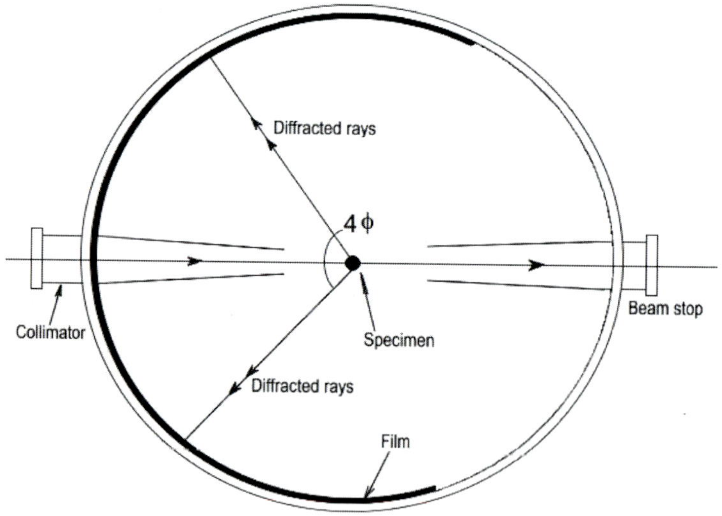

Fig. 1.4. The principle of the Unicam-type cylindrical camera

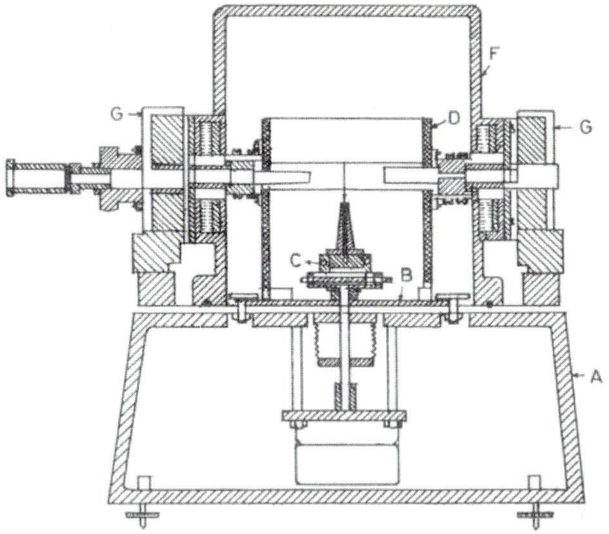

Fig. 1.5. Essential parts of the cylindrical camera

cylinder (D), the heater (E) (not shown in the figure), the evacuation chamber (F) and the film cassette (G).

As mentioned earlier, the specimen holder is different from that in the Unicam camera. Its design is shown in Fig. 1.6. The centering of the specimen is accomplished by the device shown in Fig. 1.6a (top view) and Fig. 1.6b (cross-section). The specimen is mounted in a cavity drilled in the top portion of a thin brass rod (A) which has a circular base (B). The base is rigidly fixed over a small horizontal table (C) which is fixed in a slot by means of a screw (D) with stoppers at the ends. To obtain transverse motion of table (C) another similar table (E) is fixed below (C). Table (E) is fixed with screw (F). The two screws (D) and (F) are at right angles to each other and enable the specimen to be moved linearly. A brass bush is fixed at the bottom of the lower table (E). The entire arrangement is mounted on a shaft with a provision to lock it on to the shaft or to release it. The sample filled in a thin capillary tube is fixed in the cavity of the brass rod (A) using dental cement. Centering of the specimen is checked by viewing it through a microscope fixed to the main base. The sample holder is released from the shaft and rotated to bring screw (D) to the right of the viewer. If the sample does not coincide with the vertical cross-wire in the microscope, screw (D) is manipulated suitably. The sample holder is then rotated through a right angle such that screw (F) is on the right and the specimen is again centred using screw (F). This process is repeated until no displacement is observed from the axis of rotation. The specimen holder is now locked on to the shaft. The other end of the shaft connects with

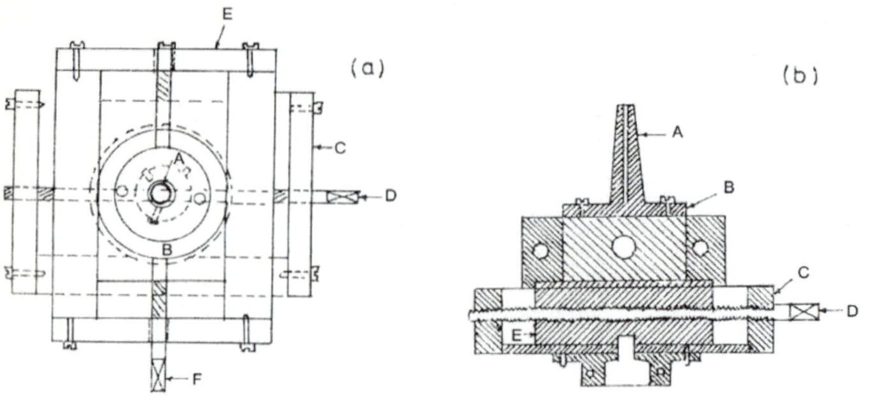

Fig. 1.6. (a) Top view, (b) cross-section of the specimen holder for the camera in Fig. 1.5

the spindle of a 1 rpm motor. When the film cassette is in position, the sample may be observed by its shadow on the fluorescent screen of the beam stop.

For determination of lattice constants at high temperature, a heating arrangement is necessary. The chief requirements of high temperature operation are the temperature stability over a period of 6–10 h and the measurement of specimen temperature. A suitable design is shown in Fig. 1.7. A hollow tube (A) of length 6 cm and diameter 1.5 cm with a slot (B) and a hole (C) at the center is brazed at its top to a circular disc (D). The slot is of width 10 mm extending around the tube over an angle of 220° and is at right angles to the axis of the tube. The hole permits the incident X-ray beam to strike the specimen and the slot allows the diffracted beam to pass out to the film. A refractory hollow tube (E) of thickness 1 cm with a similar slot and hole is compounded with the brass tube. The refractory tube snugly fits over the brass tube. The refractory tube has linear holes all round through which a coiled super-kanthal wire (G) is passed. A thin circular asbestos plate (H) with a central hole of diameter 2 cm is fixed on the lower side of the disc (D) to prevent heat losses due to conduction. The lead wires of the coils are brought out to the terminals through a small two-holed ceramic tube (I) fixed in a hole drilled in the brass disc. There is a hole at the center of the disc through which a brass bush (J) of length 2.7 cm is press-fitted into the brass tube. Two two-holed ceramic tubes (K, L) are rigidly fixed vertically in the holes drilled at the bottom of the bush. The lower ends of the ceramic tubes just enter into the gap provided for the passage of diffracted X-rays. Chromel–alumel thermocouple wires are passed through the ceramic tubes such that their junctions (M, N) lie very close to the specimen on either side of the portion of the sample exposed to X-rays.

The entire assembly is inserted in the central brass cylinder. A circular groove (F) is engraved in the disc (D) on the side facing the base of the

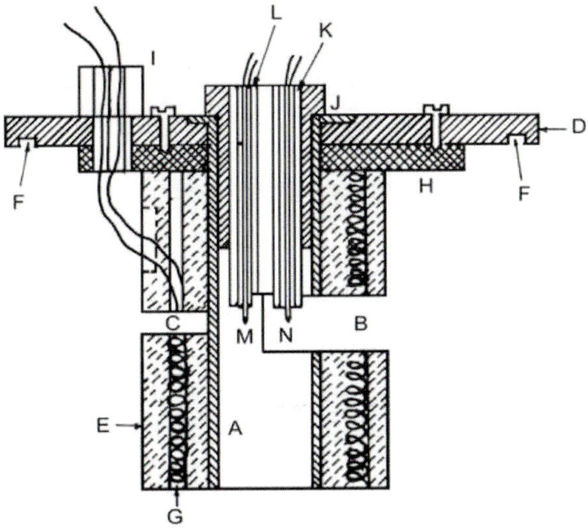

Fig. 1.7. Heating unit for the camera in Fig. 1.5

camera and fits on the top of the central brass cylinder. The disc can be rotated horizontally about the axis of rotation of the sample holder and can be fixed such that the incident X-ray beam after passing through the hole in the heater strikes the sample and diffracted X-rays reach the film arranged to receive backward reflection. The heater wire and thermocouple wires outside the heater are passed through ceramic beads. The free ends of the heater and thermocouple wires are connected to the terminals on the base plate of the camera. The thermocouples are connected to two separate millivoltmeters. The heater is fed from an AC supply through a stabilizer and a variac. A temperature of about 800°C can be attained within 20 min. The temperature stability is obtained by adjusting the output of the variac to give a constant voltage. The temperature distribution in the heater was investigated by positioning the thermocouple along the exposed length of the specimen. It was observed that the temperature remains fairly steady along the length of the specimen exposed to X-rays. The temperature recorded by the two thermocouples always agreed within 0.5°C. The constancy of temperature with time is excellent once the equilibrium is reached. The fluctuation in temperature during exposure was $< \pm 1$°C up to 400°C and $< \pm 2$°C at higher temperatures. The heater is so designed that it can be lifted in and out of the camera as a single unit during the sample mounting and alignment procedure. Also if the heater element is burnt out it can be replaced easily.

Unlike the original Unicam camera which employs Bradley–Jay film mounting, Van Arkel type of mounting is employed in this camera. With this mounting, both sides of the ring system are recorded and the ring diameters of back reflection lines can be measured without the need for knife-edge or any other calibration. A photograph of the camera in the assembled form is

Fig. 1.8. Photograph of the cylindrical camera

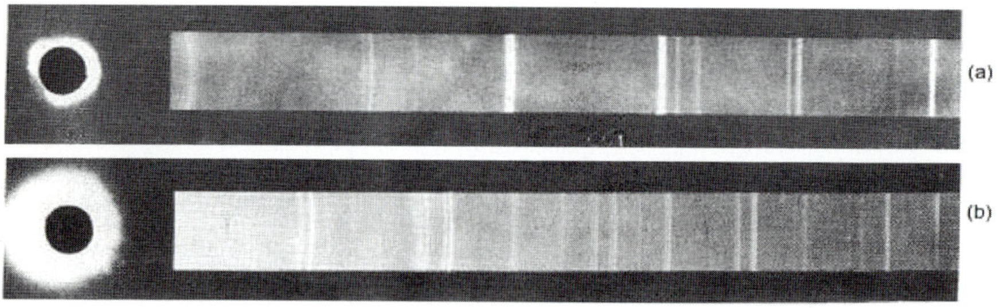

Fig. 1.9. Diffraction pattern for (a) MgO and (b) PbF_2 taken with the cylindrical camera

shown in Fig. 1.8. As will be discussed later, the error function suitable for this camera is $f(\theta) = (1/2)\ [(\cos^2\theta/\sin\theta) + (\cos^2\theta/\theta)]$.

As examples, the diffraction patterns of MgO and PbF_2 obtained with this camera are shown in Fig. 1.9. The extrapolation plots are shown in Fig. 1.10.

Symmetric Focusing Camera (SFC)

The geometry of a symmetric focusing camera is shown in Fig. 1.11. The basic principle of the design of a symmetric focusing camera is that if a divergent beam of X-rays falls on a sample spread over a cylindrical surface, the diffracted beam is focused provided the slit, the sample and the film lie on a circle called the focusing circle. When these conditions are satisfied, the reflections are sharp and intense. Further, since the effective film-to-specimen distance is large, compared to that in a Debye–Scherrer camera of the same radius, the resolution is larger.

A camera based on this principle designed by Sirdeshmukh and Deshpande [1.6] is described. The camera is 15 cm in diameter. From a carefully cut

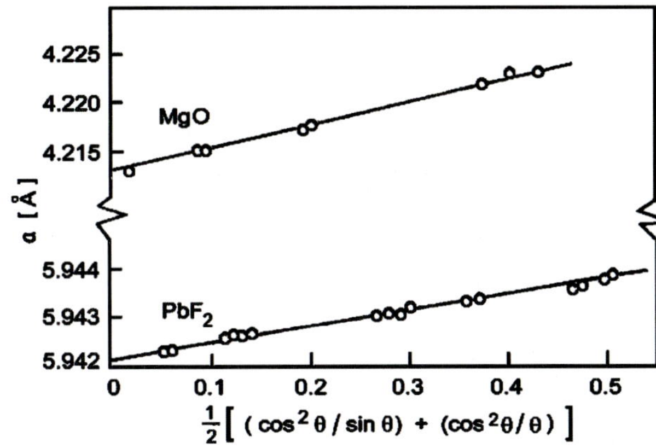

Fig. 1.10. Extrapolation plots for (a) MgO and (b) PbF_2

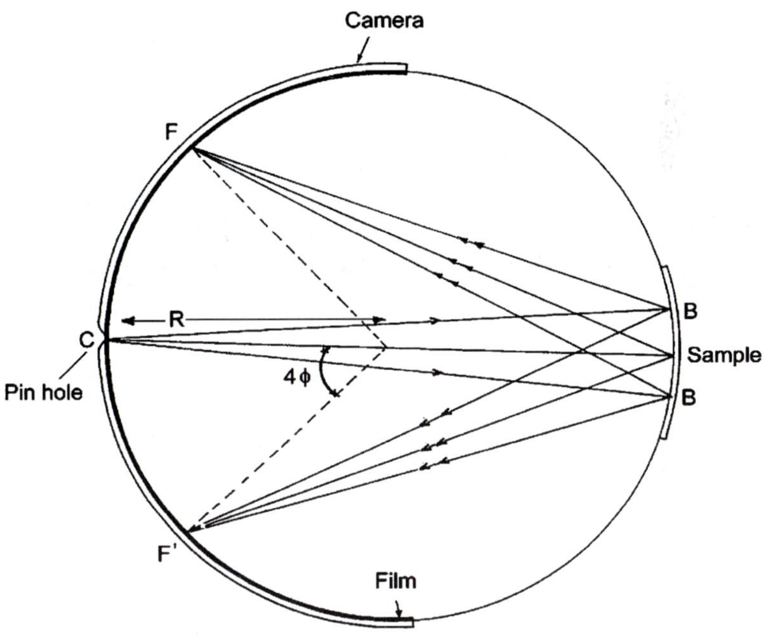

Fig. 1.11. Geometry of a symmetric focusing camera

cylinder, a minor segment is cut away. The major segment is the film-holder. It can record reflections with Bragg angles in the range of 65° onwards. The main parts of the camera are shown in Fig. 1.12. It has at its base a tripod (A) with leveling screws (B). The tripod carries a horizontal bench (C) which can rotate about a vertical axis. It can be locked onto the tripod by means of the screw (D). The horizontal bench carries an upright (E) which consists of a fixed piece and a sliding piece. The slider can be moved up and down

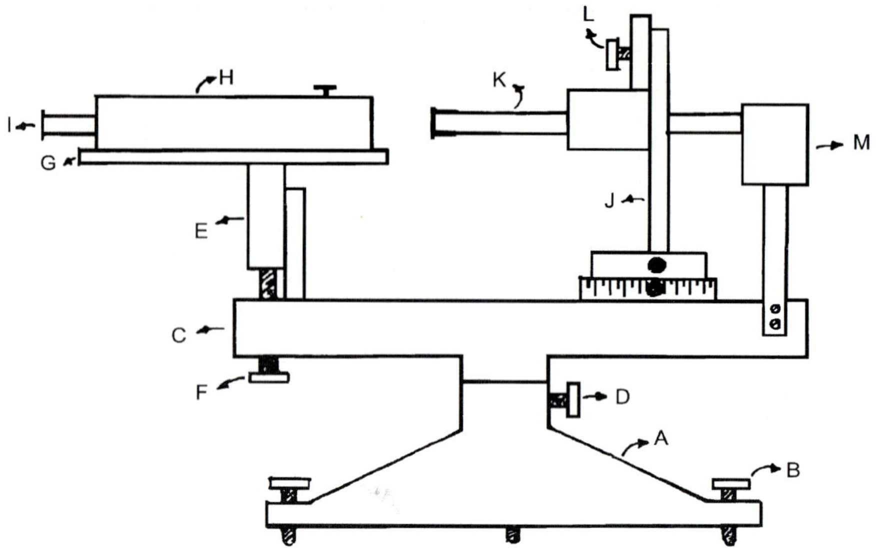

Fig. 1.12. Main parts of the symmetric focusing camera

Fig. 1.13. Diffraction pattern of aluminium using the symmetric focusing camera

by the micrometer screw (F). At the top of the sliding piece is a platform (G) which carries the film holding assembly (H) along with the collimator (I). The upright (J) is mounted over the horizontal bench. The upright can be displaced towards or away from the film-holder. There is also provision for lateral displacement of the upright. At its top the upright carries the sample holding tube (K) which can be displaced in the vertical direction by means of the screw (L). The tube is coupled to a motor (M) through an eccentric cam.

For convenience a flat sample is used instead of a sample with a cylindrical surface. The systematic errors in this camera have been analysed and it was found that the function ϕ tan ϕ is a suitable error function. Here $\phi = (\pi/2)-\theta$. A diffraction photograph of Al is shown in Fig. 1.13 and the extrapolation plot for Al is shown in Fig. 1.14. The extrapolated value of 4.0499(2) Å agrees well with the value 4.0499(1) Å obtained by Wilson [1.7].

For lattice constant measurement at higher temperatures, a tubular heater is introduced such that the sample surface is at the central region of the heater. The temperature is measured with a copper–constantan thermocouple welded at the back of the sample holder.

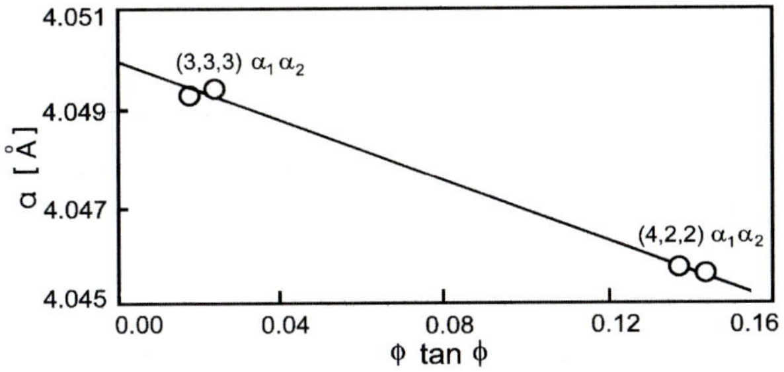

Fig. 1.14. Extrapolation plot for Al

Flat Film Camera (FFC)

The geometry of the flat film camera is shown in Fig. 1.15. As the name indicates, a flat film is employed. For accurate determination of lattice constants, the back-reflection geometry is used. Because of the limitation of size of the film, only a few reflections are recorded (Fig. 1.16). If S is the diameter of a ring and D the film-to-specimen distance, the angle θ is given by

$$\theta = (\pi/2) - \phi = [(\pi/2) - (1/2)\tan^{-1}(S/2D)]\ [180/\pi]^\circ . \qquad (1.6)$$

X-Ray Powder Diffractometer

The X-ray diffractometer is a versatile technique which facilitates at once the measurement of the Bragg angle of a reflection, its intensity and its profile.

The detection of the diffracted beam is done with a GM counter (almost obsolete now) or a proportional counter or a scintillation counter. The electronic circuitry for stability and detection is quite involved. Detailed discussion of these aspects is given in the texts mentioned at the beginning of this section. The specimen preparation is to be done with care (see Chap. 3). The geometry of the system shown in Fig. 1.17 and described later follows the treatment by Peiser et al. [1.1].

The point of divergence A is the line focus of the X-ray tube. BOB′ is the trace of the specimen. The detector C is carried on the arm pivoted at O; OA and OC are equal. The detector and the sample rotate maintaining a θ–2θ relationship so that ON, the normal to BOB′ always bisects ∠AOC. AOC is the focusing circle. The detector moves along the circle described about O. The focusing circle AOC (which has no physical reality) changes continuously as the sample and detector traverse. This is known as the Bragg–Brentano focusing system.

While the instrument uses a divergent beam, it is also necessary to limit the divergence. This is done with the help of special devices known as Soller slits. As the detector occupies some length, its movement restricts the angular region that can be explored – generally up to $\theta = 60^\circ$. For accuracy in lattice

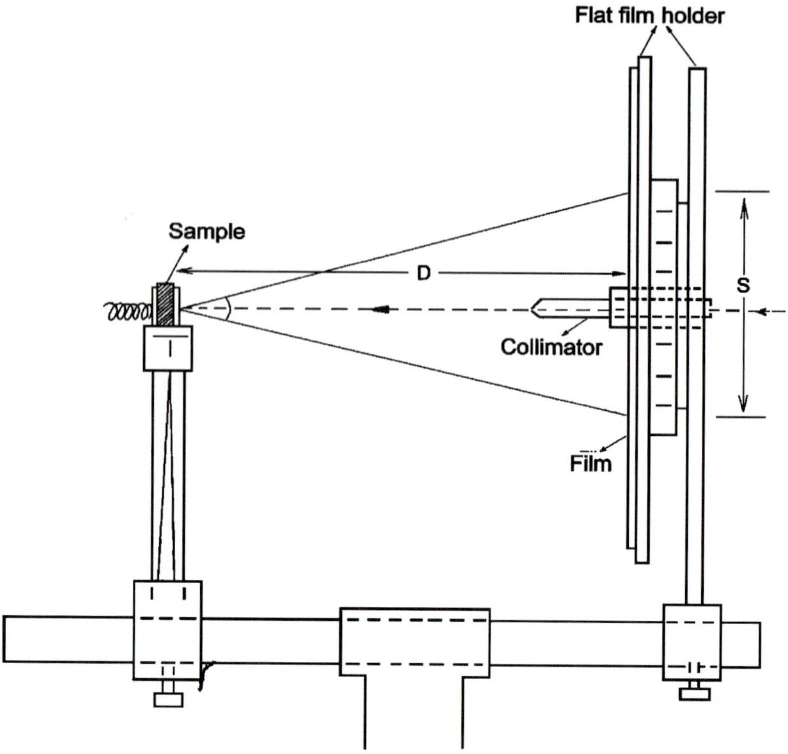

Fig. 1.15. Geometry of the flat film camera

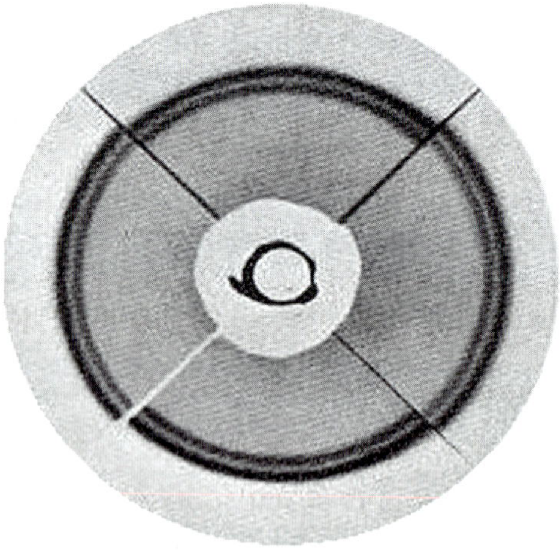

Fig. 1.16. Powder photograph of two silver gold alloys using a flat film camera

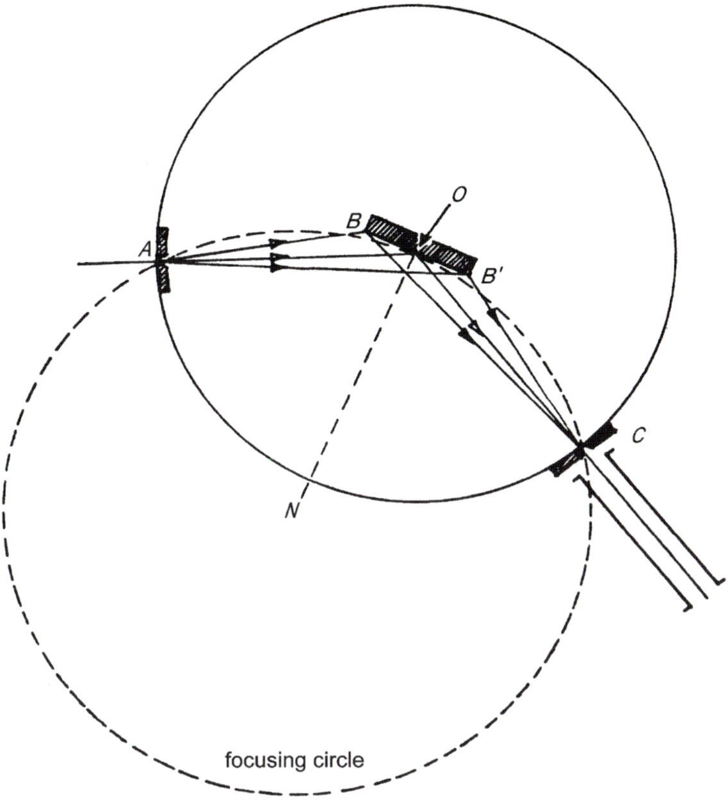

Fig. 1.17. Geometry of the X-ray powder diffractometer

constant determination, higher angles are preferred (see Sect. 1.2.3). However, because of the distances involved, the stringent geometry and the efficient detection, the XRD has a high resolution even at low angles. A typical XRD pattern is shown in Fig. 1.18.

Most of the commercial models of diffractometers are based on the principles discussed in the preceding paragraphs. However, in Bond's [1.8] diffractometer, a single crystal is employed and two wide-aperture detectors are placed in the symmetric back-reflection configuration (Fig. 1.19). The crystal is turned so that a reflection from the crystal is recorded by the two detectors in succession. A high precision is obtained by this method.

1.2.3 Strategies for Accuracy

Errors occur in the determination of lattice constants. These errors are of two types: systematic and random. Differentiation of the Bragg equation leads to

$$(\Delta d/d) = -(\cot\theta)\Delta\theta. \tag{1.7}$$

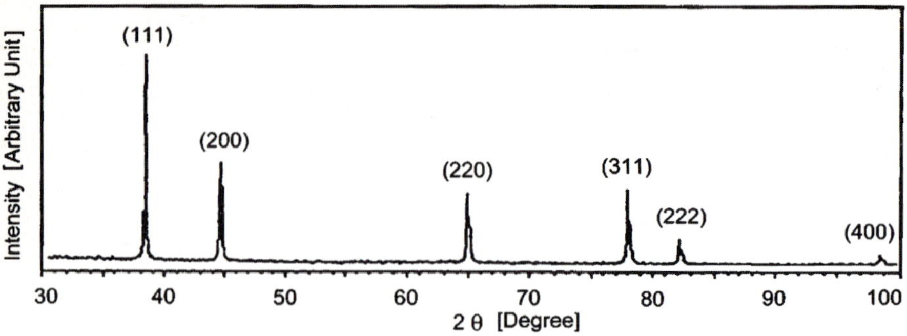

Fig. 1.18. Typical X-ray diffractogram of a powder sample (Ag)

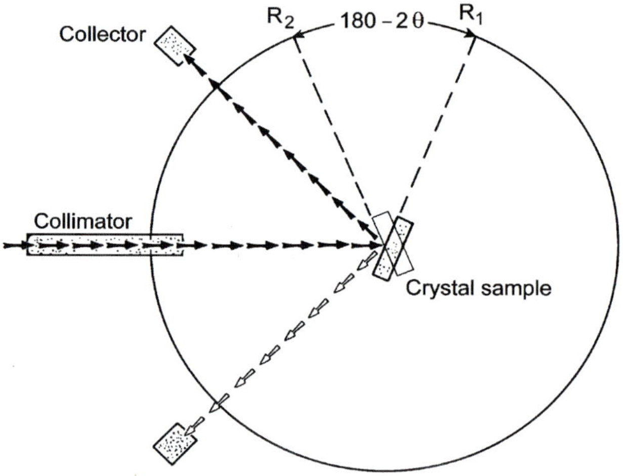

Fig. 1.19. Geometry of Bond's (single crystal) diffractometer

Thus an error $\Delta\theta$ in the measurement of the angle results in a large error $(\Delta d/d)$ at low angles and a smaller error at high angles. In fact $(\Delta d/d) \to 0$ as $\theta \to 90°$. Hence, the first strategy for accurate determination is to use high angle reflections, generally $\theta > 60°$.

But the effect of errors still remains to be attended. Systematic errors arise due to errors in geometrical parameters or physical effects. For instance, in a cylindrical camera, errors may arise due to the following causes:

1. Error in camera radius
2. Displacement of sample from camera center
3. Absorption of beam by sample
4. Divergence of beam

Table 1.1. Error functions for different techniques

	Technique	f(θ)	f(θ) at high θ
i	Cylindrical (DSC and UC) camera	$\frac{1}{2}\left[\frac{\cos^2\theta}{\theta}+\frac{\cos^2\theta}{\sin\theta}\right]$	$\cos^2\theta$
ii	Flat film camera (back reflection)	$\cos 2\phi - \cos^2 2\phi$	$\cos^2\theta$
iii	Symmetric focusing camera	$\phi\ \tan\phi$	$\cos^2\theta$
iv	Diffractometer	$\cos\theta\cot\theta$	$\cos^2\theta$

The errors have been worked out and it has been shown that $(\Delta d/d)$ is a function of the Bragg angle. Thus,

$$(\Delta d/d) = f(\theta), \tag{1.8}$$

where $f(\theta)$ is the error function. Error functions for various techniques are given in Table 1.1. It may be seen from the table that in each case $f(\theta) \to 0$ as $\theta \to 90°$. For a cubic crystal

$$\Delta d/d = \Delta a/a. \tag{1.9}$$

From (1.8) and (1.9), it follows that:

$$\Delta a = a_{\text{obs}} - a_{\text{true}} = a_{\text{true}}\ f(\theta). \tag{1.10}$$

Hence, a plot of a_{obs} obtained from each reflection against the corresponding $f(\theta)$ will be a straight line. When this line is extrapolated to $\theta = 90°$ the intercept gives a_{true}.

Random errors are minimized, first, by measuring θ repeatedly and taking the mean and, second, by drawing the extrapolation plot by least square analysis. Instead of a graphical extrapolation, Cohen [1.9] suggested error elimination by a least square processing of the data starting with the $\sin^2\theta$ values.

1.2.4 Present Level of Accuracy

A summary of the accuracy in lattice constant measurement possible with different methods is given in Table 1.2. It can be seen that an accuracy of one part in (1–5) $\times 10^4$ can be achieved with routine methods while a higher accuracy of one part in (1–2) $\times 10^5$ is possible with special methods. This is just an order less than the accuracy in X-ray wavelengths [one part in (5–10) $\times 10^5$]. Results of some studies where lattice constants have been measured with very high accuracy are given in Table 1.3.

1.3 An Overview

There is an enormous amount of information on the lattice constants of materials. This information is scattered in a vast number of papers published

Table 1.2. Accuracy of lattice parameter determination by various methods

Method	Accuracy in a
	1 part in
Rotating crystal camera	< 10,000
Weissenberg camera	< 10,000
Debye–Scherrer camera	15,000
Diffractometer (asymmetric)	15,000
Unicam camera	50,000
Focusing camera	50,000
Kossel line method	100,000
Bond's method	200,000
Uncertainty (in λ)	500,000

Table 1.3. Accurate values of lattice constants (a) of some crystals; method and accuracy given wherever mentioned in source

Crystal	Method	Temperature [°C]	a [Å]	Ref.
Al		25	4.04958	[1.10]
Au		25	4.07825	[1.10]
Cu		18	3.61496	[1.10]
NaCl		26	5.64056	[1.10]
NaBr		26	5.97324	[1.10]
KCl		25	6.29294	[1.10]
KI		25	7.06555	[1.10]
Diamond	FFC	25	3.56696(7)	[1.11]
W	DSC	25	3.16491(5)	[1.13]
Ge	Kossel	25	5.657736(8)	[1.14]
GaAs	Bond	25	5.653663(5)	[1.14]

in various journals. Among them, mention may be made of Acta Crystallographica, Proceedings of the Physical Society (London), Journal of the Physical Society of Japan, Journal of Applied Physics and Journal of Chemical Physics. Exhaustive compilations of data on lattice constants is available in [1.10, 1.15, 1.16]. In this section, some studies where the measurement of lattice constant has been used to probe various physical phenomena in solid state are discussed.

1.3.1 Characterisation of Semiconductor Materials

For semiconductor devices, material with highest purity and free from defects is required. The lattice parameter provides a useful control on device material. Silicon being the most commonly used semiconductor material, its lattice parameter has been determined with very high precision so much so that it is now used as a standard in X-ray diffraction instrumentation. Some results on the lattice parameter of silicon are given in Table 1.4. There is good internal consistency among the values of the lattice parameters obtained by various

Table 1.4. Lattice constant of silicon

Single crystal sample		Powder sample	
a [Å]	Ref.	a [Å]	Ref.
5.43108	[1.17]	5.43085	[1.12]
5.43106	[1.18]	5.43090	[1.21]
5.43107	[1.19]	5.43089	[1.22]
5.43107	[1.20]	5.43092	[1.20]

workers using single crystal samples. Similarly, there is agreement among the results obtained using powder samples But between the two sets of results (single crystal and powder), there is a consistent difference of 0.00015 Å, the value for the single crystal sample being larger. Parrish [1.12] suggested that this difference may be caused by the formation of a layer of SiO_2 formed in the process of grinding. A more plausible explanation is given by Hubbard et al. [1.20] in terms of crystal boundary effects.

Doping has significant effect on semiconductor behaviour. The addition of impurities alters the carrier concentration and hence the electrical behaviour. Tap et al. [1.23] and Gille and Schenk [1.24] studied the effect of doping PbTe with Bi and Tl ions on the lattice parameter. The addition of Bi ions reduces the lattice parameter. This is to be expected since the ionic radius of Bi^{3+} is smaller than that of Pb^{2+}. Although the radius of the Tl^{+} ions is larger than that of Pb^{2+}, the lattice parameter decreases on addition of Tl^{+} ions. This is explained on the basis of a difference in the binding between Tl and Te in relation to that between Pb and Te on the basis of complexes formed by Tl^{+} with the vacancies in PbTe.

1.3.2 Characterisation of Doped Crystals

Doping of crystals (addition of small but controlled quantities of impurities) has an effect on many physical properties of crystals. The effect is particularly prominent in optical, mechanical and electrical properties. Doping also has an effect on the lattice parameter and hence the lattice parameter can be used to characterize doped crystals.

Stott et al. [1.25] measured the lattice parameters of calcium fluoride doped with La, Tm and Y ions. The results are shown in Fig. 1.20. It is seen that the addition of rare earth ions results in an increase in the value of the lattice parameter. The per cent change in lattice constant for unit molar fraction of impurity has values 6, 6 and 13 for Y, Tm and La, respectively, and thus seems to depend on the radius of the rare earth ion. Another observation is that the value of the lattice parameter varies slightly depending on the section of the crystal boule from where the specimen is chosen. This indicates that the distribution of the impurity within the crystal is slightly inhomogeneous.

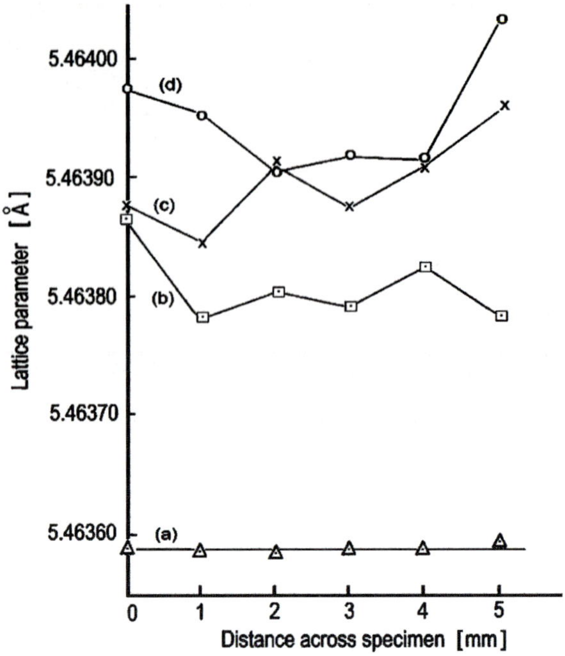

Fig. 1.20. Lattice parameters of: (**a**) undoped CaF_2; (**b**) CaF_2 containing 0.074 mol% TmF_3; (**c**) CaF_2 containing 0.088 mol% YF_3 and (**d**) CaF_2 containing 0.045 mol% LaF_3

1.3.3 Effect of Deuteration on Lattice Constants

The substitution of hydrogen by deuterium in hydrogen-containing substances is seen in lattice constant changes. Zimmerman [1.26] made measurements of lattice constant of crystals in the LiH_xD_{1-x} system. The lattice constants of LiH and LiD are 4.0831(4) and 4.0684(5) Å. The composition dependence of the partially substituted crystals was found to be linear.

Belouet et al. [1.27] studied the effect of deuteration on potassium dihydrogen phosphate (KDP) crystals. KDP is a tetragonal crystal. It can be seen from Fig. 1.21 that the 'a' parameter varies continuously with the deuteration parameter x. Further in the high x region, the measured values of 'a' show negative deviations from additivity. On the other hand, there is no systematic variation in the 'c' parameter due to deuteration.

1.3.4 Effect of Hydrogen on Lattice Parameters of Rare Earth Elements

The rare earth elements easily absorb hydrogen and this causes changes in the lattice parameter. A systematic study was made by Spedding and

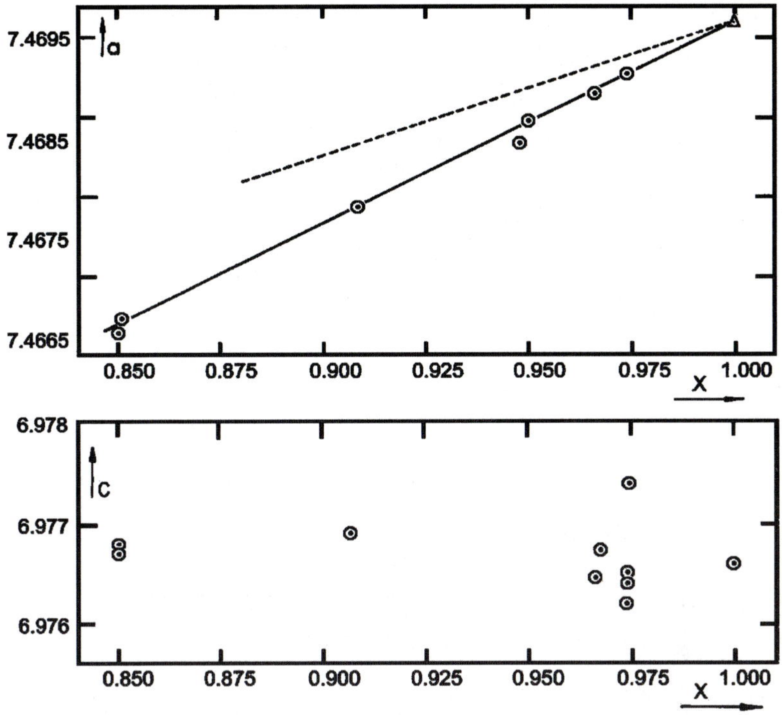

Fig. 1.21. Plot of lattice parameter a (Å) and lattice parameter c (Å) against deuteration parameter x of $KH_{2(1-x)}D_{2x}PO_4$ crystals. *Dashed line* represents additive variation of 'a'

Beaudry [1.28]. Very pure rare earth metal samples were heated in a hydrogen atmosphere until saturation and the lattice parameters were determined with the help of a Debye–Scherrer camera. The observed changes are shown in Table 1.5 and also in Fig. 1.22. It is seen that (i) the lattice parameters increase with hydrogen treatment, (ii) the increase in the 'c' parameter is more than that in the 'a' parameter and (iii) the increments in 'a' and 'c' vary from element to element.

1.3.5 Lattice Constants of Mixed Crystals

Mixed crystals are an important class of materials. Typical data on the composition variation of lattice parameters of the KI–RbI mixed crystal system taken from Van Den Bosch et al. [1.29] are shown in Fig. 1.23. The lattice parameters a_C of a mixed crystal system A_xB_{1-x} generally follow the equation:

$$a_C^n = xa_A^n + (1-x)a_B^n, \tag{1.11}$$

Table 1.5. Lattice parameters of the ABAB rare earth metals [1.28]

Rare earth metal	Lattice parameter		Increase in parameter due to hydrogen	
	a[Å]	c[Å]	a [Å]	c [Å]
Gadolinium	3.6336 ± 4	5.7810 ± 5	0.0003	0.0008
Terbium	3.6055 ± 4	5.6966 ± 6	0.0007	0.0001
Dysprosium	3.5915 ± 2	5.6501 ± 4	0.0013	0.0017
Holmium	3.5778 ± 2	5.6178 ± 3	0.0032	0.0052
Erbium	3.5592 ± 2	5.5850 ± 3	0.0052	0.0115
Thulium	3.5375 ± 4	5.5540 ± 2	0.0092	0.0216
Lutetium	3.5052 ± 4	5.5494 ± 5	0.0224	0.0598
Yttrium	3.6482 ± 2	5.7318 ± 6	0.0103	0.0383
Scandium	3.3088 ± 2	5.2680 ± 3	0.0371	0.0347

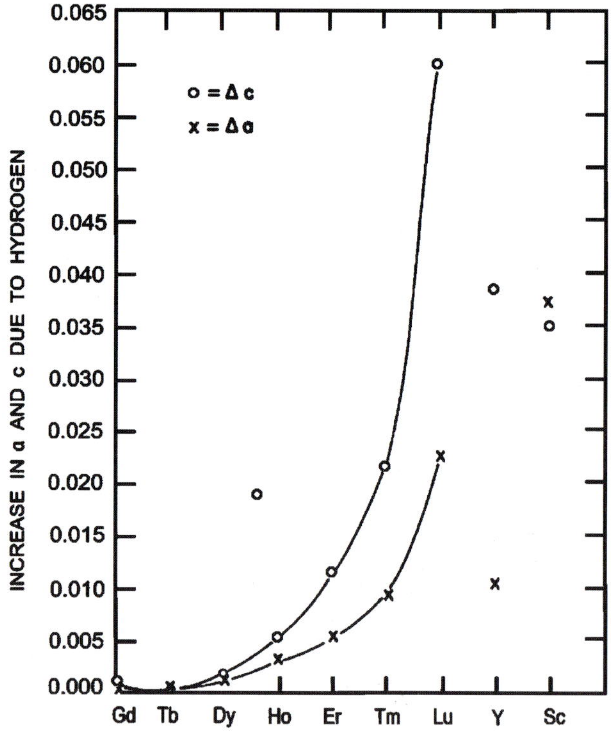

Fig. 1.22. Effect of hydrogen on the lattice parameters of rare earth metals

where a_A and a_B are the lattice parameters of the end members A and B, respectively, and x is the molar fraction of crystal A in the mixture. From the data on a number of systems, Sirdeshmukh and Srinivas [1.30] found that (1.11) with $n = 1$ provides the best description of the composition dependence

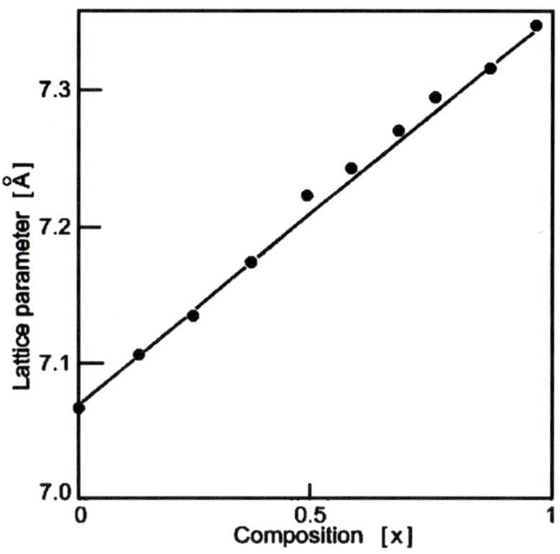

Fig. 1.23. Composition dependence of lattice parameters of $K_{1-x}Rb_xI$ mixed crystal

of lattice parameter in a mixed crystal system. Data on several mixed crystal systems will be discussed in Chap. 7.

Depending on the system, the composition of mixed crystals can be determined by potentiometric titration, polarography, atomic absorption spectroscopy or X-ray fluorescence. The lattice parameter can also be used as a means to estimate the composition. This method of characterisation has the advantage that using a small quantity of the material, lattice parameters can be measured with high accuracy.

1.3.6 Mixed Valence Effects in Lattice Constants

Mixed valence compounds have gained great prominence in recent years. The typical situation in these compounds is that the 5d bands of the rare earth ions are above the sharp 4f bands but the band gap is very narrow – often of the order of 0.5 eV. Under stimulation like pressure and temperature changes or by addition of other ions to the lattice, overlapping develops enabling the 4f electrons to participate in valence. Thus, several compounds belonging to this category show sudden changes in valence which, further, assumes a non-integral value.

In Fig. 1.24, the lattice parameter data obtained by Kaldis et al. [1.31] for the system $TmSe_xTe_{1-x}$ are shown. This system has the NaCl structure. The composition dependence of lattice parameter is very much different from that in Fig. 1.23 where it is linear. With the addition of Se, the valence of Tm suddenly changes from 2 to anywhere around 2.7. This results in the S-shaped curve. Here also, a number of parallel lines can be drawn to represent the

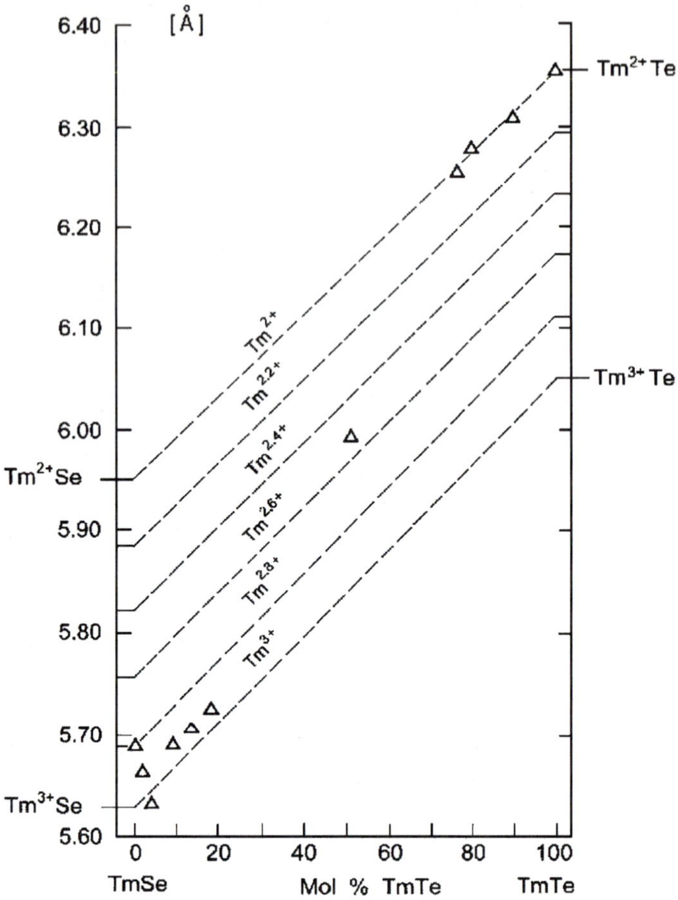

Fig. 1.24. Lattice parameters for the system $TmSe_xTe_{1-x}$

Vegard's law variation for a given value of valence. Thus the curve can be used to characterize the valence in these compounds.

Sampathkumaran and Vijayaraghavan [1.32] studied another interesting system – mixed crystals of $CeNi_2Si_2$ (which is a mixed valence crystal) and $CeCu_2Si_2$ (which is a heavy Fermion system). The lattice parameter variation with composition is shown in Fig. 1.25. Anomalous variation in the c parameter is observed at $x = 0.65$. At the same composition, anomalous variation is observed in magnetic susceptibility. It is suggested that the x-dependence of c may be useful in identifying heavy Fermion materials.

1.3.7 Temperature Variation of Lattice Constant

The temperature variation of lattice constant provides an important method for the determination of thermal expansion of crystals. The special advantage

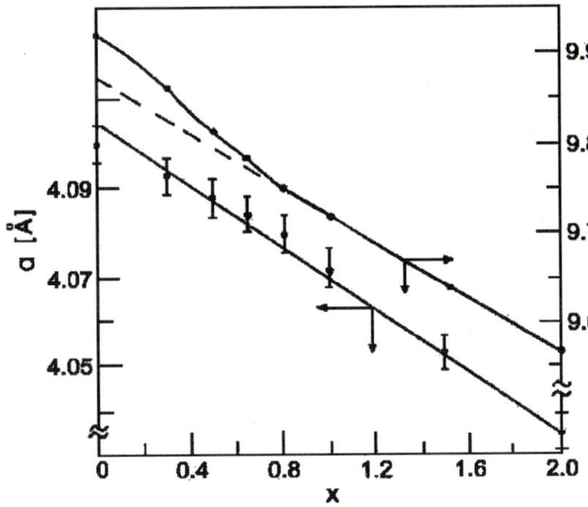

Fig. 1.25. Composition dependence of lattice parameters for the system $CeCu_{2-x}Ni_xSi_2$; dashed line indicates additive behaviour

of this method is that a very minute quantity of the material serves as sample. Further, in anisotropic crystals, the thermal expansion in several directions can be determined in a single experiment. Studies of temperature variation of lattice constants reveal many interesting phenomena like negative thermal expansion and phase transitions. Several studies of temperature variation of lattice parameters of crystals are cited in Chap. 2.

1.3.8 Pressure Variation of Lattice Parameters

By enclosing the sample in a pressure medium or in a diamond anvil press, the changes in lattice constant can be studied. From these changes, the linear compressibilities can be evaluated. A minute crystal or a small quantity of material in the powder form serves as sample. This method and results on several materials are quoted in Chap. 8.

1.3.9 Effect of Magnetic Field on Lattice Constant

There is very little work on this aspect. Kida et al. [1.33] studied the effect of magnetic field on the 'c' lattice constant of Dy which is antiferromagnetic with $T_N = 179$ K and ferrimagnetic with $T_C = 91$ K. The study was carried out by making measurements on the (006) reflection from a single crystal using X-rays.

The variation of 'c' with magnetic field is shown in Fig. 1.26. There is a discontinuous change in 'c' in the experiments conducted at 130 and 150 K. This 'gap' in c-parameter vanishes at 170 K. The observations have been explained in terms of exchange integrals.

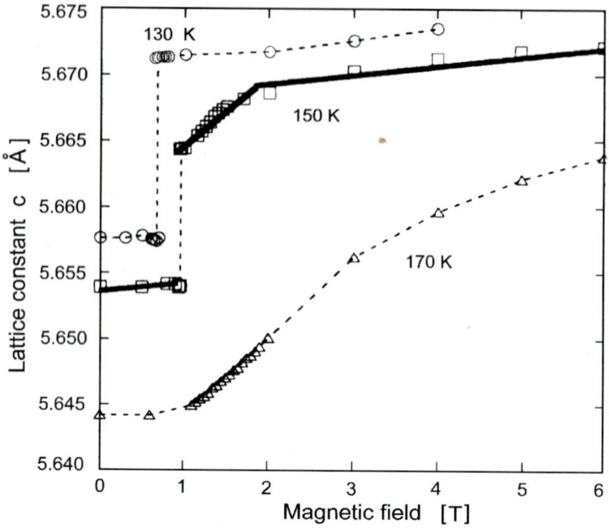

Fig. 1.26. Variation of lattice parameter c of Dy with magnetic field at different temperatures

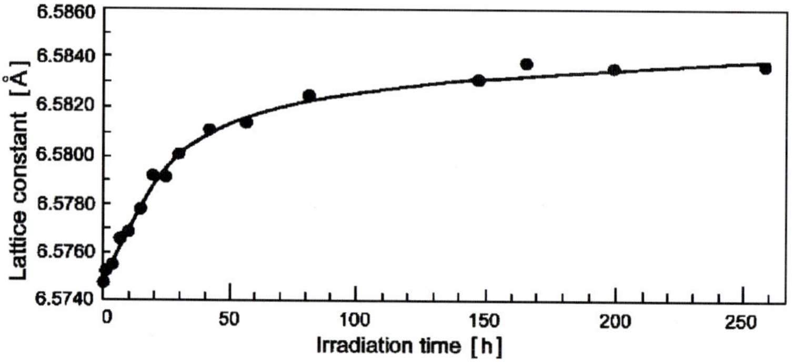

Fig. 1.27. Lattice parameter as a function of γ-irradiation time for $NaClO_3$

1.3.10 Radiation Damage

Irradiation of solids by high energy radiation affects several physical properties. Irradiation effects are seen in lattice parameter values also.

Stapien et al. [1.34] studied the damage in $NaClO_3$ crystals irradiated with γ-rays from a ^{60}Co source. The variation in the value of lattice parameter as a function of irradiation time is shown in Fig. 1.27. There is a fast increase in lattice parameter during the first 50 h of irradiation. Thereafter, the rate of increase slows down and the lattice parameter reaches a saturation value. There is considerable evidence to show that on irradiation the ClO_3 ion breaks into a number of species including gaseous oxygen and ozone. The strains produced in the lattice by these radiolysis products result in the lattice parameter variations.

Zircon is a mineral of gem quality. It is a tetragonal crystal. It has been observed that zircon samples from different locations show a considerable range of physical properties. This applies to the lattice parameter values also. These variations cannot be explained in terms of differences in chemical composition. It is known that zircon samples invariably contain uranium and thorium as impurities. It has been suggested that the differences in lattice parameters are due to effects of irradiation by α-particles originating in the radioactive disintegration of uranium and thorium. That it is so can be seen from the plots of the lattice parameters a and c as a function of the α activity (Fig. 1.28). These results are taken from the work of Holland and Gottfried [1.35]. The plots are smooth curves. The observed variation could be due to a combination of causes like ionisation by the α particles and displacement of atoms by recoil nuclei.

The interesting result of this study of lattice parameters of zircon is that since lattice parameters are found to correlate with α-activity and since the latter correlates with the age of the mineral, it may be possible to use the lattice parameter as a measure of the age in zircon samples.

1.3.11 Effect of Particle Size on Lattice Constant

A smaller particle size shows two effects on X-ray (or electron) diffraction patterns. First, there is a broadening of the powder diffraction lines and, second, there is a measurable change in the lattice constant. A detailed electron diffraction investigation by Boswell [1.36] on a few metals and some alkali halides clearly established that the lattice constant decreases with decreasing particle size. This is further confirmed by numerous subsequent studies. Typical results obtained by Boswell are given in Table 1.6. The decrease in lattice constants with decrease in particle size is consistent with the theory of surface effects in crystals [1.37].

1.3.12 Lattice Constants and Point Defects in Crystals

Lattice parameter determination yields interesting information regarding point defects in crystals. It is well known that the formation of vacancies in metals or Schottky defects in ionic crystals leads to an increase in bulk volume as new atomic layers are formed on the crystal surface. This means that the bulk density (ρ_m) decreases. No such change takes place when Frenkel defects are formed. On the other hand, the existence of pure interstitials results in an increase in bulk density.

The density of a crystal can be calculated from the lattice parameter from the relation:

$$\rho_{\mathrm{x}} = Mn\,/N_{\mathrm{A}}\,a^3, \qquad (1.12)$$

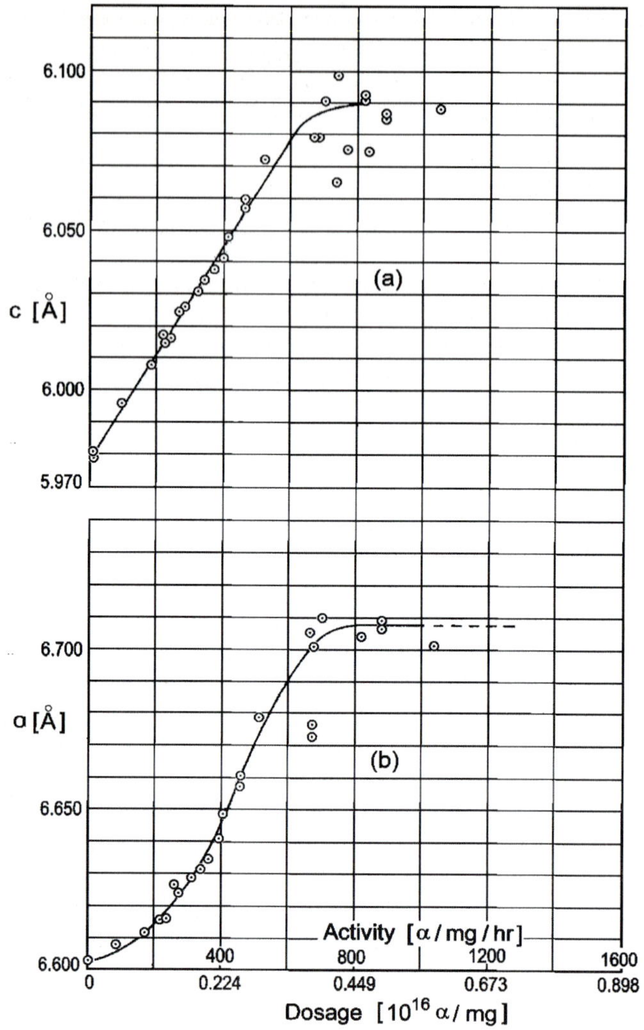

Fig. 1.28. Variation of lattice parameters c and a of zircon with α-activity

where M is the molecular weight, n the number of formula units in the unit cell, N_A Avogadro's number and a the lattice constant. ρ_x is called the X-ray density or ideal density. Its value is not affected by the presence of a small number of defects. From the above discussion, we have the following relations between ρ_m and ρ_x:

Schottky defects $\rho_m < \rho_x$
Frenkel defects $\rho_m = \rho_x$
Interstitials $\rho_m > \rho_x$

Table 1.6. Change of lattice constant with particle size [1.36]

Gold		NaCl	
Particle size [Å]	Lattice constant [Å]	Particle size [Å]	Lattice constant [Å]
14	3.983	48	5.614
18	3.993	75	5.622
33	4.059	80	5.626
40	4.062	150	5.630

Table 1.7. Value of lattice constant (a), ρ_x, ρ_m and concentration of Schottky defects (n_S) for the $Na_xK_{1-x}Cl$ mixed crystal system [1.38]

x	a [Å]	ρ_x [g cm^{-3}]	ρ_m[g cm^{-3}]	n_S [%]
0	6.2916	1.9881	1.9880	0
0.1	6.2354	1.9982	1.9964	0.09
0.3	6.1185	2.0217	2.0117	0.49
0.5	5.9913	2.0538	2.0368	0.82
0.7	5.8571	2.0922	2.0683	1.14
0.9	5.7156	2.1363	2.1321	0.20

Thus, from a comparison of ρ_m and ρ_x it is possible to infer the type of defect present and also to estimate its concentration.

As an example, the results obtained by Barrett and Wallace [1.38] on the $K_xNa_{1-x}Cl$ system are reproduced in Table 1.7. It can be seen that there are Schottky defects in this system. Further, the concentration of the defects (n_S) is maximum in the composition range 0.5–0.7.

It may be mentioned that some authors have used a different approach. Thus, Straumanis [1.39] calculated the molecular weight M_x from (1.12) using the measured density and compared it with the standard value of M. On the other hand, Semiletov [1.40], compared the measured lattice constants of III–VI semiconductors with those calculated from covalent radii and drew information about vacancies and interstitials.

1.3.13 Lattice Constant Variations due to Dislocations

The stress fields associated with dislocations produce local variations in lattice parameter. The effect gets averaged out and cannot be detected with routine methods. However, if a sample can be scanned with an X-ray beam, these variations can be observed. In Fig. 1.29, the experimental set up used by Rozgonyi et al. [1.41] is shown. Here the computer-controlled diffractometer positions the detector at the maximum of the Bragg peak as the specimen is scanned and the value of the Bragg angle is recorded as a function of some distance on the sample. An X-ray topograph (XRT) is also recorded and the two records are superposed (Fig. 1.30). It is seen that the value of the Bragg angle is constant up to a certain point (E) and then it decreases. This region is a strained region and the strain can be computed from the variation in lattice

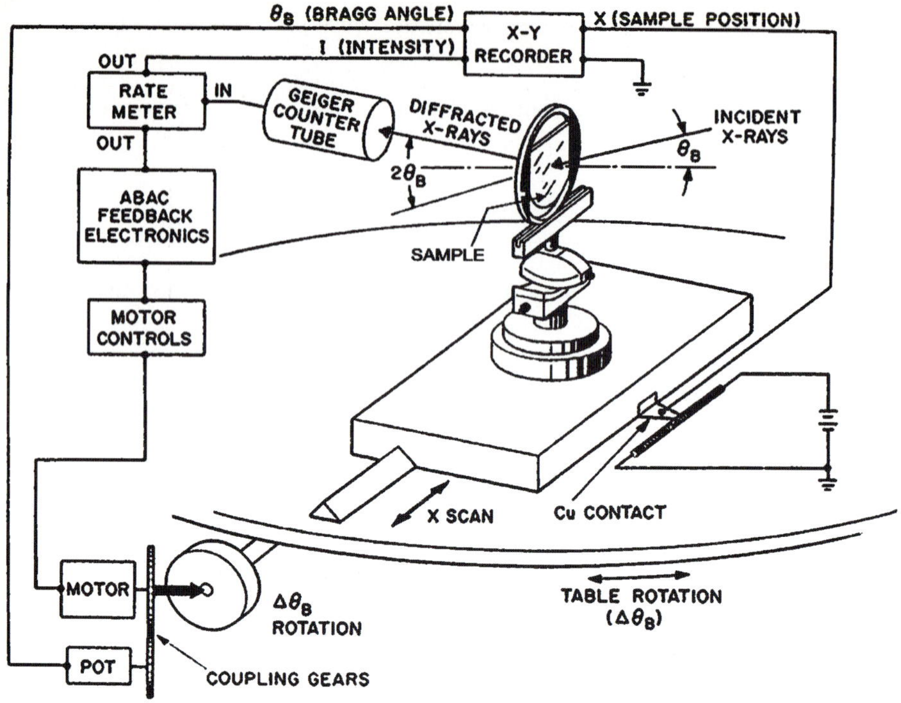

Fig. 1.29. Diagram of automatic Bragg angle control (ABAC) to study variations in lattice constant with dislocations

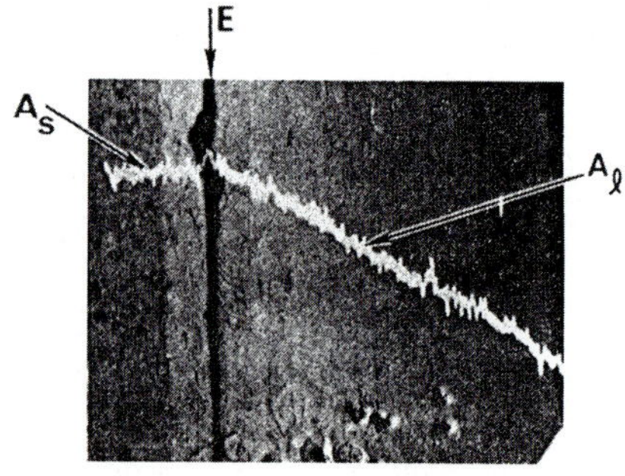

Fig. 1.30. ABAC trace superimposed on XRT for a $Ga_{0.66}Al_{0.34}As/GaAs$ sample. A_s and A_l are, respectively, ABAC trace for the substrate alone and the layer/substrate combination. E is the edge of the LPE layer

parameter. These variations are obviously due to the stress produced by the dislocations. Fig. 1.30 relates to a section of LPE layers of $Ga_{1-x}Al_xAs_yP_{1-y}$ on GaAs substrate. This information is very important in the performance control of heterojunctions.

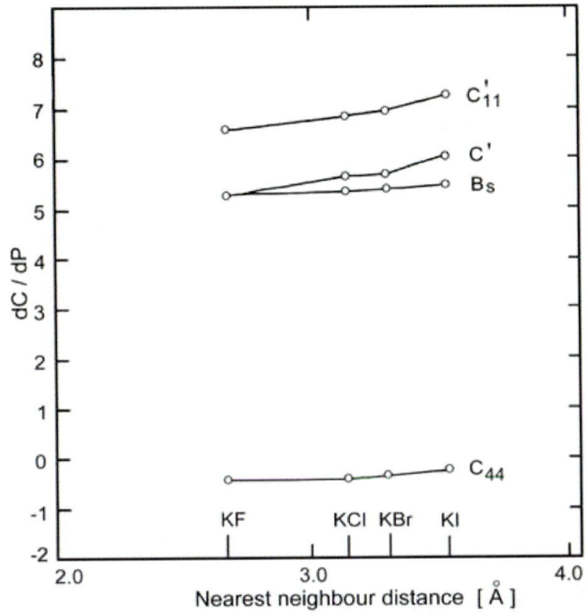

Fig. 1.31. Plots of pressure derivatives of elastic constants vs. nearest neighbour distance of alkali halides

1.3.14 Lattice Constant as a Scaling Parameter

A number of physical properties of crystals scale with the lattice constant i.e., they vary smoothly with the lattice constant. In Figs. 1.31–1.33 plots of the pressure derivatives, temperature derivatives of elastic constants and the surface energy, of some alkali halides are shown with the nearest neighbour distance as the variable. Such correlations are useful either in estimating unknown values by interpolation or in understanding the nature of the property.

1.4 Some of our Results

1.4.1 Lattice Parameters – Data Generation

Using the techniques described in Sect. 1.2, the lattice parameters of several cubic and a few tetragonal and hexagonal crystals have been determined. The lattice parameter results are a by-product of the experiments aimed at thermal expansion measurements and Debye–Waller factor studies. These values are given in Tables 1.8 and 1.9. The data obtained by camera techniques are systematically more accurate than those obtained from diffractometric measurement. This is due to the inherent differences in the two techniques. Most

Table 1.8. Lattice constant (a) and X-ray density (ρ_x) of some cubic crystals at room temperature

Structure	Crystal	Temperature [°C]	a^a [Å]	ρ_x[g cm^{-3}]	Method^b	Ref.
FCC	Al	28	4.0499(2)	2.598	SFC	[1.42]
	Pb	28	4.9505(2)	11.339	SFC	[1.42]
NaCl	NaCl	29	5.6407(2)	2.163	UC	[1.43]
	RbCl	26	6.5916(2)	2.832	FPC	[1.42]
	RbBr	27	6.8927(2)	3.357	FPC	[1.42]
	RbI	27	7.3466(2)	3.557	FPC	[1.42]
	MgO	26	4.2124(2)	3.581	SFC	[1.42]
	CdO	32	4.6954(3)	8.238	UC	[1.43]
	CaS	26	5.6951(2)	2.594	FPC	[1.42]
	PbS	26	5.9358(1)	7.599	SFC	[1.42]
	MnS	28	5.222(2)	4.058	XRD	[1.44]
	EuS	30	5.953(1)	5.793	XRD	[1.45]
	TmSe	28	5.703(2)	8.876	XRD	[1.44]
	SmS	28	5.962(2)	5.717	XRD	[1.44]
	SmSe	27	6.190(2)	6.4212	XRD	[1.44]
	SmTe	27	6.593(2)	6.4391	XRD	[1.44]
	ZrC	28	4.698(2)	6.612	XRD	[1.46]
	NbC	28	4.471(2)	7.796	XRD	[1.46]
	HfC	28	4.636(2)	12.697	XRD	[1.46]
	TaC	27	4.454(2)	14.503	XRD	[1.46]
	VN	29	4.135(2)	6.101	XRD	[1.46]
	HfN	29	4.510(3)	13.936	XRD	[1.46]
CsCl	CsBr	28	4.295(2)	4.459	XRD	[1.44]
	CsI	27	4.567(2)	4.528	XRD	[1.44]
	TlCl	27	3.842(2)	7.021	XRD	[1.44]
	TlBr	28	3.986(2)	7.453	XRD	[1.44]
	Gd–Zn	29	3.600(2)	7.923	XRD	[1.44]
	Gd–Mg	29	3.812(2)	5.442	XRD	[1.44]
	NH_4Cl	27	3.8763(2)	1.525	FPC	[1.42]
	NH_4Br	28	4.0603(2)	2.429	FPC	[1.42]
ZnS	CdTe	28	6.480(1)	5.858	XRD	[1.47]
CaF_2	CaF_2	31	5.4650(2)	3.177	UC	[1.43]
	SrF_2	28	5.7982(2)	4.280	SFC	[1.42]
	BaF_2	27	6.1940(2)	4.900	SFC	[1.42]
	CdF_2	30	5.398(3)	6.351	XRD	[1.46]
	PbF_2	37	5.9421(2)	7.762	UC	[1.43]
	EuF_2	30	5.835(4)	6.350	XRD	[1.46]
Miscellaneous	FeS_2	28	5.4165(3)	5.014	SFC	[1.42]
	$Sr(NO_3)_2$	20	7.7798(2)	2.985	DSC	[1.48]
	$NaBrO_3$	28	6.7072(2)	3.321	SFC	[1.49]
	$Eu_3Fe_5O_{12}$	30	12.488(2)	6.323	XRD	[1.45]

[a] *Accuracy*: The figure in parenthesis denotes uncertainty in the last decimal place.
[b] *Methods*: DSC – 11.46 cm Debye–Scherrer camera; FFC – Flat film camera; SFC – 15 cm symmetric focusing camera; UC – 19 cm Unicam camera; XRD – X-ray powder diffractometer

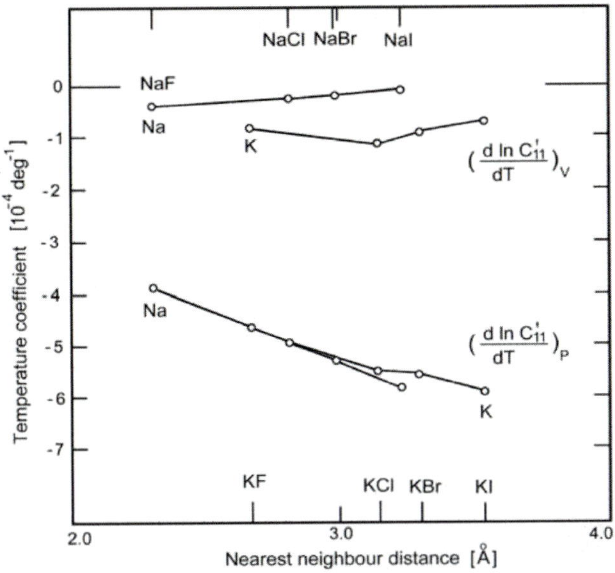

Fig. 1.32. Plots of temperature derivatives of elastic constants; $C'(C_{11}-C_{12})/2$, $C'_{11}(C_{11}+C_{12}+2C_{44})/2$, $B_S = (C_{11}+2C_{12})/3$ vs nearest neighbour distance of alkali halides

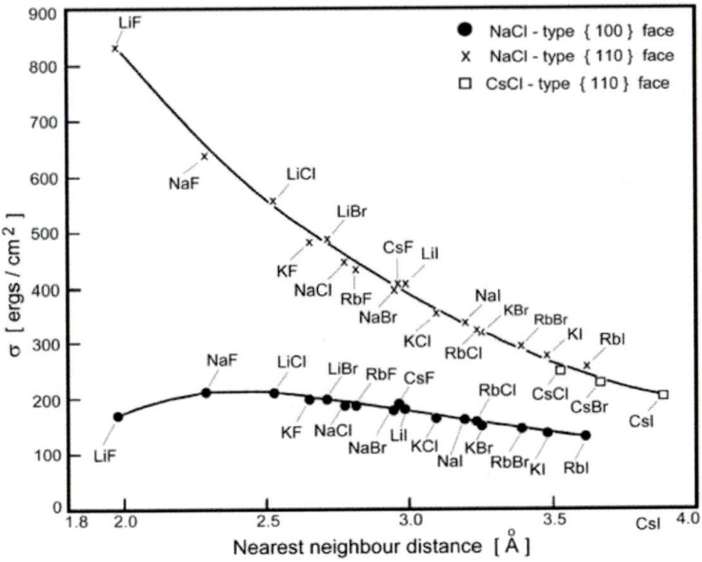

Fig. 1.33. Plots of surface energy σ vs. nearest neighbour distance of alkali halides

Table 1.9. Lattice parameters (a, c) and X-ray density of some tetragonal and hexagonal crystals (notation for accuracy and methods same as in Table 1.8)

Crystal class	Temperature [°C]	Crystal	a [Å]	c [Å]	X-ray density	Method	Ref.
Tetragonal	25	Sn	5.8318(3)	3.1819(3)	7.285	FFC	[1.42]
	26	KDP	7.4528(4)	6.9683(4)	2.335	FFC	[1.42]
Hexagonal	26	ZnO	3.2495(2)	5.2064(3)	5.119	SFC	[1.42]
	25	PbI_2	4.5562(4)	6.9830(4)	6.097	FFC	[1.42]

of the values agree within limits of errors with accurate values reported by others. Some of the values in Table 1.8 are now discussed.

In the case of $Sr(NO_3)_2$, the value of 7.7794 Å [1.48] differed considerably from the value of 7.81 Å reported earlier by Vegard [1.50]. The densities calculated from these two values using (1.12) are 2.930 and 2.985 [$g\,cm^{-3}$]. Comparison of these values with the experimental value of 2.986 indicates that our value of 'a' [1.48] is more reliable.

The value of 4.2124 Å for the lattice constant of MgO is based on measurements on a 'coarse' sample of MgO powder supplied by the US Bureau of Mines. The USBM also supplied a 'fine' sample of the same purity. Measurements on this second sample yielded a lower value of 4.2117(2) Å. This difference in the values for the two samples is consistent with the general observation that the lattice constant decreases with the grain size of the sample.

PbS is thermally stable and can be obtained in the pure form. Its powder photograph has several reflections at high angles making an accurate measurement of lattice constant possible. Razik [1.51] proposed PbS as a standard for precision lattice parameter measurement. The value of 5.9358(1) Å for the lattice constant of PbS [1.42] agrees well with the value 5.9350(7) Å obtained by Razik [1.51].

1.4.2 Lattice Constant as a Scaling Parameter

Keyes [1.52] showed that the elastic constants of crystals with ZnS structure scale with their lattice constants. Mitra and Marshall [1.53] showed that the compressibility of alkali halides scales with the lattice constant. In our work, we have observed that the lattice constant scales with (i) the Debye temperature and (ii) the hardness.

Nagaiah et al. [1.54] obtained a linear plot between the Debye temperature and the reciprocal of the lattice constant of some rare earth garnet crystals. They used this plot (Fig. 1.34) to estimate the Debye temperatures of samarium iron garnet, holmium iron garnet and lutecium iron garnet crystals from their lattice constants. Similarly, Gopi Krishna et al. [1.55] calculated the Debye temperatures of some rare earth monochalcogenides from known data on elastic constants and obtained a linear plot between θ and a^{-1}.

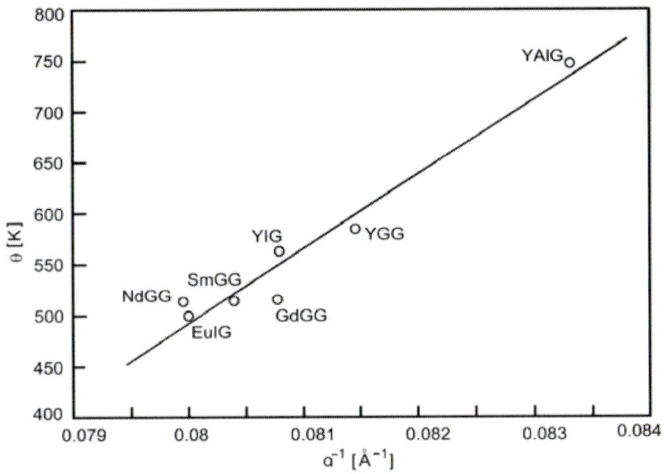

Fig. 1.34. Plot of Debye temperatures of some rare earth garnets against the reciprocal of lattice constant (a)

Regarding the hardness, Thirmal Rao and Sirdeshmukh [1.56] and Sirdeshmukh et al. [1.57–1.59] showed that the lattice constant correlated with the hardness of alkali halides with NaCl structure, divalent chalcogenides with NaCl structure, the rare earth garnets and the cesium halides, respectively. The plot for the cesium halides is shown in Fig. 1.35. Typically, the hardness decreases with increasing lattice constant and the a vs. H plots are smooth curves. The variation of the two parameters, one decreasing and the other increasing is linked to the weakening of the interatomic binding. Details are given in the Chap. 4.

1.4.3 Temperature Variation of Lattice Constant

The lattice constant generally increases as the temperature increases. This variation provides an important method for the determination of the thermal expansion coefficient of crystals. The temperature variation of the lattice parameters of several crystals has been investigated. As an example, the temperature variation of the lattice constants of CdF_2 and PbF_2 [1.60] is shown in Fig. 1.36. Results on other crystals are given in Chap. 2.

1.4.4 Radiation Induced Changes in Lattice Constant of $NaBrO_3$

Hussain et al. [1.49] measured the lattice constant of $NaBrO_3$ subjected to different dosages of γ radiation. The lattice constant was determined with the help of a symmetric focusing camera; filtered Cu radiation was used.

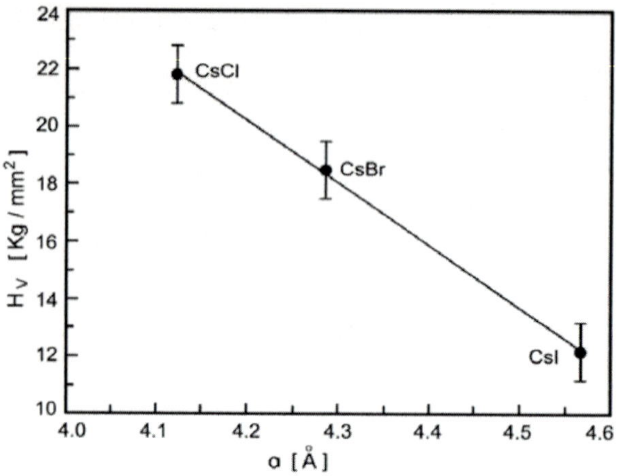

Fig. 1.35. Plot of microhardness (H_V) of cesium halides against the lattice constant (a)

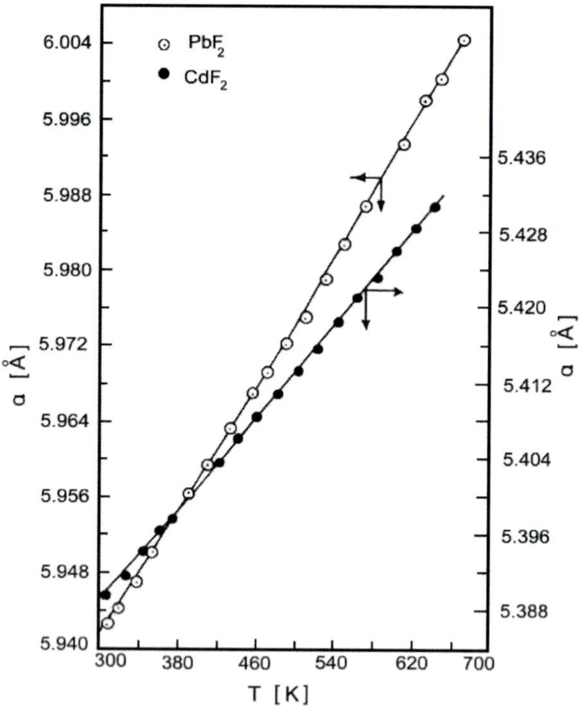

Fig. 1.36. Temperature variation of the lattice constants of CdF_2 and PbF_2

Table 1.10. Lattice constant a of $NaBrO_3$ at different γ-irradiation doses

Dosage [10^6 rad]	a [Å]	Dosage [10^6 rad]	a [Å]
0.00	6.7072(2)	12.42	6.7092(2)
2.48	6.7074(2)	14.84	6.7104(2)
5.11	6.7075(2)	17.32	6.7106(2)
7.45	6.7081(2)	19.94	6.7109(2)
10.01	6.7090(2)	22.49	6.7114(2)

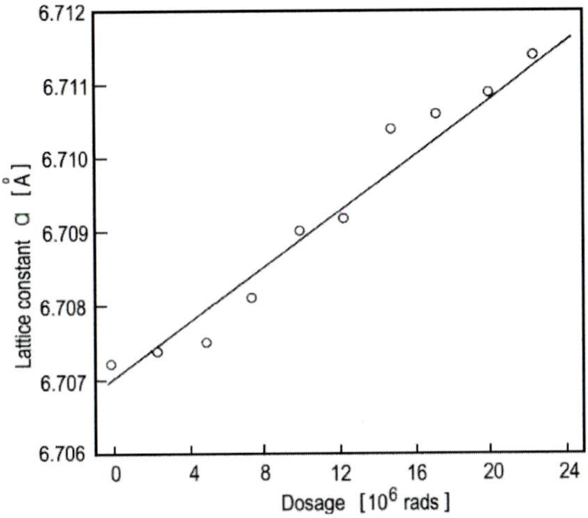

Fig. 1.37. Lattice constant vs. γ-irradiation dosage for $NaBrO_3$

The γ-irradiation was carried out with a 1 kCi ^{60}Co source at a dose rate of 1.38×10^5 rad h^{-1}.

The lattice constants are given in Table 1.10. The dependence of the lattice constant of $NaBrO_3$ on irradiation is shown in Fig. 1.37. It is seen that in $NaBrO_3$ as well as $NaClO_3$ (see Sect. 1.3.10), the lattice constant increases with γ-ray dosage. This radiation induced lattice expansion is attributed to the creation of radiolysis products like BrO^{-1}, BrO^{-2}, $Br_2O_3^-$, etc. in $NaBrO_3$ (and similar ions in $NaClO_3$).

1.4.5 Lattice Constants of Mixed Crystals

The lattice constants of mixed crystals are a function of the composition. Generally, the composition dependence is linear. In some cases, on closer analysis, a slight positive deviation from linearity is observed. The composition

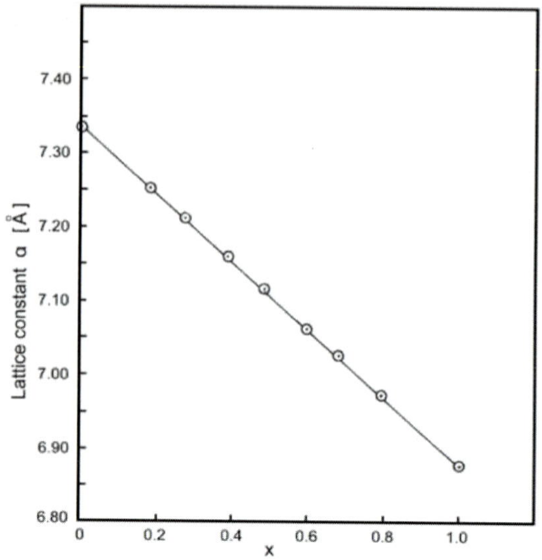

Fig. 1.38. Lattice constant vs composition (x) for the $RbBr_xI_{(1-x)}$ mixed crystal system

dependence of the lattice parameter of some mixed crystal systems has been studied. As an example, the composition dependence of lattice constants in the $RbBr_xI_{(1-x)}$ system observed by Kumara Swamy et al. [1.61] is shown in Fig. 1.38. Results on other systems are given in Chap. 7.

2

Thermal Expansion

2.1 Introduction

The thermal expansion of solids is a basic physical property representing the dimensional changes in a solid induced by a change in temperature. It is of technical importance as it determines the thermal stability of a crystal. The thermal shock resistance of crystals depends on the thermal expansion [2.1]. The thermal expansion characteristics decide the choice of material for construction of metrological instruments. In modern epitaxial device technology, lattice mismatch is an important factor; this is related to the thermal expansion behaviour [2.2]. In nuclear fuel technology, the thermal expansion is a deciding factor in the choice of container material [2.3]. The knowledge of thermal expansion is necessary in the experimental determination of the temperature variation of elastic constants, refractive index, dielectric constants and photoelastic constants. Thermal expansion data are also needed in the conversion of C_P into C_V.

The thermal expansion of a solid is a consequence of the anharmonic nature of the atomic vibrations of the atoms. When a solid is heated, the atoms vibrate with increased amplitudes. The anharmonic increase in the amplitudes results in the displacement of the effective mean positions of the atoms. As a consequence, the volume of the solid increases. The intimate connection between the anharmonicity of vibrations and the expansion coefficient can be established by assuming a simple model, the monatomic one-dimensional lattice [2.4].

If the equilibrium separation of the atoms increases by x, the change in potential energy may be represented approximately by

$$U(x) = a_1 x^2 - a_2 x^3, \tag{2.1}$$

where a_1 and a_2 are constants. Here, the first term represents harmonicity and the second the anharmonicity of atomic vibrations. Using Boltzmann statistics, we obtain for the average value of x

$$\bar{x} = 3a_2 k_{\mathrm{B}} T / 4a_1^2, \tag{2.2}$$

where k_B is the Boltzmann constant and T the absolute temperature. The linear coefficient of expansion α is given by

$$\alpha = (\mathrm{d}\bar{x}/\mathrm{d}T)/r_0 = (3a_2 k_B/4a_1^2 r_0), \quad (2.3)$$

where r_0 is the interatomic distance at the initial temperature. According to (2.3), the expansion coefficient is a constant. This is because of the classical expression for the energy used in (2.2). If the quantum expression is employed, one gets [2.5]

$$\alpha = (\mathrm{d}\bar{x}/\mathrm{d}T)/r_0 = (3a_2 k_B/4a_1^2 r_0)(h\nu/k_B T)^2 \exp(-h\nu/k_B T), \quad (2.4)$$

where the terms have their usual significance. From (2.4), it can be seen that the thermal expansion coefficient is directly related to the anharmonic term in the expression for potential energy; there would be no thermal expansion in the absence of anharmonicity. It can also be seen from (2.4) that the coefficient of thermal expansion is temperature dependent as against (2.3) which gives a temperature independent quantity. Further (2.4) predicts that the coefficient of thermal expansion decreases as the temperature decreases and finally approaches zero, as the temperature tends to zero in conformity with the third law of thermodynamics.

The thermal expansion behaviour of a solid is closely related to the interatomic forces in the solid and its structure. The expansion coefficient is very low for covalent crystals and quite large for Van der Waal solids, the metallic and ionic solids having intermediate values of expansion coefficients. Krishnan [2.6] pointed out that in ionic crystals the thermal expansion depends on the valency of the ions. Megaw [2.7] gave a quantitative relation between the coefficient of expansion and the electrostatic share, q. The relation is

$$\alpha\, q^2 = \text{constant}. \quad (2.5)$$

The electrostatic share is defined as the ratio of the valency of the ion to the number of ions of the opposite sign, which are its immediate neighbours.

The thermal expansion behaviour varies considerably with the structure even when the constituent ions are the same. This can be seen from the expansion of sphalerite and wurtzite [2.8], rutile and anatase [2.9] and the tetragonal and hexagonal phases of GeO_2 [2.10, 2.11].

The difference in the interatomic binding in different directions has considerable influence on the magnitude of thermal expansion anisotropy. The anisotropy of expansion in layer and anti-layer structures is discussed by Wooster [2.12] in terms of the intra-layer and inter-layer bonding. In complex structures, where the constituents are groups of atoms, the change with temperature in the shape of these groups and their orientation affects the thermal expansion. This effect has been observed in the expansion of ammonium dihydrogen phosphate [2.13], potassium iodate [2.14] and in several other crystals [2.15].

A theory of thermal expansion was proposed by Gruneisen [2.16]. Gruneisen defined a parameter, γ, which expresses the volume dependence of the frequencies of vibration. The Gruneisen parameter is given by the expression

$$\gamma = -(\mathrm{d}\log\nu)/\mathrm{d}\log V), \tag{2.6}$$

where ν is the frequency of atomic vibrations and V the volume of the solid. Gruneisen finally derived the equation:

$$\alpha = (\gamma\,\psi\,C_V)/3\,V. \tag{2.7}$$

Thus, the Gruneisen parameter relates the linear coefficient of thermal expansion α to the isothermal compressibility ψ, the specific heat at constant volume C_V and the molar volume V.

Gruneisen assumes the same value for the constant γ for all frequencies. As such, (2.6) may also be written as

$$\gamma = -(\mathrm{d}\log\nu_{\mathrm{m}})/\mathrm{d}\log V), \tag{2.8}$$

where ν_{m} is the Debye characteristic frequency. In a more elaborate way, each mode of vibration ν_i has its corresponding Gruneisen parameter γ_i and also makes its contribution C_i to the specific heat. The thermal expansion coefficient α is now given by

$$\alpha = (\psi/3V)\sum_i \gamma_i C_i. \tag{2.9}$$

The comparison of α calculated from the lattice vibration spectrum with the experimental values provides a good check on the lattice dynamical model [2.17, 2.18].

2.2 Experimental Methods

2.2.1 General

The coefficient of thermal expansion α is defined as $(1/\Delta T)(\Delta l/l)$ where Δl is the increase in length of a sample of length l corresponding to an increase ΔT in temperature. For most solid materials, α is in the range $10\text{–}60 \times 10^{-6}(\mathrm{K}^{-1})$. The crystal samples used in measurements are generally a few mm thick. As thermal expansion is temperature dependent, measurements have to be made over small temperature intervals of a few degrees at a time. Thus, sensitivity of the order of $10^{-4}\text{–}10^{-6}$ in $(\Delta l/l)$ is generally required. Sometimes for detecting smaller changes, a higher sensitivity is required. This demand has led to development of a number of experimental techniques for measurement of thermal expansion of crystals. These various methods may be grouped as (i) optical methods, (ii) capacitance methods, (iii) diffraction methods, (iv) dilatometric methods and (v) other methods. Several of these methods have been discussed in [2.6, 2.19, 2.20]. Some of the important methods will be briefly discussed.

2.2.2 Optical Methods

Most optical methods make use of the phenomenon of interference. Fizeau [2.21, 2.22] was the first to set up an interferometer for the measurement of thermal expansion. The sample was a crystal block with one face polished. A glass plate was supported above this face. A beam of approximately monochromatic light travelled down the system and reflections occurred from the lower surface of the glass plate and the upper surface of the specimen. When these two surfaces were at a small angle, the two reflected beams resulted in a set of straight line fringes. Displacement of the crystal surface due to heating resulted in the displacement of the fringe system. From the shift of fringes the thermal expansion can be calculated.

Several modifications of Fizeau interferometers have appeared [2.23–2.27]. The main difference between these versions and that of Fizeau are (i) the use of quartz or glass optical flats, (ii) use of partially metallized plates, (iii) use of three small pieces of the sample or a cylindrical sample, (iv) use of a laser to give sharp fringes and (v) the use of a camera or other recording system.

Frazer and Hollis-Hallett [2.28] and Meincke and Graham [2.29] employed a Fabry–Perot etalon for the measurement of thermal expansion. The specimen was in the form of a cylinder 2 in. long and 1 in. in diameter with an axial hole half inch in diameter. The ends were in the form of three feet with polished co-planar optically flat surfaces. This sample was sandwiched between two half-silvered optical flats. A beam of monochromatic light passes through the etalon to the opposite side resulting in a fringe pattern. From a continuous monitoring of the intensity of the central interference fringe as the etalon is warmed, the thermal expansion coefficient can be obtained.

Kirby [2.30] modified an Abbey–Pulfrich interferometer for thermal expansion measurement. He devised a holder such that a large single crystal is held inside an etalon. The design of the holder is such that experiments can be done on the same crystal sample placed in different orientations in the etalon, thus enabling the measurement of thermal expansion in different directions. In yet another modification, Aurora et al. [2.31] used a Jamin interferometer in conjunction with a laser for measuring the thermal expansion of some superionic conductors.

2.2.3 Capacitance Methods

In this method a heterodyne oscillator is used. The plate separation of one capacitance is controlled by the sample thickness and the other capacitance is variable. When the sample expands on heating, it results in the change in the capacitance of the first condenser which in turn alters the frequency. Either the change in frequency Δf is measured, or the change ΔC in capacitance C in the second condenser necessary to restore the original frequency is measured. It can be shown that $2\Delta f/f = \Delta C/C = \Delta l/l$. Hence the expansion coefficient can be calculated.

Bijl and Pullan [2.32] were the first to use this method. Several capacitance dilatometers with minor modifications have been proposed [2.33, 2.34]. A major refinement was introduced by Carr et al. [2.35] who designed a three terminal capacitance cell. Two such capacitances are connected in a transformer bridge and the small changes in capacitance are measured by means of an A.C. bridge. The overall sensitivity of this method is very high ($\Delta l/l \sim 10^{-10}$). White [2.36] used this set-up for the measurement of thermal expansion of a number of solids at very low temperature. Jones and Richards [2.37] discuss transducer designs for sensitive capacitance micrometry. Ema et al. [2.38] have designed a capacitance dilatometer with a sensitivity of ($\Delta C/C \sim 10^{-7}$); they used this arrangement to measure the changes in the expansion coefficient of sodium nitrite through the transition.

2.2.4 Diffraction Methods

Bragg's law of diffraction is

$$d = \lambda/(2\sin\theta), \tag{2.10}$$

where λ is the wavelength, θ the Bragg angle and d the interplanar spacing. This law provides the basis for the diffraction methods of determining thermal expansion. A change in temperature of a crystal causes a change in the value of d, which is observed as a change in θ. Thus from the changes in the Bragg angles, the expansion coefficients can be obtained. However, actually, instead of finding changes in the d values of the individual lines, the lattice constants are determined accurately by using all the available lines and the thermal expansion is calculated from the temperature variation of lattice constants. In terms of the lattice constant 'a' and the change Δa corresponding to the temperature change ΔT, α becomes $(1/\Delta T)(\Delta a/a)$.

The accurate determination of lattice parameters from X-ray diffraction is now a highly developed field. Excellent treatment of this topic is available in treatises [2.39–2.41]. Single crystal methods are used rarely. The powder method is more commonly used. A number of cameras based on different geometries have been proposed. The geometries generally employed in powder cameras are (i) flat film camera, (ii) cylindrical camera and (iii) focusing cameras. Some of these cameras are described in Chap. 1.

Cameras with low temperature and high temperature facilities have been designed by many workers and are also commercially available. The well-known Unicam high temperature camera with 19-cm diameter has been employed by many workers for lattice parameter studies up to 1,000°C. The lattice parameters can be determined, generally, with an accuracy of ±0.0002 Å. This accuracy has been enhanced in some cases by using special techniques [2.41]. Another X-ray diffraction technique is the X-ray diffractometer wherein the diffraction pattern is recorded with the help of a detector. This technique is also discussed in Chap. 1.

2.2.5 Dilatometric Methods

Push-rod dilatometry is one of the techniques of measurement of thermal expansion. Several push-rod dilatometers are commercially available. Fused quartz dilatometers are simple and comparatively inexpensive. In a typical dilatometer the sample is enclosed in a fused quartz tube. The end of the sample is in contact with a sensitive dial-gauge. The composite holder (quartz tube and sample) is placed in a heater and length changes are directly read on the dial-gauge. Janson and Sjoblom [2.42] describe such a dilatometer.

A push-rod dilatometer was constructed and was used by Rao [2.43] for studies of some crystals. The instrument is described in some detail since it is simple in design, inexpensive and made from indigenously available components. The main parts of the push-rod dilatometer and the procedure for its use are discussed in this section.

The Push-Rod Assembly

The rods used in the present experimental set-up are made of fused silica. They are cylindrical in shape and their surface is satin-glazed. The rod assembly is made up of two parts. Three rods of equal length form the base of the assembly and a set of four short rods makes its upper part. The three base rods rest on two metal stands fixed to a wooden base. The metal stands are provided with horizontal projections on either side. The two projections facing each other have three grooves made in each of them to hold the rods. The exterior projections are flat so as to support the dial-gauge micrometers. The rod assembly as seen from the front is shown in Fig. 2.1a. The long rod, designated BR is one of the base rods fixed to the metal stands. The two shorter rods placed on the base rods rest lengthwise in the space provided by the adjacent base rods. The sample is sandwiched between the two short rods wherein the rod marked FR is the fixed rod and MR the movable rod. The shaded portion marked S shows the position of the sample. The movable rod MR is in communication with the dial-gauge DG. The position of the heater is shown in the figure by H.

The differential arrangement of the rod assembly, along with the positions of the specimen and the standard reference material, as seen from above is presented in Fig. 2.1b. The shaded portions, S_1 and S_2, represent the two specimens, i.e., the sample and the standard reference material. FR_1 and FR_2 are the two fixed short rods, fixed to the base rods by means of an adhesive applied at the end away from the heater. MR_1 and MR_2 are the two movable rods which are in communication with the two dial-gauge indicators DG_1 and DG_2 kept on the two exterior projections of the metal stand. The position of the heater with respect to the two samples is shown by H. When heated, the specimen under study communicates its dilation to the micrometer kept on the metal stand on one side of the assembly while the standard reference

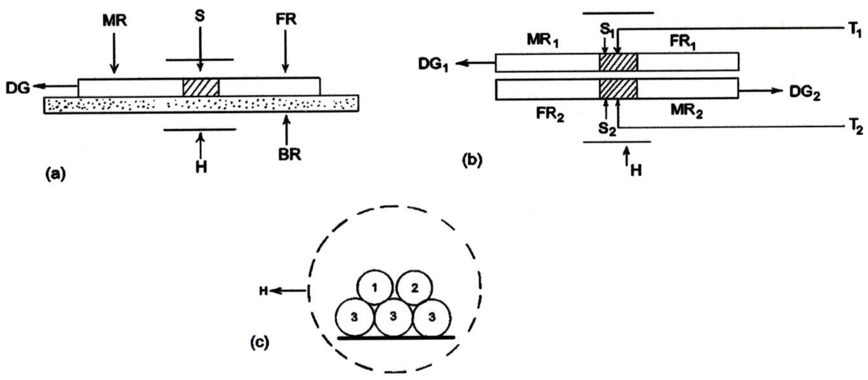

Fig. 2.1. (**a**) Front view, (**b**) top view and (**c**) cross-sectional view of the rod assembly

material communicates its dilation to the micrometer on the other side. The positions of the two thermocouples, T_1 and T_2 are as shown in the figure.

The cross-sectional view of the rod assembly is given in Fig. 2.1c. The circles marked 1 and 2 represent the cross-sections of the short rods whereas the circles marked 3 represent the base rods. The cross-section of the heater as seen from one side of the assembly is indicated by the dotted circle.

The Dial-Gauge Micrometers

When the sample is heated, it expands and pushes the silica rod which is in communication with the dial-gauge indicator. The push-rod is in contact with the arm of the dial-gauge which passes through a small hole drilled in the metal stand just above the horizontal projection. Two dial-gauge micrometers, one for measuring the expansion of the standard reference material and the other for measuring the expansion of the specimen, are used.

The construction and working of the dial-gauge have been described by Cook and Rabinowicz [2.44]. The dial-gauge consists of levers and gears having two toothed wheels one moving over the other. The input displacement pushes an arm or shaft which communicates the displacement to a toothed wheel. This toothed wheel moves another toothed wheel of lesser radius in the opposite direction. A light pointer needle is attached radially to the second toothed wheel. The needle moves on a graduated circular scale. The displacement of the arm of the dial-gauge is magnified and the output may directly be read on the circular scale.

In the present design, the dial indicators are Mitutoya circular dial-gauges made in Japan. In this, the circular scale has two semi-circular parts each marked 0–100 divisions, each division corresponding to 0.001 cm. One-fifth of each division could be made out visually. Thus, this micrometer is capable of measuring length changes to within two parts in 10^5. It may be mentioned that for most crystals the relative expansion $\Delta L/L$ for a rise of 1°C is of the order of 10^{-5} or more.

The Heating Arrangement

The heater H shown in Fig. 2.1 is a cylindrical tube and is made of asbestos. The heating coil, a super-kanthal wire, was wound uniformly over the tube over a length of about 6 cm. The heater coil was fed by AC supply from a dimmerstat through a TEMPO temperature control. The temperature distribution along the heater tube was investigated by placing the hot junction of the thermocouple at different distances from an open-end. The samples and the hot junctions of the respective thermocouples are placed in the region of uniform temperature.

Measurement of Temperature

The temperature of the samples was measured with copper–constantan thermocouples. The thermocouples are placed as shown in Fig. 2.1b. The hot junction of the thermocouple was kept very close to the sample, both being located at the middle of the heater. Both the thermocouples are connected to a sensitive galvanometer through a commutator arrangement. As the specimen under study and the standard material are located in the same region of the heater, the temperatures of both are nearly the same. However, occasionally a slight difference of a degree or two was observed. But, as will be discussed in a later section, this does not affect the results.

The temperature of the specimen could be measured with an accuracy of 0.5°C. However, the true error may be ±1°C, which is taken into consideration while estimating the error in thermal expansion coefficients.

Experimental Procedure

The sample is mounted in its position and is sandwiched between the two short rods, one fixed to the base rods and the other free to slide. The measurements are made simultaneously on the specimen under investigation as well as on the NBS copper sample kept by its side as shown in Fig. 2.1b. The dilations are recorded at various temperatures for the sample and the standard substance. Although the expansion of the small portion of the silica rods inside the heater is very small compared to the expansion of the samples, this effect is eliminated by the following procedure.

Let ΔR_1 and ΔR_2 be the observed dilations recorded in the dial-gauges in communication with the specimen under study, say KCl, and with the standard sample, respectively. These include the dilations ΔC_1 and ΔC_2 of the silica rods inside the heater. As very nearly the same length of the silica rods is heated with both the samples, we may write $\Delta C_1 = \Delta C_2$. Let ΔL_1

and ΔL_2 be the true dilations of KCl and the standard material, respectively. Then

$$\Delta R_1 = \Delta L_1 + \Delta C_1, \tag{2.11}$$

and

$$\Delta R_2 = \Delta L_2 + \Delta C_2. \tag{2.12}$$

As

$$\Delta C_1 = \Delta C_2,$$
$$\Delta R_1 - \Delta R_2 = \Delta L_1 - \Delta L_2$$

and

$$\Delta L_1 = \Delta R_1 - \Delta R_2 + \Delta L_2. \tag{2.13}$$

In (2.13) the factor $(\Delta L_2 - \Delta R_2)$ corresponds to the difference between the correct dilation of the standard sample and its observed dilation due to any expansion of the silica rods. This difference, when added to the observed dilation for the specimen under study, gives the corrected dilation. The correction factor $(\Delta L_2 - \Delta R_2)$ is a very small quantity when compared to ΔL_1, even at high temperatures. Thus, any slight difference of temperature between the specimen and the standard material will give rise to only a negligible effect on the value of the correction term $(\Delta L_2 - \Delta R_2)$.

A copper standard SRM-736 was obtained from the US National Bureau of Standards. Kirby and Hahn [2.45] gave values of relative expansions for the copper standard SRM-736 in the NBS certificate of analysis. A curve was drawn between these relative expansions and temperature. From this curve, values of relative expansion were obtained for a reference temperature of 30°C. The values of $\Delta L/L_{30}$ thus obtained were plotted against temperature. The correct dilations of any sample of copper SRM-736 can be determined from this curve, if its length is known.

The relative expansions $\Delta L/L_{30}$ of the specimen crystal are determined after correcting with reference to the expansion of the SRM-736 copper sample at the appropriate temperatures as discussed in the preceding paragraph. Here, ΔL is the length change and L_{30} is the length of the crystal at room temperature measured with a micrometer screw gauge capable of measuring lengths of the order of 0.001 cm. The length is then corrected for 30°C using the room temperature expansion coefficient and the relative expansions, $\Delta L/L_{30}$, are found. A smooth curve is drawn between temperature (t°C) and relative expansion. The true coefficients of expansion are determined from this curve at various temperatures from the slopes of the curve at those temperatures. These are the observed coefficients of linear expansion of the crystal.

The experimental arrangement was standardized by measuring the thermal expansion of a single crystal specimen of potassium chloride. From a least square analysis of the observed data, the thermal expansion coefficient of KCl is given by

$$\alpha = (36.66 \times 10^{-6} + 3.391 \times 10^{-8} t + 1.029 \times 10^{-12} t^2)/°\mathrm{C}, \tag{2.14}$$

Table 2.1. Comparison of linear thermal expansion coefficient (α) of KCl by different methods

Temperature [K]	$\alpha[10^{-6}\,\mathrm{K}^{-1}]$				
	X-ray [2.46]	X-ray [2.47]	Interferometer [2.48]	X-ray [2.49]	Dilatometer [2.43]
300	36.4	35.0	37.1	37.1	37.6
350	38.4	–	39.0	38.8	39.3
400	39.7	38.2	41.0	40.5	41.0
450	41.1	–	42.9	42.2	42.7
500	42.8	42.9	44.9	43.9	44.4
550	45.6	–	46.8	45.6	46.1

where t is the temperature in °C. In Table 2.1 a comparison is made between the thermal expansion coefficients of KCl thus obtained [2.43] and those measured by other workers. It can be seen from Table 2.1 that the agreement between the results of Rao [2.43] and those available in literature is good. Considering the uncertainty in the measurement of length change and in the determination of temperature, the uncertainty in the values of the coefficient of thermal expansion obtained from this set-up is estimated as 5%.

2.2.6 Other Methods

Nielsen and Leipold [2.50] measured the thermal expansion of magnesium oxide with the help of tele-microscopes. A 3 in. long specimen was heated in an induction furnace. Through a window in the furnace, the displacement of the tip of the sample was measured with a tele-microscope having a filar micrometer eye-piece.

Jaakkola et al. [2.51] designed a hydrostatic weighing apparatus for precision measurement of volumes. The solid to be studied formed one arm of a sensitive analytical balance. The suspended solid was kept in an enclosure which was thermally insulated and vibration free. Further, a temperature control of 0.002 K was achieved. The heating was done by immersing the solid in a liquid whose density is known.

Foster and Finnie [2.52] developed a method suitable for crystals with very small coefficients of expansion. This method employs a single frequency He–Ne laser. The beam from the laser is split. One beam passes through an interferometer which includes the sample. The laser length is modulated by a signal which modulates the output frequency. By mixing this with the unmodulated beam the change is calibrated in terms of the laser length. In the actual experiment, the laser length changes because of the expansion of the crystal. The expansion is obtained from the change in frequency combined with the calibration.

Shrivastava and Joshi [2.53] proposed a method for the measurement of thermal expansion from a knowledge of the stress coefficient of electrical resistance. Two identical wires are used in the experiment, one is stretched and the other is heated. The change in resistance of both the wires is measured and the change in length of the heated wire is calculated in terms of the change in length due to stress. Obviously, this method is useful only for conducting materials available in the form of wires.

2.3 An Overview

This overview is confined to some salient contributions dealing with experimental techniques, theories and studies of whole families of crystals.

2.3.1 Some Novel Experimental Techniques

The commonly employed techniques for measurement of thermal expansion have been discussed in Sect. 2.2. A few other methods have been proposed and used for limited work. These are briefly discussed here because of their novelty.

An Ultrasonic Method

Mantysalo [2.54, 2.55] developed an ultrasonic method which uses an ultra-high-frequency (UHF) acoustic transmission probe and an UHF spectrometer. The amplitude modulated UHF oscillator and a signal amplifier function as a continuous wave spectrometer. When the UHF energy is fed into a transmission line and then transferred to the resonance probe, a transducer converts it into ultrasonic energy. These longitudinal ultrasonic waves propagated through a crystal sample cause a flow of heat and a dissipation of energy. In other words, the ultrasonic wave amplitude is attenuated. A second transducer converts the ultrasonic energy into UHF signal.

Due to thermal expansion, the crystal expands and resonance occurs resulting in standing waves. The result is an oscillating curve of the ultrasonic attenuation as a function of temperature (Fig. 2.2). The period of these oscillations (in units of temperature) is

$$\Delta T = \lambda/(2L_T\alpha) \tag{2.15}$$

and

$$\lambda = v_T/\nu \ , \tag{2.16}$$

where λ is the wavelength of the UHF wave, L_T the length of the specimen at temperature TK, ν the UHF frequency and v_T the velocity of the UHF wave. When the temperature difference is small, L_T can be replaced by its room temperature value L_0 and (2.15) and (2.16) reduce to

$$\alpha = v_T/(2L_0\,\nu\,\Delta T). \tag{2.17}$$

Mantysalo used this method to determine the coefficient of expansion of Li and obtained a value of 47×10^{-6} K^{-1}.

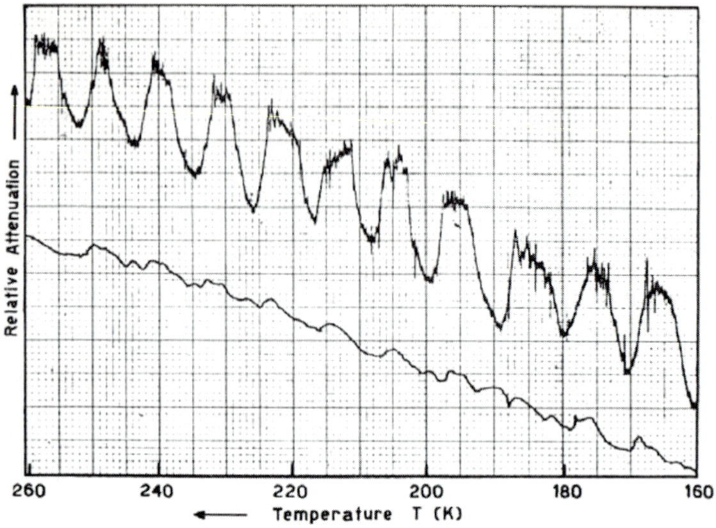

Fig. 2.2. Recorder traces of the ultrasonic attenuation in a lithium single crystal as a function of temperature. (For the upper curve the acoustical surfaces are nearly parallel. For the lower curve the parallelism is poor.)

Holographic Method

Heflinger et al. [2.56] developed a holographic method for the determination of thermal expansion. Their set-up uses a He–Ne laser or a Ruby laser. A reference beam and another beam (both from the same source) reflected from the object are superposed to give a hologram. Further the superposition of two holograms with the imposed temperature change occurring between the two exposures results in a fringe pattern.

The fringes appear straight (Fig. 2.3a) when the two exposures occur at the same temperature (i.e., no thermal expansion). This is so irrespective of the shape of the object. On the other hand, the fringes develop curvature (Fig. 2.3b) when there is a temperature difference. The authors describe a simple procedure of reading the fringe characteristics from which α is calculated. This is a no-contact method and is applicable to objects of any shape.

γ-Ray Attenuation Method

A new technique for the thermal expansion of isotropic solids was proposed by Drotning [2.57]. This method utilizes γ-ray attenuation to measure thermal expansion. His set-up is shown in Fig. 2.4. A γ-ray beam passes through the sample and is detected with a scintillation detector. The mass attenuation coefficient σ is determined from the relation

$$I(T) = I_0(T)\exp[-\sigma\,\rho(T)\,l(T)], \qquad (2.18)$$

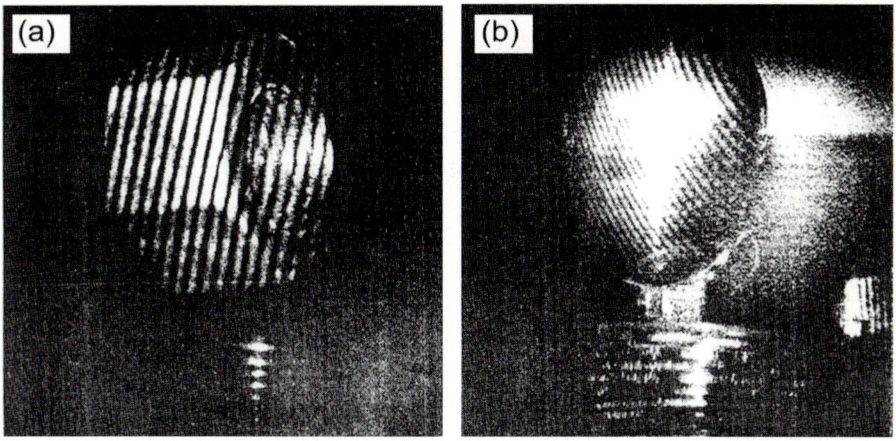

Fig. 2.3. (**a**) Straight fringes for an object in the absence of temperature change and (**b**) curved fringes for an object heated through 9°C

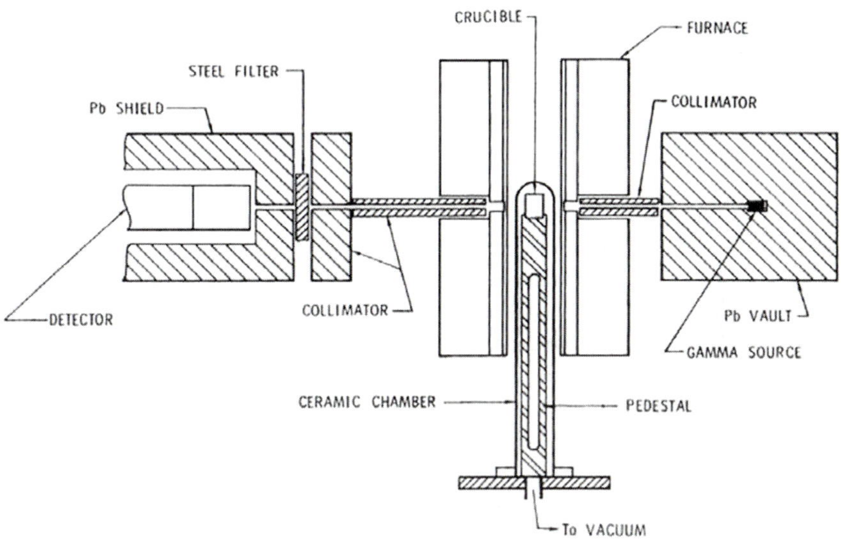

Fig. 2.4. Schematic diagram of gamma attenuation experimental apparatus

where $I(T)$ and $I_0(T)$ are the γ-ray intensities at temperature T measured after passage through the apparatus with and without the sample. $\rho(T)$ and $l(T)$ are the density and length of the material at temperature T.

Quantities z and x are defined as:

$$z = \frac{\log[I(T_1)I_0(T_2)/I(T_2)I_0(T_1)]}{\sigma\, \rho_1\, l_1}, \tag{2.19}$$

$$x = (\Delta T)\alpha, \tag{2.20}$$

x is obtained by solving

$$6x^3 + 3x^2 - 2x - z = 0. \tag{2.21}$$

The procedure, then, involves determination of $I_0(T)$ and $I(T)$ as a function of temperature and independent measurements of ρ_1 and $l_1$at some temperature T_1 (e.g. room temperature). The unique features of this method are that it is a non-contact probe and it can be used for measuring the thermal expansion of a sample in the solid state as well as the molten state through the melting temperature. On the other hand, the method is limited to determination of isotropic thermal expansion.

2.3.2 Experimental Data on Thermal Expansion of Crystals

The data that has been generated on the thermal expansion of crystals is so enormous that it is not possible even to give a bibliography. We can only give a few important sources and refer to some individual contributions where whole families of crystals have been studied.

Krishnan [2.6] includes a chapter on thermal expansion in his book. Deshpande and Mudholker [2.58] discussed analytical methods to obtain thermal expansion coefficients from X-ray data. The book on thermal expansion by Krishnan et al. [2.20] is more comprehensive and contains discussions of experimental techniques, theories of thermal expansion and a large section on experimental data. The most comprehensive data compilation is given by Touloukian et al. [2.59] in which almost all possible experimental data on a very large number of solids are reproduced.

While considering individual contributions, the pioneering work of Fizeau [2.21] deserves mention. He developed the optical interferometric technique which has been used by successive generations of researchers, with due modifications and improvements. Besides, he reported results on a very large number of crystals (for numerical results and references, see [2.6, 2.20]). Many of the crystals studied by him have been studied again by using recent techniques but there is hardly a case where his data has been faulted. Fizeau's results have stood the test of time.

White [2.60] set up a 3-terminal capacitance dilatometer and established stringent procedures for the measurement of thermal expansion at very low temperatures (30–2 K). He used his method to determine the thermal expansion of a variety of solids like cubic, trigonal and hexagonal metals, elemental semiconductors, alkali halides and fluorite type oxides (for references see [2.20]).

Sharma [2.61] made systematic measurements of thermal expansion of several crystals at high temperatures. Bailey and Yates [2.24] determined the thermal expansion of the alkaline earth fluorides at low temperature. Accurate measurements of the thermal expansion of mineral crystals were made by

Falzone and Stacey [2.62]. The thermal expansion of binary oxides with fluorite, antifluorite and rutile structures has been reviewed by Taylor [2.63]. The results on measurements of thermal expansion of crystals with ADP, scheelite, rutile and calcite structures have been analysed vis-à-vis their structural features by Rao and Deshpande [2.64]. A systematic study of the thermal expansion of crystals with the A-15 structure has been made by Somi Reddy and Suryanarayana [2.65, 2.66].

2.3.3 'Invar'

Invar stands for "invariable". In a study of the physical properties of Fe–Ni alloys, Guillaume [2.67] and his colleagues at the International Bureau of Standards (located in France) were led to the discovery of an alloy "the temperature coefficient of which is practically zero". The composition of this alloy is 64.4% Fe and 35.6% Ni and the actual value of thermal expansion is 1.2 ($10^{-6}\,{}^{\circ}C^{-1}$).

This near-zero expansion of Invar found applications in drawing geodesic base lines, in railway signal transmission, in designing compensated clocks and watches and in making standards of length. An interesting application of INVAR is in detecting minute movements of the Eiffel tower due to atmospheric temperature fluctuation.

Guillaume was awarded the 1920 Nobel Prize in physics.

2.3.4 Thermal Expansion of Inert Gas Solids

The inert gas solids are important as they are systems which closely approximate to the Lennard-Jones 6–12 potential which is suitable for Van der Waal solids. The measurement of their thermal expansion presents difficulties as the entire experiment, starting with the growth of crystals, their orientation and the recording of X-ray diffraction photographs, has to be done at very low temperatures, typically in the range 3–20 K. The range of temperature over which measurements are made being limited, a high accuracy is required. In their X-ray diffraction work Simmons and coworkers (references in Table 2.2) used a large film-to-specimen distance (~50 cm) which permitted a high accuracy of ±0.00008 Å in the lattice constant determination. However, the difficulties are to some extent compensated by the large value of the expansion coefficients, which are typically 10 times those of metals and inorganic crystals.

The values of thermal expansion coefficient at 20 K for the inert gas solids are given in Table 2.2. Barron [2.68] theoretically constructed a universal curve for the temperature variation of the Gruneisen constant for the inert gas solids according to which the Gruneisen constant varies from 2.85 to 3.00. The values calculated from experimental data are of this order.

Table 2.2. Values of thermal expansion coefficient (αp of inert gas solids at a temperature close to 20K

Inert gas solid	$\alpha[10^{-6}\,°\mathrm{C}^{-1}]$	Ref.
Ar	185	[2.69]
Ne	1334	[2.70]
Kr	157	[2.71]
Xe	116	[2.72]

2.3.5 Correlations of Thermal Expansion with other Physical Properties

The thermal expansion coefficient correlates with several other physical properties of solids. Some of these correlations are:

1. The product of the coefficient of expansion and the melting point is a constant. This is known in literature as Lindemann's formula. Application to various families of crystals showed that the constant varies from family to family. Van Uitert et al. [2.73] showed that $\alpha\ t_{\mathrm{m}}$ (t_{m} being in °C) has an average value of 0.027 for the alkali halides.
2. Touloukian et al. [2.59] plotted the expansion coefficient of the alkali halides against the refractive index and obtained a linear plot.
3. Hannemann and Gatos [2.74] proposed the empirical relation:

$$\alpha = C\psi, \tag{2.22}$$

where C is a constant and ψ the compressibility. Hanneman and Gatos drew α vs. ψ plots for several metal groups. Not all the plots were linear as predicted by (2.22).

4. Askil [2.75] plotted α for metals against E_{diff} the activation energy for diffusion. He obtained a hyperbolic plot which can be represented by

$$\alpha\ E_{\mathrm{diff}} = \text{constant}. \tag{2.23}$$

2.3.6 Thermal Expansion and Vacancies in Solids

Solids contain vacancies in thermal equilibrium with the lattice. If N is the total number of sites and ΔN the number of vacancies, the fractional concentration ($\Delta N/N$) increases exponentially with temperature. The measurement of thermal expansion provides an important method for a study of thermally generated defects.

It has been mentioned that the thermal expansion of a solid can be measured by making measurements of length changes ($\Delta L/L$) in a bulk sample or the lattice constant changes ($\Delta a/a$) caused by a change in temperature.

For a given temperature interval, these two values should be equal. However, if both the techniques are highly accurate, a slight but significant difference is observed with $(\Delta L/L) > (\Delta a/a)$. This is because in the bulk (or single crystal) sample, the vacancies form new surface layers. The difference $[(\Delta L/L) - (\Delta a/a)]$ increases with temperature and is directly related to the concentration of defects as given by

$$(\Delta N/N) = 3\,[(\Delta L/L) - (\Delta a/a)] = Ae^{-E_{\mathrm{f}}/k_{\mathrm{B}}T}, \tag{2.24}$$

where A is a constant, E_{f} the energy of formation of monovacancies, k_{B} the Boltzmann constant and T the absolute temperature. Thus a plot of log $[(\Delta L/L) - (\Delta a/a)]$ and T^{-1} will be a straight line with a slope of $(E_{\mathrm{f}}/k_{\mathrm{B}})$.

Simmons and Balluffi [2.76–2.79] made careful measurements of $(\Delta L/L)$ and $(\Delta a/a)$ on samples of Al, Ag, Au and Cu. The sample for the length measurements was a rod (~50 cm). The ΔL measurements were made with filar micrometer microscopes by observing the shift of two indentation marks on the rod. A single crystal grain was identified on the surface of the rod. This was made the X-ray single crystal sample from which back reflection photographs were recorded. The sample rod was kept in a furnace (~300 cm long) which provided a sufficiently large zone where the temperature was constant within ±0.2 °C. As an example, the plot of $(\Delta L/L)$ and $(\Delta a/a)$ at different temperatures for Al is shown in Fig. 2.5 and the corresponding plot of log $3[(\Delta L/L) - (\Delta a/a)]$ vs. T^{-1} is shown in Fig. 2.6. The values of E_{f}, the energy of formation of monovacancies for the metals studied by Simmons and Balluffi are given in Table 2.3.

2.3.7 Effect of Gross Defects on Thermal Expansion

Effect of Dislocations

In the preceding section, the difference between the dilatometric and lattice expansions in cubic crystals was discussed. In non-cubic crystals, in place of (2.24), we get

$$\begin{aligned}(\Delta N/N) &= 2\,\Delta_a(T) + \Delta_c(T)\\ &= Ae^{-E_{\mathrm{f}}/k_{\mathrm{B}}T}\end{aligned} \tag{2.25}$$

Here,

$$\Delta_a(T) = (\Delta L/L)_a - (\Delta a/a) \tag{2.26}$$

and

$$\Delta_c(T) = (\Delta L/L)_c - (\Delta c/c). \tag{2.27}$$

Again, a plot of log $[2\,\Delta_a(T) + \Delta_c(T)]$ vs. T^{-1}, is a straight line with slope $E_{\mathrm{f}}/k_{\mathrm{B}}$.

Nowick and Feder [2.80] theoretically examined the role of dislocations on the difference between the dilatometric and lattice expansions in hcp crystals. They arrived at the following results:

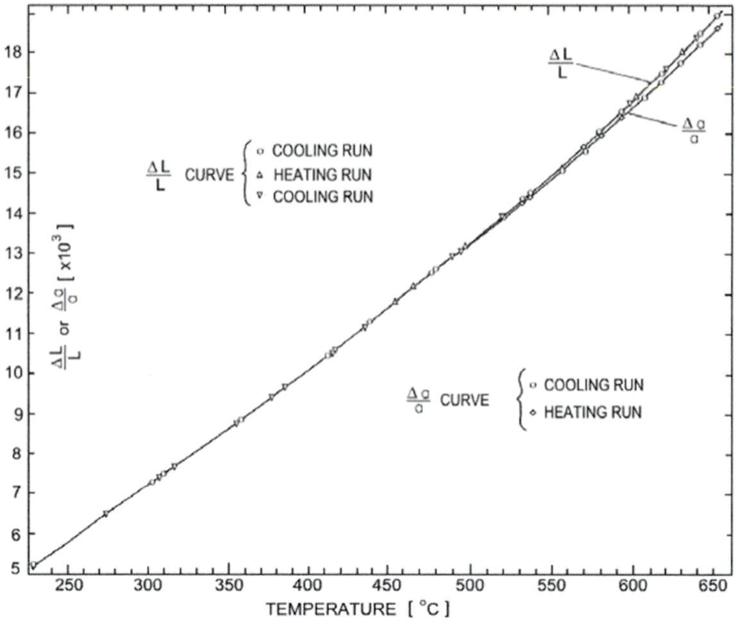

Fig. 2.5. Plot of $\Delta a/a$ and $\Delta L/L$ against temperature for aluminium

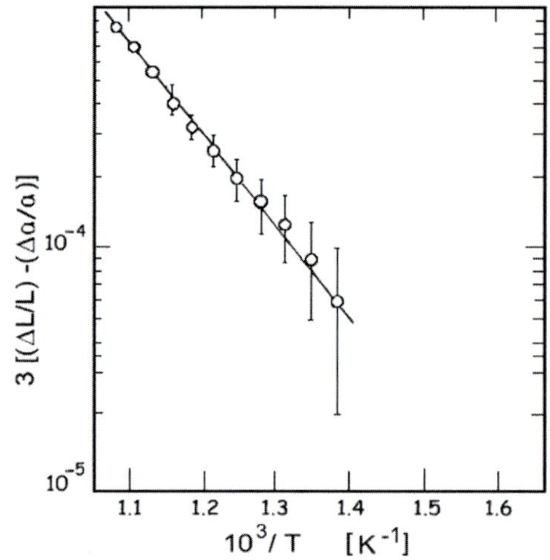

Fig. 2.6. Plot of log 3 $[(\Delta L/L)–(\Delta a/a)]$ against T^{-1} for aluminium

(a) As the dislocations act as sources and sinks for vacancies, the ratio $(\Delta c/\Delta a)$ is dependent on the mechanism of dislocation climb.
(b) For diffusion-limited climb, $(\Delta c/\Delta a)$ is independent of temperature.
(c) For climb-rate limitation, $(\Delta c/\Delta a)$ is a function of temperature.

Table 2.3. Values of formation energy (E_{f}) of monovacancies for some metals obtained from thermal expansion measurements

Metal	E_f [eV]	Ref.
Al	0.76	[2.76]
Ag	1.09	[2.77]
Au	0.94	[2.78]
Cu	1.17	[2.79]

Table 2.4. Theoretical values of $(\Delta c/\Delta a)$ for different values of ϕ

Plane	$\cos^2\phi$	$(\Delta c/\Delta a)$
$\{01\bar{1}0\}$	1	2
$\{11\bar{2}2\}$	0.78	1.3
$\{10\bar{1}1\}$	0.83	1.4

(d) The ratio $(\Delta c/\Delta a)$ is related to the distribution of dislocations and is, hence, sample-dependent.
(e) Feder and Nowick [2.81] showed theoretically that the ratio $(\Delta c/\Delta a)$ is related to the orientation (ϕ) of the dislocation (Burgers vector) with the normal to the slip plane. Values calculated by them for some typical cases are given in Table 2.4.

Feder and Nowick [2.81] verified these predictions by making very accurate dilatometric and lattice expansion measurements on cadmium crystals. They observed that $(\Delta c/\Delta a)$ is temperature independent in Cd which indicated diffusion-limited climb. In contrast, Janot et al. [2.82] found that $(\Delta c/\Delta a)$ was strongly dependent in Mg which indicates a climb-rate limited mechanism. Lastly, Feder and Nowick [2.81] found that $(\Delta c/\Delta a)$ was, indeed, sample dependent with values of 2.3 and 0.7 for different samples of Cd.

Effect of Mosaic Block Readjustment

Generally, the thermal expansion values obtained by bulk methods turn out to be larger than those obtained by the lattice (X-ray) method. As mentioned in the preceding section, this difference is attributed to the effect of thermally generated point defects. However, cases do occur where the lattice values are larger than the bulk values. The thermal expansion of sodium chlorate ($NaClO_3$) is an example. The coefficients of linear thermal expansion of $NaClO_3$ obtained by Sharma [2.83] by the optical interferometric method employing single crystal samples and those obtained by Deshpande and Mudholker [2.84] by the X-ray powder diffraction method are shown in Fig. 2.7. It is clearly seen that the expansion coefficient by the bulk method is less than that by the lattice method by about 15% at 200°C.

To explain these differences, Deshpande and Mudholker [2.84] invoked the effect of defects like dislocations and voids in the single crystal. They suggested

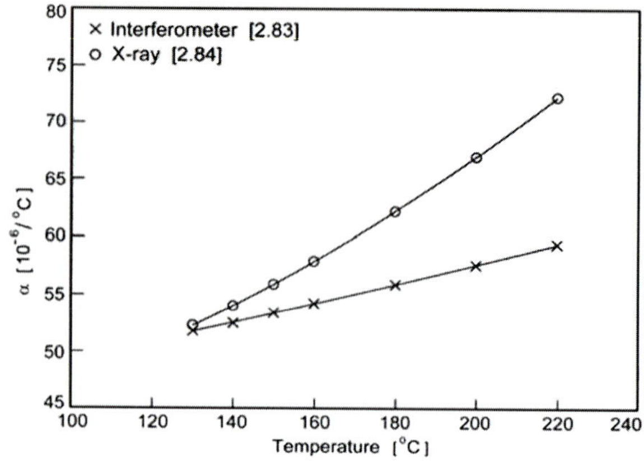

Fig. 2.7. Thermal expansion coefficient (α) of $NaClO_3$ against tempreration (t) by the interferometric and X-ray methods

that due to these defects, a part of the lattice expansion of the individual mosaic domains is accommodated in the interspaces between the domains and is not shown up in macroscopic measurements. Calling this effect $\alpha_{\text{readjustment}}$ and denoting the lattice expansion by α_{lattice} and that due to point defects by α_{Schottky}, Deshpande and Mudholker proposed the relation

$$\alpha_{\text{macroscopic}} = \alpha_{\text{lattice}} + \alpha_{\text{Schottky}} - \alpha_{\text{readjustment}}. \tag{2.28}$$

According to this relation, depending on the relative values of the last two terms, $\alpha_{\text{macroscopic}}$, the value determined by bulk methods, may be more or less than α_{lattice}.

Effect of Cracks and Voids

Graphite has a very good cleavage parallel to the layers and crystallites of graphite have a large number of cracks and voids. The thermal expansion coefficient of polycrystalline graphite measured by dilatometric method is less than the ideal isotropic value $(2\alpha_a + \alpha_c)/3$ where α_a and α_c are the linear coefficients of expansion in the a and c directions. Another way to obtain the isotropic value is to consider the orientation distribution of the crystallites. This is given by

$$\alpha = R\alpha_c + (1 - R)\alpha_a, \tag{2.29}$$

where R is an integral representing the orientation distribution of the crystallites [2.85, 2.86]. But the experimental value is still less than the value calculated in this way. Price and Bohras [2.87] modified (2.29) to:

$$\alpha = R\alpha_c + (1 - R)f\alpha_a, \tag{2.30}$$

Table 2.5. Effect of γ-irradiation on the thermal expansion coefficient α of $NaClO_3$

Irradiation [10^6 rad]	$\alpha[10^{-6}\,°C^{-1}]$
0	45.51
5.0	45.72
7.5	46.73
9.0	47.06
10.5	47.63
12.0	46.75
14.0	46.26
17.0	45.85

where f is a constant called 'accommodation factor'. This factor is introduced to take into account the accommodation of the crystalline expansion into the inter-crystallite spaces. Equation (2.30) gives better agreement with experimental data than (2.29).

2.3.8 Effect of Irradiation

Merriam et al. [2.88] made measurements on X-ray irradiated NaCl. They employed a novel photoelastic method to determine the volume changes. They made two types of measurements: (a) the fractional volume change ($\Delta V/V$) caused by irradiation and (b) the fractional increase in thermal expansion ($\Delta\alpha/\alpha$). Merriam et al. found that (i) there is an increase in thermal expansion, (ii) the fractional increase in thermal expansion is a linear function of the fractional volume change and (iii) the fractional increase in expansion coefficient is typically less than 1% and is about 14 times the concentration of radiation-induced defects as measured by the volume expansion.

Stapien-Damm et al. [2.89] measured the lattice constant of $NaClO_3$ irradiated by γ-rays from a ^{60}Co source. The expected change in lattice constant being small, the highly accurate Bond's method was used. The values of the linear coefficient of thermal expansion of $NaClO_3$ at different γ-ray doses are given in Table 2.5. It is seen that α increases from 45.5 to 47.6 ($10^{-6}\,°C^{-1}$) up to a dosage of 10.5 (10^6 rad) and then decreases reaching the normal (unirradiated) value at a dosage of 17 (10^6 rad). The effect is attributed to strains created by the release of radiolysis products.

2.3.9 Surface Thermal Expansion

Many properties (Debye temperatures and hardness for instance) show a difference when measured at the surface and in the bulk of a crystal. The thermal expansion has also been measured close to the surface. Nesterenko et al. [2.90] used LEED to determine the temperature variation of lattice parameters of Si, Ge and GaAs. Finzel et al. [2.91] used atomic beam diffraction (ABD) to measure the thermal expansion of NaF and LiF. Watanabe et al. [2.92] studied the temperature dependence of lattice parameters of LiF using ABD. In ABD, neutral atoms are used with energies in the range of 10–100 MeV. ABD probes only the upper surface layer. The lattice parameter a is calculated from the relation

$$a = 2\lambda/(\sin\theta_{\mathrm{f}} - \sin\theta_{\mathrm{i}}), \tag{2.31}$$

where θ_{f} and θ_{i} are the incident and scattered angles. The temperature variation of lattice parameters of LiF observed by Watanabe et al. [2.92] for the surface layers and for the bulk crystal by Swanson and Tatge [2.93] are shown in Fig. 2.8. The values of the ratio $\alpha_{\mathrm{surface}}/\alpha_{\mathrm{bulk}}$ are given in Table 2.6. In all the cases studied, the surface thermal expansion is much higher than the bulk expansion. It may be mentioned that theoretical calculations of the surface thermal expansion coefficients [2.94] also show that $\alpha_{\mathrm{surface}}$ is larger than α_{bulk}.

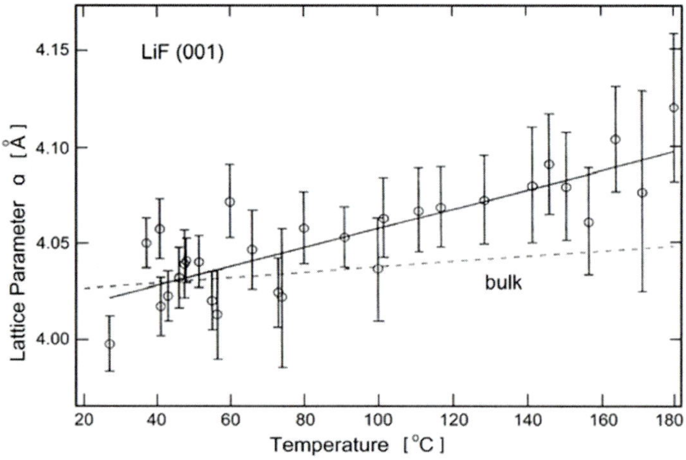

Fig. 2.8. Temperature dependence of the lattice parameter of the uppermost surface layer of LiF. The dotted line indicates the bulk lattice parameter

Table 2.6. Values of $(\alpha_{\mathrm{surface}}/\alpha_{\mathrm{bulk}})$ for some crystals

Crystal	$(\alpha_{\mathrm{surface}}/\alpha_{\mathrm{bulk}})$	Ref.
Si	8	[2.90]
Ge	4	[2.90]
GaAs	3	[2.90]
NaF	2	[2.91]
LiF	2	[2.91]
	4	[2.92]

Table 2.7. Values of the volume coefficient of expansion $\beta = (1/V)(\mathrm{d}V/\mathrm{d}T)$ at different pressures

Crystal	$\beta[10^{-4}\,°\mathrm{C}^{-1}]$					Ref.
	Experimental results (up to 30 kbar)					
P (kbar) →	0	10	20	30		
NaCl structure						
LiF	0.96	0.93	0.87	0.79		[2.95]
NaCl	1.169	0.968	0.841	0.756		[2.96]
	Experimental results (up to 80 kbar)					
P (kbar) →	0	20	40	60	80	
LiF	0.96	0.89	0.86	0.83	0.80	[2.97]
NaF	0.94	0.84	0.79	0.75	0.73	[2.97]
KF	0.97	0.84	0.78	–	–	[2.97]
CsCl structure						
CsCl	1.38	0.85	0.54	–	–	[2.97]
High pressure CsCl phase						
KF	(1.78) Extrapolated from high pressure values		1.04	0.85	0.70	[2.97]
	Theoretical results (up to 40 kbar)					
P (kbar) →	0	10	29	30	40	
NaCl structure						
LiCl	1.32	1.07	0.90	0.78	0.69	[2.98]
LiBr	1.50	1.15	0.94	0.79	0.68	[2.98]
LiI	1.80	1.25	0.96	0.78	0.66	[2.98]
NaBr	1.26	0.96	0.77	0.65	0.56	[2.98]
NaI	1.37	0.96	0.74	0.60	0.51	[2.98]

2.3.10 Pressure Variation of Thermal Expansion

With the development of high-pressure X-ray diffraction techniques, it has become possible to study the pressure variation of thermal expansion of solids. Boehler and Kennedy [2.95, 2.96] made length measurements whereas Yagi [2.97] made unit cell volume measurements at high pressures. Results on the measurement of the volume expansion coefficient β at high pressures for some alkali halides (in the NaCl and CsCl phases) are given in Table 2.7.

Yagi [2.97] pointed out that the ratio (β/ψ) is independent of pressure, ψ being the compressibility. Kumar [2.98] calculated the pressure variation of α for some alkali halides from the relation:

$$\beta(P) \;=\; \beta(0)\;\left[1+(\delta\psi)+(P-P_0)\right]^{-1}, \tag{2.32}$$

where δ is the Anderson–Gruneisen parameter. His calculated values agreed with experimental values for NaCl and LiF. Hence, his calculated results for a few other alkali halides (for which experimental data are not available) are included in Table 2.7.

2.3.11 Theories of Thermal Expansion

Gruneisen [2.16, 2.99] was the first to propose a full-fledged theory of thermal expansion. A detailed elaboration of Gruneisen's theory is given in [2.100, 2.101]. The main results are (i) the introduction of a constant γ defined by

$$\gamma = -\mathrm{d}\log\nu/\mathrm{d}\log V, \tag{2.33}$$

where ν is a lattice frequency and V, the volume and (ii) the derivation of

$$\alpha = C_V/3Q[1 - KE_{\mathrm{th}}/Q)]^2, \tag{2.34}$$

where C_V is the specific heat, $E_{\mathrm{th}} = \int_0^T C_V\,\mathrm{d}T$, $Q = V/\gamma\psi$ and K is a constant.

Gruneisen [2.99] applied this equation to several crystals. Subsequently, the equation has been used in different forms to analyse the thermal expansion data on metals by Hume-Rothery [2.102] and on alkali halides by Fitschmeister [2.103].

The constant γ was shown to be equal to

$$\gamma = 3\alpha\, V/\psi\, C_V, \tag{2.35}$$

where ψ is the compressibility, C_V the specific heat, α the linear expansion coefficient and V the volume. γ was introduced by Gruneisen as a constant independent of temperature. But calculations of γ from (2.35) using experimental data at different temperatures revealed that it is not so. In fact, γ has a low temperature limit (γ_0) and a high temperature limit (γ_∞).

Most of the later developments in theories of thermal expansion are related to a more sophisticated definition of γ. With the development of lattice dynamics, it was clear that a mode Gruneisen constant γ_i should be associated with each mode of vibration. The mean γ is given by

$$\gamma = \sum \gamma_i C_i \Big/ \sum C_i, \tag{2.36}$$

where C_i is the Einstein specific heat function for the ith mode. Different models were assumed for the lattice vibration spectrum and so γ could be evaluated for each model.

While much work has been done on different aspects of γ (its temperature dependence, its volume dependence, its dependence on the lattice dynamical model, its directional dependence), the mean value of γ remains the focal point in all theoretical work on thermal expansion. It is rightly known after its innovator as the Gruneisen constant.

Table 2.8. Negative minimum values of α and the temperature at which the negative minimum value is observed [2.104, 2.105]

System	T [K]	$\alpha[10^{-6}\mathrm{K}^{-1}]$
IV–IV		
Si	~80	−0.77
Ge	28	−0.4
III–IV		
GaSb	30	−1
GaAs	30	−0.5
InSb	~30	−1.6
AlSb	~38	−1.0
II–VI		
ZnTe	28	−0.68
ZnS	28	−0.32
ZnSe	32	−0.69
CdTe	28	−3.3
I–VIII		
CuCl	32	−0.33
Alkali halides		
RbCl	4.4	−0.02
RbBr	7.6	−0.58
RbI	7.6	−0.55

Table 2.9. Examples of systems having negative expansion in one direction; α_c, α_a expansion coefficients in c and a directions respectively. Data for metals [2.106] and for others [2.64]

System	$\alpha_c[10^{-6}\,\mathrm{K}^{-1}]$	$\alpha_a[10^{-6}\,\mathrm{K}^{-1}]$
As	15	−2.0
Graphite	28	−1.2
Se	−4.5	69
Te	−2.3	30
In	−7.7	55
FeF_2	−0.1	16.6
RuO_2	−1.4	6.9
CrO_2	−15.0	18.7
$CaCO_3$	25.0	−3.68
$CdCO_3$	19.0	−2.13

2.3.12 Negative Thermal Expansion

The negative thermal expansion displayed by some solids is an interesting phenomenon. At very low temperatures, some crystals with diamond, zinc blende and alkali halide structures have an overall negative thermal expansion. This information is given in Table 2.8. Several anisotropic crystals have a negative thermal coefficient in one of the principal directions even at ordinary temperatures. Some examples of this type are given in Table 2.9.

In anisotropic crystals, the observed thermal expansion in any direction is the difference between the pure thermal expansion in that direction and the Poisson's contraction caused by expansion in other orthogonal directions. If the Poisson contraction is larger than the pure thermal expansion, the net observed thermal expansion in that direction is negative.

This qualitative explanation does not explain the overall negative expansion in crystals mentioned in Table 2.8. For these as well as for the anisotropic crystals showing negative expansion in one direction, recourse has to be made to the role of the mode Gruneisen constants. A crystal has several mode Gruneisen constants γ_i and the mean Gruneisen constant γ is the weighted mean of the mode Gruneisen constants. The thermal expansion coefficient α is given by

$$\alpha = (\psi\, C_V/3V)\gamma = (\psi\, C_V/3V)(\Sigma\gamma_i C_i/\Sigma C_i), \tag{2.37}$$

where C_i is the Einstein specific function for the ith mode of vibration and the other quantities are already defined. Blackman [2.107] made the important suggestion that in some crystals, some of the γ_i's may be negative and if their contribution dominates in the summation in (2.37), then the resulting value of α may be negative.

Gerlich [2.108] evaluated the mode γ's for crystals with diamond and ZnS structures from data on the pressure variation of elastic constants. Talwar et al. [2.104] calculated the γ_i's from lattice dynamical calculations. They obtained negative values for some γ_i's and observed that α calculated from (2.37) using these γ_i's is indeed negative.

2.3.13 Anisotropy of Thermal Expansion

The coefficient of linear thermal expansion of a crystal is a second rank tensor. Consequently, the number of principal expansion coefficients is 1 for cubic, 2 for tetragonal, hexagonal and trigonal, 3 for orthorhombic, 4 for monoclinic and 6 for triclinic crystals. Quantitatively, the anisotropy of thermal expansion differs from crystal to crystal and the theoretical explanation of the observed anisotropy in expansion is difficult.

Gruneisen and Goens [2.109] modified Gruneisen's [2.99] theory. By introducing directional Gruneisen constants (γ_a and γ_c) and directional Debye temperatures (θ_a and θ_c) for hcp elements, they derived the following expressions:

$$\alpha_a = (S_{11} + S_{12})q_{xx} + S_{13}q_{zz}, \tag{2.38}$$

$$\alpha_c = 2S_{13}q_{xx} + S_{33}q_{zz}, \tag{2.39}$$

where the S's are elastic constants and

$$q_{xx} = \frac{3R}{V}\gamma_a C\,(\theta_a/T)\,, \tag{2.40}$$

$$q_{zz} = \frac{3R}{V}\gamma_c C\left(\theta_c/T\right), \tag{2.41}$$

where the C's are Debye functions for the specific heats. With these equations, Gruneisen and Goens could account for the anisotropy in the thermal expansion of Zn including the negative value of α_a at low temperatures. In fact, Gruneisen and Goens theoretically predicted that α_a will have negative values for Cd at low temperatures. Such negative values were later confirmed experimentally by McCammon and White [2.110]. Riley [2.111] extended Gruneisen and Goens' formulation to high temperatures to explain the anisotropic thermal expansion of graphite.

Pastine [2.112], on the other hand, introduced anisotropy in the lattice potential of a tetragonal crystal in the following way:

$$\phi_j = -A\left(1 + \varepsilon\, P_z\right)/R_j^m + B/R_j^n, \tag{2.42}$$

where A, B, ε, m and n are constants, R_j the interatomic distance and $P_z = -(1/2)(1 - 3\cos\theta_j)$, θ_j being the angle between R_j and the $z-$axis. Pastine finally obtained an expression for the anisotropy ratio α_a/α_c in terms of the elastic constants and the directional Gruneisen constants. Pastine was able to get values for α_a/α_c in agreement with experimental values for the thermal expansion of In and γ-Mn.

2.4 Some of our Results

2.4.1 Coefficients of Thermal Expansion – Data Generation

The coefficients of thermal expansion of 26 crystals (22 cubic, 2 tetragonal and 2 hexagonal) have been determined at elevated temperatures. For these measurements a commercial Seifert flat film camera and two cameras fabricated by us, one focusing and the other of the Unicam type, were used. The design of the latter two cameras is discussed in Chap. 1. For six crystals a push-rod dilatometer (described in Sect. 2.2.5) was used.

In each case the lattice constant 'a' (or the length change Δl) was plotted as a function of the temperature t. From the smooth curve the values of $\mathrm{d}a/\mathrm{d}t$ (or $\mathrm{d}l/\mathrm{d}t$) were determined at different temperatures and the coefficient of thermal expansion $\alpha = (1/a_0)(\mathrm{d}a/\mathrm{d}t)$ or $(1/l_0)(\mathrm{d}l/\mathrm{d}t)$ was evaluated. Here a_0 and l_0 are the lattice constant or length of the sample at the starting temperature. These α values were fitted by least squares method to the equation

$$\alpha = (a_1 + a_2 t + a_3 t^2). \tag{2.43}$$

The values of α at room temperature [α (RT)] and the values of the constants a_1, a_2 and a_3 in (2.43) for the crystals studied are given in Tables 2.10 and 2.11. Some results derived from these data are discussed in the subsequent sections.

Table 2.10. Coefficient of thermal expansion at room temperature [α (RT)] and constants a_1, a_2 and a_3 in (2.43) for some cubic crystals [Methods: XC –X -ray camera; XRD –X-ray diffractometer; DLM –Dilatometer]

Crystal	Temperature range [°C]	Method	α(RT) [10^{-6} °C^{-1}]	a_1 [10^{-6} °C^{-1}]	a_2 [10^{-8} °C^{-1}]	a_3 [10^{-10} °C^{-1}]	Ref.
NaCl structure							
RbCl	30–200	XC	38.13	–	–	–	[2.113]
RbBr	30–140	XC	37.43	38.00	−1.54	4.13	[2.113]
RbI	40–160	XC	39.7	40.23	−3.81	7.46	[2.114]
MgO	20–260	XC	10.2	9.73	2.53	−0.20	[2.115]
CaS	40–280	XC	11.4	11.23	1.93	0.18	[2.114]
PbS	30–260	XC	19.0	18.81	0.74	–	[2.116]
EuS	20–700	XRD	14.3	14.16	0.32	–	[2.117]
TiC	30–300	DLM	6.17	5.95	1.10	–	[2.43]
CsCl structure							
NH_4Cl	30–55	XC	58.5	–	–	–	[2.118]
NH_4Br	30–130	XC	57.9	58.96	−11.98	35.00	[2.113]
CaF_2 structure							
CaF_2	30–330	XC	18.35	18.28	0.24	–	[2.119]
SrF_2	30–340	XC	18.04	17.98	0.23	–	[2.119]
BaF_2	30–480	XC	19.04	18.93	0.42	–	[2.119]
CdF_2	30–380	XC	21.77	18.94	0.52	–	[2.120]
PbF_2	35–380	XC	25.43	25.00	1.56	–	[2.120]
EuF_2	20–200	DLM	14.70	14.23	−0.01	2.88	[2.121]
	25–140	XRD	15.20	10.30	19.10	–	[2.117]
	140–240	XRD	−29.00	–	–	–	[2.117]
Miscellaneous structures							
V_3Si	20–300	DLM	8.54	8.15	1.98	−0.23	[2.122]
$Bi_4(GeO_4)_3$	40–300	DLM	7.40	6.01	3.38	–	[2.123]
$Bi_4(SiO_4)_3$	40–300	DLM	7.40	6.01	3.38	–	[2.123]
$Bi_{12}SiO_{20}$	30–300	DLM	11.83	10.71	5.78	−1.01	[2.43]
$Eu_3Fe_5O_{12}$	20–700	XRD	10.40	–	–	–	[2.117]
FeS_2	30–360	XC	8.78	8.67	0.38	–	[2.119]

2.4.2 USBM Inter-Laboratory Project on Thermal Expansion of MgO

The United States Bureau of Mines (USBM) conducted an inter-laboratory programme on thermal expansion of MgO. The USBM prepared pure MgO powder and samples from a single batch were supplied to several participating laboratories. The participants were asked to communicate the results on basic measurements (‘a’ or ‘l’, the length of the brickets) at different temperatures to the USBM where data processing was done. The results for all participating laboratories were published by Campbell [2.1]. Here, only such results are discussed which have a relevance to our contribution.

Table 2.11. Coefficient of thermal expansion at room temperature [α (RT)] and constants a_1, a_2 and a_3 in (2.43) for some tetragonal and hexagonal crystals along the a and c directions (method: XC)

Crystal	Temperature range [°C]	Direction	α(RT) [10^{-6} °C^{-1}]	a_1 [10^{-6} °C^{-1}]	a_2 [10^{-8} °C^{-1}]	a_3 [10^{-10} °C^{-1}]	Ref.
Tetragonal							
Sn (β)	35–150	a	38.13	14.64	6.00	−0.58	[2.124]
		c	30.90	28.14	9.36	0.03	[2.124]
KH_2PO_4	40–140	a	18.80	10.10	21.68	−2.62	[2.125]
		c	35.90	28.73	17.65	1.83	[2.125]
Hexagonal							
ZnO	20–250	a	6.50	–	–	–	[2.126]
		c	3.20	–	–	–	[2.126]
PbI_2	25–115	a	31.00	26.36	16.50	–	[2.127]
		c	23.60	15.69	26.30	–	[2.127]

1. Fig. 2.9 shows a plot of the per cent expansion [$(\Delta a/a) \times 100$] as a function of temperature. Our results (Laboratory R) are shown along with results from five other laboratories. All the six laboratories used X-ray camera techniques. What is of interest is that our results are consistent with the results obtained by several other participants.
2. The USBM had supplied two samples of equal purity but differing in grain size. Measurements were made on both the samples. At room temperature, the lattice constant for the fine sample (4.2117 Å) was found to be less than that for the coarse sample (4.2124 Å). However, this relative difference in lattice constant continued throughout. The coefficients of thermal expansion obtained for the two samples are given in Table 2.12. It can be seen that the values for the two samples agree within limits of experimental errors. Since the difference in lattice constant for two particle sizes is related to surface forces, the lack of difference in the thermal expansion coefficients for the two samples indicates that the surface forces are not (or are only slightly) temperature dependent.

2.4.3 Aspects of Gruneisen Theory

Gruneisen's theory has been discussed earlier. Some of our results on thermal expansion are analyzed in the light of Gruneisen's theory.

The Thermal Gruneisen Constant

The Gruneisen constant is defined as

$$\gamma = -d \log \nu / d \log V \tag{2.44}$$

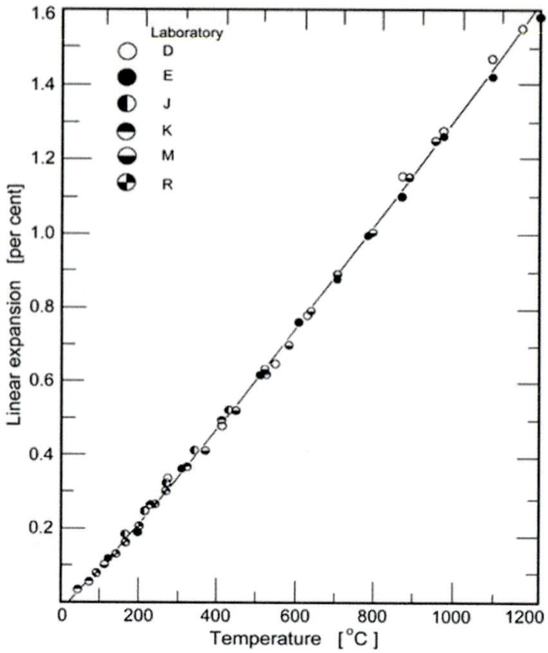

Fig. 2.9. Linear expansion values of magnesium oxide

Table 2.12. Coefficient of thermal expansion (α) of coarse and fine samples of MgO supplied by USBM

Temperature [°C]	$\alpha[10^{-6}\,°C^{-1}]$	
	Coarse sample	Fine sample
20	10.2	10.3
60	11.2	11.1
100	12.1	11.9
140	12.9	12.6
180	13.6	13.2
220	14.3	13.8
260	15.0	14.9

where ν is a lattice frequency and V the lattice volume. Thus γ represents the volume dependence of lattice frequencies. Phenomenologically, the Gruncisen constant is given by

$$\gamma_{\mathrm{th}} = \frac{3\alpha V}{\psi C_V} \tag{2.45}$$

where α is the coefficient of linear expansion, ψ is the compressibility, C_V the molar specific heat and V the molar volume. This value is called the thermal

Table 2.13. Room temperature values of γ_{th}

Crystal	γ_{th}	Ref.
Sn	2.01	[2.124]
As	1.44	[2.128]
Sb	1.18	[2.128]
Bi	1.11	[2.128]
Te	1.28	[2.128]
Se	1.59	[2.128]
ZnO	0.64	[2.129]
BeO	1.19	[2.129]
ZnS	0.63	[2.129]
CdS	0.61	[2.129]
MgF_2	1.02	[2.130]
$ZrSiO_4$	1.08	[2.131]
$Bi_4(GeO_4)_3$	1.18	[2.123]
$Bi_4(SiO_4)_3$	1.88	[2.123]
EuF_2	1.17	[2.121]

Gruneisen constant (γ_{th}) to distinguish it from values estimated from other crystal properties.

γ_{th} has been evaluated for several crystals. For some of them [Sn, $Bi_4(SiO_4)_3$, $Bi_4(GeO_4)_3$ and EuF_2], values of α from Table 2.10 are used. All other data are taken from literature. These values of γ_{th} are given in Table 2.13.

The values of γ_{th} are generally in the range 1–3. For ionic crystals γ_{th} is generally in the range 1–2. Srinivasan [2.132] pointed out purely empirically that γ_{th} has smaller values for covalent and partially covalent crystals. Based on Srinivasan's [2.132] suggestion, it is concluded that ZnO and MgF_2 are ionic and BeO, ZnS, CdS are partially covalent. Further, on the same basis, Rao et al. [2.123] concluded that $Bi_4(GeO_4)_3$ is less ionic than $Bi_4(SiO_4)_3$.

Gruneisen Constant from Other Properties

The Gruneisen constant can also be calculated by other methods. Two of these are (a) from pressure variation of elastic moduli and (b) from interatomic potentials. The results obtained from these methods are discussed in Chap. 6.

Temperature Variation of Thermal Expansion from Gruneisen's Theory

Several authors have analysed data on the temperature variation of thermal expansion of metals and the alkali halides on the basis of Gruneisen's theory. Several AB_2 type crystals have been studied [2.119] in our laboratory and the results have been analysed in the context of Gruneisen's theory.

Gruneisen's theory leads to the following equations:

$$\alpha = (QC_V/3(Q - KE_{\text{th}})^2, \tag{2.46}$$

$$Q = V/\gamma\psi, \tag{2.47}$$

$$K = \gamma + (2/3) \tag{2.48}$$

and

$$E_{\text{th}} = \int_0^T C_V \, \mathrm{d}T \tag{2.49}$$

The value of K is obtained from γ_{th} and C_V from known values of the Debye temperature θ. The value of Q is obtained by normalizing the experimental and calculated values of α at 300 K. Using the values of Q and K, α is calculated for temperatures in the range 300–700 K. From this, $\Delta\alpha$, the difference in the values of α at 300 and 700 K is calculated. This value ($\Delta\alpha$) is then compared with experimental values of $\Delta\alpha$ from different sources. The input parameters for (2.46) and the $\Delta\alpha$ values are given in Table 2.14. It can be seen that the experimental $\Delta\alpha$ values obtained by Hussain [2.119] are closer to those estimated from (2.46) than most of the other experimental values.

2.4.4 Studies of Some Anomalous Phenomena

Lead Sulphide

The temperature variation of lattice constant of lead sulphide (Galena, PbS) was studied by Deshpande [2.144] at elevated temperatures. Deshpande [2.144] found an anomalous change in the rate of increase in 'a' at about 100°C. The thermal expansion coefficient which was $24.4 \times 10^{-6}\,°\mathrm{C}^{-1}$ at room temperature decreased to $16.3 \times 10^{-6}\,°\mathrm{C}^{-1}$ at 100°C and then increased again. Deshpande [2.144] suggested that his observations needed a confirmation by a more systematic study.

Deshpande [2.144] made measurements with a flat film camera which recorded only two reflections. Further, the lattice constants were determined

Table 2.14. Values of $\Delta\alpha$ (increase in α over the range 300–700 K)

	Input for (2.46)			$\Delta\alpha[10^{-6}\,\mathrm{K}^{-1}]$	
Crystal	$Q(10^5)$	K	θ[K]	Calc.(2.46)	Experimental
CdF_2	2.957	2.96	328	3.2	2.1 [2.119], 17.0 [2.133], 9.0 [2.134]
PbF_2	2.422	2.74	237	6.3	6.2 [2.119], 28.1 [2.135]
CaF_2	2.961	2.41	513	5.2	0.9 [2.119], 45 [2.136], 5.4 [2.137], 7.7 [2.138]
SrF_2	3.215	2.29	380	3.3	0.9 [2.119], 6.2 [2.139]
BaF_2	3.194	2.36	282	2.9	1.7 [2.119], 10.5 [2.140], 1.8 [2.141]
FeS_2	5.613	2.13	645	1.9	1.5 [2.119], 5.9 [2.142], 2.3 [2.143]

Table 2.15. Lattice constant 'a' and coefficient of expansion "α" of PbS at elevated temperatures

t[°C]	a [Å]	t[°C]	$\alpha[10^{-6}\,°C^{-1}]$
26	5.9358	30	19.0
28	5.9362	40	19.1
72	5.9416	60	19.2
88	5.9435	80	19.4
101	5.9452	100	19.5
116	5.9470	120	19.7
138	5.9494	140	19.9
170	5.9535	180	20.1
195	5.9569	220	20.4
227	5.9610	260	20.8
243	5.9628		
271	5.9668		

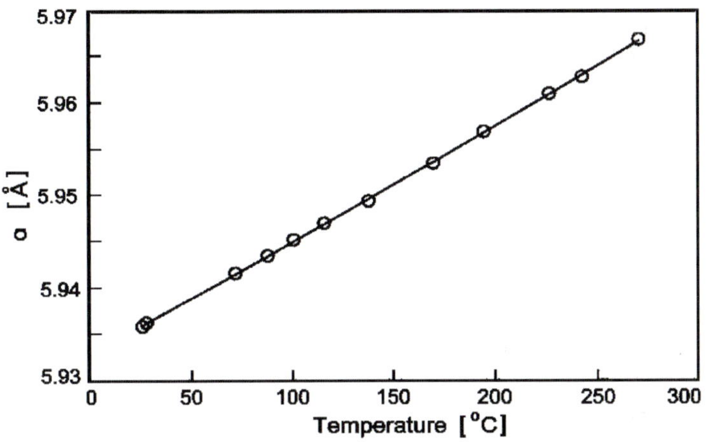

Fig. 2.10. Plot of lattice constant (a) versus temperature of PbS

only at five elevated temperatures up to 250°C. Sirdeshmukh and Deshpande [2.116] carried out more systematic measurements with a symmetric focusing camera which recorded five well-resolved doublets. Further, measurements were made at eleven elevated temperatures. The lattice constants obtained by Sirdeshmukh and Deshpande [2.116] are given in Table 2.15 and are shown in Fig. 2.10. It can be seen that the temperature variation is smooth. The coefficient of thermal expansion calculated from the data of Sirdeshmukh and Deshpande [2.116] is also given in Table 2.15 and shown in Fig. 2.11. Here also, the variation of α is smooth and there is no indication of any anomaly.

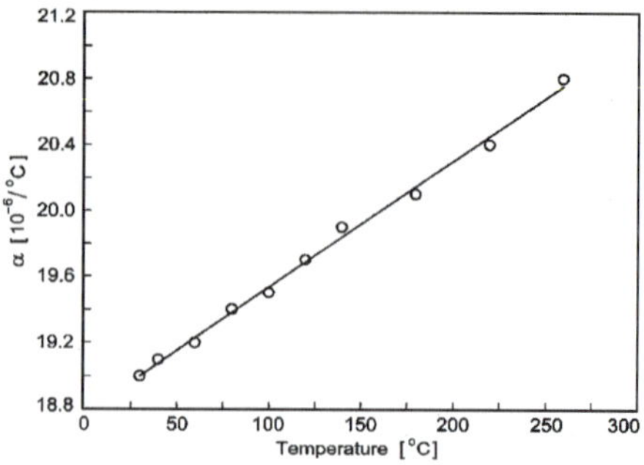

Fig. 2.11. Temperature variation of thermal expansion (α) of PbS

β Tin

The α–β transition in tin at low temperatures is well documented. A β–γ transition has been conjectured at a temperature of about 100 °C on the basis of studies of properties like malleability and thermoelectricity of tin-based alloys; a rhombic structure has been suggested for this γ phase.

Deshpande and Sirdeshmukh [2.145] carried out a detailed X-ray study with the help of a symmetric focusing camera making measurements at close temperatures. The study did neither reveal any new reflections other than those for the β phase nor any abnormal broadening of reflections. Further, the plots of the lattice parameters and the corresponding coefficients of expansion as a function of temperature were smooth over the temperature range where the transition was suspected. Thus no evidence of a structural phase transition was observed.

Europium Fluoride

An X-ray study of EuF_2 at elevated temperatures was carried out by Hussain and Sirdeshmukh [2.117] using a powder X-ray diffractometer. The temperature variation of the lattice constant for temperatures up to about 240°C is shown in Fig. 2.12. Up to about 140°C, the temperature variation of the lattice constant is normal. The coefficient of thermal expansion calculated for this temperature range is given in Table 2.10. Above 140°C, the lattice constant shows a decrease with temperature. The angular shift of the (331) reflection is shown in Fig. 2.13 which clearly shows a change in sign of the expansion coefficient. The expansion coefficient for the temperature range 140–240°C is -29×10^{-6}/°C. The reflections associated with the CaF_2 structure persist right up to 240°C. Other crystals with CaF_2 structure do not show any such anomaly. Although some crystals like $CaCO_3$, ADP and $NaNO_2$ show negative expansion in one direction and some crystals with ZnS structure show a negative overall expansion at low temperatures, no crystal is known to have

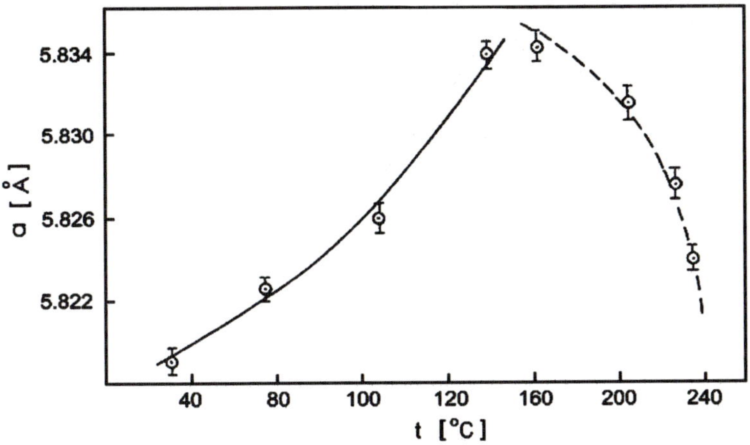

Fig. 2.12. Temperature variation of the lattice constant of EuF_2; dashed curve, guide to the eye

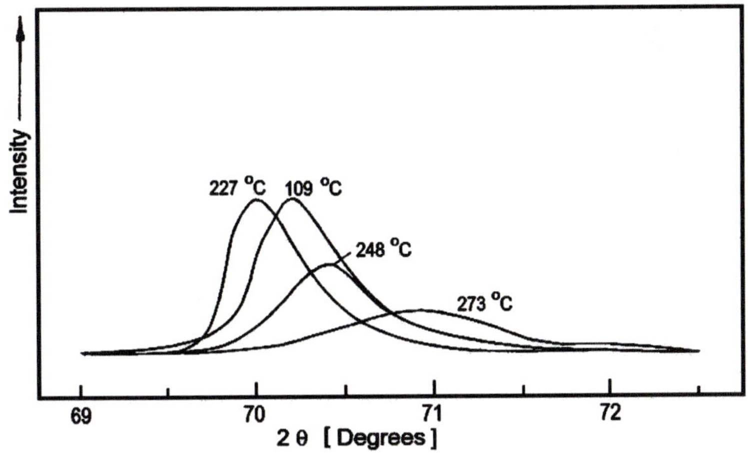

Fig. 2.13. The (331) reflection of EuF_2 at different temperatures

an overall negative expansion coefficient at elevated temperatures. An independent study of thermal expansion of this crystal and also studies of other properties are desirable.

2.4.5 Empirical Relations

Relation Between Thermal Expansion Coefficient and Debye Temperature

The Lindemann formula

$$\theta = c(T_{\mathrm{m}}/MV^{2/3})^{1/2} \tag{2.50}$$

has been used by several authors to estimate the Debye temperatures θ. T_m, M and V are the melting temperature, the molecular weight and the molar volume, respectively; c is a constant. This formula is particularly useful when the Debye temperature is not available from other properties like specific heats or elastic constants. Deshpande and Sirdeshmukh [2.146] pointed out that for binary crystals it is more appropriate to use the reduced mass $\mu = M_1 M_2/(M_1 + M_2)$ where M_1 and M_2 are the masses of the two atoms. One limitation of the Lindemann formula is that it cannot be applied to substances which decompose and hence, do not have a congruent melting point. Lindemann has also proposed a relation between the thermal expansion coefficient (α) and the melting point. This is given by:

$$\alpha\, T_m = c', \tag{2.51}$$

where c' is another constant.

Combining these two relations, Sirdeshmukh [2.147] obtained the relation:

$$\theta^2 \alpha\, \mu\, a^2 = c'', \tag{2.52}$$

where a is the lattice constant and c'' is yet another constant. The values of c'' for some crystals with the NaCl structure are given in Table 2.16. c'' has a mean value of 115.4×10^7 for this group of crystals. Using this value and the known values of α, μ and a for CaS, Sirdeshmukh [2.147] estimated the value of θ for CaS, which is not known from other methods. This value is given in Table 2.16 in parenthesis.

Relation Between Thermal Expansion Coefficient (α) and the Effective Ionic Charge (q^*)

While proposing a theory of the dielectric constant of ionic crystals, Szigeti [2.148] introduced the concept of the effective ionic charge (q^*). This is the ratio of the reduced charge ze* of an ion to its ideal ionic charge ze ($q^* = ze^*/ze$).

Sirdeshmukh [2.149] observed an empirical relation between the effective ionic charge (q^*) and the coefficient of thermal expansion (α) of the alkali

Table 2.16. Values of the constant c'' in (2.52) for some crystals with NaCl structure

Crystal	μ	a[Å]	$\alpha[10^{-6}\,\mathrm{K}^{-1}]$	θ [K]	$c'' = \theta^2 \alpha\, \mu\, a^2 \times 10^{-7}$
RbCl	25.06	6.591	38.1	171	133.4
RbBr	41.30	6.892	38.9	129	127.4
RbI	51.07	7.346	40.7	99	110.2
MgO	9.65	4.212	11.2	729	101.4
PbS	27.77	5.936	20.3	229	104.4
CaS	17.81	5.695	12.2	(403)	

Table 2.17. Values of α, q^* and $\alpha(q^*)^2$ for alkali halides

Crystal	$\alpha[10^{-6}\,K^{-1}]$	q^*	$\alpha(q^*)^2[10^{-6}\,K^{-1}]$
LiCl	44	0.74	24
LiBr	50	0.71	25
LiI	59	0.67	27
NaCl	40	0.76	23
NaBr	43	0.71	22
NaI	48	0.67	22
KCl	38	0.81	25
KBr	40	0.78	24
KI	45	0.68	21
RbCl	36	0.85	26
RbBr	38	0.78	23
RbI	43	0.70	21

halides. The relevant data on α and q^* are given in Table 2.17. It is observed that a gradation exists between α and q^*, one increasing as the other decreases. On careful examination, it is found that a simple relation exists between α and q^* viz:

$$\alpha(\,q^*)^2 = \text{constant.} \tag{2.53}$$

The values of $\alpha(q^*)^2$ are given in Table 2.17.

Sirdeshmukh [2.149] gave a qualitative explanation of this observation in terms of Megaw's [2.7] concept of ionic share. If q is the ionic share, Megaw [2.7] proposed the relation

$$\alpha\, q^2 = \text{constant.} \tag{2.54}$$

According to (2.54), α should be the same for a family of crystals like the alkali halides for which $q = e/n$, n being the coordination number. That it is not so is seen from Table 2.17 where α has a range of values from 38 to 59. Extending the concept of the effective ionic charge, the electrostatic share q becomes q^*e/n; substitution in (2.54) leads to (2.53).

The data in Table 2.17 are reproduced from Sirdeshmukh [2.149]. Inclusion of the alkali halides not included in Table 2.17 and use of more recent data on α and q^* (Sirdeshmukh et al. [2.105]) may slightly alter the values of $\alpha(q^*)^2$ but the conclusions remain unaffected.

Relation Between Thermal Expansion Coefficient (α) and the Compressibility (ψ)

Hannemann and Gatos [2.74] proposed an empirical relation between the thermal expansion coefficient (α) and the compressibility (ψ). The Hannemann–Gatos relation is given by

$$\alpha = k_1\psi, \tag{2.55}$$

where k_1 is a constant. It predicts a linear α vs. ψ plot. Hannemann and Gatos [2.74] drew the α vs. ψ plots for fcc metals, the alkali metals and the bcc metals. While two of the plots were linear, as predicted by (2.55), the third was non-linear. Sirdeshmukh [2.150] drew the α vs. ψ plot for another crystal family viz. the alkali halides and found that the data points were scattered all over the plot with no indication of any curve connecting them.

Sirdeshmukh [2.150] suggested a modified relation:

$$\alpha V = k_2 \psi, \tag{2.56}$$

where V is the volume and k_2 is another constant. The (αV) vs. ψ plots for three metal families considered by Hannemann and Gatos [2.74] as well as the alkali halides are shown in Figs. 2.14 –2.17. It can be seen that for all the four

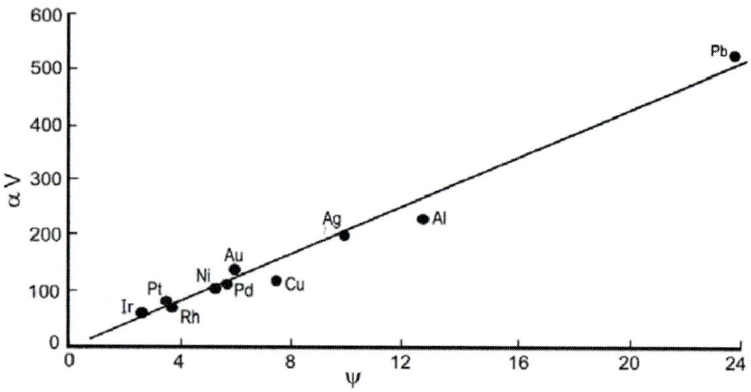

Fig. 2.14. $\alpha V - \psi$ plot for fcc metals (α in units of $10^{-6}\,\mathrm{C}^{-1}$, ψ in units of $10^{-12}\,\mathrm{cm^2 dyne^{-1}}$ and V in $\mathrm{cm^3 g \cdot atom^{-1}}$)

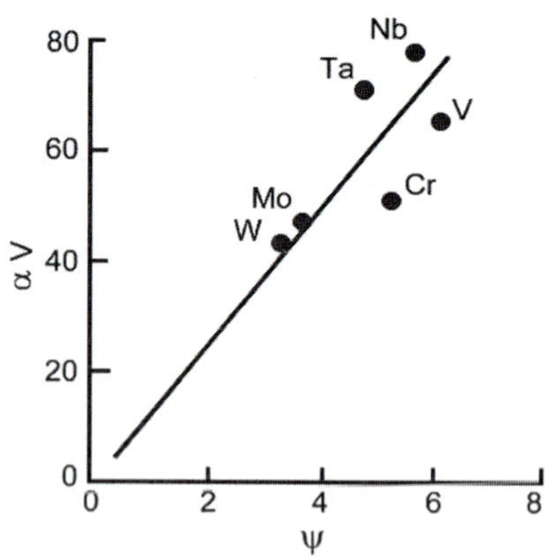

Fig. 2.15. $\alpha V - \psi$ plot for bcc metals (units as in Fig. 2.14)

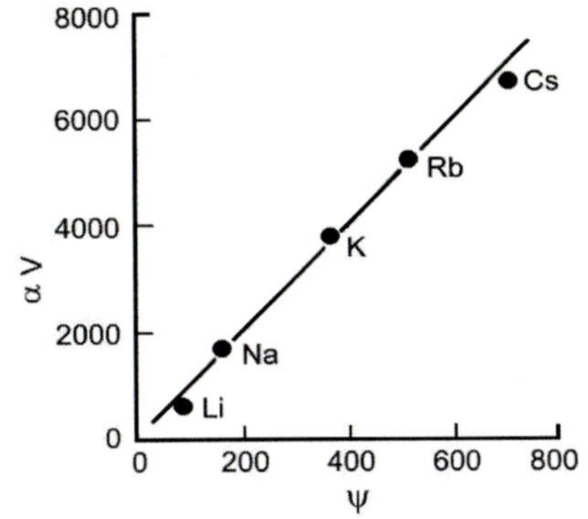

Fig. 2.16. $\alpha V - \psi$ plot for alkali metals (units as in Fig. 2.14)

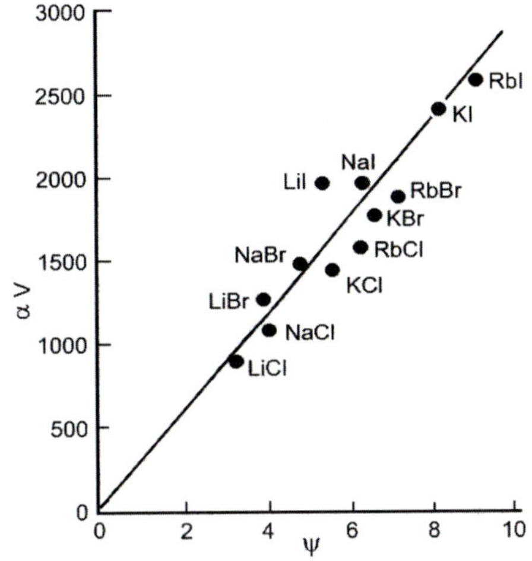

Fig. 2.17. $\alpha V - \psi$ plot for alkali halides (units as in Fig. 2.14)

families, linear plots are obtained as predicted by (2.56). Thus (2.56) appears to represent the relation between α and ψ better than (2.55).

Sirdeshmukh [2.150] pointed out that the empirical relation (2 56) follows from the expression for the coefficient of thermal expansion in terms of the specific heat C_V and the Gruneisen constant γ given by

$$\alpha V = (\gamma C_V \psi/3) \tag{2.57}$$

if it is assumed that the product γC_V is constant. Examination of data on γ and C_V for a large number of crystals shows that this is a reasonable approximation.

3

Debye–Waller Factors of Crystals

3.1 Introduction

Soon after the discovery of diffraction of X-rays by crystals by Laue, Debye undertook a theoretical study of the effect of temperature on the intensity of diffracted X-rays. Through this theory he introduced a factor known in X-ray literature as the Debye–Waller factor. Apart from providing a correction for the temperature effect on intensities, the Debye–Waller factor is interrelated to several physical properties and has emerged as an important solid state parameter. Considerable work is now available in literature on the theoretical as well as experimental aspects of the Debye–Waller factor.

3.2 Brief Outline of the Debye–Waller Theory

The effect of thermal vibration on the intensities of diffraction of X-rays in crystals has been considered by Debye [3.1], Waller [3.2] and later workers [3.3–3.8]. The theory as presented here follows the treatment of the subject by Warren [3.9]. The most familiar effect of thermal vibrations is the reduction of intensities of the reflections from crystal planes by a factor e^{-2M} derived on the basis of the Debye–Waller theory. In addition to reducing the intensities of the Bragg reflections, the thermal vibrations produce a diffuse intensity.

Consider a monatomic cubic crystal. The vibrations of the different atoms in a crystal are not independent of one another. The coupling between the displacements of the atoms may be expressed by representing the thermal vibrations in the crystal in terms of a set of elastic waves propagating in several directions, the amplitudes and wavelengths of the individual waves being different. At any instant the displacement of a particular atom is the sum of the displacements due to each of the elastic waves.

The periodic nature of the lattice sets a lower bound to the wavelength of the elastic waves propagating in the lattice. Waves shorter than twice the repeat distance along a crystal direction are not possible. An elastic wave may be represented by a wave vector $\mathbf{g}$, whose magnitude equals the reciprocal of the wavelength of the wave and, whose direction is along the wave normal. The lattice nature of the crystal restricts all possible $\mathbf{g}$ vectors to the first Brillouin zone.

The finite size of the crystal further restricts the possible $\mathbf{g}$ vectors. For a crystal containing N atoms, the number of $\mathbf{g}$ vectors possible is also N, the terminal points of these vectors being uniformly distributed throughout the first Brillouin zone. To each of these vectors there correspond three independent modes of vibration limiting the total number of waves to $3N$. The displacements of the lattice points due to thermal vibrations may be represented by the sum of the displacements due to the $3N$ independent elastic waves, these constituting the normal coordinates of the system.

The vector displacement $\mathbf{u}_n$ of the atom n due to the lattice waves is given by

$$\boldsymbol{u}_n = \sum_{gj} a_{gj} \mathbf{e}_{gj} \cos(\omega_{gj} t - 2\pi \mathbf{g} \,.\, \mathbf{r}_n - \delta_{gj}), \tag{3.1}$$

where $\mathbf{r}_n (= n_1 \mathbf{a}_1 + n_2 \mathbf{a}_2 + n_3 \mathbf{a}_3)$ is the position vector of the nth atom in the crystal, $\mathbf{e}_{gj}$ ($j = 1, 2, 3$) a unit vector along the direction of vibration associated with the wave vector $\mathbf{g}$, $a_{g\,j}$ the amplitude of the wave, ω_{gj} the circular frequency and δ_{gj} an arbitrary phase factor accounting for the fact that in an actual crystal, the phases of the elastic waves vary rapidly and arbitrarily with time, so that no phase relations are possible among the waves.

For an undisturbed lattice composed of identical atoms, the X-ray diffracted intensity expressed as a fraction of the waves scattered by a single electron is given by

$$I = f^2 \sum_m \sum_n \mathrm{e}^{\mathrm{i}k\,\mathbf{S}\,.\,(\mathbf{r}_m - \mathbf{r}_n)}, \tag{3.2}$$

f being the atomic scattering factor, $\mathbf{S} = (\mathbf{s} - \mathbf{s}_0)$, where $\mathbf{s}$ and $\mathbf{s}_0$ are the unit vectors along the diffraction direction and the incident beam direction, respectively, and $k = 2\pi/\lambda$, λ being the wavelength of X-rays. The vector $\mathbf{S}$ is perpendicular to the reflecting planes and its magnitude $|\mathbf{S}| = 2 \sin\theta$, where 2θ is the angle of scattering. Due to thermal vibrations, the atoms undergo small displacements, $\mathbf{r}_n \to \mathbf{r}_n + \mathbf{u}_n$ and the intensity at any instant is given by

$$I = f^2 \sum_m \sum_n \mathrm{e}^{\mathrm{i}k\,\mathbf{S}\,.\,(\mathbf{r}_m - \mathbf{r}_n)} \mathrm{e}^{\mathrm{i}k\,\mathbf{S}\,.\,(\mathbf{u}_m - \mathbf{u}_n)}. \tag{3.3}$$

Writing

$$P_{mn} = k\mathbf{S}.(\mathbf{u}_m - \mathbf{u}_n) \tag{3.4}$$

the time average of the intensity is given by

$$I = f^2 \sum_m \sum_n \mathrm{e}^{\mathrm{i}k\,\mathbf{S}\,.\,(\mathbf{r}_m - \mathbf{r}_n)} < \mathrm{e}^{\mathrm{i}P_{mn}} > . \tag{3.5}$$

For small values of P_{mn}

$$< e^{iP_{mn}} >= e^{-<P_{mn}^2>/2} . \qquad (3.6)$$

Using (3.1) and (3.4) and remembering that the phases of the waves are completely independent (resulting in the vanishing of cross terms), we may write

$$\frac{1}{2} < P_{mn}^2 >= \frac{1}{2} \sum_{gj} (k\mathbf{S} \, . \, \mathbf{e}_{gj})^2 < a_{gj}^2 > [1- < \cos \, 2\pi g \, .(\mathbf{r}_m - \mathbf{r}_n) >] . \qquad (3.7)$$

The first term in this equation is given by

$$\tfrac{1}{2} \sum_{gj} (k \; \mathbf{S}. \, \mathbf{e}_{gj})^2 < a_{gj}^2 > = \sum_{gj} G_{gj} = 8\pi^2 (\sin\theta/\lambda)^2 \sum_{gj} < a_{gj}^2 >_s , \qquad (3.8)$$

where $< a_{gj}^2 >_s$ is the component of the mean square amplitude in the direction of $\mathbf{S}$. Since the mean square displacement due to a wave is half the mean square amplitude, $< u_s^2 >$ the mean square of the component of displacement parallel to the direction of $\mathbf{S}$ is given by

$$< u_s^2 > = \tfrac{1}{2} \sum_{gj} < a_{gj}^2 >_s . \qquad (3.9)$$

Hence

$$\sum_{gj} G_{gj} = 16\pi^2 < u_s^2 > (\sin^2\theta/\lambda^2) = 2M, \qquad (3.10)$$

where

$$\begin{aligned} M &= 8\pi^2 < u_s^2 > (\sin^2\theta/\lambda^2) \\ &= B \, (\sin^2\theta/\lambda^2) \end{aligned} \qquad (3.11)$$

with $B = 8\pi^2 < u_s^2 >$. For a cubic crystal, the mean square displacement $< u_s^2 >$ is the same along the three perpendicular directions so that the total mean square displacement $< u^2 >$ is equal to $3 < u_s^2 >$. Then

$$B = 8\pi^2 < u_s^2 > = (8\pi^2/3) < u^2 > . \qquad (3.12)$$

The intensity of diffraction by a crystal is therefore

$$I = f^2 e^{-2M} \sum_m \sum_n e^{ik \, \mathbf{S} \, . \, (\mathbf{r}_m - \mathbf{r}_n)} \, e^{\sum_{gj} G_{gj} \cos \, 2\pi \mathbf{g} \, . \, (\mathbf{r}_m - \mathbf{r}_n)} . \qquad (3.13)$$

Since $\sum_{gj} G_{gj} \cos \, 2\pi\mathbf{g} \, . \, (\mathbf{r}_m - \mathbf{r}_n)$ is small, the intensity may be written as

$$I = f^2 e^{-2M} \sum_m \sum_n e^{ik \, \mathbf{S} \, . \, (\mathbf{r}_m - \mathbf{r}_n)} + f^2 e^{-2M} \sum_m \sum_n e^{ik \, \mathbf{S} \, . \, (\mathbf{r}_m - \mathbf{r}_n)} \qquad (3.14)$$

$$\sum_{gj} G_{gj} \cos \, 2\pi \, \mathbf{g} \, . \, (\mathbf{r}_m - \mathbf{r}_n).$$

The first term represents the ordinary Bragg intensity reduced by the factor e^{-2M}. The second term represents the first-order temperature diffuse scattering (TDS). The factor $f\ e^{-M} \sum_n e^{(2\pi\,i/\lambda)\ \mathbf{S}\,.\,\mathbf{r}_n}$ represents the structure factor F_T of the lattice including the effect of thermal vibrations. For a crystal containing more than one kind of atom in the unit cell, the structure factor is expressed as

$$F_T = \sum_n f_n\ e^{-M_n} e^{(2\pi i/\lambda)\ \mathbf{S}\,.\,\mathbf{r}_n}, \tag{3.15}$$

where f_n is the atomic scattering factor for atom of type n and M_n is the corresponding temperature factor defined by

$$M_n = 8\pi^2 < u_{ns}^2 > (\sin\theta/\lambda)^2 \tag{3.16}$$

$< u_{ns}^2 >$ being the mean square displacement of atom of type n in the direction of **S**.

The mean square amplitude of vibration may be expressed in terms of the kinetic energy of thermal vibration. The kinetic energy is given by

$$E_{\text{kin}} = \tfrac{1}{2} \sum_n m\dot{u}_n^2, \tag{3.17}$$

where m is the mass of the atoms in the crystal. The mean total energy $< E >$ is twice the mean kinetic energy and using (3.1) is given by

$$\begin{aligned} < E > &= (1/2)Nm \sum_{gj} \omega_{gj}^2 < a_{gj}^2 > \\ &= \sum_{gj} < E_{gj} >, \end{aligned} \tag{3.18}$$

where $< E_{gj} >$ is the average energy per wave and N is the number of lattice points in the crystal. If the elastic waves are treated as harmonic oscillators, the energy per wave on the basis of the quantum theory is

$$< E_{gj} > = \ [(e^{h\nu_{gj}/k_B T} - 1)^{-1} + (1/2)]\ \ h\nu_{gj}, \tag{3.19}$$

where $\omega_{gj} = 2\pi\nu_{gj}$ and h and k_B are the Planck and Boltzmann constants, respectively. In terms of the average energy per wave, the mean square amplitude is given by

$$< a_{gj}^2 > = \ 2 < E_{gj} > /Nm\omega_{gj}^2. \tag{3.20}$$

The temperature factor $2M$ can now be evaluated. Using (3.8), (3.10), (3.19) and (3.20), $2M$ can be expressed as

$$2M = (h/4\pi^2 Nm) \sum_{gj} (k\ \mathbf{S}.\mathbf{e}_{gj})^2 [(e^{h\nu_{gj}/k_B T} - 1)^{-1} + (1/2)]\ /\nu_{gj}. \tag{3.21}$$

Three aspects of (3.21) may be noted. First, for the evaluation of M, a knowledge of the complete vibration spectrum of the solid is necessary. Second, a monatomic solid has been assumed so far. For a non-monatomic solid a temperature factor is associated with each atom. Third, in non-cubic crystals, the temperature factor is anisotropic.

It may be noted that the factors e^{-2M}, M or B have all been referred to by various authors as the temperature factor due to thermal displacements of the atoms or the Debye–Waller factor [3.8–3.10]. To avoid ambiguity, e^{-2M}, M or B factor will be referred to as the temperature correction, temperature factor and the Debye–Waller factor, respectively.

Equation (3.21) takes a useful form if the Debye model is assumed for a solid. A crystal is assumed by Debye to behave as a continuous solid with regard to the propagation of elastic waves. Following Debye, the waves are assumed to be pure transverse and pure longitudinal, the velocities of these waves being independent of wavelength and direction. Though waves of the same wavelength do not necessarily travel with the same speed in different directions even in a cubic crystal, average values for the velocities of the longitudinal and transverse waves may be taken to obtain a fair approximation to M.

Since the number of $\mathbf{g}$ vectors is large, the summation over g in (3.21) is facilitated by replacing it by an integration over a sphere of radius $g_{\rm m}$, where $g_{\rm m}$ represents the maximum possible value for g. The density of lattice points in such a sphere is $N/(4/3)\pi {g_{\rm m}}^3$. The number of lattice points in an element of volume for values of g between g and $(g+{\rm d}g)$ is $(4\pi g^2{\rm d}g)\ [N/(4\pi/3){g_{\rm m}}^3]$. Since for a wave of given type j, the vibration direction $\boldsymbol{e}_{gj}$ takes all directions with equal probability relative to $\mathbf{S}$, we may write

$$\begin{aligned} <(k\mathbf{S}.\mathbf{e}_{gj})^2> &= k^2\left|\mathbf{S}^2\right| <\cos^2(\mathbf{S},\mathbf{e})> \\ &= (4\pi\sin\theta/\lambda)^2/3. \end{aligned} \tag{3.22}$$

Equation (3.21) may now written as

$$2M = \left[\frac{h}{12\pi^2 Nm}\right]\left(\frac{4\pi\sin\theta}{\lambda}\right)^2 \sum_j \int_0^{g_{\rm m}} \left[\frac{1}{{\rm e}^{h\nu_{gj}/k_{\rm B}T}-1}+\frac{1}{2}\right]\frac{3Ng^2{\rm d}g}{\nu_{gj}g_{\rm m}^3}. \tag{3.23}$$

For a wave of given type j, $\nu_j = V_j g$, where V_j is the average velocity for each kind of wave. Changing the integration variable, (3.23) can be written as

$$2M = \left[\frac{4h}{m}\right]\left(\frac{\sin\theta}{\lambda}\right)^2 \sum_j (1/\nu_{{\rm m}j}^3) \int_0^{\nu_{{\rm m}j}} \left[\frac{1}{{\rm e}^{h\,\nu/k_{\rm B}T}-1}+\frac{1}{2}\right]\nu\,{\rm d}\nu. \tag{3.24}$$

Writing

$$\phi(x) = (1/x)\int_0^x y\,{\rm d}y/({\rm e}^y-1), \tag{3.25}$$

where $y = (h\nu/k_\mathrm{B}T)$ and $x_j = (h\nu_{mj}/k_\mathrm{B}T)$, (3.24) becomes

$$2M = (4k_\mathrm{B}T/m)(\sin\theta/\lambda)^2 \sum_j (1/\nu^2_{\mathrm{m}j})\,[\phi\,(x_j) + (x_j/4)]. \qquad (3.26)$$

An average value of the expression under the summation in (3.26) for different values of j is used to evaluate $2M$. The function $[\phi(x) + (x/4)]$ can be shown to have almost equal values for the transverse and longitudinal waves at not too low temperatures. Debye's theory of specific heats defines a characteristic temperature $\theta = (h\nu_\mathrm{m}/k_\mathrm{B})$. Since $\nu_{\mathrm{m}j}$ depends on the velocity of the waves, θ for the longitudinal and transverse waves cannot be the same. Instead, a mean θ_M is defined for use in (3.26), given by

$$x = \theta_\mathrm{M}/T$$

and

$$(3/\theta^2_\mathrm{M}) = (1/\theta^2_\mathrm{l}) + (2/\theta^2_\mathrm{t}), \qquad (3.27)$$

where the subscripts 'l' and 't' referring to the longitudinal and transverse waves, respectively. The mean Debye temperature defined by (3.27) differs from θ_D in Debye's theory of specific heats where

$$(3/\theta^3_\mathrm{D}) = (1/\theta^3_\mathrm{l}) + (2/\theta^3_\mathrm{t}). \qquad (3.28)$$

With these approximations, we obtain

$$2M = 2B(\sin\theta/\lambda)^2 = (12h^2/mk_\mathrm{B}\theta_\mathrm{M})\left[\frac{\phi(x)}{x} + \frac{1}{4}\right](\sin\theta/\lambda)^2. \qquad (3.29)$$

3.3 Experimental Procedures

In this section, the procedures for the evaluation of the Debye–Waller factors and the X-ray Debye temperatures from the measurement of X-ray diffraction intensities are discussed in detail. The procedures followed in our work are described; however, they have a general relevance to all work in this field.

3.3.1 Measurement of Integrated Intensity

Details of Diffractometric Procedure

The details of the diffractometric method are given here. Measurements were made with a Phillips PW 1051 diffractometer using Cu $K\alpha$ radiation, filtered to eliminate $K\beta$ radiation. In several cases, an argon-filled G.M. counter was used to collect the intensity data. In our later work, an NaI (Tl) scintillation counter was used. The integrated intensities were recorded on a strip chart recorder. The X-ray tube was usually operated at 34–36 kV and 15–17 ma. The width of the receiving slit was 0.2 mm. For angles $2\theta < 80°$ a 1° divergent slit and for $2\theta > 80°$ a 4° divergent slit were used. A scanning speed of 1/4° per minute was adopted. A time constant of 4 s was chosen for the counting circuit. A 2θ scan was adopted. An initial adjustment with a standard silicon sample ensured that the error in 2θ was not greater than 0.1°. The chart speed was adjusted to be 800 mm h^{-1}. The counter was operated at 1,650 V. For the GM counter used, the dead time was 150 µs. The diffractometer incorporates a high degree of stabilization in the X-ray generator and receiving circuits, minimizing the errors due to variations in intensity of the source or in the efficiency of the receiving circuits, due to fluctuations in the line voltage. This was confirmed by checking the integrated intensity of a peak after an interval of about 8 h. The variation was always found to be negligible.

Sample Preparation

The starting material was reduced to a fine powder by slow grinding in an agate mortar for periods ranging from 30 min to an hour (the grinding time varying with the sample chosen), so that almost the whole of the powder passed through a 325 mesh screen. Reduction of particle size by slow grinding effectively minimizes the preferred orientation errors [3.11]. It also reduces primary extinction to a negligible level [3.12]. This was found to hold in the present investigation since no noticeable reduction in intensities of the low angle reflections could be observed. The intensities were therefore not corrected for extinction although a method of correcting for extinction in powders has been proposed by Mitra and Chattopadhyay [3.13].

The sample mounting is done by the conventional method of packing the powder in a rectangular sample holder. This consists of a rectangular sheet of aluminium $4 \times 2.5 \times 0.2$ cm in size, with a rectangular hole of dimensions 2×1 cm. A glass plate is bound to the sample holder with tape. The powder is then sifted into the rectangular cavity so formed. The powder is made to fill the cavity evenly by tamping with a glass plate or with a spatula. The excess powder is removed by means of a razor blade. Some more powder is added on to the surface of the sample and lightly pressed with a glass plate, the excess powder being once again removed. This process is repeated three to four times until a nearly plane surface is obtained. The effect of prolonged grinding on the effective temperature factor, due to the introduction of strains in the sample has been studied by Inagaki et al. [3.14]. The presence of strains in a sample affects the width of the Bragg peak, the width increasing with increase of strain. However, for the substances chosen in the present study, grinding for about an hour or less was often sufficient. Further, from the method adopted for sample mounting, strain likely to be introduced as with pelleting under high pressure was least probable. This was also supported by the absence of unduly broad peaks in the diffractograms.

Some other sample characteristics affecting the intensities are porosity, surface effects and preferred orientation. In accurate intensity measurement, it is necessary either to correct for the effects of these factors or to choose the experimental conditions so as to minimize the effects of these factors. The present mode of sample preparation was conducive to minimum preferred orientation. However, since the samples were not packed at high pressures, the sample densities may differ from the true density of the specimen. The reduction in intensity due to the average structure of the specimen, termed the porosity effect P_0, is uniform, so that P_0 may be included in the scale factor in relative intensity measurements [3.15]. The surface roughness effect P_s due to granularity in the surface layer depends on the Bragg angle only at low 2θ values. In a typical case, Suortti [3.15] quotes an error of 2 % in the intensity of a reflection at $2\theta = 20°$ due to the surface roughness. However, the measured Bragg peaks for the samples studied generally have values of $2\theta > 20°$, so that a neglect of the granularity correction is not totally unjustified. It may be mentioned here that the choice of particle size and procedure of sample preparation is very much similar to that followed by Blattner et al. [3.16] who observe that under these conditions, the effects due to extinction and preferred orientation are minimized.

An accurate value of the absorption correction is essential for absolute intensity measurements. For a flat powder specimen the absorption correction is not a function of the Bragg angle and hence can be included in the scale factor for relative intensity measurements.

Measurement of Intensities

The intensities were measured on a relative basis. Filtered radiation was used and a crystal monochromator was not used. Togawa [3.17] has pointed out that 'the relative intensity method with use of filtered radiation has an advantage because the reflection intensity is stronger than with a monochromator beam' although it has the disadvantage of a slightly higher background. The line profiles of the Bragg peaks were recorded with the help of a strip chart recorder. The counting rate was adjusted so that statistical counting errors were about 2%. The peaks were scanned for a range of values 2θ covering two degrees on either side of the Bragg peak, so that the background could be determined with minimum ambiguity. All measurable peaks corresponding to the range of 2θ values between 20° and 140° were scanned. The relative intensities were determined by measuring the area under a Bragg peak. A minimum of six independent measurements were made for each peak and their average taken as a measure of the intensity. The intensities measured in this way generally agreed within 2%, and in no case did the disagreement exceed 5%. The agreement between the relative values for the different peaks was good. The trace of a typical peak is shown in Fig. 3.1.

A straight-line background scheme was adopted by drawing a common tangent to the line profile on either side of the Bragg peak. The ratios of the peak-to-background intensities were high so that standard errors due to

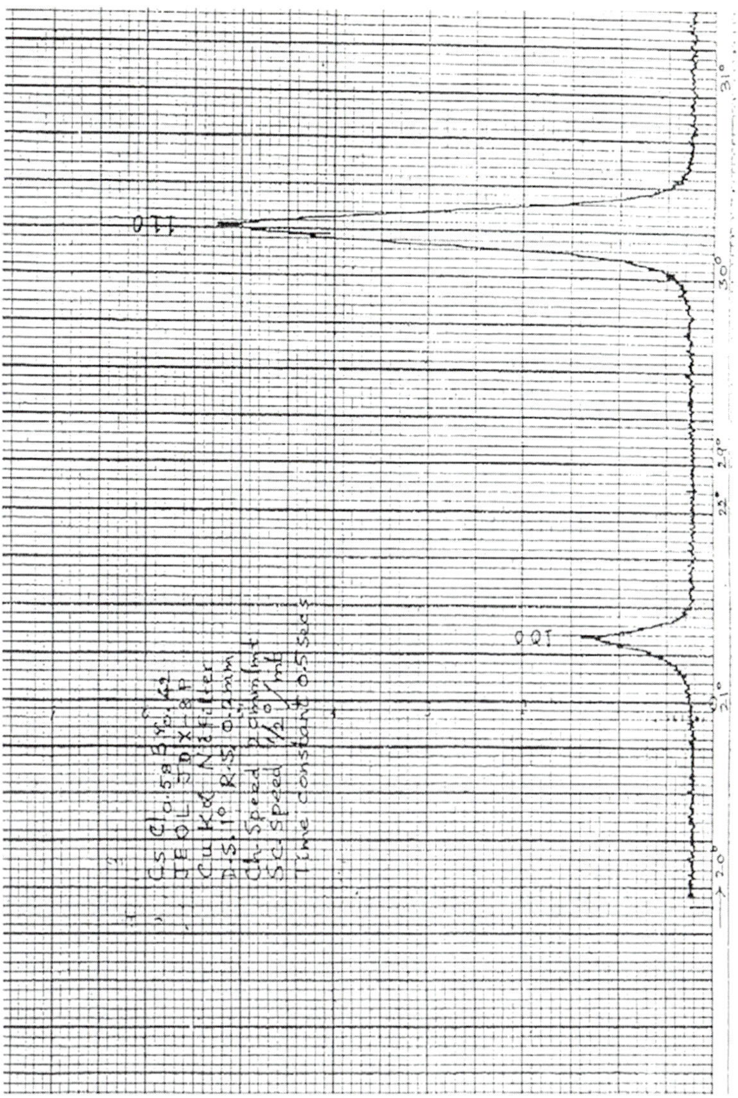

Fig. 3.1. Trace of (100) and (110) reflections from a powder sample ($CsCl_{0.58}Br_{0.52}$)

background radiation were low. Since the α_2 component of the Cu $K\alpha$ radiation cannot be eliminated, consideration of the relative contributions of the α_1 and α_2 components to be in the ratio 2:1 enabled the intensity due to the α_1 component to be used in the final analysis.

The measured intensities were corrected for temperature diffuse scattering (TDS) by the method of Chipman and Paskin [3.18]. According to this method, the correction factor β for TDS is given by

$$\beta = qa_0(\cos\theta/\lambda)QB(\sin\theta/\lambda)^2, \tag{3.30}$$

where a_0 is the lattice parameter, Q the length of the 2θ scan in radians, θ the Bragg angle, λ the wavelength and the factor q is $(\pi/3)^{1/3}$ for the fcc lattice and $(2\pi/3)^{1/3}$ for the bcc lattice. The value of B, used to estimate the correction β is obtained from an initial least square treatment of the intensity data.

The correction for dead time to the measured intensities was applied by an approximate method proposed by Chipman [3.19]. Since the count rate continually varies as a peak is scanned, the true intensity is given by

$$\int N_\mathrm{t}\,\mathrm{d}t = \int [N_0/(1-N_0\tau)]\mathrm{d}t, \tag{3.31}$$

where N_t and N_0 correspond to the instantaneous values of the true and measured count rates respectively. For a Gaussian peak Chipman obtains to a first approximation $[1+(\tau N_\mathrm{p}/\sqrt{2})]$ for the ratio between the true and measured integrated intensities, τ being the dead time of the counting apparatus and N_p is the count rate at the peak.

The integrated intensity I_0 is obtained from the measured intensity $I_0{}'$ from the relation

$$I_0 = \frac{I_0{}'}{(1+\beta)}\left(1+\frac{\tau N_p}{\sqrt{2}}\right) = I_0{}'(1+\psi)/(1+\beta), \tag{3.32}$$

where $\psi = \tau N_\mathrm{p}/\sqrt{2}$ and β represent the dead time and TDS corrections, respectively, (3.30) and (3.31). Finally, the integrated intensity I_0 of a Bragg reflection from a powder sample for unpolarised radiation [3.19] is given by

$$I_0 = \left[\frac{P\,r_\mathrm{e}^2\,l\,t\,\lambda^3}{64\pi\,R^2\mu_\mathrm{l}\,\omega\,V_\mathrm{a}^2}\right]\left[\frac{1+\cos^2 2\theta}{\sin^2\theta\,\cos\theta}\right] J\,F_\mathrm{T}^2. \tag{3.33}$$

Here, I_0 is the duly corrected experimental intensity (3.32). In (3.33), P is the total power in the primary beam, r_e the classical electron radius, R the specimen-to-receiving slit distance, V_a the volume of the unit cell, μ_l the linear absorption coefficient, ω the angular velocity of the receiving slit, λ the wavelength, l and t the length and width of the receiving slit, J the multiplicity factor and F_T the structure factor. The factor $(1+\cos^2 2\theta)/\sin^2\theta\cos\theta$ is the Lorentz-polarisation factor (LP), θ being the Bragg angle.

The measurement of intensities can be done either on an absolute basis or a relative basis. Absolute intensity measurement can, again, be done from preliminary measurements of the intensity of the incident beam with the help of attenuators or by obtaining a scale factor whereby the relative intensities of the substance examined are brought to an absolute scale. This is achieved by comparison of the intensity of a particular reflection of the substance under study with the intensity of a reflection from a standard substance.

3.3.2 Analysis of Intensity Data

Determination of Debye–Waller Factors

From (3.33) the observed intensity I_0 may be written as

$$I_0 = C\ (\mathrm{LP})\ J F_{\mathrm{T}}{}^2, \tag{3.34}$$

where C is called the scale factor. From (3.15) the structure factor is given by

$$F_{\mathrm{T}}^2 = \left[\sum f_i \mathrm{e}^{-B_i\,(\sin\theta/\lambda)^2} \cos 2\pi\ (hx_i + ky_i + lz_i)\right]^2 + \left[\sum f_i \mathrm{e}^{-B_i\,(\sin\theta/\lambda)^2} \sin\ 2\pi\ (hx_i + ky_i + lz_i)\right]^2, \tag{3.35}$$

where h, k, l are Miller indices and $x_i,\ y_i,\ z_i$ are the position coordinates of the ith atom; the summation includes all atoms in the unit cell. For crystals with simple structures, (3.35) gets simplified. The structure factors for some simple structures are given in Table 3.1.

The Debye–Waller factors can be determined by three methods:

(a) By a least squares refinement of (3.34) and (3.35).

Table 3.1. Structure factor square (F^2) for some simple structures

Structure	Indices	F^2
1. fcc	h,k,l all odd or all even	$[4f\mathrm{e}^{-B(\sin\theta/\lambda)^2}]^2$
	$h+2k=3n$, l even	$[4f^2]\mathrm{e}^{-(4\pi\sin\theta/\lambda)^2[\phi(u,\psi)]}$
2. hcp	$h+2k=3n\pm1$, l odd	$[3f^2]\mathrm{e}^{-(4\pi\sin\theta/\lambda)^2[\phi(u,\psi)]}$
	$h+2k=3n\pm1$, l even	$[f^2]\mathrm{e}^{-(4\pi\sin\theta/\lambda)^2[\phi(u,\psi)]}$ $\phi = <u_{\mathrm{c}}>^2\cos^2\psi - <u_{\mathrm{a}}>^2\sin^2\psi$ $\cos\psi = (la/c)\{(4/3)(h^2 + hk + k^2) + (la/c)^2\}^{-1/2}$ $B = (8\pi^2/3)\{2<u_{\mathrm{a}}^2> + <u_{\mathrm{c}}^2>\}$
3. NaCl (AB)	h,k,l all even	$[4f_{\mathrm{A}}\mathrm{e}^{-B_{\mathrm{A}}(\sin\theta/\lambda)^2} + 4f_{\mathrm{B}}\mathrm{e}^{-B_{\mathrm{B}}(\sin\theta/\lambda)^2}]^2$
	h,k,l all odd	$[4f_{\mathrm{A}}\mathrm{e}^{-B_{\mathrm{A}}(\sin\theta/\lambda)^2} - 4f_{\mathrm{B}}\mathrm{e}^{-B_{\mathrm{B}}(\sin\theta/\lambda)^2}]^2$
4. CsCl (AB)	$h+k+l$ even	$[f_{\mathrm{A}}\mathrm{e}^{-B_{\mathrm{A}}(\sin\theta/\lambda)^2} + f_{\mathrm{B}}\mathrm{e}^{-B_{\mathrm{B}}(\sin\theta/\lambda)^2}]^2$
	$h+k+l$ odd	$[f_{\mathrm{A}}\mathrm{e}^{-B_{\mathrm{A}}(\sin\theta/\lambda)^2} - f_{\mathrm{B}}\mathrm{e}^{-B_{\mathrm{B}}(\sin\theta/\lambda)^2}]^2$
5. ZnS (AB)	$h+k+l=4\,n$	$[4f_{\mathrm{A}}\mathrm{e}^{-B_{\mathrm{A}}(\sin\theta/\lambda)^2} + 4f_{\mathrm{B}}\mathrm{e}^{-B_{\mathrm{B}}(\sin\theta/\lambda)^2}]^2$
	$h+k+l=4\,n+2$	$[4f_{\mathrm{A}}\mathrm{e}^{-B_{\mathrm{A}}(\sin\theta/\lambda)^2} - 4f_{\mathrm{B}}\mathrm{e}^{-B_{\mathrm{B}}(\sin\theta/\lambda)^2}]^2$
	$h+k+l=4\,n+1$	$[4f_{\mathrm{A}}\mathrm{e}^{-B_{\mathrm{A}}(\sin\theta/\lambda)^2}]^2 + [4f_{\mathrm{B}}\mathrm{e}^{-B_{\mathrm{B}}(\sin\theta/\lambda)^2}]^2$
6. $CaF_2(AB_2)$	$h+k+l=4\,n+1$	$[4f_{\mathrm{A}}\mathrm{e}^{-B_{\mathrm{A}}(\sin\theta/\lambda)^2}]^2$
	$h+k+l=4\,n$	$[4f_{\mathrm{A}}\mathrm{e}^{-B_{\mathrm{A}}(\sin\theta/\lambda)^2} + 8f_{\mathrm{B}}\mathrm{e}^{-B_{\mathrm{B}}(\sin\theta/\lambda)^2}]^2$
	$h+k+l=4\,n+2$	$[4f_{\mathrm{A}}\mathrm{e}^{-B_{\mathrm{A}}(\sin\theta/\lambda)^2} - 8f_{\mathrm{B}}\mathrm{e}^{-B_{\mathrm{B}}(\sin\theta/\lambda)^2}]^2$

(b) For certain structures, the structure factor F_T takes a simple form (see Table 3.1). In such cases, semigraphical–semianalytical procedure can be used. As an example, we shall consider crystals with zinc blende structure. The structure factors, as given in Table 3.1, are

$$F_T{}^2 = 16\left[f_A e^{-B_A(\sin\theta/\lambda)^2} + f_B e^{-B_B(\sin\theta/\lambda)^2}\right]^2 \quad \text{for} \quad h+k+l=4n, \tag{3.36}$$

and

$$F_T{}^2 = 16\left[f_A e^{-B_A(\sin\theta/\lambda)^2} - f_B e^{-B_B(\sin\theta/\lambda)^2}\right]^2 \quad \text{for} \quad h+k+l=4n+2. \tag{3.37}$$

We may now introduce relative structure factors F_i' [3.9] as

$$F_i' = [I_0/(\text{LP})\,J]^{1/2}\,. \tag{3.38}$$

The plot of F_i' vs. $(\sin\theta/\lambda)$ is a smooth curve for reflections in (3.36) and (3.37). Typical curves are shown in Fig. 3.2. From these curves, we can choose values of F_{4n}' and F_{4n+2}' at a given value of $(\sin\theta/\lambda)$. The sum and difference of these values give quantities f_A^0 and f_B^0 defined as follows:

$$f_A^0 = \left(F_{4n}' + F_{4n+2}'\right) = C_1 f_A e^{-B_A(\sin\theta/\lambda)^2} \tag{3.39}$$

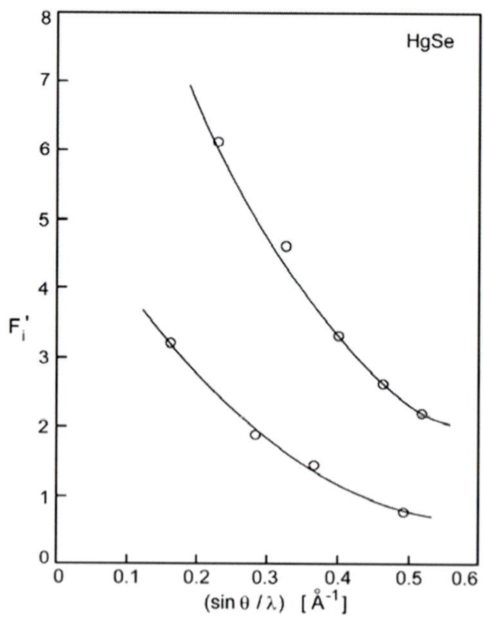

Fig. 3.2. Plot of relative structure factor F_i' vs $(\sin\theta/\lambda)$ for HgSe

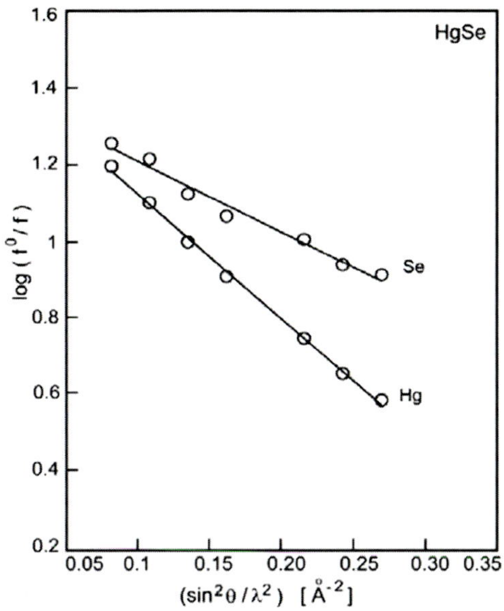

Fig. 3.3. Plot of log $(f_A{}^0/f_A)$ and log $(f_B{}^0/f_B)$ vs $(\sin\theta/\lambda)^2$ for HgSe

and

$$f_{\mathrm{B}}^0 = \left(F'_{4n} - F'_{4n+2}\right) = C_2 f_{\mathrm{B}} \mathrm{e}^{-B_{\mathrm{B}}\,(\sin\theta/\lambda)^2}. \tag{3.40}$$

Equations (3.39) and (3.40) may be written in logarithmic form as

$$\log\,(f_{\mathrm{A}}^0/f_{\mathrm{A}}) = \log C_1 - B_{\mathrm{A}}\,(\sin\theta/\lambda)^2 \tag{3.41}$$

and

$$\log\,(f_{\mathrm{B}}^0/f_{\mathrm{B}}) = \log C_2 - B_{\mathrm{B}}\,(\sin\theta/\lambda)^2. \tag{3.42}$$

Typical plots representing (3.41) and (3.42) are shown in Fig. 3.3. From these plots, B_{A} and B_{B} and the constants C_1 and C_2 can be evaluated; C_1 and C_2 are expected to be equal. The mean Debye–Waller factor $\overline{B}$ is obtained from

$$\overline{B} = \frac{m_{\mathrm{A}} B_{\mathrm{A}} + m_{\mathrm{B}} B_{\mathrm{B}}}{m_{\mathrm{A}} + m_{\mathrm{B}}}, \tag{3.43}$$

where m_{A} and m_{B} are masses of atoms A and B, respectively.

(c) If B_{A} and B_{B} are expected to be close to each other (for instance, if the two atoms do not differ much in mass or if the number of reflections is not sufficient to permit determination of B_{A} and B_{B} separately), a mean value $\overline{B}$ is assumed for both atoms. Then (3.34), (3.36) and (3.37) yield

$$I_0 = C\;(\mathrm{LP})\;J(f_{\mathrm{A}} \pm f_{\mathrm{B}})^2 \mathrm{e}^{-2B(\sin\theta/\lambda)^2}. \tag{3.44}$$

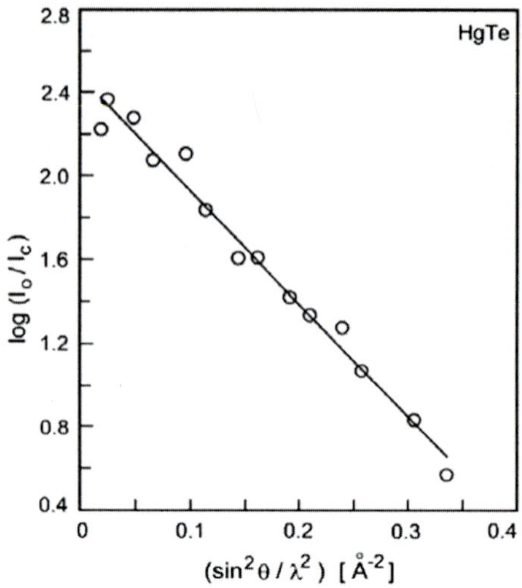

Fig. 3.4. Plot of log (I_0/I_c) vs $(\sin\theta/\lambda)^2$ for HgTe

A plot of log $(I_0\ /\ I_c)$ vs. $(\sin\theta/\lambda)^2$ yields the value of $\overline{B}$ with

$$I_c = C\ (\mathrm{LP})\, J(f_A \pm f_B)^2 . \tag{3.45}$$

A typical plot is shown in Fig. 3.4. From the plot, $\overline{B}$ can be evaluated. Values of the Lorentz-polarisation factor (LP) and the multiplicity factor J can be taken from standard books on X-ray crystallography. The values of the atomic scattering factors are taken from Cromer and Waber [3.20] and are corrected for anomalous dispersion [3.21].

Determination of Debye Temperature and Amplitudes of Vibration

The value of the characteristic temperature θ_M corresponding to the B value so obtained is calculated from

$$B = (6h^2/\overline{m}k_B\theta_M)\left[\frac{\phi(x)}{x} + \frac{1}{4}\right], \tag{3.29}$$

where $\overline{m}$ is the average mass of the atoms, $x = \theta_M/T$, T is the absolute temperature and the function $\phi(x)$ is defined by (3.25). In (3.29), $\phi(x)$ is a function of θ_M/T. In order to evaluate θ_M from the experimental value of B, iterative procedures are in use. A value of θ_M is assumed to evaluate the rhs of (3.29) and this is compared with the observed B. If there is a difference, the value of θ_M is adjusted and the process repeated until there is agreement between the values of B thus calculated and the experimental value. This procedure has been followed by Baldwin and Tompson [3.22], Walford [3.23] and Naidu and Houska [3.24] among others. However, as will be shown below, such iterative procedures are not necessary.

Equation (3.29) can be written as

$$W(x) = \left[\frac{\phi(x)}{x^2} + \frac{1}{4x}\right] = \left(\frac{B\overline{m}k_{\mathrm{B}}T}{6h^2}\right). \tag{3.46}$$

Values of the function $W(x)$ have been tabulated by Benson and Gill [3.25] for a wide range of values of x for small increments, so that the Debye temperature can be easily determined from the mean Debye–Waller factor without recourse to an iterative procedure.

From (3.11) the mean square displacement of the ions is related to the mean B value by

$$B = (8\pi^2/3) < u^2 > . \tag{3.47}$$

Evaluation of Errors

The errors quoted in the subsequent sections in the values of the X-ray Debye characteristic temperature θ_{M}, the Debye–Waller factor B and the rms amplitude of vibration $< u^2 >^{1/2}$ refer to the random errors in the recorded intensities. The standard error in the value of B is estimated from the relation

$$\Delta B = 1/2\left[\frac{n\sum|e_Y|^2}{(n-2)\,(n\sum X^2 - (\sum X)^2)}\right]^{1/2}. \tag{3.48}$$

Here n refers to the number of reflections included in the least square analysis, $Y = \log(I_0/I_c)$, $X = (\sin^2\theta/\lambda^2)$ and $|e_Y|$ refers to the magnitude of the difference between Y_{meas} and $Y_{\mathrm{calc.}}$, the latter being obtained from the equation corresponding to the least squares line.

The error in B may be used to evaluate the corresponding errors in θ_{M} and the rms amplitude of vibration. The error in θ_{M} is obtained by differentiating the equations $x = (\theta_{\mathrm{M}}/T)$ and $B = (6h^2/\overline{m}k_{\mathrm{B}}T)W(x)$, whence

$$\Delta\theta_{\mathrm{M}} = \frac{W(x)}{\Delta W(x)}\frac{\Delta B}{B}T\Delta x. \tag{3.49}$$

The fractional change $\Delta W(x)/W(x)$ corresponding to the change Δx can be estimated from the values of $W(x)$ tabulated by Benson and Gill [3.25]. The standard error in the rms amplitude of vibration is given by

$$\Delta < u^2 >^{1/2} = (3/8\pi^2 B)^{1/2}(\Delta B/2). \tag{3.50}$$

The errors calculated in this manner refer only to random errors. As discussed earlier, the systematic errors arising due to preferred orientation, extinction, porosity and surface effects have been sought to be minimized under the conditions chosen for the measurement of intensities. The neglect of these factors may possibly give errors larger in magnitude than those quoted by 2–3%.

3.3.3 Other Methods

The method of determination of B and θ_M from intensities measured at a single temperature is discussed in earlier sections. For sake of completeness, a couple of other methods are discussed in this section.

θ_M From Bragg Intensities at Different Temperatures

The Bragg intensity is a function of temperature. The intensity $I(T)$ of a Bragg reflection at temperature Tcan be written as

$$I(T) = \text{constant } (LP)JF^2\mathrm{e}^{-2M}. \tag{3.51}$$

If $I(T_1)$ and $I(T_2)$ are the intensities at temperatures T_1 and T_2,

$$\log\,[I(T_1)/I(T_2)] = \log\,[(LP)_1F_1^2/(LP)_2F_2^2] - 2M(T_1) + 2M(T_2) \tag{3.52}$$

In view of. (3.29), (3.53) becomes

$$\begin{aligned}\log[I(T_1)/I(T_2)] = \log\,&[(LP)_1F_1^2/(LP)_2F_2^2]\\ &-(12h^2/mk_B\lambda^2)[T_1\sin^2\theta_1\{\phi(x_1)+(x_1/4)\}/\theta^2_{M_1}\\ &-T_2\sin^2\theta_2\{\phi(x_2)+(x_2/4)\}/\theta^2_{M_2}].\end{aligned} \tag{3.53}$$

If one assumes that over small temperature increments the variation of θ_M is small, and that $\theta_{M_1} = \theta_{M_2} = \theta_M$ (Av.), then θ_M (av.) can be found from the equation

$$\theta_M{}^2(\text{Av.}) = \frac{(12h^2/mk_B)\;[T_2\phi(x_2) - T_1\phi(x_1)](\sin^2\theta/\lambda^2)}{[\log I(T_1)/I(T_2)]}. \tag{3.54}$$

Since the function $\phi(x)$ depends on θ_M, the calculation of θ_M from (3.54) involves an iterative procedure. The reliability of the approximation in (3.54) depends on the smallness of the temperature interval chosen over which θ_M is assumed independent of temperature. Since the expression for θ_M involves the logarithm of the intensity ratio, the intensities must be known accurately. This sets a limit on the optimum value of the temperature increment.

Debye–Waller Factors From the Intensity of Temperature Diffuse Scattering

Another method where both the atomic scattering factor and the Debye temperature may be simultaneously obtained for an element was suggested by Borie [3.26]. This involves the determination of the integrated intensities of the Bragg peaks and also the contribution to the background from the temperature diffuse scattering. The contribution to the background intensity from the temperature diffuse scattering is given by

$$I_{\text{TDS}} = f^2[1-\mathrm{e}^{-2M}]\;G(\sin\theta/\lambda), \tag{3.55}$$

where the function $G(\sin\theta/\lambda)$ is given in [3.9, 3.27, 3.28]. Thus from TDS, $f^2[1-\mathrm{e}^{-2M}]$ is obtained as a function of $(\sin\theta/\lambda)$ while from the Bragg intensities f^2e^{-2M} is obtained as a function of $(\sin\theta/\lambda)$. The sum of these two curves gives f^2 and their ratio yields M and hence θ_{M}. Borie [3.26] used this method to obtain f, B and θ_{M} for copper.

3.4 An Overview

From the earlier sections, it can be seen that the determination of the Debye–Waller factors has at least three applications viz. (a) they provide the temperature corrections for observed intensities, (b) they can be compared with lattice dynamical calculations to obtain information about the soundness of the model and (c) they yield values of the X-ray Debye temperature which, by itself, is an important solid state parameter. Further, the Debye–Waller factor and the Debye temperature correlate with several other physical properties. In view of this importance, there has been considerable activity in the field. In this section, we consider some of the important contributions from various groups and laboratories.

3.4.1 Earlier Work of Historical Importance

Soon after the theoretical work of Debye and Waller, experimental work was undertaken to verify theoretical predictions. James [3.29] and James and Furth [3.30] measured the intensities of reflections of NaCl at high and low temperatures, respectively, and made only qualitative comparisons with theory. Waller and James [3.31] reinterpreted the measurements of James [3.29] and James and Furth [3.30] taking into account the diatomic nature of NaCl. Further measurements and verification of the Debye–Waller theory was carried out by James and Brindley [3.32] for KCl and James et al. [3.33] for Al. Brindley [3.34] measured the intensities of reflexions for NaF, NaCl, LiF and KCl and calculated the mean square amplitudes of vibrations from the definition of the Debye–Waller factor. Shonka [3.35] measured the intensities of NaF and determined B and θ_{M} from the Debye–Waller theory. Brindley and Ridley [3.36] applied temperature corrections to the measured intensities for Al, Cu and KCl and obtained experimental values of the atomic scattering factors, and compared them with theoretical values. Ribner and Wollan [3.37] and Brindley and Ridley [3.38] determined B and θ_{M} for MgO from the measured X-ray diffraction intensities.

3.4.2 Experimental Values of Debye–Waller Factors at Room Temperature

The principle of determination of the Debye–Waller factors and the X-ray Debye temperature has been discussed in Sect. 3.2. There is a large amount

Table 3.2. Experimental room-temperature B values of NaCl

B_{Na} [Å^2]	B_{Cl} [Å^2]	Method	Ref.
1.54	1.25	X-ray, single crystal	[3.31]
1.25	1.12	X-ray, single crystal	[3.44]
1.25	1.00	X-ray, single crystal	[3.45]
1.29	1.15	X-ray, powder	[3.46]
1.63	1.42	Neutron, single crystal	[3.47]
1.639 ± 0.037	1.326 ± 0.023	X-ray, single crystal	[3.48]
1.84	1.44	X-ray, powder	[3.49]
1.89	1.61	X-ray, single crystal	[3.50]
1.81 ± 0.03	1.49 ± 0.03	X-ray, powder	[3.51]

of data on the room temperature values of these parameters. B-values for a variety of crystals are given in International Tables for X-ray Crystallography [3.39]. Recently, compilations of data on Debye–Waller factors and the associated Debye temperatures have been published by Butt et al. [3.40] for cubic elements, Butt et al. [3.41] for cubic compounds and Gopi Krishna and Sirdeshmukh [3.42] for hcp elements. Several aspects of the Debye–Waller factors of alkali halides have been discussed by Sirdeshmukh et al. [3.43].

NaCl, NaF, KCl and MgO are substances for which numerous measurements have been made. Here, we shall examine the results on NaCl in detail to bring out the problems in Debye–Waller factor measurements. The various results on NaCl are given in Table 3.2. A look at the data in the table reveals that there is wide variation in the values from report to report. For B_{Na}, the reported values vary from 1.25 to 1.89 Å^2 – a variation of about 50%. For B_{Cl}, the variation is from 1.00 to 1.61 Å^2 – again, a variation of about 60%.

Several factors contribute to the uncertainties in the measured B values. These include the uncertainty in intensity measurements, background correction, correction for thermal diffuse scattering and choice of atomic scattering factors. Groenewegen and Huiszoon [3.52] mention that variations in the range 0.03–0.06 and 0.04–0.20 Å^2 arise in B because of different methods of TDS correction employed in powder and single crystal measurements, respectively. Lawrence [3.53] attributed the large difference in his B-values for MgO and those reported by Togawa [3.17] to the difference in the angular range of reflexions used in the two studies. Besides, errors may be introduced by the state of perfection and strains in the sample.

A closer look at the table indicates that the B values are systematically lower in the earlier studies than in the more recent measurements. This may be due to improvements in techniques of intensity measurements and also improvements in methods of applying TDS correction.

3.4.3 Effect of Choice of Atomic Scattering Factors on Measured *B*-values

In the determination of the Debye–Waller factors, the atomic scattering factors constitute an important input. Atomic scattering factors are available for

Table 3.3. Debye–Waller factors and R factor for different states of V and N in VN

	State	B_V [Å^2]	B_N [Å^2]	R
(a)	V^0N^0	0.367	0.318	2.83
(b)	V^+N^-	0.295	0.224	2.84
(c)	$V^{+2}N^{-2}$	0.311	0.321	1.10
(d)	$V^{+3}N^{-3}$	0.456	0.337	1.13

different ionic states and also for different atomic models. The atomic scattering factors have to be chosen with care while determining the Debye–Waller factors.

The Debye–Waller factors of VN determined by Hosoya et al. [3.54] employing atomic scattering factors appropriate to the (a) V^0, N^0, (b) V^{+1}, N^{-1}, (c) V^{+2}, N^{-2} and (d) V^{+3}, N^{-3} states are shown in Table 3.3. Hosoya et al. [3.54] found that the choice of atomic scattering factors had an effect not only on the magnitudes of the B values but also on their relative values ($B_V < B_N$ in case (c) and otherwise in other cases). The R factor was also calculated for each case. From the lowest value of R, it was concluded that the V^{+2} N^{-2} state is probably the true electronic state for VN.

Stewart [3.55] analysed X-ray intensities for diamond powder specimens using atomic scattering factors obtained from the Hartree–Fock and molecular carbon models and obtained 0.20 and 0.172 Å^2 for the Debye–Waller factor.

Dawson and Howard [3.56] used atomic scattering factors for Li^0, F^0 and Li^{+1}, F^{-1} states in analyzing intensity data for LiF and found only a marginal effect on the resulting B-values.

3.4.4 Debye–Waller Factor for a Real Crystal

From the dynamical theory of X-ray diffraction applicable to thick and perfect crystals, we get $I \propto F\mathrm{e}^{-M}$ whereas from the kinematical theory applicable to thin and mosaic crystals, we get $I \propto F^2\mathrm{e}^{-2M}$. The expressions given in Sect. 3.2, thus, assume that the sample is mosaic.

Parthasarathi [3.57] pointed out that since a real crystal may be intermediate between a perfect and imperfect crystal, its Debye–Waller factor may be e^{-pM} where $1 < p < 2$. He also suggested that the value of p could be used as a measure of the degree of perfection.

Batterman [3.58] found that the temperature factor is $\mathrm{e}^{-1.3M}$ for a germanium crystal. Batterman [3.59] verified experimentally that the temperature factor is e^{-M} for a perfect Si crystal and e^{-2M} for Si powder samples. Annaka [3.60] found that the temperature factor is $\mathrm{e}^{-1.3M}$ for NaCl. Batterman and Strock [3.61] obtained e^{-M} for a perfect CaF_2 crystal and $\mathrm{e}^{-1.6M}$ for a crystal which was not characterized for perfection. For powder samples and for single crystals rendered mosaic by quenching, the temperature factor may be assumed to be e^{-2M}.

3.4.5 Debye Temperatures of Thin Films and Fine Particles

The X-ray Debye temperature can be measured from intensities measured either at a single temperature or at different temperatures. In the former method, the atomic scattering factors are directly used whereas in the latter, ratios of intensities are used which nearly cancels out the atomic scattering factors.

Using the single temperature method, Carpenter [3.62] obtained a value of 1549 K for θ_M for diamond. On the other hand, Post [3.63] obtained a value of 1,990 K for θ_M for diamond using the method of temperature variation of intensities. In view of the difference in the two methods discussed in the preceding paragraph, Barron et al. [3.64] concluded that the difference in the two θ_M values could be due to some inadequacy in the atomic scattering factors used by Carpenter [3.62]. Schoening and Vermuelen [3.65] made measurements of θ_M on diamond powders of different particle sizes using the temperature variation method and obtained a value of 1,880 and 1,500 K for θ_M for the fine (6 µm) and coarse (780 Å) powders, respectively. These different values have been obtained without involving the atomic scattering factors. Thus the suggestion of Barron et al. [3.64] that atomic scattering factors for C need revision is contradicted.

Mitra and Chaudhari [3.66] determined θ_M for Cu and Ag films of different thickness using X-ray and electron diffraction methods and found that θ_M is much lower for thin films than for bulk samples.

Blakeley [3.67] refers to a number of low energy electron diffraction studies of Debye–Waller factors as a function of the electron energy. Electron beams of different voltages probe surface layers of different thicknesses. The general conclusion from these studies is that $(\overline{u^2})_{\text{surface}} \Big/ (\overline{u^2})_{\text{bulk}}$ is about 2.

Ohshima and Harada [3.68] made X-ray measurements on fine particles of Cu, Ag and Au in the range 60–2,000 Å. Their results are given in Table 3.4. In all the cases the Debye temperature decreased as particle size increased. This variation is attributed to a softening of the thermal vibrations in the surface of the fine particles.

Considering a particle of radius r to consist of a core and a shell of thickness Δr (Fig. 3.5) and associating Debye temperatures θ_c and θ_s with the core and the shell, it can be shown that the Debye temperature θ_P of a particle is given by

$$\theta_P^{-2} = \theta_c^{-2} + (3\Delta r/r)[\theta_s^{-2} - \theta_c^{-2}]. \tag{3.56}$$

Thus, a plot of θ_P^{-2} and r^{-1} should be linear; such linear plots are shown in Fig. 3.6. θ_c can be obtained from the intercept of the plot. θ_s is estimated from the slope of the plot by assuming that the shell thickness Δr is of the order of the nearest neighbour distance. Values of θ_P, θ_s and θ_c are given in Table 3.4 along with θ_M values for bulk samples. The θ_c values are expectedly close to θ_M values.

Table 3.4. Values of $\theta_P, \theta_s, \theta_c$ and θ_M

Specimen		θ_P [K]	θ_s [K]	θ_c [K]	θ_M [K] [3.69, 3.70]
Metal	Size [Å]				
Cu	160(30)	284(9)	165(25)	306(7)	322(22)
	400(50)	292(13)			
	560(70)	303(11)			
Ag	130(20)	187(8)	120(12)	216(6)	212(7)
	190(30)	204(9)			
	565(80)	204(9)			
	2000(500)	217(6)			
Au	60(10)	134(5)	98(7)	165(5)	177(5)
	115(30)	145(4)			
	150(40)	144(3)			
	230(50)	161(4)			
	380(70)	158(6)			

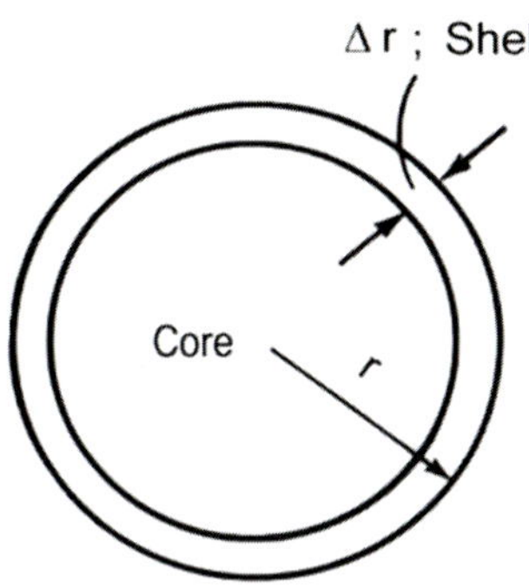

Fig. 3.5. The 'shell' and 'core' of a particle of radius r

3.4.6 Effect of Lattice Strain on B

Inagaki et al. [3.14] measured the intensities of X-ray reflexions for several powdered materials ground for different durations. They found that the slope of the log (I_o/I_c) vs. $(\sin^2\theta/\lambda^2)$ plot (which yields the value of B) changes substantially with the time of grinding; the magnitude of the effect was different in different substances.

In a subsequent paper Inagaki et al. [3.71] determined the Debye–Waller factor B and also the lattice strain for samples ground for different times. The lattice strain was determined from the linear plots between $(\beta\cos\theta/\lambda)$ and $(\sin\theta/\lambda)$ using the Scherrer equation:

$$(\beta\cos\theta/\lambda) = (K/L) + 2\varepsilon_1\,(\sin\theta/\lambda). \tag{3.57}$$

Here β is the half-width of a reflexion, K the shape factor (~1), L the particle size and ε_1 the lattice strain. Inagaki et al. [3.71] found that the B vs. ε_1 plots (Fig. 3.7) are smooth but differ for different substances.

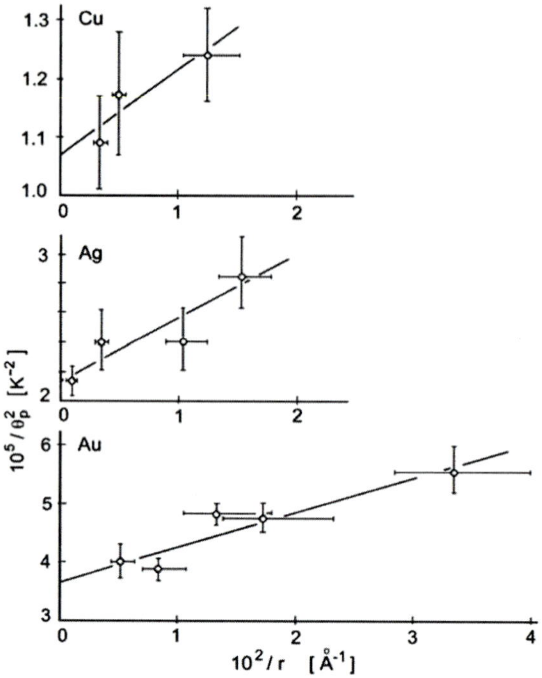

Fig. 3.6. Plot of θ_P^{-2} vs. r^{-1}

Inagaki et al. [3.71] were the first to systematically study the effect of grinding (and the resulting strain) on the determination of Debye–Waller factors. They produced strain in several inorganic compound powders (CaF_2, TiO_2, CdO, graphite and $BaTiO_3$) by grinding and determined the mean Debye–Waller factors. Inagaki et al. [3.71] found that in all the materials studied by them, the measured Debye–Waller factor (B_{eff}) increased with increasing strain (ε_l). Further they found that above $\varepsilon_l = 0.2 \times 10^{-2}$, B_{eff} tends to saturate. The limiting (saturation) value of B_{eff} is almost twice that of the starting material. Graphite, however, was an exception where B_{eff} continued to increase without any tendency to saturate up to the maximum value of strain produced. Thus, lattice strain was shown to cause a large effect on the measured values of the Debye–Waller factors of the materials studied.

3.4.7 Anisotropy of Debye–Waller Factors

At a gross level, the Debye–Waller factor has the same symmetry as the crystal itself. Thus, while B is isotropic for the cubic symmetry, there are two principal Debye–Waller factors for the tetragonal, trigonal and hexagonal crystals, three for the orthorhombic, four for the monoclinic and six for the triclinic systems.

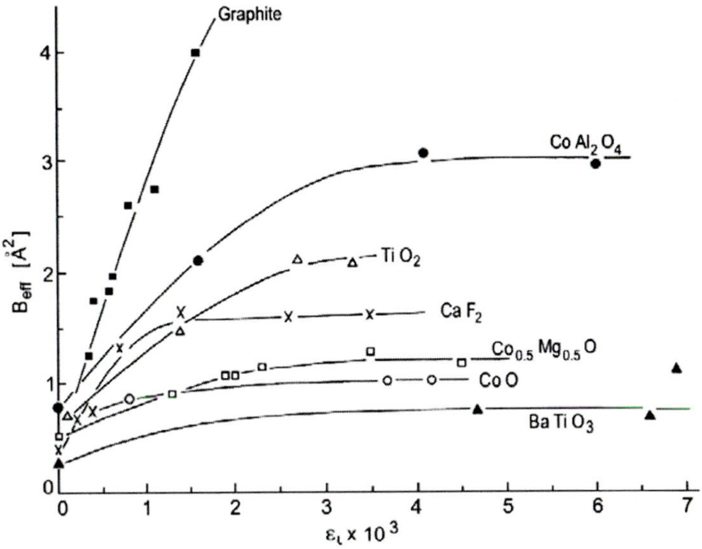

Fig. 3.7. Plot of effective Debye–Waller factor B vs. lattice strain (ε_1)

Gilbert and Lonsdale [3.72] measured the Debye–Waller factor for several reflexions for urea (tetragonal) and found that the direction-dependence in this crystal follows the rule $B = a + b\cos^2\phi$ where a and b are the Debye–Waller factors in the principal directions and ϕ is the angle between normals to the (hkl) and (001) planes.

At a finer level, the temperature factor is anisotropic for each atom. This anisotropy is represented by a thermal vibration ellipsoid given by

$$2M = \beta_{11}h^2 + \beta_{22}k^2 + \beta_{33}l^2 + 2\beta_{12}hk + 2\beta_{23}kl + 2\beta_{31}lh \tag{3.58}$$

for each atom. By standard procedures, this is reduced to

$$2M = 8\pi^2[\overline{u_1^2}l_1^2 + \overline{u_2^2}l_2^2 + \overline{u_3^2}l_3^2](\sin^2\theta/\lambda^2), \tag{3.59}$$

where u_i and l_i are the thermal vibration amplitude and direction cosine of the ith axis of the ellipsoid with respect to the scattering vector. For purposes of evaluating the mean Debye–Waller factor from the thermal vibration ellipsoid parameters, McWhan et al. [3.73] suggest

$$B_{\text{average}} = 8\pi^2\overline{(u^2)_{av}} = 8\pi^2(1/k)\sum_k M_k\overline{u_k^2}, \tag{3.60}$$

where $\overline{u_k{}^2}$ is the average for the ellipsoid of the kth atom. However, such measurements are generally made while performing structure determination of crystals containing a large number of atoms.

3.4.8 Pressure Variation of θ_M

There is very little experimental work on this aspect. The X-ray Debye temperature of Al was studied by Matsumoro et al. [3.74] at pressures up to 6 GPa using synchrotron radiation. At each pressure, the intensities were measured at 45°C and 264°C.

If $I(T_1)$ and $I(T_2)$ are the intensities at temperatures T_1 and T_2, then starting with the equation

$$I(T) = C\,(LP)\,JF^2 e^{-2M}, \tag{3.61}$$

we get

$$\log\,[I(T_2)/I(T_1)] = 2\,\Delta B\,(\sin\theta/\lambda)^2 + C. \tag{3.62}$$

From a plot of log $[I(T_2)/(T_1)]$ vs. $(\sin\theta/\lambda)^2$ (Fig. 3.8) ΔB is determined. θ_M is calculated from

$$\theta_M{}^2 = (6h^2/mk_B)\,(\Delta T/\Delta B). \tag{3.63}$$

This θ_M value is associated with the mean of the temperatures T_1 and T_2 and with the pressure at which $I(T_1)$ and $I(T_2)$ are measured. This whole procedure is repeated at different pressures. The resulting values of θ_M at different pressures are shown in Fig. 3.9. The parameter V/V_0 is used instead of P as the variable.

The observed pressure dependence of θ_M of Al can be fitted to the equation:

$$\theta_M = (\theta_M)_0 \exp\{[\gamma(p) - \gamma(0)]/A\}, \tag{3.64}$$

$$\gamma(p) = \gamma(0)(V/V_0)^A, \tag{3.65}$$

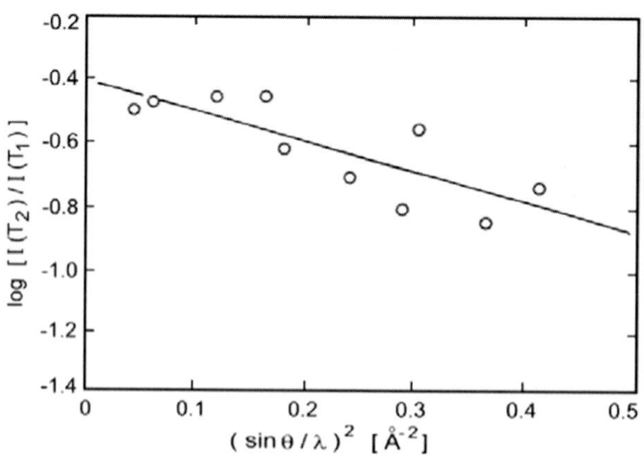

Fig. 3.8. Plot of log $I(T_2)/I(T_1)$ vs. $(\sin\theta/\lambda)^2$

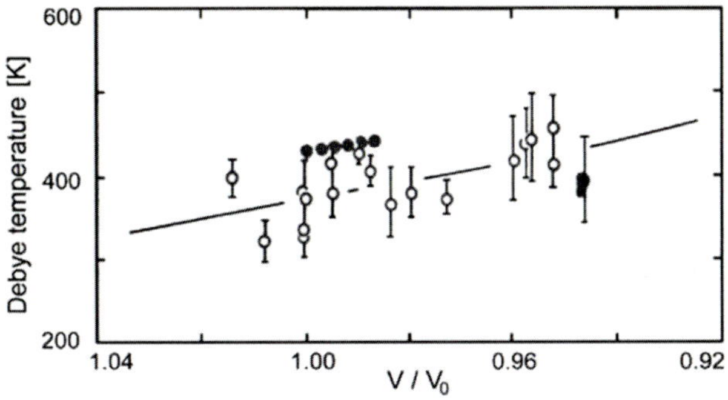

Fig. 3.9. Plot of θ_M vs. V/V_0

where $\gamma(p)$ and $\gamma(0)$ are the Gruneisen constants at a given pressure and zero pressure respectively. A is a constant $\sim$1. From the data, $\gamma(0)$ and $(\theta_{\mathrm{M}})_0$ are obtained as 3 and 367 K which agree with other independent measurements.

3.4.9 Temperature Variation of B and θ_{M}

The temperature variation of the Debye–Waller factors of some alkali halides has been reported by Bastow et al. [3.75]. Fig. 3.10 is a typical representation of the temperature variation of the Debye–Waller factors of the alkali halides. Mention may be made of the determination of the B-values for KCl and NaCl at temperatures close to the melting point by Viswamitra and Jayalaxmi [3.76]. Debye–Waller factors of some of the alkali halides have been determined at low temperatures ($\sim$80 K); these are listed by Sirdeshmukh et al. [3.43]. Systematic measurements of Debye–Waller factors of several crystals with CaF_2 structure have been made up to high temperatures using neutron diffraction [3.77–3.79].

The temperature variation of θ_{M} of Al, Cu and Pb was studied by Owens and Williams [3.80]. Wilson et al. [3.81] made measurements of θ_{M} of Ni as a function of temperature. The temperature variation of θ_{M} of several alkali halides at high temperatures has been studied by Pathak and his group and their results have been listed by Sirdeshmukh et al. [3.43]. Results for the Rb halides are shown in Fig. 3.11 as typical examples.

3.4.10 Anharmonic Effects in Debye–Waller Factors and Debye Temperature

The theory of anharmonic effects in Debye–Waller factors and Debye temperatures has been discussed in different (but equivalent) forms in [3.82–3.86]. According to the treatment by Willis [3.86], the temperature factor is:

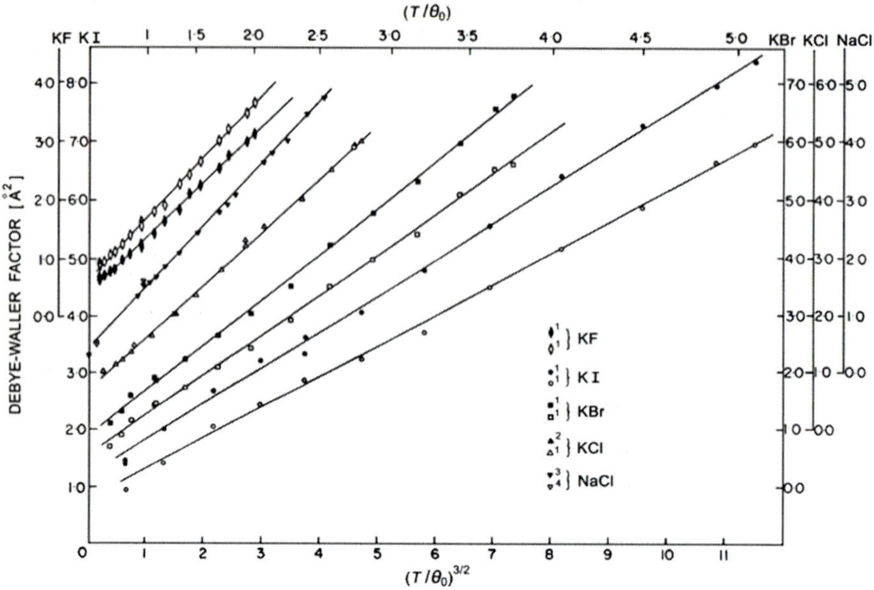

Fig. 3.10. Temperature variation of Debye–Waller factors ($[T/\theta_{\mathrm{D}}]^{3/2}$ is used as the variable)

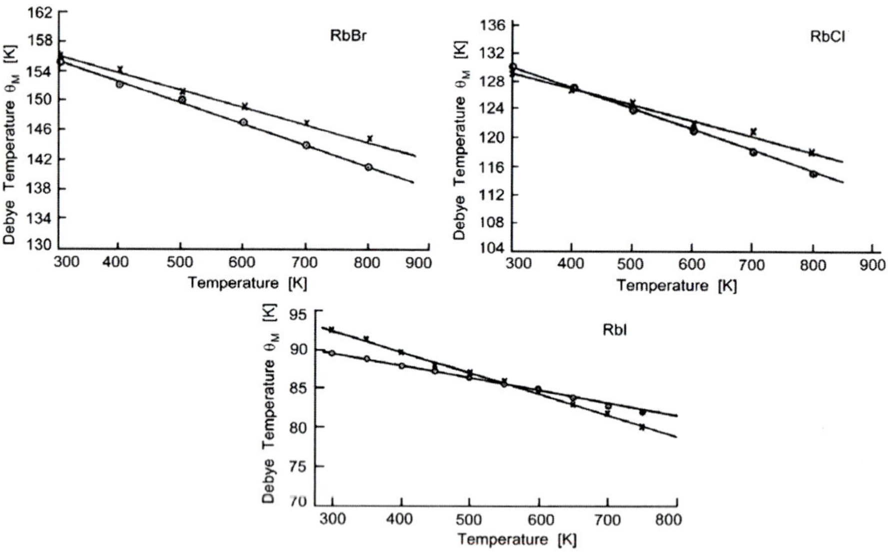

Fig. 3.11. Plots of X-ray Debye temperature (θ_{M}) vs temperature (T) for RbCl, RbBr and RbI

$$
\begin{aligned}
2M = {} & (2\pi/a_0)^2(h^2+k^2+l^2)(1/\alpha_0)k_\mathrm{B}T + (2\pi/a_0)^2(h^2+k^2+l^2) \\
& \times (2\gamma\beta/\alpha_0 k_\mathrm{B})(k_\mathrm{B}T)^2 - (2\pi/a_0)^2(h^2+k^2+l^2)(20\gamma_0/\alpha_0^3)(k_\mathrm{B}T)^2 \\
& + (2\pi/a_0)^4(h^2+k^2+l^2)(2\gamma_0/\alpha_0^4)(k_\mathrm{B}T)^3 \\
& - (2\pi/a_0)^4[h^2k^2+k^2l^2+l^2h^2-(1/3)h^2-(1/3)k^2-(1/3)l^2] \\
& \times (12\delta_0/5\alpha_0^4)(k_\mathrm{B}T)^3 .
\end{aligned} \tag{3.66}
$$

Here a_0 is the lattice constant, γ the Gruneisen constant, β the volume expansion coefficient and k_B the Boltzmann constant. α_0, γ_0 and δ_0 are anharmonicity parameters.

The first term is the harmonic term, the second is the thermal expansion correction to the harmonic term (quasiharmonic term), the third and fourth terms are isotropic anharmonic terms and the fifth term is the anisotropic anharmonic term. The fourth and fifth terms are negligible. Equation (3.66) can finally be reduced to:

$$B(T) = B^\mathrm{h}(T)[1 + (2T\beta\gamma) - 20Tk_\mathrm{B}(\gamma_0/{\alpha_0}^2)]. \tag{3.67}$$

Here $B^\mathrm{h}(T)$ is the harmonic B-factor given by

$$B^\mathrm{h}(T) = 8\pi^2 k_\mathrm{B}T/\alpha_0. \tag{3.68}$$

To compare experimental results with theory, we use the parameter Y given by

$$Y = (\lambda/\sin\theta)^2 \log\left[I(T)/I(T_0)\right] = 2\left[B(T_0) - B(T)\right]. \tag{3.69}$$

Further, (3.67) may be expressed in terms of the Debye temperature as

$$[1/\theta_\mathrm{M}(T)]^2 = (mk_\mathrm{B}^2/3\hbar^2\alpha_0)[1 + T(2\beta\gamma - 20k_\mathrm{B}\gamma_0/\alpha_0^2)] \tag{3.70}$$

The terms α_0 and γ_0 can be obtained by fitting experimental data at some temperatures.

The experimental data on Al, Na and (α) iron have been interpreted in terms of this theory by Jyoti Prakash et al. [3.87]. The Y–T plots and the θ_M–T plots are shown in Figs. 3.12 and 3.13, respectively. The agreement is, expectedly, better when the anharmonic term is included.

Similar comparisons of experimental data with the anharmonic theory have been made for KCl and BaF_2 by Willis [3.86], for KBr by Sureshchandra et al. [3.88], for Ni, Mo and W by Sneh [3.89] and for NaCl, KCl and KBr by Shepard et al. [3.90] with their new data.

3.4.11 Debye–Waller Factors from Lattice Dynamics

The Debye–Waller factor is related to the vibration spectrum of the crystal lattice (Sect. 3.2). The vibration spectrum of a crystal can be calculated assuming

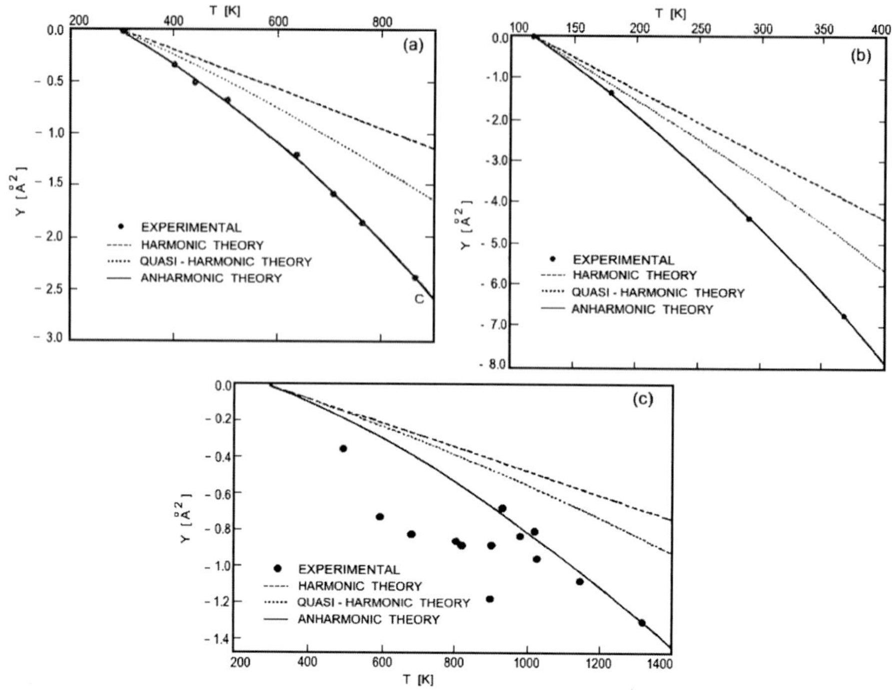

Fig. 3.12. Plots of the parameter Y (3.69) vs. temperature T : (a) aluminium, (b) sodium and (c) α-iron

different models. Once the vibration spectrum is evaluated the Debye–Waller factor can be calculated.

Comparison of such calculated values with experimental values can serve two purposes. If the experimental values are accurate and well-established and if calculated values from different models differ, the comparison will help to choose the better model. On the other hand, if calculated values from different models are consistent among themselves and if the experimental values differ widely, the comparison will help to sift the more acceptable value.

In Table 3.5, examples of both types are given. In MgO, calculated values from two models are close but experimental values from two sources are very different. By comparison, one can conclude that the experimental values by Togawa [3.17] are inferior to the values by Lawrence [3.53]. On the other hand, in LiF, experimental values from two sources are close but calculated values from two models differ considerably. Here, it may be concluded that the shell model [3.93] is better than the rigid ion model [3.94].

Drawing such conclusions is not always so straight forward. As an example, we shall consider the results for NaCl. The lattice dynamical values from different models are collected in Table 3.6. Buyers and Smith [3.97] have given a detailed discussion of the factors which introduce uncertainties in the lattice

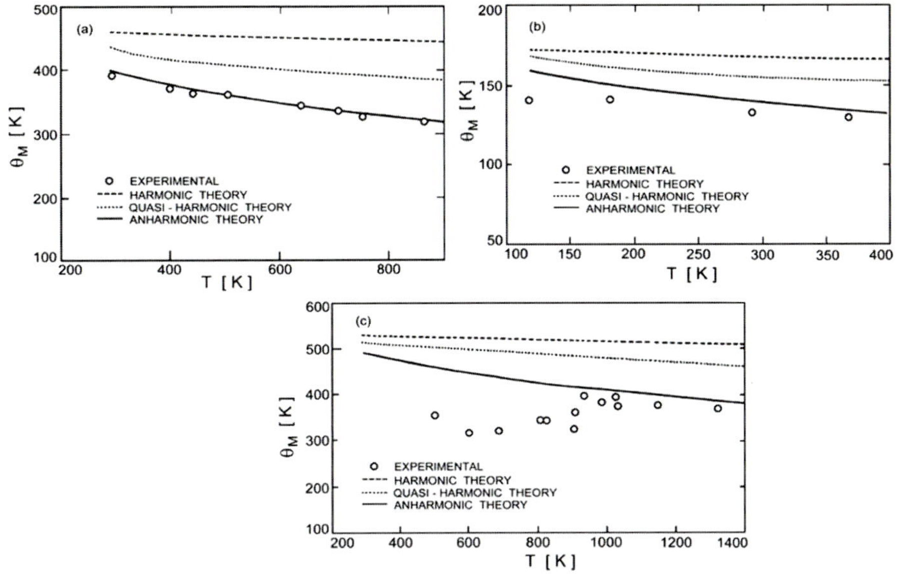

Fig. 3.13. Plots of X-ray Debye temperature (θ_{M}) vs. temperature (T) for (a) aluminium, (b) sodium and (c) α-iron

Table 3.5. Experimental and theoretically calculated Debye–Waller factors (B_1, B_2) of some crystals

Substance	B_1 [Å^2]	B_2 [Å^2]	Model/method	Ref.
MgO	0.30	0.33	Shell model	[3.91]
	0.315	0.341	8-parameter 3-body force model	[3.92]
	0.31	0.34	Powder X-ray diffraction	[3.53]
	0.24	0.19	Powder X-ray diffraction	[3.17]
LiF	0.91	0.73	Shell model	[3.93]
	1.53	0.57	Rigid ion model	[3.94]
	1.10	0.67	Powder X-ray diffraction	[3.45]
	1.01	0.68	Powder X-ray diffraction	[3.95]

dynamical values. Some of these are (a) inadequacies in the assumed model, (b) uncertainties in the experimental input data, (c) computational procedures and (d) anharmonicity correction.

With all these uncertainties, it is noted that the model-to-model differences in the B-values are much less than the differences in the experimental values (given in Table 3.2). From a comparison with experimental data available at that point of time, Buyers and Smith [3.97] concluded that "theories predict Debye–Waller factors that are higher than experimental determinations". However, if we compare the theoretical values with the more recent experimental values (Table 3.2), this conclusion has to be revised. The experimental values, particularly those for B_{Na}, are larger than the theoretical

Table 3.6. Debye–Waller factors B_{Na} and B_{Cl} for NaCl calculated at 295 K from lattice dynamical models

	Model	B_{Na} [Å^2]	B_{Cl} [Å^2]	Ref.
(1)	Ionic deformation model	1.45	1.48	[3.96]
(2)	Karo and Hardy's deformation model	1.53	1.46	[3.97]
(3)	Shell model	1.556	1.348	[3.98]
(4)	5-parameter rigid ion model	1.63	1.59	[3.52]
(5)	11-parameter breathing shell model	1.585	1.316	[3.99]

values. Considering the possibilities of differences in experimental values, it may be concluded that there is fair agreement between experimental and theoretical values irrespective of the model used. Hence, whenever there is lack of data (e.g. CsF) or very little data (e.g. RbF), lattice dynamical calculation would be useful. References for several sources of lattice dynamical calculations of Debye–Waller factors of alkali halides are given by Sirdeshmukh et al. [3.43]. A critical comparison of Debye–Waller factors of zinc blende type crystals calculated from several models is given by Reid [3.100].

3.4.12 Debye–Waller Factors and Melting

According to the theory of melting by Lindemann [3.101], a crystalline solid melts when the ratio of the square of the amplitude of vibration $(\overline{u^2})$ and the interatomic distance r attains a certain value. This ratio $(\overline{u^2}/r)$ is called the Lindemann parameter. It is obvious that determination of the Debye–Waller factors close to the melting point either experimentally or theoretically provides a method to estimate the Lindemann parameter.

For the alkali halides, values of $(\overline{u^2})$ close to the melting point have been reported only for NaCl and KCl by Viswamitra and Jayalaxmi [3.76]. They obtained values ~0.17 for the Lindemann parameter. Kushwaha [3.102] calculated $(\overline{u^2})$values close to the melting point lattice-dynamically and obtained a mean value of 0.16 for the Lindemann parameter. Vetelino et al. [3.103] calculated $(\overline{u^2})$ for several zinc-blende-type crystals lattice-dynamically at high temperatures. The mean value of the Lindemann parameter from their results is 0.25. Thus, in general, the Lindemann parameter is a constant for a family of related solids but its value may differ from family to family.

3.4.13 Debye–Waller Factors and Temperature Dependence of Band-gap in Semiconductors

Band-gaps in semiconductors change with temperature. This temperature variation is known experimentally for many semiconductors. Generally, the temperature coefficient is negative i.e., the band-gap energy decreases as the temperature increases.

A theory for calculating the energy bands as a function of temperature was proposed by Brooks and Yu [3.104]. In these calculations, the Debye–Waller factors are used as input parameters. Using this procedure, Tsay et al. [3.105] calculated the temperature coefficients of energy gaps for several III–V semiconductors. For these calculations, they used the Debye–Waller factors calculated theoretically by Vetelino et al. [3.103]. Comparing their results with experimental data, Tsay et al. [3.105] concluded that the agreement between theory and experiment is as good as the agreement between different experimental results.

Generally, the Debye–Waller factor of a heavier ion is smaller than that of a lighter ion. But in PbTe, Keffer et al. [3.106, 3.107] observed experimentally that $B_{\mathrm{Pb}} > B_{\mathrm{Te}}$. Interestingly, a lattice dynamical calculation by the same authors also showed that $B_{\mathrm{Pb}} > B_{\mathrm{Te}}$. Earlier, Prakash [3.108] had observed from absorption edge experiments that $\mathrm{d}E_g/\mathrm{d}T$ is positive in PbTe. Keffer et al. [3.106, 3.107] calculated $\mathrm{d}E_g/\mathrm{d}T$ for PbTe using the Brooks–Yu method. For these calculations they used the Debye–Waller factors determined by them as input parameters. They obtained a value of 2×10^{-5}Ry K^{-1} for $\mathrm{d}E_g/\mathrm{d}T$ compared to the value of 4×10^{-5}Ry K^{-1} obtained by Prakash [3.108] from experiments. Keffer et al. [3.107] showed that a part of this difference can be explained in terms of uncertainties in the values of the Debye–Waller factors.

3.4.14 Debye Temperature in an Antiferromagnetic Transition

Chromium undergoes an antiferromagnetic transition at a Neel temperature of about 313 K. The calculation of Debye temperatures from elastic constants measured around this temperature by Wilson et al. [3.81] indicated a discontinuity of 3 K in the Debye temperature. Wilson et al. [3.81] also determined the X-ray Debye temperatures in the range 100–500 K by the method of temperature variation of intensities but the accuracy of their results was not sufficient to establish the existence of a discontinuity in θ_{M} at the Neel point.

Koumelis [3.109] decided to restudy the problem by determining the change in θ_{M} rather than θ_{M} itself. For this purpose, the X-ray diffraction intensities from a powder sample were accurately measured at 310.2 and 313.2 K and the change in Debye temperature $\Delta\theta_{\mathrm{M}}$ was calculated from

$$(\Delta I/I) = (12h^2/mk_{\mathrm{B}})(\sin\theta/\lambda)^2[(1/4) - (\mathrm{e}^x - 1)^{-1} + (3T^2/\theta_{\mathrm{M}}^2)\int_0^x (\xi \mathrm{d}\xi)/(\mathrm{e}^\xi - 1)](\Delta\theta_{\mathrm{M}}/\theta_{\mathrm{M}}). \quad (3.71)$$

From his experiment, Koumelis [3.109] obtained $\Delta\theta_{\mathrm{M}} = 7.9 \pm 0.8$ K. This value, though of the same order, is larger than the value calculated from elastic data.

Florias and Koumelis [3.110] repeated the measurement, this time using a single crystal. Their new experiments yielded $\Delta\theta_{\mathrm{M}} = 3.6 \pm 1.1$ K which is in better agreement with the elastic-θ results. Florias and Koumelis [3.110]

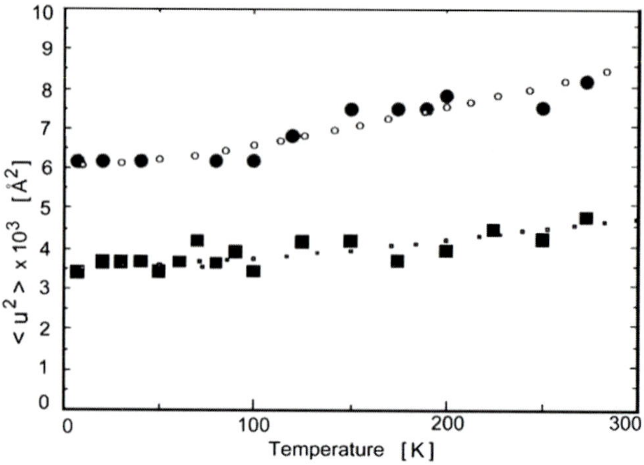

Fig. 3.14. Plots of $\overline{u}^2$ vs. T for encapsulated (closed circles) and bulk particles (closed squares) of TaC

attributed the difference from earlier result to possible plastic deformation effect on the powder sample used.

The studies of Cr, besides being important with respect to the antiferromagnetic transition, are also an example of persistent and planned efforts to improve the accuracy in θ_M measurements.

3.4.15 Nano Effect on Debye–Waller Factor and Debye Temperature

Yosida [3.111] studied the effect of encapsulation of TaC particles in carbon nanotubes on the Debye–Waller factor and X-ray Debye temperature. X-ray diffraction experiments were made in the range 7–273 K on TaC using bulk particles and particles encapsulated in carbon nanotubes. From TEM images, the encapsulated particles were found to have a size of 72 Å in comparison of 3,000 Å for the bulk particles.

Considerable differences were observed between the thermal parameters of the two samples. Thus, at 273 K, the $(\overline{u^2})^{1/2}$ values were 0.090 and 0.06 Å for the encapsulated and bulk particles, respectively. The X-ray Debye temperatures were found to be 340 and 489 K for the encapsulated and bulk samples, respectively. The temperature variation of B (or $(\overline{u^2})$) as shown in Fig. 3.14 is also different being more for the encapsulated than for the bulk samples.

3.4.16 Energy of Defect Formation from Debye Temperature

It was mentioned earlier that the Debye temperature is related to several other physical properties. These relations may be used to estimate the Debye temperature from a known physical property or the property may be estimated from the known value of the Debye temperature. As an example, we shall consider the relation between the Debye temperature and the formation energy of defects in crystals.

The concentration of defects in a solid at a given temperature depends upon the value of the formation energy. The formation energy E_d is, thus, an important parameter describing the defect state of the crystal. The main defect in metals is a vacancy whereas in the case of ionic crystals, like the alkali halides, the Schottky defects exist in pairs. E_d represents the formation energy for a vacancy in metals and for a Schottky pair in ionic crystals. The formation energy can be estimated experimentally from the data on ionic conductivities and diffusion [3.112] for ionic crystals and from thermal expansion for metals. The experimental values are available only for some metals and ionic crystals. The formation energy can also be calculated on the basis of the interionic potential [3.112]. As experimental values of the formation energy are not available for several crystals attempts have been made by several workers to establish empirical relations between the formation energy and other physical properties. That the formation energy of defects is related to the Debye temperature through an equation like

$$\theta_\mathrm{D} = \text{constant} \left(E_\mathrm{d} \Big/ M V^{2/3} \right)^{1/2} \tag{3.72}$$

was shown by Mukherji [3.113] empirically and by March [3.114], Glyde [3.115] and Tewary [3.116] from different theoretical approaches. Here M is the molecular weight and V, the molar volume. Glyde [3.115] further pointed out that in these calculations the X-ray Debye temperature should be used rather than the Debye temperature from the specific heats. Sastry and Mulimani [3.117] verified the relation in the case of the alkali halides with the NaCl structure. Pathak and Trivedi [3.118] verified the same relation for the alkali halides with the NaCl structure employing Debye temperatures obtained from X-ray data.

3.4.17 Effect of Electronic Environment on Debye–Waller Factor

The values of (B/a^2) for some alkali halides are given in Table 3.7. The most prominent feature in these values observed by Linkoaho [3.51] is that (B/a^2) of an individual ion increases when the number of electrons, n_e, of its companion ion increases. This dependence is observed to be quite linear in the case of Cl ion and is reproduced in Fig. 3.15. There is no regularity of (B/a^2) values in the case of the cation. Cooper and Rouse [3.122] have pointed out that the relative mean square amplitudes of the fluorine ions in CaF_2, SrF_2 and BaF_2 also show the same trend as observed by Linkoaho [3.51]. No theoretical justification has been made for the validity of this relationship. It is desirable to examine whether this relationship is generally valid.

Table 3.7. Values of the relative mean square amplitudes of vibration for some alkali halides; B_1 for cation, B_2 for anion

Substance	$[B_1/a^2] \times 10^2$	$[B_2/a^2] \times 10^2$	Reference for B values
LiCl	8.81 ± 0.38	4.47 ± 0.38	[3.119]
NaCl	5.69 ± 0.11	4.68 ± 0.11	[3.51]
KCl	5.25 ± 0.13	5.20 ± 0.13	[3.120]
RbCl	5.27 ± 0.23	5.64 ± 0.23	[3.121]

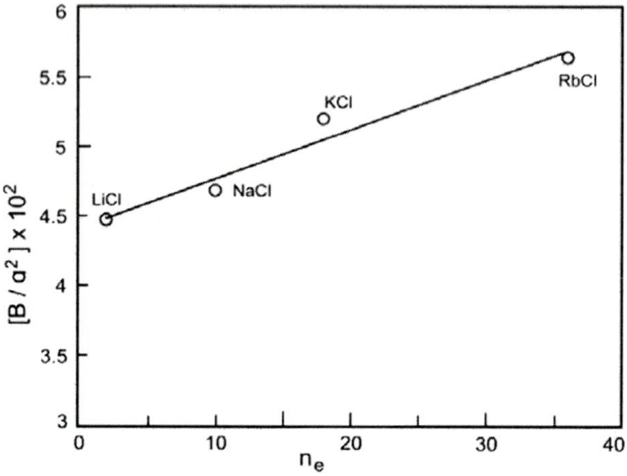

Fig. 3.15. Plot of (B/a^2) of the chlorine ion vs. the electron number (n_e) for the alkali ion for some alkali chlorides

3.4.18 Debye–Waller Factor of Mixed Crystals

Wasastjerna [3.123] studied the Debye–Waller factors of single compositions of KCl–KBr and KCl–RbCl mixed crystals. Beg et al. [3.124] determined the Debye–Waller factor of $K_{0.5}Rb_{0.5}F$ mixed crystals. In all these cases, it was observed that the Debye–Waller factor for intermediate compositions exceeds the value for the end members. The Debye–Waller factors of $Sm_{0.7}Y_{0.3}S$ and $Eu_{0.8}Y_{0.2}S$ mixed crystals have been measured by Dernier et al. [3.125]. The values are anomalous and have been explained in terms of the mixed valence effect.

It should also be noted here that the disorder in mixed crystals contributes to the effective (measured) Debye–Waller factor. In order to obtain the true Debye temperature, the measured Debye–Waller factor has to be corrected for a static component.

3.4.19 Debye–Waller Factors of Protein Structures

In the earlier sections, we have mostly discussed 'small crystals' i.e. crystals containing a small number of atoms in the unit cell. When the number of atoms is large, the simple analytical methods discussed earlier are not useful and recourse is made to the least squares refinement of (3.35).

Proteins are important 'large' molecules and the determination of their structures was a landmark in X-ray Crystallography resulting in the award of the Nobel Prize to Dorothy Hodgkins. From the standpoint of Debye–Waller factor determinations also, it may be some sort of a record that in the process of determination of the structure of 2 Zn pig insulin, Baker et al. [3.126] estimated the Debye–Waller factors of more than 1,100 atoms. An analysis of these Debye–Waller factors led to some information about atomic motion in the crystal. Fig. 3.16 shows plots of $\overline{U}$, the rms displacements, against r, the distance of each residue from the centre of mass. A correlation of these parameters means that the structure is vibrating about its centre. The plots show that (a) a correlation exists between $\overline{U}$and r (b) that the correlation shows a larger spread for the side chains and (c) that the best correlation exists for molecules 1 and 2 with the dimer and hexamer centroids.

Recently, Lenin et al. [3.127] investigated whether the B-values can be correlated to the mutability of residues in globular proteins and whether an assessment can be made of the degree of change in the Bvalues of structurally equivalent residues in the course of evolution. For this purpose Lenin et al. [3.127] considered available data on haemoglobin structures, trypsin structures and triosephosphate isomerase structures. For each class, Lenin et al. [3.127] correlated the smoothened B-value with (a) the aligned residue number, (b) smoothened Dayhoff's scores and (c) the pairwise correlation coefficients of B-values were also correlated with pairwise Dayhoff's scores. These correlation plots for haemoglobin are shown in Figs. 3.17–3.19. The plot between B-values and the aligned residue number (Fig. 3.17) consists of a number of bumps. The correlations in Figs. 3.18 and 3.19 show considerable scatter. The plots for the other two groups are very similar. Lenin et al. [3.127] concluded that there is no systematic variation in B as the amino acid replacement increases. This indicates that the protein flexibility is conserved during the course of evolution.

3.5 Some of our Results

In view of the importance of the Debye–Waller factor as a solid state parameter, a comprehensive programme of study of various aspects of the X-ray determination of the Debye–Waller factors and Debye temperatures was undertaken in our laboratory. The systems studied belong to the fcc, hcp, NaCl, CsCl, ZnS and CaF_2 structures. The results are discussed in this section.

3.5.1 Debye–Waller Factors – Data Generation

Experimental values of Debye–Waller factors are given in Tables 3.8–3.13. In the case of hcp metals, the Debye–Waller factors in the 'a' and 'c' directions

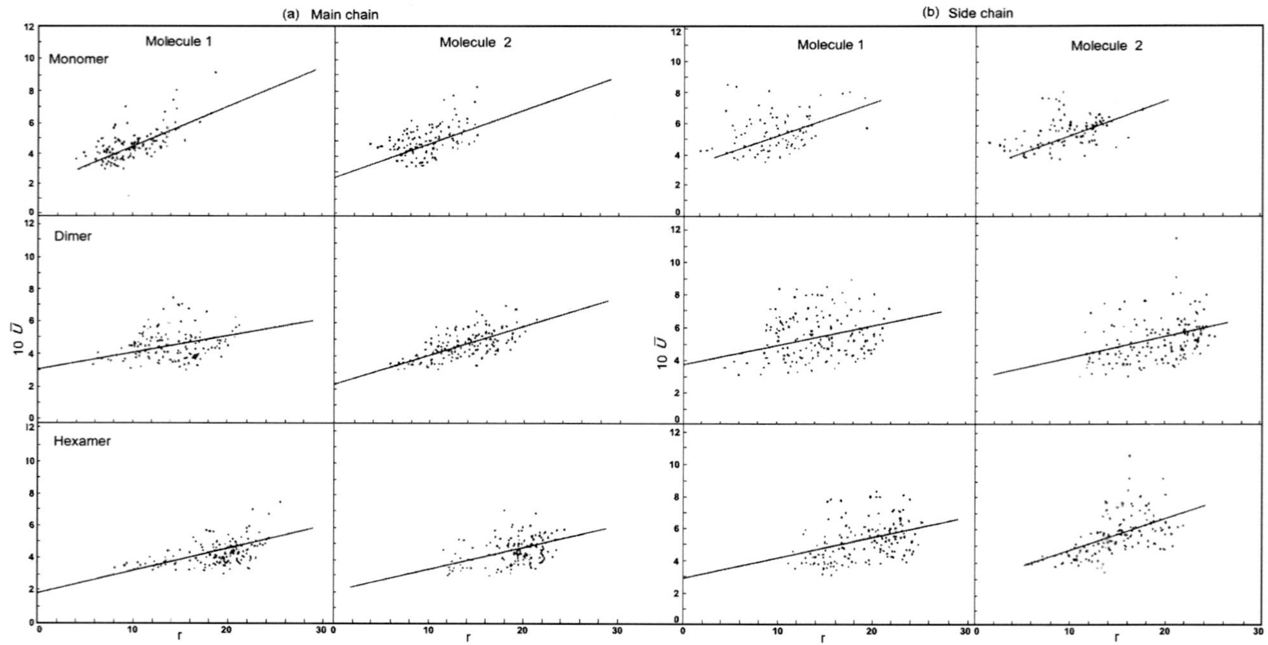

Fig. 3.16. Plots of $\overline{U}$ against the radial distance r from the centre of mass of (a) the main-chain atoms for molecule 1, left, and molecule 2, right, as the monomer, dimer and hexamer (b) of the side-chain atoms in the same order

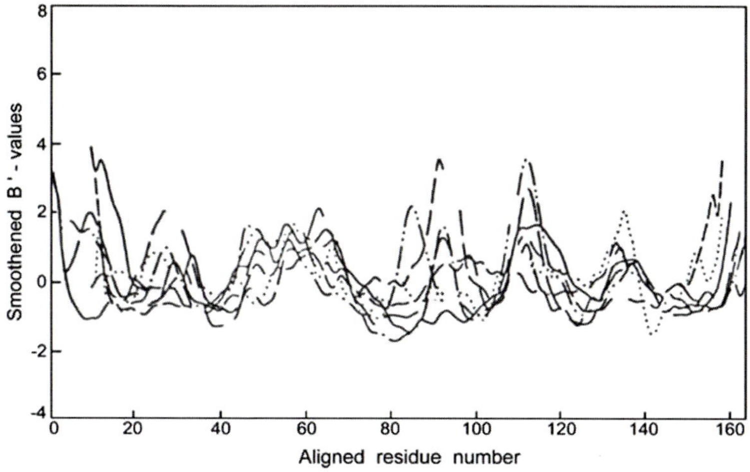

Fig. 3.17. Plot of Debye–Waller factor B vs. the aligned residue number for haemoglobin

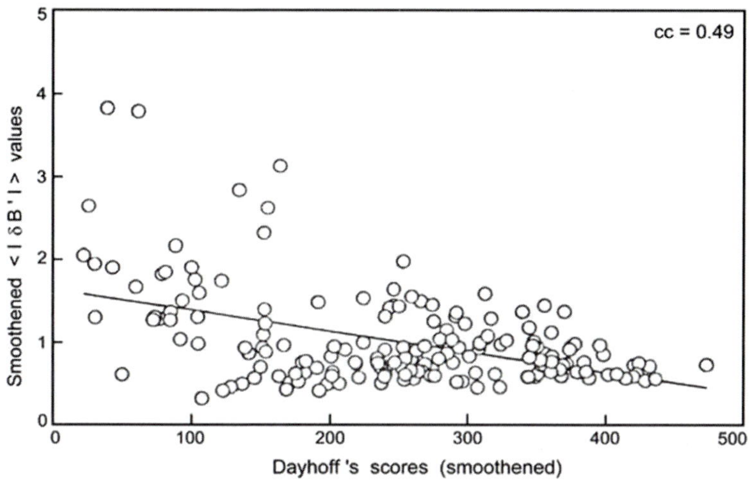

Fig. 3.18. Plot of smoothened B values vs. smoothened Dayhoff's scores for haemoglobin

are given besides the mean Debye–Waller factor. In the other systems, the mean Debye–Waller factor is given. However individual atomic Debye–Waller factors are given wherever such determination was possible.

3.5.2 Debye–Waller Factors and Mass Ratio

There has been some controversy with regard to the relative values of the Debye–Waller factors of atoms in a crystal. Brindley [3.34] observed that

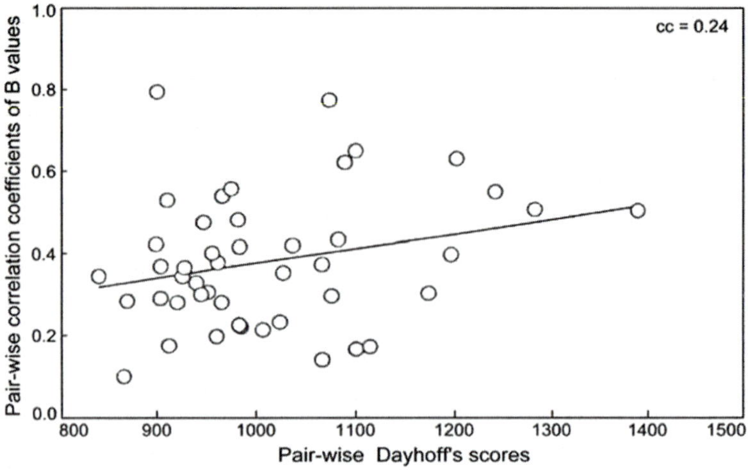

Fig. 3.19. Plot of pairwise correlation coefficient of B vs. pairwise Dayhoff's scores for haemoglobin

Table 3.8. Debye–Waller factor B of some fcc elements at room temperature; figures in parenthesis in Tables 3.8–3.13 indicate uncertainties in the last two digits

Crystal	B [Å^2]	Ref.
Al	0.88 (03)	[3.128]
Yb	1.65 (15)	[3.129]

in NaCl the lighter ion has the larger amplitude of vibration. The same observation was made by Menz [3.153] from experimental data on Debye–Waller factors for a few alkali halides obtained from electron diffraction. In fact, Menz proposed an inverse proportionality between the ratios of Debye–Waller factors and the mass ratios. However considering data on Debye–Waller factors for a larger number of alkali halides, Linkoaho [3.51] observed that such a proportionality does not exist. Huiszoon and Groenewegen [3.154] pointed out that in AgCl [3.155] and PbTe [3.107] the heavier atoms have a higher Debye–Waller factor. All these crystals have the NaCl structure. Similar results on some crystals of other structures have also been reported. These are HgSe [3.156], $AuGe_2$ [3.157], $Pb(NO_3)_2$ [3.158] and TlCl [3.159].

Huiszoon and Groenewegen [3.154] showed, with the aid of lattice dynamics, that the Debye–Waller factors become mass-independent at temperatures above the Debye temperature. Feldman [3.160] by an independent analysis came to the same conclusion. But Scheringer [3.161] reconsidered this problem and pointed out that at higher temperatures, the mass dependence of the Debye–Waller factors is weakened and not eliminated. As such, the amplitudes of vibration should show a mass dependence, however feeble. Further,

Table 3.9. Directional Debye–Waller factors B_a and B_c, and mean Debye–Waller factor $\overline{B}$ of some hcp elements at room temperature

Crystal	B_a [Å^2]	B_c [Å^2]	$\overline{B}$ [Å^2]	Ref.
Er	0.57 (07)	0.73 (07)	0.62 (07)	[3.130]
Dy	0.81 (08)	0.89 (07)	0.84 (08)	[3.131]
Gd	0.80 (01)	0.88 (02)	0.83 (01)	[3.131]
Lu	0.86 (06)	0.97 (06)	0.90 (06)	[3.131]
Y	0.83 (02)	0.80 (03)	0.82 (02)	[3.131]
Sc	0.72 (01)	0.73 (01)	0.72 (01)	[3.132]
Tb	0.67 (04)	0.71 (04)	0.68 (04)	[3.132]
Ti (α)	0.54 (01)	0.48 (06)	0.52 (03)	[3.133]
Zr	0.52 (01)	0.51 (04)	0.52 (02)	[3.133]
Ru	0.14 (02)	0.15 (01)	0.14 (02)	[3.133]
Tm	0.82 (01)	0.84 (01)	0.83 (01)	[3.133]
Hf	0.41 (02)	0.41 (01)	0.41 (02)	[3.133]

Scheringer suggested the desirability of a statistical analysis of Debye–Waller factors vis-a-vis the atomic mass covering a large number of crystals.

Following Scheringer's suggestion, the data on the ratios of the Debye–Waller factors for a large number of crystals with NaCl structure are collected in Table 3.14. These include some of the results obtained in this laboratory. The mass ratios m_1/m_2 (m_1 always being the heavier of the two masses) are also given. It is seen that in most cases $B_2/B_1 > 1$.

Before we analyse these data, we shall discuss the three cases pointed out by Huiszoon and Groenewegen [3.154] viz. MgO, AgCl and PbTe. In the case of MgO, Huiszoon and Groenewegen [3.154] quote the experimental results of Sanger [3.162]. But there are several later reports, particularly that of Lawrence [3.53] where the reported values of the Debye–Waller factors lead to a ratio $B_2/B_1 = 1.11$. Lattice dynamical calculations by Sanger [3.162] also support this value for the ratio B_2/B_1. Again, in the case of AgCl, Huiszoon and Groenewegen [3.154] quote the results of Korhonen and Linkoaoho [3.155]. Later measurements by Srinivas and Sirdeshmukh [3.140] yield a value of $B_2/B_1 > 1$. Lattice dynamical values by Groenewegen and Huiszoon [3.52] support this value. Thus MgO and AgCl cannot be considered as violating the empirical mass-ratio rule and, hence, the values from later reports are included in Table 3.14.

With regard to the third case mentioned by Huiszoon and Groenewegen [3.154] viz. PbTe, the anomalous value of $B_2/B_1 < 1$ quoted by Huiszoon and Groenewegen [3.154] from the report of Keffer et al. [3.106] has been confirmed by independent experimental measurements by Noda et al. [3.163] and Nagaiah [3.145]. Further this ratio ($B_2/B_1 < 1$) is also supported by lattice dynamical calculations by Keffer et al. [3.107]. Hence, it has to be considered as an established case where the mass ratio rule fails.

Table 3.10. Atomic Debye–Waller factors (B_A, B_B, and mean Debye–Waller factors $\overline{B}$ of crystals with NaCl structure (suffix A refers to first atom in formula and B to second atom)

Crystal	B_A [Å^2]	B_B [Å^2]	$\overline{B}$ [Å^2]	Ref.
LiCl	1.99 (16)	0. 62 (01)	1.13 (11)	[3.134]
LiBr	2.52 (25)	0.90 (02)	1.14 (12)	[3.134]
NaCl	1.52 (10)	1.15 (15)	1.56 (11)	[3.134]
NaBr	1.55 (05)	1.14 (04)	1.67 (10)	[3.134]
NaI	2.63 (08)	1.81 (05)	2.24 (19)	[3.134]
KCl			1.93 (08)	[3.135]
KBr			2.23 (05)	[3.136]
KI			2.99 (13)	[3.137]
RbCl			2.18 (08)	[3.138]
RbBr			2.89 (18)	[3.139]
RbI			3.36 (14)	[3.139]
AgCl	2.13 (03)	2.38 (05)	2.19 (03)	[3.140]
AgBr			2.14 (06)	[3.140]
MnS			0.90 (09)	[3.141]
CdO			0.84 (03)	[3.141]
TmSe			0.90 (17)	[3.142]
SmS			1.56 (15)	[3.142]
SmSe			1.29 (24)	[3.142]
SmTe			1.10 (29)	[3.142]
EuS	0.39 (11)	1.12 (28)		[3.143]
EuTe			1.43 (20)	[3.143]
EuSe	1.22 (13)	1.37 (20)		[3.144]
PbS	1.36 (10)	0.78 (03)	1.28 (08)	[3.145]
PbTe	1.41 (20)	1.06 (15)	1.28 (20)	[3.145]
TaC	0.21 (01)	0.47 (03)	0.23 (14)	[3.146]
ZrC	0.19 (01)	0.37 (05)	0.21 (02)	[3.146]
HfC	0.19 (03)	0.38 (05)	0.20 (04)	[3.146]
NbC	0.20 (02)	0.37 (02)	0.22 (02)	[3.146]
HfN	0.37 (02)	0.80 (05)	0.40 (02)	[3.146]
VN	0.38 (02)	0.47 (08)	0.40 (03)	[3.146]

Values of Debye–Waller factors with $B_2/B_1 < 1$ have been reported for PbS and PbSe by Noda et al. [3.163] and for NiO and CoO by Meisalo and Inkinen [3.164]. Out of these, the PbS results have been qualitatively confirmed by the experimental results of Nagaiah and Sirdeshmukh [3.165] and also lattice dynamically by Gnanasoundari and Ramachandran [3.166]. Thus this is another case where the violation of the mass ratio rule is confirmed. Results on PbSe, NiO and CoO need independent experimental confirmation as well as theoretical support.

While it is now established that the ratio B_2/B_1 is, in general, greater than unity, it remains to be seen whether there is any relationship between

Table 3.11. Mean Debye–Waller factor $\overline{B}$ of some crystals with CsCl structure

Crystal	$\overline{B}$ [Å^2]	Ref.
CsCl	1.87 (10)	[3.147]
CsBr	2.32 (24)	[3.148]
CsI	2.59 (12)	[3.148]
TlCl	2.79 (08)	[3.148]
TlBr	2.72 (09)	[3.148]
KRS-5		
$TlBr_{0.46}I_{0.54}$	2.73 (06)	[3.149]
KRS-6		
$TlCl_{0.46}Br_{0.54}$	3.03 (10)	[3.150]
NH_4Cl	1.82 (16)	[3.148]
NH_4Br	2.60 (22)	[3.148]
Gd-Zn	2.66 (16)	[3.151]
Gd-Mg	3.70 (46)	[3.151]

Table 3.12. Atomic Debye–Waller factors $B_{\mathrm{A}}, B_{\mathrm{B}}$, and mean Debye–Waller factors $\overline{B}$ of crystals with sphalerite (zinc blende) structure (formula AB) at room temperature; Ref. [3.152]

Crystal	B_{A} [Å^2]	B_{B} [Å^2]	$\overline{B}$ [Å^2]
GaAs			0.80 (10)
InP	1.19 (08)	1.76 (20)	1.27 (10)
InSb	1.60 (23)	1.00 (16)	1.29 (19)
ZnTe	1.16 (09)	0.99 (02)	
CdTe			1.86 (16)
HgSe	3.25 (15)	1.83 (18)	2.85 (16)
HgTe			2.70 (11)

Table 3.13. Atomic Debye–Waller factors $B_{\mathrm{A}}, B_{\mathrm{B}}$, and mean Debye–Waller factor $\overline{B}$ of some crystals with fluorite structure (formula AB_2) at room temperature; Ref. [3.146]

Crystal	B_{A} [Å^2]	B_{B} [Å^2]	$\overline{B}$ [Å^2]
SrF_2	0.55 (07)	1.04 (10)	0.70 (08)
BaF_2	0.63 (08)	0.93 (09)	0.70 (08)
CdF_2	0.54 (12)	1.16 (19)	0.70 (13)
$PbF_2(\beta)$	0.50 (04)	1.31 (11)	0.63 (05)
EuF_2	0.78 (27)	1.53 (23)	0.93 (26)

the values of m_1/m_2 and B_2/B_1. The values of m_1/m_2 and B_2/B_1 are plotted in Fig. 3.20. It is observed that the data points are scattered about a curve. The limits of scattering are indicated by the shaded region of width of 0.4 in B_2/B_1. Within these limits of scattering, the trend is that B_2/B_1 increases as m_1/m_2 increases. up to about $m_1/m_2 \approx 8$ and thereafter it saturates to $B_2/B_1 \approx 2$.

Table 3.14. Mass ratio (m_1/m_2) and the ratio (B_2/B_1) for crystals with NaCl structure

	Crystal	m_1/m_2	B_1	B_2	B_2/B_1	Ref.
1.	LiF	2.74	1.05	0.65	1.616	[3.41]
2.	LiCl	5.11	1.99	0.62	3.210	[3.134]
3.	LiBr	11.51	2.52	0.90	2.80	[3.134]
4.	NaF	1.21	0.91	0.91	1.00	[3.41]
5.	NaCl	1.54	1.52	1.15	1.322	[3.134]
6.	NaBr	3.47	1.55	1.14	1.360	[3.134]
7.	NaI	5.52	2.63	1.81	1.453	[3.134]
8.	KCl	1.10	2.17	2.16	0.995	[3.41]
9.	KBr	2.04	2.36	2.38	0.992	[3.41]
10.	RbCl	2.41	2.14	2.27	1.061	[3.41]
11.	MgO	1.52	0.30	0.34	1.133	[3.53]
12.	CaO	2.51	0.35	0.44	1.257	[3.167]
13.	SrO	5.48	0.43	1.93	4.488	[3.167]
14.	BaO	8.58	0.79	1.19	1.506	[3.167]
15.	MnO	3.43	0.38	0.45	1.184	[3.164]
16.	EuS	4.74	0.39	1.12	2.872	[3.143]
17.	EuSe	1.93	1.22	1.37	1.123	[3.144]
18.	TaC	15.07	0.21	0.47	2.268	[3.146]
19.	HfC	14.86	0.19	0.38	1.973	[3.146]
20.	NbC	7.74	0.20	0.37	1.858	[3.146]
21.	ZrC	7.60	0.19	0.37	1.943	[3.146]
22.	VN	3.64	0.38	0.47	1.249	[3.146]
23.	HfN	12.74	0.37	0.80	2.160	[3.146]
24.	AgCl	3.04	2.13	2.38	1.117	[3.140]
25.	PbS	6.46	1.36	0.78	0.574	[3.145]
26.	PbTe	1.62	1.41	1.06	0.752	[3.145]
27.	PbSe	2.62	1.43	1.13	0.792	[3.163]
28.	NiO	3.67	0.63	0.40	0.635	[3.164]
29.	CoO	3.68	0.51	0.47	0.922	[3.164]
30.	CdO	7.03	0.64	0.81	1.266	[3.168]

EuS, LiCl, SrO, CdO and BaO have $m_1/m_2 > 1$ and also $B_2/B_1 > 1$. But their data points show very large deviations from the curve (beyond the shaded region). The experimental results for these crystals need a redetermination.

3.5.3 Comparison of Experimental Results with Lattice Dynamical Results

As mentioned earlier, the Debye–Waller factors can be calculated from lattice-dynamical models. The comparison of such values with experimental values throws light on the reliability of the model and/or the experimental data. In Table 3.15, the experimental values of Debye–Waller factors obtained by us for some crystals are compared with calculations from lattice dynamical models.

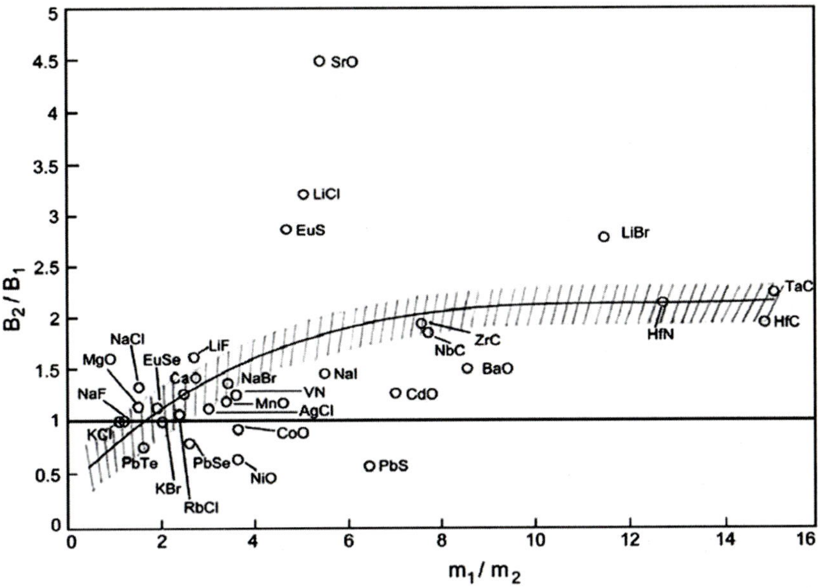

Fig. 3.20. Plot of mass ratio m_1/m_2 vs. the ratio B_2/B_1

From a comparison, the following conclusions may be drawn:

(1) In general, there is an order-of-magnitude agreement between the experimental and theoretical values.
(2) In the rubidium halides, the experimental values show a better agreement with 11-parameter shell model calculations by Govindarajan [3.170] using neutron inelastic scattering data than with other models.
(3) In the other alkali halides, the experimental values agree more with values reported by Kushwaha [3.102] based on the 7-parameter bond-bending model than with other models.
(4) In AgCl, there is a fair agreement between experimental values and those reported by Groenewegen and Huiszoon [3.52] based on a 5-parameter rigid ion model.
(5) In PbS, calculations have been made by Gnanasoundari and Ramachandran [3.166] from a simple shell model. The calculated value agrees well with the experimental value for Pb. However, for S, the agreement is not so good.
(6) For the transition metal carbides, lattice dynamical values are available from Feldman [3.172]. There is very good agreement between the experimental and theoretical values for the metal atoms. The agreement for the carbon atoms is not so good and it worsens in the sequence ZrC–TaC. The difference may be due partly to lesser accuracy in the experimental values for the C atom and partly to the limitation of the model.

Table 3.15. Comparison of experimentally determined Debye–Waller factors [in Å^2] with lattice dynamical models; experimental values from Table 3.10

Method	B_A	B_B	B_A	B_B				
(a) Sodium halides	NaBr		NaI					
Expt.	1.55	1.14	2.63	1.81				
11-parameter shell model [3.169]	1.89	1.55	2.41	1.96				
7-parameter bond-bending model [3.102]	1.75	1.70	2.47	2.21				
(b) Rubidium halides	RbCl	RbBr	RbI					
	$\overline{B}$	$\overline{B}$	$\overline{B}$					
Expt.	2.18	2.89	3.36					
11-parameter shell model, neutron data [3.170]	2.331	2.708	3.265					
11-parameter shell model, elastic constant [3.169]	1.992	2.361	2.909					
Deformation shell model [3.171]	2.168	2.451	3.003					
7-parameter bond-bending model	1.983	2.343	2.900					
	B_A	B_B						
(c) AgCl								
Expt.	2.13	2.38						
5-parameter rigid ion model [3.52]	2.13	2.15						
(d) PbS								
Expt.	1.36	0.78						
Simple shell model [3.166]	1.42	1.07						
(e) Transition metal carbides	ZrC		NbC		HfC		TaC	
	B_A	B_B	B_A	B_B	B_A	B_B	B_A	B_B
Expt.	0.19	0.37	0.20	0.36	0.19	0.37	0.21	0.47
Double-shell model [3.172]	0.18	0.29	0.20	0.26	0.16	0.27	0.19	0.24

3.5.4 Anisotropy of Debye–Waller Factors

The relations between the anisotropy in physical properties have been discussed by Wooster [3.173] and Boas and Mackenzie [3.174]. These discussions do not include the anisotropy of the Debye–Waller factors.

The Debye–Waller factors are anisotropic in anisotropic crystals. Thus, hexagonal crystals have two principal Debye–Waller factors B_a and B_c associated with the 'a' and 'c' directions. The values of B_a and B_c for several hexagonal close-packed (hcp) metals are given in Table 3.16.

In hcp crystals, the ideal c/a ratio is 1.633. Such a crystal will be nearly isotropic. A larger c/a ratio indicates weakening of interatomic forces in the 'c' direction. Since the Debye–Waller factor (or atomic amplitude) reflects the interatomic bonding, the B_c/B_a ratio is expected to be > 1 for crystals with $c/a > 1$ while $B_c/B_a \approx 1$ for crystals with c/a close to 1.633.

Watanabe et al. [3.176] noted this feature in their study of Mg and Cd. Further, they plotted a curve between B_c/B_a and c/a with data for only four crystals available at that time. Now that experimental values are available for a larger number of crystals, this aspect can be examined anew. In Table 3.16, the B_c/B_a values and c/a values for seventeen hcp crystals are given. The same data are shown in Fig. 3.21. The c/a value is significantly different from 1.633 only in the case of Cd and Zn. The B_c/B_a values are larger than 1 by a significant amount in these two crystals. In all other cases, $c/a \approx 1.63$ and B_c/B_a values are also $\sim$1. No significance can be given to the cases where B_c/B_a is slightly larger than 1 while c/a values are slightly less than 1.63.

Table 3.16. c/a ratio and the ratio B_c/B_a for hcp metals (c/a values from [3.175]; B_c/B_a values for Be, Cd, Zn, Ho and Mg from [3.42] and rest from Table 3.9)

Crystal	c [Å]	a [Å]	c/a	B_c [Å^2]	B_a [Å^2]	B_c/B_a
Be	3.5833	2.2866	1.566	0.54	0.60	0.9
Cd	5.61	2.97	1.89	3.04	1.15	2.64
Dy	5.67	3.58	1.58	0.89	0.81	1.10
Er	5.59	3.55	1.57	0.73	0.57	1.28
Gd	5.79	3.63	1.59	0.88	0.80	1.10
Ho	5.61	3.58	1.57	0.97	0.96	1.00
Lu	5.55	3.50	1.58	0.97	0.86	1.12
Mg	5.21	3.21	1.62	1.58	1.34	1.18
Sc	5.27	3.31	1.59	0.73	0.72	1.00
Tb	5.69	3.60	1.58	0.71	0.67	1.06
Y	5.73	3.64	1.57	0.80	0.83	0.96
Zn	4.95	2.66	1.86	2.04	1.23	1.65
Ti (α)	4.68	2.95	1.58	0.48	0.54	0.88
Zr	5.15	3.23	1.59	0.51	0.52	0.98
Ru	4.28	2.70	1.58	0.15	0.14	1.07
Tm	5.55	3.53	1.57	0.84	0.82	1.02
Hf	5.05	3.19	1.58	0.41	0.41	1.00

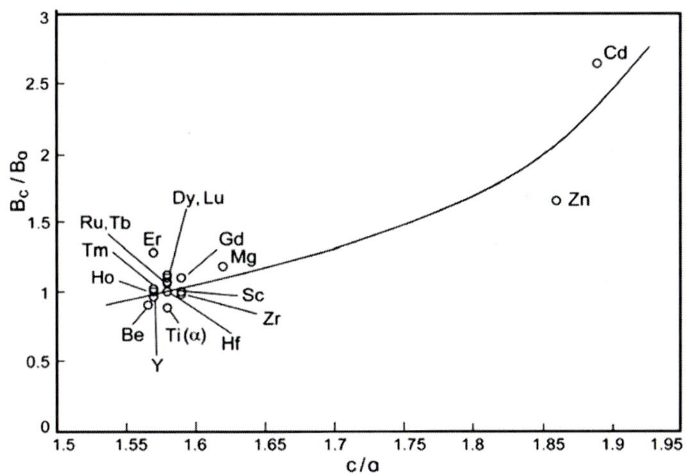

Fig. 3.21. Plot of B_c/B_a vs. axial ratio (c/a) for hcp metals

This is obviously due to the errors in the Debye–Waller factors. The figure shows a clear trend with B_c/B_a increasing with c/a. There is a gap in data from $c/a = 1.65$ to $c/a = 1.85$. Data points in this range will help to firmly establish the plot. This would be possible when the Debye–Waller factors of hcp alloys with intermediate values of c/a are studied.

3.5.5 Effect of Strain on Debye–Waller Factors

Inagaki et al. [3.71] showed that lattice strains affect the measured Debye–Waller factor. All the materials studied by Inagaki et al. [3.71] were inorganic compounds. It was considered worthwhile to examine this effect in other materials like metals and semiconductors. The rare earth metal Yb [3.177] and the semiconductor material CdTe [3.178] were chosen for study.

The CdTe and Yb filings were subjected to slow grinding for varying periods. At different stages of grinding, the lattice strain and mean Debye–Waller factor were determined. The lattice strain was determined from the half-widths of X-ray diffraction lines by using the Scherrer equation:

$$(\beta \cos\theta/\lambda) = (K/L) + 2\varepsilon_l(\sin\theta/\lambda), \tag{3.73}$$

where β is the half-width, K the shape factor, L the particle size and ε_l the strain. The slope of the linear plot between $(\beta\cos\theta/\lambda)$ and $(\sin\theta/\lambda)$ gives $2\varepsilon_l$.

The variation of B with ε_l for CdTe and Yb is shown in Figs. 3.22 and 3.23. In both the materials, B increases with ε_l. In CdTe the variation of B with ε_l shows the same trend as in the materials studied by Inagaki et al. [3.71]. B increases with ε_l and then tends to saturate at $\varepsilon_l = 0.8 \times 10^{-3}$. This is much lower than the strain at which saturation begins in the work of Inagaki et al. [3.71]. On the other hand, in Yb, the variation of B with ε_l is more like that observed by Inagaki et al. [3.71] for graphite; B increases monotonously with ε_l without attaining saturation.

Since the interest is in obtaining accurate values of the Debye–Waller factors, Gopi Krishna and Sirdeshmukh [3.177] and Sirdeshmukh et al. [3.178] suggested the extrapolation of the B vs. ε_l plot to $\varepsilon_l = 0$ to get the strain-corrected value of the Debye–Waller factor. The values of the Debye–Waller factor at maximum strain, at initial strain and at zero strain are given in

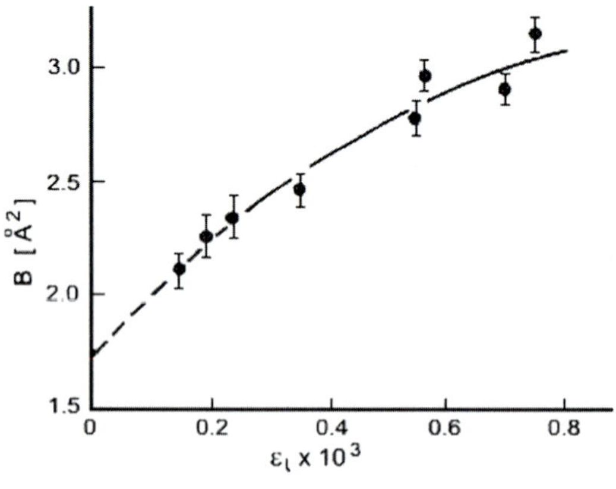

Fig. 3.22. Variation of Debye–Waller factor B vs. the strain ε_l for CdTe

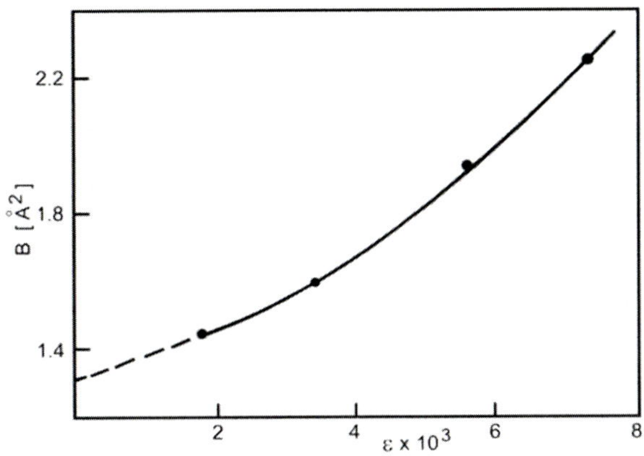

Fig. 3.23. Variation of Debye–Waller factor B vs. the strain ε_l for Yb

Table 3.17. Values of the Debye–Waller factor (B) at different strains

	B [Å^2]			
	Max. strain	Initial strain	Zero strain	Other values
CdTe	3.0	2.1	1.75	1.90 [3.179]
Yb	2.2	1.45	1.24	1.65 [3.129]

Table 3.17. The values of B reported by earlier workers are larger than those for the strain-free values. It is quite possible that the samples used on those studies were affected by strain.

Taking these results along with those of Inagaki et al. [3.71] it may be concluded that the lattice strain introduced in the process of sample preparation has an effect on the measured values of the Debye–Waller factors in all types of materials. It is suggested that when the Debye–Waller factors are determined from X-ray diffraction intensities, it is desirable to make an estimate of the strain and, should it be large, to make a suitable correction.

3.5.6 Effect of Atomic Scattering Factors on B

From (3.34)–(3.45), it can be seen that for the determination of the Debye–Waller factor, the theoretical values of the scattering factors have to be used. Scattering factors are available in literature for neutral atoms and for atoms in different valence states. As mentioned earlier, Dawson and Howard [3.56] found a small but significant difference in the Debye–Waller factors of LiF determined with scattering factors for Li^0F^0 and $Li^{-1}F^{-1}$ states. Differences of a larger magnitude were observed by Hosoya et al. [3.54] in the Debye–Waller factors of VN determined with scattering factors for the V^0N^0, $V^{+1}N^{-1}$, $V^{+2}N^{-2}$ and $V^{+3}N^{-3}$states.

There is some controversy regarding the bonding in CdO. From a comparison of the calculated and observed values of the lattice energies and nearest neighbour distances De Noble [3.180] concluded that CdO is the most ionic of all the cadmium chalcogenides. The ionicity value of Phillips [3.181] for CdO is the lowest among crystals with NaCl structure. From their X-ray studies, Linkoaho et al. [3.168] obtained ionic charges +3 ± 0.5 and –2 ± 0.5 for the cadmium and oxygen ions, respectively. While these values are not exactly what would be expected if CdO were completely ionic, Linkoaho et al. concluded that CdO forms an ionic crystal. On the other hand, from the X-ray intensities, Linkoaho et al. [3.168] calculated a parameter represented by $\mathrm{d}Z/R^2\mathrm{d}R$. This parameter is proportional to the charge density around the cation and the anion. A low value of this parameter at the radius of best separation is an indicator of high ionicity whereas a high value indicates accumulation of charge between the atoms and consequently a covalent nature of the bond. Linkoaho et al. [3.168] have given the values of this parameter for a few crystals with NaCl structure. CdO has the highest value for this parameter suggesting a departure from ionicity. The nature of the bonding in PbS is similarly not clear. It has a low hardness [3.182]) and an effective ionic charge 0.8 characteristic of a highly ionic crystal [3.183]. However, other properties of PbS like the optical properties, suggest a departure from ionic bonding [3.184].

There is meager work on the physical properties of MnS. Not much information is available on the electronic state of the atoms in this substance. From an X-ray study of MnO, Kuriyama and Hosoya [3.168] concluded that the Mn and O atoms are in the state of divalent ions. Wells [3.184] makes the general remark that an M-S bond is more covalent than an M–O bond where M is a metal atom.

In view of the uncertainties regarding the valence states of atoms in these substances, calculations of the mean Debye–Waller factors, the mean amplitudes of vibration and the Debye temperature were carried out by Subhadra and Sirdeshmukh [3.141] using scattering factors for the two extreme states viz. neutral atoms and doubly charged ions. The results of these calculations are summarized in Table 3.18. Although there is a small difference in the values of R-factor for the two sets of calculations. the two sets of scattering factors yield values of B and θ which agree within limits of estimated errors. It may be mentioned that the substances studied contain atoms with fairly large numbers of electrons and the use of atomic or ionic scattering factors in these systems does not seem to have a noticeable effect on the derived values of the Debye–Waller factor.

3.5.7 Debye–Waller Factors and the Electronic Environment

Linkoaho [3.51] observed an effect of the electronic environment of an ion on its thermal vibration amplitude. This was shown in the form of a linear plot between B/a^2 and n_e ('B' Debye–Waller factor of atom, 'a' lattice constant, 'n_e' electron number of the neighbouring atom). Linkoaho's observation was confined to a few alkali halides.

Table 3.18. Experimental results for PbS, MnS and CdO for the two electronic states

Substance	Electronic state for scattering factors	B[Å^2]	θ [K]	R
CdO	(Cd^0O^0)	0.84 (04)	255 (6)	0.008
	($Cd^{+2}O^{-2}$)	0.83 (03)	255 (5)	0.006
PbS	(Pb^0S^0)	1.37 (13)	145 (7)	0.018
	($Pb^{+2}S^{-2}$)	1.40 (13)	143 (6)	0.018
MnS	(Mn^0S^0)	0.81 (11)	318 (22)	0.019
	($Mn^{+2}S^{-2}$)	0.90 (10)	299 (16)	0.016

Table 3.19. Relative mean square amplitudes of vibration for carbides with the NaCl structure (B_1 for metal atom, B_2 for carbon atom)

Substance	n_e	$[B_1/a^2] \times 10^2$	$[B_2/a^2] \times 10^2$	Ref. for B values
TiC	22	1.296	1.641	[3.185]
ZrC	40	0.852	1.692	[3.146]
NbC	41	0.991	1.806	[3.146]
HfC	72	0.897	1.745	[3.146]
TaC	73	1.053	2.389	[3.146]

Table 3.20. Values of the relative mean square amplitudes of vibration in some CaF_2 type crystals (B_1 for anion, B_2 for the F ion)

Substance	n_e	$[B_1/a^2] \times 10^2$	$[B_2/a^2] \times 10^2$	Ref. for B values
CaF_2	18	1.699	2.340	[3.122]
SrF_2	38	1.647	3.060	[3.146]
CdF_2	46	1.854	3.992	[3.146]
BaF_2	54	1.643	2.425	[3.146]
EuF_2	61	2.302	4.510	[3.146]
PbF_2	80	1.418	3.729	[3.146]

This approach is extended to the transition metal carbides and the fluorite type crystals for which the Debye–Waller factors have been determined. The relevant data are shown in Tables 3.19 and 3.20 and the plots are shown in Figs. 3.24 and 3.25. The plots are linear and they show that (B/a^2) for C and F increases with n_e of the corresponding atoms. These results are consistent with the observations of Linkoaho [3.51].

3.5.8 Debye–Waller Factors in Mixed Crystals

It was mentioned in Sect. 3.4.16 that the Debye–Waller factor in mixed crystals is larger than that of the pure crystals. It was also pointed out that there is only limited work. Systematic measurements on several mixed crystal systems have been carried out by the authors and their coworkers and the composition dependence studied. These results will be discussed in Chap. 7.

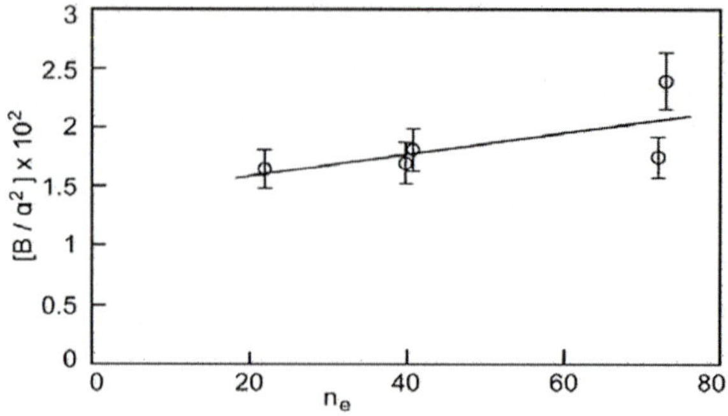

Fig. 3.24. Plot of (B/a^2) of the carbon atom vs. the electron number (n_e) for the metal atom in some carbides

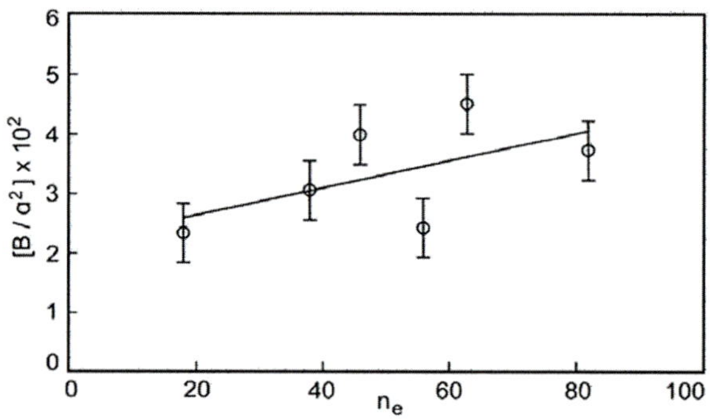

Fig. 3.25. Plot of (B/a^2) for the fluorine ion vs. the electron number (n_e) for the metal ions in some CaF_2 type crystals

3.5.9 X-ray Debye Temperatures Derived from Debye–Waller Factors

The procedure for deriving the X-ray Debye temperature θ_M from the mean Debye–Waller factor $\overline{B}$ has been discussed in Sect. 3.3. The Debye temperatures derived from Debye–Waller factors given in Tables 3.8–3.13 are given in Table 3.21. This data on Debye temperatures is analyzed in the following sections.

3.5.10 Comparison of θ from Different Methods

The Debye temperature θ is given by $h\nu_D/k$ where ν_D is either a characteristic frequency or an average frequency. θ values can be derived from physical properties which depend on the vibrational spectrum of the solid. Blackman [3.186], Herbstein [3.10] and Gschneidner [3.187] have critically examined the θ values derived from different physical properties of a solid. Salter [3.188] showed that,

Table 3.21. X-ray Debye temperatures (θ_M) of crystals derived from mean Debye–Waller factors given in Tables 3.8–3.13. The Debye temperatures derived from specific heats (θ_D) and elastic constants (θ_E) are also given

	Crystal	θ_M [K]	θ_D [K]	θ_E [K]
(a) fcc				
1.	Al	387 (7)	390	403
2.	Yb	109 (5)		
(b) hcp				
1.	Er	182 (10)	163	186
2.	Dy	160 (8)	158	177
3.	Gd	165 (1)	155	171
4.	Lu	148 (5)	166	184
5.	Y	219 (3)	214	245
6.	Sc	330 (2)		
7.	Tb	178 (7)		
8.	Ti (α)	391 (22)		
9.	Zr	267 (20)		
10.	Ru	495 (24)		
11.	Tm	155 (10)		
12.	Hf	217 (20)		
(c) NaCl type				
1.	LiCl	387 (2)	477	394
2.	LiBr	267 (14)	419	249
3.	NaCl	278 (8)	288	322
4.	NaBr	202 (6)	240	224
5.	NaI	144 (6)	203	167
6.	KCl	218 (3)	232	230
7.	KBr	162 (2)	172	170
8.	KI	118 (3)	156	131
9.	RbCl	157 (5)	168	169
10.	RbBr	121 (4)	134	129
11.	RbI	99 (3)	106	102
12.	AgCl	149 (2)	161	146
13.	AgBr	131 (2)	144	137
14.	MnS	299 (16)		419
15.	CdO	255 (5)		
16.	TmSe	176 (6)	200	258
17.	SmS	155 (7)	269	268
18.	SmSe	153 (14)	206	
19.	SmTe	151 (20)		
20.	EuS	205 (16)		276
21.	EuTe	134 (10)		189
22.	EuSe	153 (9)		232
23.	PbS	143 (6)	225	229
24.	PbTe	127 (19)	125	177
25.	TaC	459 (40)	616	572
26.	ZrC	617 (33)	649	690
27.	HfC	461 (65)	549	552
28.	NbC	637 (37)	761	740
29.	HfN	291 (7)		

Table 3.21. Continued

	Crystal	θ_M [K]	θ_D [K]	θ_E [K]
30.	VN	527 (27)		
(d) CsCl type				
1.	CsCl	148 (4)	174	159
2.	CsBr	118 (6)	130	136
3.	CsI	101 (2)	120	115
4.	TlCl	101 (2)		125
5.	TlBr	95 (2)		114
6.	KRS-5 ($TlBr_{0.46}I_{0.54}$)	90 (1)		
7.	KRS-6 ($TlCl_{0.7}Br_{0.3}$)	97 (2)		
8.	NH_4Cl	268 (11)	285	270
9.	NH_4Br	164 (7)		177
10.	Gd-Zn	108 (3)		
11.	Gd-Mg	101 (6)		
(e) ZnS (zinc blende) type				
1.	GaAs	246 (15)	360	348
2.	InP	194 (8)	425	303
3.	InSb	151 (11)	240	207
4.	ZnTe	185 (4)	201	225
5.	CdTe	124 (6)	200	163
6.	HgSe	93 (3)	212	152
7.	HgTe	88 (2)	112	142
(f) CaF_2 type				
1.	SrF_2	348 (20)	418	380
2.	BaF_2	294 (17)	360	282
3.	CdF_2	317 (29)		328
4.	PbF_2	260 (10)		219
5.	EuF_2	243 (27)		329

in general, the Debye temperatures obtained from one experimental method cannot usually be used to describe some other physical property. Differences between the values based on different properties have to be expected from the fact that the contribution to these properties is not always from the same region of the vibration spectrum of a solid. However, since Debye temperatures from X-ray measurements and specific heats both involve averages over the entire vibrational spectrum, comparison between these values could be undertaken to a certain extent, though differences between these have to be expected due to the different methods of averaging involved in the two cases. This was first pointed out by Zener and Bilinsky [3.189], who showed that the ratio θ_M/θ_D (the subscripts M and D referring to the X-ray and specific heat values, respectively) can be expressed as a function of the Poisson's ratio of the substance. Barron et al. [3.190] have shown that this method of relating the two Debye temperatures is not strictly correct. This is supported by Herbstein [3.10]. Gschneidner [3.187] observes that even after correcting for this factor the agreement between θ_M and θ_D is in general poor and for a majority of substances studied $\theta_M < \theta_D$, both values corresponding to room temperature.

It is of interest to know the position regarding the relation between θ_M and θ_E (from elastic constant data). Blackman [3.186] observes that θ_E and θ_D are equal only at sufficiently low temperatures. Gschneidner [3.187] also carried out a comparison of θ_E and θ_d at room temperature. He observed that while equality between θ_D and θ_E does not generally hold at room temperature, no systematic pattern could be observed between the two (i.e. approximately equal numbers of values of θ_E are larger and smaller than θ_D). As such Gschneidner concludes that $\theta_D \approx \theta_E$ at room temperature. Since $\theta_D > \theta_M$ as already pointed out, $\theta_E > \theta_M$. This is the general conclusion drawn by Blackman [3.186], Herbstein [3.10], Gschneidner [3.187] and subsequent workers.

Our θ_M values are compared with θ_E and θ_D values wherever available. These values are also given in Table 3.21. Since the present measurements relate to room temperature, for the most part θ_E and θ_D values derived from room temperature data are chosen. In the case of crystals for which room temperature data are not available, the low temperature θ values are included.

It can be seen from the values in Table 3.21 that the present X-ray values conform to the trend mentioned earlier viz., $\theta_M < \theta_D$ or θ_E. A more detailed statistical analysis of the data is given in Table 3.22. It is very clear that θ_M and θ_D, are very close only in a few (about 10 %) cases. In a majority of cases (85 %), it is found that $\theta_M < \theta_D$. This is in agreement with the observation made by Gschneidner [3.187] in the case of metals. A comparison of the θ_M and θ_E, values shows that these values are close (within ± 5 K) in only a few cases (~10%). In a majority of cases (80%), $\theta_M < \theta_E$. This is in agreement with the observations of Blackman [3.186], Herbstein [3.10] and Gschneidner [3.187].

3.5.11 A modified Expression for the X-ray Debye Temperature (θ_M)

The expression for the Debye–Waller factors in terms of the Debye model (given in Sect. 3.2) is derived for a monatomic solid. This expression was initially applied to monatomic solids. It was soon extended to diatomic crystals like KCl with the justification that the two atoms have nearly equal masses and hence it can be approximated to a monatomic crystal with a mass equal to the average mass of the two atoms. In course of time, the procedure has been extended to diatomic and even triatomic crystals irrespective of whether the masses are equal or not.

Table 3.22. Statistical comparison of θ values

Values compared	Total no.	Equal within $\pm$ 5 K	$\theta_M < \theta_D, \theta_E$
θ_M, θ_D	42	5	35
θ_M, θ_E	51	6	41

The question of applicability of this expression to polyatomic crystals has been critically examined by Horning and Staudenmann [3.191]. Their final result is that (3.29) for monatomic solids gets modified in the case of polyatomic solids to

$$B = (6h^2/pmk_B\theta'_M)\ [(\phi(x)/x) + (1/4)]\,. \qquad (3.74)$$

Here, p is the number of atoms per lattice point (2 for NaCl, 3 for CaF_2, etc.) and m is the average mass. The modification is applicable at higher temperatures ($T > \theta_M$). Horning and Staudenmann [3.191] suggested that θ_D should be compared to θ'_M rather than θ_M. According to them, θ_D may still differ from θ'_M.

Horning and Staudemann [3.191] calculated θ'_M from reported values of θ_M for some crystals having the NaCl, CsCl and ZnS structures and compared them with θ_D values. There was better agreement between θ'_M and θ_D compared with that between θ_M and θ_D. We have applied this modification to some crystals with NaCl structure studied in this work. The results are shown in Table 3.23. While $\theta_M < \theta_D$, the modification overshoots θ_D, i.e. $\theta'_M > \theta_D$ in several cases. The e.s.d.'s for ($\theta_D - \theta_M$) and (θ_D–θ'_M) are comparable.

3.5.12 Energy of Defect Formation from Debye Temperatures

It was shown in Sect. 3.4 that the formation energy of defects (E_d) can be estimated from the Debye temperature using a formula of the type

$$\theta_M = \text{constant}\left(E_d\Big/MV^{2/3}\right)^{1/2}. \qquad (3.75)$$

We have extended these calculations to the crystals for which we have determined the Debye temperatures. The results are presented structurewise:

Table 3.23. Comparison of θ_M and θ'_M with θ_D for some diatomic crystals; θ_M and θ_D from Table 3.21

Crystal	θ_M [K]	θ'_M [K]	θ_D [K]	$\lvert\theta_D - \theta_M\rvert$ [K]	$\lvert\theta_D - \theta'_M\rvert$ [K]
EuS	205	290	262	57	28
EuSe	153	216	176	23	60
EuTe	134	189	140	6	49
RbCl	157	222	186	29	36
RbBr	121	171	135	14	36
RbI	99	140	115	16	25
NaCl	278	393	288	10	105
NaBr	202	286	240	32	46
NaI	144	204	203	59	1
LiCl	387	547	477	90	70
LiBr	267	378	419	52	41
KI	118	167	156	39	11
e.s.d				42	47

Table 3.24. Comparison of the value of energy of Schottky defect formation E_d for CsCl type compounds

Substance	Values of E_d [eV per pair]		
	From (3.75)	From experiment	From calculations by Murthy and Murti [3.193]
CsCl	1.86	1.86	1.86
CsBr	1.56	2.00	1.74
CsI	1.58	1.90	1.46
NH_4Cl	1.65	0.81	1.33
NH_4Br	1.24	–	1.17
TlCl	1.03	1.30	1.27
TlBr	1.16	–	1.20

Crystals with CsCl structure

Using a value of 151 K for θ_M for CsCl [3.192] and the experimental value for E_d quoted by Murthy and Murti [3.193], a value of 5,002 is obtained for the constant. The values of E_d for some crystals with CsCl structure are given in Table 3.24. The table also gives the values of the formation energy calculated by Murthy and Murti [3.193]. It can be seen from the data that the values obtained from the above relation agree reasonably with values obtained from rigorous theoretical calculations.

With the exception of NH_4Cl, there is fair agreement between the experimental values and the values calculated here. Kroger [3.194] pointed out that NH_4Cl is a mixed conductor and if the electronic and protonic contributions to the electrical conductivity are not taken into account, a spuriously low value is likely to be obtained for the vacancy formation energy.

Crystals with CaF_2 Structure

From the Born model, the formation energies were calculated for CaF_2 by Franklin [3.195]. The formation energy is 2.7 ± 0.4 eV for an anion Frenkel pair, 7.5 ± 0.8 eV for a cation Frenkel pair and 5.1 ± 0.7 eV for a Schottky trio. These calculations are in agreement with the experimental results of Ure [3.196]. Thus the anion Frenkel (AF) pair is the more predominant defect in CaF_2 type crystals unlike the alkali halides in which the Schottky pair is predominant. Using the value of formation energy for the anion Frenkel pair for CaF_2 and the Debye temperature of CaF_2 calculated from the B values obtained by Cooper and Rouse [3.122] the value of the constant in (3.75) comes out to be 3,550. Using this constant the formation energies of the anion–Frenkel pair are calculated for the other crystals. The results are tabulated in Table 3.25.

The values obtained in here agree reasonably well with the values 2.39 eV for SrF_2 and 1.92 eV for BaF_2 obtained by Catlow and Norgett [3.197] from shell model calculations. For BaF_2 the formation energy of the anion Frenkel

Table 3.25. Formation energies for the anion Frenkel pair (E_d) in CaF_2-type crystals

Substance	E_d [eV per AF pair]	
	Equation (3.75)	From [3.197, 3.198]
CaF_2	2.7	
SrF_2	2.48	2.39
CdF_2	2.13	
BaF_2	2.82	1.92 1.5
EuF_2	1.84	
PbF_2	2.82	

Table 3.26. Values of energy of formation of neutral metal atom vacancies (E_d) for ZnS type crystals

Crystal	E_d [eV per vacancy]	Crystal	E_d [eV per vacancy]
GaAs	2.85	CdTe	1.49
InP	2.13	HgSe	0.88
InSb	1.88	HgTe	1.07

defect is obtained as 1.5 eV by Mizuta et al. [3.198] which is lower than the calculated value. Rao [3.199] has attributed the difference between the theoretically calculated and experimental values of linear thermal expansion coefficients of some CaF_2 type crystals to the formation of defects. The disagreement was found to be the most in the case of EuF_2 among the substances studied. From this, he suggested that the formation energy of defects may be the least for EuF_2 among the fluorides with the CaF_2 structure. The calculated values given in Table 3.25 also show that the value of the formation energy for EuF_2 is the least among the fluorides.

ZnS-Type Crystals

Bailly [3.200] has pointed out that in the semiconducting compounds with ZnS structure, the covalently bound chalcogen atoms do not form defects and the predominant defects are neutral metal vacancies. Balaiah [3.147] calculated E_d for these crystals from (3.75).

A value of 2,900 was obtained for the constant using the value $E_d = 2.85$ eV for GaAs. The calculated values of E_d are given in Table 3.26. The E_d values are of the same order as that for GaAs.

Metals

For metals Tewary [3.116] gave a value of 32.8 for the constant in (3.75). Using this value and the θ_M values given in Table 3.21, the E_d values for some hcp metals were estimated by Gopi Krishna [3.130]. The values estimated are given in Table 3.27.

Table 3.27. Values of the energy of vacancy formation E_d for rare earth metals

Metals	Values of E_d [eV]	
	Experiment	Equation (3.75)
Yb	–	0.676
Gd	–	1.165
Dy	–	1.096
Er	1.721	1.430
Lu	–	0.965
Y	1.439	1.154
Ho	–	0.930

The estimated values agree reasonably well with the experimental values 1.721 eV for Er and 1.439 eV for Y quoted by Dariel [3.201]. The experimental values for other rare earth metals are not available for comparison but the estimated values could be taken as reasonable in view of the fair agreement in cases where experimental data are available.

4

Hardness

4.1 Introduction

"Hardness is like the storminess of the sea, easily appreciated but not readily measured (or defined)" [4.1]

"Hardness is a measure of the resistance (of the material) to permanent deformation or damage" [4.2]

It "(Hardness) is not a single property but rather a whole complex of mechanical properties and at the same time a measure of intrinsic bonding of the material" [4.3]

"(There are) clear connections between chemical bonding, hardness and dislocation mobilities" [4.4]

"The precise definition (of hardness) depends entirely on the method of measurement" [4.5]

"Resistance to the movement of dislocations will determine the hardness of the material" [4.6]

The above quotes are an indication of the complexity of the concept of hardness. However, if we choose the last two, it is possible to see how hardness is related to one or other of several factors. For this purpose, one may consider the commonest method of determining hardness. This consists of indenting a solid surface by a loaded indenter of a definite geometrical shape and measuring the contact area between the indenter and the material. The ratio of the load and the contact area is the experimental definition of hardness.

Assuming that the hardness is determined by the indentation method, let us consider what happens when a loaded indenter is pressed against the surface of the material. This is shown in Fig. 4.1. When the indenter is pressed, it creates dislocations. As the indentation continues, newly created dislocations push the dislocations created earlier. These dislocations move in some preferred directions in the crystal depending on the slip system. This motion of dislocations is obstructed by at least four factors, viz.:

1. Impurities and precipitates present in the material (impurity hardening)
2. Other dislocations present in the material (dislocation hardening)

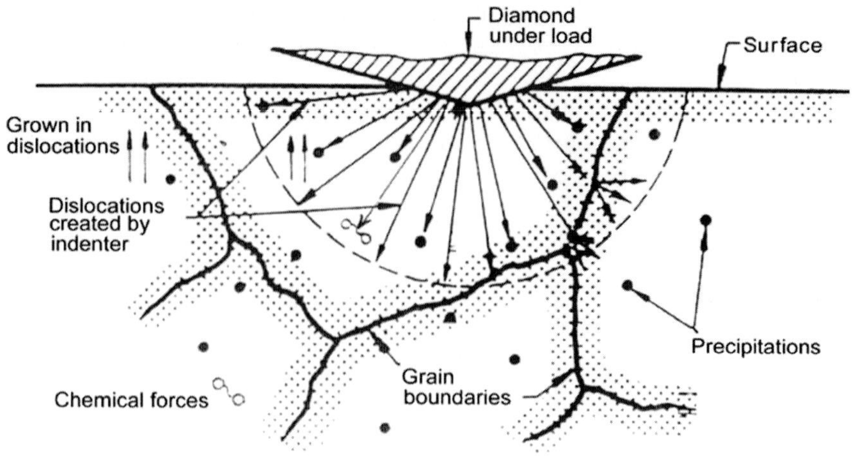

Fig. 4.1. Schematic of resistance to motion of dislocations

3. Grain boundaries (grain boundary hardening)
4. The chemical forces in the lattice of the material (intrinsic hardness)

Thus, the hardness of a pure (free from impurities) and well-annealed (free from excessive number of dislocations and grain boundaries) crystal represents the intrinsic hardness which reflects the bond strength.

The low-load indentation method of hardness testing is popular because it offers several advantages viz.:

1. It is simple and does not consume much time
2. It can be used on samples with small areas, e.g. small crystals
3. It can be used on brittle materials which cannot withstand large loads, e.g. glasses.

4.2 Experimental Methods

4.2.1 General

Several methods are available for the measurement of hardness. Shaw [4.7] lists the following categories of methods:

1. *Scratch tests* in which one merely observes whether one material is capable of scratching another. The Mohs and file hardness tests are of this type.
2. *Plowing tests* in which a blunt element (usually diamond) is moved across a surface under controlled conditions of load and geometry and the width of the groove is the measure of hardness. The Bierbaum test is of this type.

3. *Rebound tests* in which an object of standard mass and dimensions is bounced from the test surface and the height of rebound is taken as a measure of hardness. The Shore Scleroscope is an instrument of this type.
4. *Damping tests* in which the change in amplitude of a pendulum having a hard pivot resting on the test surface is the measure of hardness. The Herbert pendulum test is of this type.
5. *Cutting tests* in which a sharp tool of given geometry is caused to remove a chip of standard dimensions.
6. *Abrasion tests* in which a specimen is loaded against a rotating disk and the rate of wear is taken as a measure of hardness.
7. *Erosion tests* in which sand or abrasive grain is caused to impinge upon the test surface under standard conditions and loss of material in a given time is taken as the measure of hardness. Hardness of grinding wheels is measured thus.
8. *Static indentation tests* in which a ball, cone or pyramid is forced into a surface and the load per unit area of impression is taken as the measure of hardness. Brinell, Vickers, Rockwell, Monotron and Knoop tests are of this type.

Of these, the indentation method is most commonly employed. As mentioned, indentation hardness is defined as the ratio of the load to the surface area of the indentation. Indentation hardness measurements can, in principle, be carried out at fairly high loads ($\sim$100 kg). But, for most materials, it is convenient to make measurements at low loads $<$200 g. This low-load hardness is called microhardness. With specially designed instruments, measurements can be made at very low loads (0.01–0.5 g). The hardness resulting from such measurements is called ultra hardness. For thin films and coatings, the recently developed method of nanoindentation is employed.

4.2.2 Leitz–Wetzlar Mini-Load 2 Microhardness Tester

The machine is shown in Fig. 4.2. The essential parts are the stage, a facility for pre-selection of the indentation-site, the indenter, the load, an automatic arrangement for slow release of the load, timer and a micrometer eye-piece for measurement of the impression. In the Leitz Mini-Load 2 microhardness tester loads of 5, 10, 15, 25, 50, 100 and 200 g can be applied. The basic force of the indenting unit or the dead weight of the machine is 5 g $\pm$ 50 mg. The maximum diagonal length which can be measured by this machine is 220 μm.

The procedure described below is adopted for both the Vickers and the Knoop hardness measurements alike. Samples of a minimum size of about $3 \times 3 \times 1\,\mathrm{mm}^3$ are chosen for the microhardness measurements. Several indentations are made taking care to see that neighbouring indentations are apart by a distance of 3–4 times the size of the indentations. At each load, about 20 indentations are made and the mean of the measured diagonal lengths is used to evaluate the hardness numbers. The method of histogram [4.5] is employed

Fig. 4.2. The Leitz–Wetzlar hardness tester

to evaluate the average value of the diagonal lengths. The duration of the indentation (the loading and the hold time or the indenting and the dwelling time) at any load applied is 25 s and is maintained constant. The error in the measurement of the diagonal length of an indentation impression in general may be estimated to be about 0.25–0.5 μm. The computed hardness numbers are accurate to $\pm 1\,\mathrm{kg\,mm^{-2}}$.

The Vickers diamond pyramidal indenter has a square base with $136° \pm 30'$ pyramidal angle. The faces of the indenter meet at a sharp point and are optically flat. An indentation made with a Vickers indenter is shown in Fig. 4.3. Vickers hardness number (H_V) is expressed in terms of the contact area of the indenter and is computed from the relation,

$$H_V = 1854.4\,(P/d^2) \text{ in } \mathrm{kg\,mm^{-2}}, \qquad (4.1)$$

where P is the load applied in g and d the diagonal length of the indentation impression measured in μm.

The Knoop hardness indenter is an elongated pyramid such that the angles between the long and short faces are $172° \pm 30'$ and $130°$ respectively. The

Fig. 4.3. The impression of a Vickers indentation

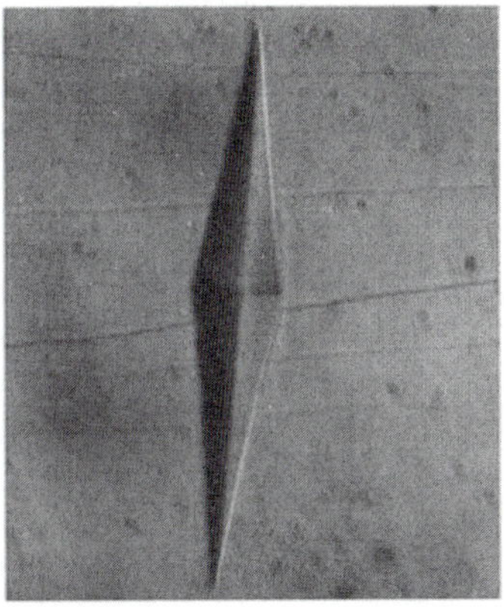

Fig. 4.4. The impression of a Knoop indentation

Knoop indenter is more blunt than the Vickers indenter and hence gives a shallower impression. The Knoop indentation is in the form of a parallelogram with one diagonal seven times longer than the shorter diagonal. An indentation made with a Knoop indenter is shown in Fig. 4.4. The Knoop hardness number (H_K) is expressed in terms of the load and the projected area and is computed from the relation,

$$H_K = 14230\,(P/d^2) \text{ in } \text{kg}\,\text{mm}^{-2}, \tag{4.2}$$

where P is the load in g and d the length of the indentation along the long axis of the indenter measured in μm.

4.2.3 Shimadzu Dynamic Ultra Hardness Tester DUH 202

This instrument is essentially a Vickers tester with diamond pyramidal indenter. Some parameters of interest are: loading range 0.01–200 g; measurable

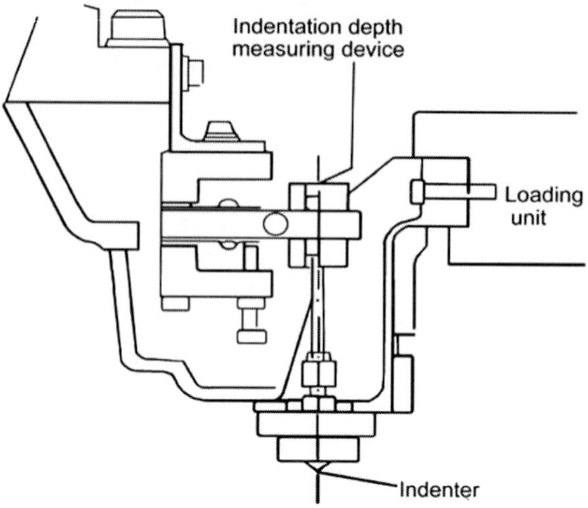

Fig. 4.5. Indentation depth measuring device (DUH)

depth 0–15 μm; minimum unit of depth measurement 0.001 μm; error in dynamic hardness 30 kg mm^{-2} at ultraloads and 10 kg mm^{-2} at microloads.

There are distinct differences between this instrument and conventional instruments. First, the loading is done with an electromagnetic force. Second, the hardness is determined in the conventional way by measuring the diagonal length of the impression under a microscope or by measuring the depth of the impression by means of a differential transformer (Fig. 4.5). Third, as the electromagnetic force is being applied, the depth is measured continuously at every point of the descent of the indenter i.e. at each value of the applied force; it is thus a dynamic measurement. At ultraloads, hardness can be measured only from the value of the depth h as the diagonal length is too small to be measured with the microscope. The dynamic ultrahardness H_{DU} is given by

$$H_{\mathrm{DU}} = 37.835\,(P/h^2) \text{ in kg mm}^{-2}. \tag{4.3}$$

Actually the computerized instrument yields data on P, h and H_{DU} along with a P vs. h plot. Thus one can study the depth variation of hardness.

4.2.4 Nanoindentation

Nanoindentation is used for measuring hardness at very small length scales. The commonly used indenter for nanoindentation is the Berkovich diamond indenter which is a triangular diamond pyramid. In this, the lateral dimension of the contact impression is approximately seven times the depth. So measurements are possible for contact impressions with lateral dimensions 0.1 μm. A detailed procedure for the determination of mechanical properties

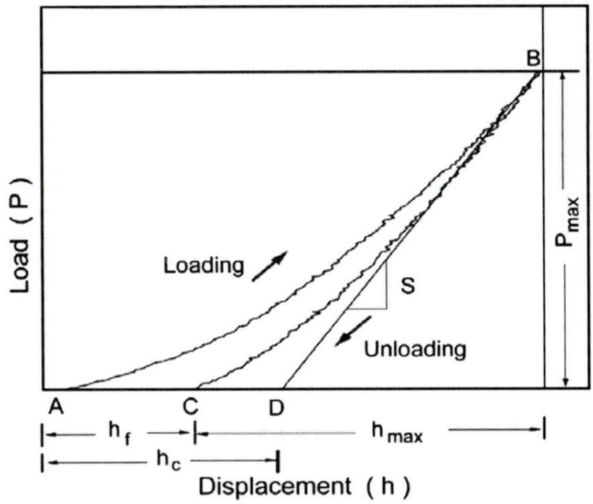

Fig. 4.6. A schematic representation of load vs. displacement data for an indentation experiment

of very thin coatings and films using a nanoindentation technique was first described by Oliver and Pharr [4.8].

The principle consists of the continuous measurement of force and displacement as an indenter of known geometry is pressed into the sample material. The force is usually applied via an electromagnetic coil and the current in the coil determines the load. The displacement is measured, in most cases, by a capacitive sensor. The indenter load is increased from zero to a certain maximum value P_{max} and then the load is gradually decreased to zero. During the loading and unloading, the load and displacement data are automatically stored in a computer. A typical load vs. indentation depth curve is shown in Fig. 4.6. From the load and displacement (indentation depth) data, the hardness (H) of the material can be calculated. In general, the loading data are influenced more by the plastic properties of the material and the unloading data by elastic properties. From Fig. 4.6 the loading part is AB, whereas unloading part is BC. The initial part (i.e., close to P_{max}) of the unloading curve is a straight line and then it becomes curved on further reduction in the load due to the elastic recovery of the indentation depth. On complete removal of the load, the residual displacement is h_f. The unloading curve for ceramics is of the shape shown in Fig. 4.6, with only the initial unloading part following a straight line. In the case of soft metals, there is very little recovery of the indentation depth on removal of the load. Therefore, the unloading curve for metals is a straight line down to about 5% of the maximum applied load. If there is no recovery of the indentation, the unloading curve will follow the line BD. AD represents the contact depth of indentation and is denoted by h_c. Because of the elastic recovery of the indentation depth on load removal, $h_c > \mathrm{h_f}$.

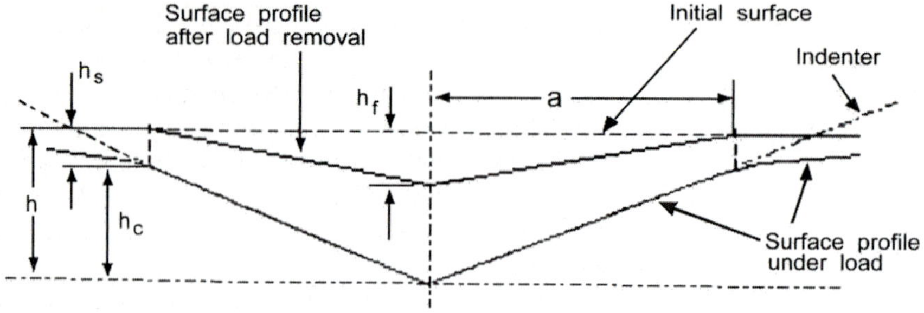

Fig. 4.7. Schematic representation of a section through an indentation, showing various quantities used in the analysis

Oliver and Pharr [4.8] also proposed a method for the analysis of the nanoindentation load vs. displacement data. Fig. 4.7 shows a cross-section of an indentation and identifies the parameters used in the analysis. At any time during loading, the total displacement h is written as $h = h_s + h_c$, where h_c is the vertical distance along which contact is made (hereafter called the contact depth) and h_s is the displacement of the surface at the perimeter of the contact. At peak load, the load and displacement are P_{max} and h_{max}, respectively, and the radius of the contact circle is a. Upon unloading, the elastic displacements are recovered, and when the indenter is fully withdrawn, the final depth of the residual hardness impression is h_f.

The hardness is defined as the mean contact pressure underneath the indenter at maximum load (P_{max}.),

$$H = P_{max}/A_c. \tag{4.4}$$

The projected contact area A_c is determined from the area function of the indenter tip $A_c = f(h_c)$. The contact depth is related to the maximum indentation depth h_{max} by:

$$h_c = h_{max} - \varepsilon\, P_{max}/S. \tag{4.5}$$

Here S is the contact stiffness, defined as an increment in load divided by the resulting increment in displacement in the absence of plastic deformation; $\varepsilon = 0.75$ for Berkovich diamond indenter. The relationship used to describe the unloading data for stiffness measurement is

$$P = \alpha(h - h_f)^m, \tag{4.6}$$

where P is the load, α and m are constants. The unloading curve can be differentiated analytically to determine the slope at maximum load:

$$S = (\mathrm{d}P/\mathrm{d}h)_{P=P_{max}}. \tag{4.7}$$

Using (4.5), the contact depth is calculated and is used to determine the projected contact area A_c from the area function. The nanohardness is obtained by substitution of A_c in (4.4).

4.2.5 Relative Hardness Measurement

Hardness testing machines are generally expensive. The design of a simple inexpensive instrument designed by Kishan Rao and Sirdeshmukh [4.9] which facilitates measurement of relative hardness of crystals from measurements of lengths of the indentation dislocation rosette (IDR) is described.

When a crystal is indented and subsequently etched, an array of dislocation etch pits is observed. This array is termed as IDR. Hopkins et al. [4.10] observed that the length of the IDR is related to the load on the indenter. Inabe et al. [4.11] observed that the relation between the load and the length of the IDR is exactly similar to that between the load and the diagonal length of the Vickers indentation impression.

Gilman [4.12] observed that in crystals with the diamond structure and ZnS structure, the critical shear stress for glide of dislocations is related to the hardness through the relation

$$\tau = kH, \tag{4.8}$$

where k is a constant and τ is the shear stress. From a study of the IDR in silicon produced by a spherical indenter, Hu [4.13] showed that

$$l = cP^{1/2}\tau^{-1/2}, \tag{4.9}$$

where l is the length of the arm of the IDR, P, the load applied and c is a constant. Combining (4.8) and (4.9), we get

$$l = AP^{1/2}H^{-1/2}, \tag{4.10}$$

where A is a new constant. This equation relates the length of the IDR to the hardness. Although (4.8) and (4.9) have been proposed for silicon, similar equations may hold for other systems also and hence (4.10) may have general validity. Equation (4.10) predicts an l^2–P relation for a given crystal. Further, if two similar crystals having hardness H_1 and H_2 are indented at loads resulting in rosettes of lengths l_1 and l_2, (4.10) leads to

$$H_2 = (l_1/l_2)^2 H_1, \tag{4.11}$$

Thus, if H_1 is known, H_2 can be estimated. This is the principle of the design of the instrument.

Figure 4.8 shows the line diagram of the instrument. It consists of two pillars A and B. A brass bar F is attached to the pillar A with ball bearings such that it is free to move about in a vertical plane. D is the sample holder fixed to the pillar B; it can be raised or lowered using a rack-and-pinion arrangement O. A fine steel pin S is fixed at one end of the bar F, which carries a screw E at its other end with which the bar is balanced. When a load P is applied on the pin, the bar is locked in the horizontal position using a screw G. The crystal X placed on the sample holder is raised until the crystal

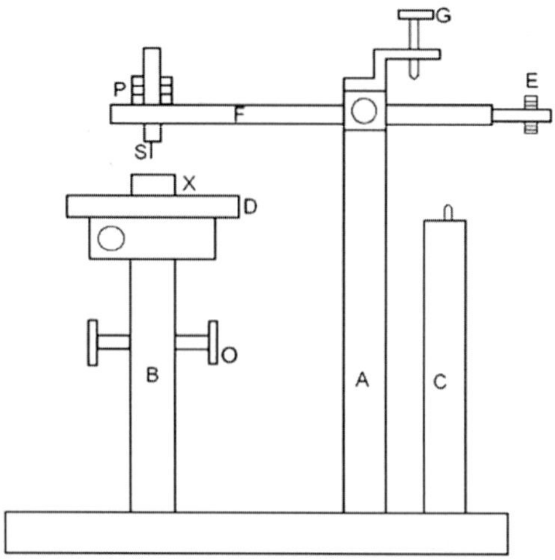

Fig. 4.8. Line diagram of the indenter

surface is close to the tip of the pin. Indentation is made by slowly releasing the locking screw until the pin touches the surface of the crystal. An electrical arrangement C, results in the switching off of a bulb when there is a contact between the pin and the crystal. All the indentations are made for 5 s. After indentation, the crystals are etched with suitable etchants and the crystal surface is observed under a microscope. The lengths of the rosette arms are measured with a micrometer eye-piece.

Using the instrument, indentations have been made on freshly cleaved surfaces of pure NaCl, doped NaCl, KCl, KBr and LiF. The crystals have been etched with known etchants. Fig. 4.9 shows the IDR pattern of pure NaCl. The l^2 values are plotted against the load P for pure NaCl (Fig. 4.10). A linear relationship is seen as predicted by (4.10). Hu [4.13] observed a similar relationship in silicon indented with a spherical indenter. These results show that the linear relationship is independent of the shape of the indenter or the crystal system. The relative microhardness of crystals was estimated in terms of the hardness of pure NaCl for which a value of 22.1 kg mm^{-2} [4.14] was used. The results are given in Table 4.1. The values obtained agree well with those obtained by others using sophisticated instruments; the gradation in the hardness values of LiF, KCl and KBr is the same in the data obtained in the present work and in data from literature. The instrument has been tested for the alkali halides for hardness values up to 100 kg mm^{-2}. The hardness of a Sr^{2+}-doped sample of NaCl was measured by this method and also directly with a Vickers indenter. These values agree closely and indicate hardening due to doping.

For accurate results, the indentation of the standard crystal (NaCl) and the sample crystals has to be made under identical loading conditions. Care has also to be taken to see that the indenter needle is exactly normal to

Table 4.1. Relative hardness values [kg mm^{-2}]

Crystal	Hardness		
	IDR method	Other values	Ref.
LiF	99.8	103	[4.14]
KCl	8.4	10.2	[4.14]
		13.1	[4.15]
KBr	7.6	10	[4.14]
		7.4	[4.16]
NaCl doped with $SrCl_2$	36.6	35.1	Direct measurement with Vickers hardness tester for the same sample

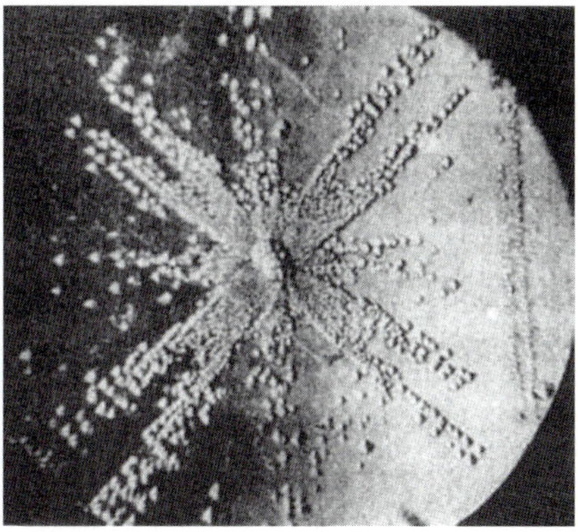

Fig. 4.9. Indentation rosette pattern on (100) faces of NaCl

the crystal surface at the position of contact. Otherwise, the rosette develops asymmetry and consequently the results will be in error. If these precautions are taken, the accuracy of the results depends on the accuracy of measurement of rosette arms and the value of the hardness of the standard. Keeping these factors in view and the scatter existing in the hardness values of NaCl, the errors in the hardness values obtained by this method are estimated to be ±5%.

The results obtained with this instrument compare well with those obtained with commercial, sophisticated hardness testing machines. This method is limited to crystals which are isomorphous and for which etchants are well established. The method should prove useful for measurement of changes in hardness of a crystal, due to any treatment given to the crystal. Some possible areas of applications are: (a) temperature variation of hardness, (b) radiation hardening, (c) impurity hardening and (d) annealing.

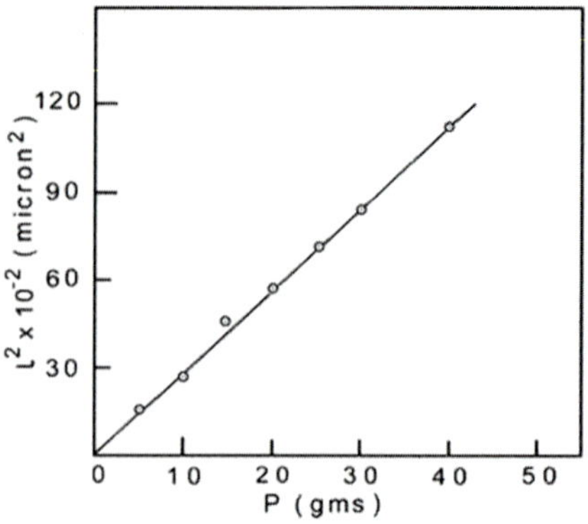

Fig. 4.10. Plot of square of rosette length (l^2) against load (P) for NaCl

4.3 An Overview

4.3.1 General

For quite some time, only metallurgists and mineralogists were interested in hardness. However, in the last several decades, physicists and chemists also have involved themselves in studies of hardness. Discussion of various aspects of hardness and data on a large number of materials are found in books [4.1,4.3,4.5], compilations [4.17–4.19] and also in several articles strewn over many scientific journals.

This section is an overview of the work done in several laboratories. It is not possible to touch upon all aspects of the available work. The overview is therefore, confined to inorganic crystals.

4.3.2 Load Variation of Hardness

The Vickers hardness is given by the relation

$$H_{\mathrm{V}} = 1854.4\ (P/d^2)\ \text{in}\ \ \mathrm{kg\,mm^{-2}}. \tag{4.1}$$

According to this equation the diagonal length of the impression changes with the applied load such that the ratio of P/d^2 is constant and the hardness has a constant value. However, in actual work when the hardness is measured at different loads, the hardness is found to show a dependence on the load. The load dependence is found to vary from material to material and from experiment to experiment. Typically, four types of results have been reported. These are shown in Fig. 4.11.

In 'a' type variation, the hardness is constant with respect to load. Such a behaviour presumes an ideal instrument response and an ideal material response. Such load-independent behaviour has been observed by several investigators [4.20–4.23].

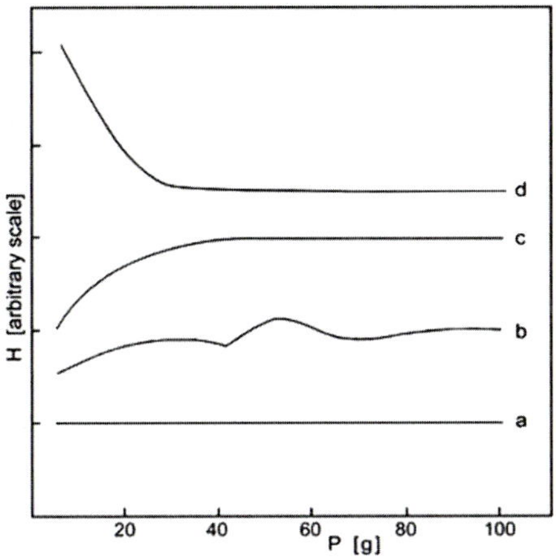

Fig. 4.11. Types of load variation of hardness

In the 'b' type variation, the load–hardness curve consists of kinks and maxima and minima. Such behaviour has been observed in some organic crystals [4.24–4.27]. These abrupt changes in the load vs. hardness curves are attributed to the activation of slip processes triggered at certain loads. This type of load dependence has also been observed in some polymeric materials [4.28–4.30]. In these materials such load variation has been attributed to chain–chain slipping.

In the 'c' type variation, the hardness increases with increasing load and at higher loads (above 25 g or so) the load variation becomes negligibly small. Load variation of this type has been observed in several studies [4.31–4.33]. Such load-dependence can arise due to a defect in the diamond indenter (chisel-tip indenter), or due to an elastic recovery of the depth of the impression [4.31] or more commonly due to a systematic positive loading error. This loading error (W) can be estimated by drawing a P vs. d^2 plot; the negative intercept of the linear plot on the load axis gives the loading error W. This correction is to be added to the applied load. The true hardness is now given by

$$H_{\mathrm{V}} = 1854.4\,(P + W)/d^2, \tag{4.12}$$

The most commonly observed load variation is of the 'd' type. Here, the hardness has a relatively larger value at low loads. The hardness shows a steep decrease with increasing loads up to about 25 g and thereafter the variation of hardness is slight. Mott [4.5] quotes several reports where such behaviour has been observed. Pratap and Hari Babu [4.34] in their study of ammonium halide crystals and Kotru [4.35] in his study of rare-earth compound crystals observed a similar load dependence.

According to the approach of Hayes and Kendall [4.36] this type of load variation can be due to the sample exerting a Newtonian pressure on the loaded indenter. This resultant pressure (W) is a function of the material being tested and represents the minimum applied load to cause an indentation as the load (W) will allow no plastic deformation. With this approach (4.1) may be written as

$$P - W = K\,d^2, \tag{4.13}$$

The $P - d^2$ plot in such a case has a positive intercept on the P-axis. This intercept represents the value of W. With the applied-load P reduced by the term 'W', the hardness is calculated from

$$H_V = 1854.4\,(P - W)/d^2. \tag{4.14}$$

It is load independent and represents the true value.

4.3.3 Solid Solution Hardening

It is observed that, generally, the hardness of a mixed crystal is much larger than that of either pure member of the system. There is considerable interest in this phenomenon. The utility of a device material is limited by its poor mechanical strength and solid solution hardening affords a method of enhancing the strength.

Subba Rao and Hari Babu [4.15,4.37] reported the composition dependence of the hardness of mixed crystals of KCl–KBr, KCl–KI, KCl–NaCl and KBr–KI systems. In all the cases, the composition dependence was non-linear with the mixed crystals in the equimolar region having a hardness much larger than that of either end member. These authors suggested a model which includes contributions from the lattice and from the disorder caused by the presence of ions of different sizes. This model qualitatively accounts for the observed composition variation. Shrivastava [4.38] calculated the hardness of several alkali halide mixed crystals using a model that considers the effect of the internal stress field of elastic interaction through dislocations.

Plendl [4.39] pointed out that the hardness of mixed crystals depends on three factors: (a) the ionic–covalent binding ratio $\psi\,(I)$, (b) overlap of atomic shells $\psi\,(A)$ and (c) the deformation of atomic polyhedra $\psi\,(D)$. The hardness is given by the function $\psi\,(B)$ which is the product of $\psi\,(I)\,\psi\,(A)^{-2}\,\psi\,(D)$. The result compared very well with the experimental observations on CaF_2–SrF_2 mixed crystals.

4.3.4 Impurity Hardening

Doping is the incorporation of impurities with a small concentration into the crystal. The impurity generally occupies a substitutional position. In a few cases, the impurity goes into interstitial positions. The impurity is generally of a higher valency than the ions in the host lattice. Impurities generally increase the hardness of a crystal.

Dryden et al. [4.40] studied the hardness of several alkali halide crystals doped with divalent ions. The host crystals were NaCl, KCl and LiF and the dopant ions were Ca^{2+}, Sr^{2+}, Ba^{2+}, Mg^{2+} and Mn^{2+}. The crystals hardened with doping and the increase in hardness was proportional to $C^{2/3}$ where C is the concentration. It may be noted that these authors did not measure the hardness directly but measured the critical resolved shear stress which is a measure of hardness.

Chin et al. [4.41] determined the yield stress as well as hardness of NaCl, NaBr, KCl and KBr crystals doped with Ca^{2+}, $Sr^{2+}$and Ba^{2+} ions. They found that the yield stress and hardness are related as follows:

$$\tau = c\,(H - H_0), \tag{4.15}$$

where τ and H are the yield stress and hardness of a doped crystal and H_0 the hardness of the pure crystal; c is a constant. Chin et al. [4.41] found that the increase in hardness (and yield stress) is proportional to $C^{1/2}$ where C is the concentration of divalent ions. However, the increase is otherwise independent of the species. Gilman [4.42] attempted a theoretical explanation of these observations in terms of an ionic model.

Pratap and Hari Babu [4.34] observed increase in hardness of NH_4Cl crystals doped with Cu^+ ions. Ascheron et al. [4.23] observed increase in hardness of InP on doping with Sn. They also quote earlier results on impurity hardening in Si and Ge.

4.3.5 Dislocation Hardening

The movement of a dislocation is impeded by other dislocations present in the crystal. Thus, a larger dislocation density implies a larger hardness. Chen and Hendrickson [4.43] measured the hardness of as-grown silver crystals and crystals damaged by severe scratching (higher dislocation density); they found that the hardness was larger for the latter than the former. Further, they mapped the dislocation density on the surface of a single crystal. They also mapped local hardness variation on the same crystal and observed a definite correlation between the hardness and the dislocation density. Lastly, they determined the dislocation density and the hardness of several silver crystals and found that crystals with larger dislocation density were harder.

4.3.6 Radiation Hardening

The hardness is affected by irradiation. In the following, the effects of photon irradiation and particle irradiation are discussed separately.

Photon irradiation effect on hardness

Earlier work on radiation hardening of alkali halides is summarized in Table 4.2. Aerts et al. [4.44] made a detailed study of the effect of X-ray irradiation on the hardness of NaCl. By measuring the hardness on two parts

Table 4.2. Summary of hardening studies of alkali halides

Crystals [Ref.]	Radiation employed	Property studied	Results
LiF, LiCl, LiBr, NaF, NaBr, NaCl, NaI, KF, KCl, KBr, KI, RbCl, CsF [4.45, 4.46]	γ-rays	Flow stress	Hardening $\propto n_F{}^{1/2}$ (n_F: F-centre concentration)
NaCl [4.44]	X-rays	Hardness	Quantitative (hardening increasing with time of irradiation)
NaCl [4.47]	X-rays	Hardness	Hardening $\propto n_F{}^{1/2}$
KCl, KBr [4.48]	X-rays	Hardness	Qualitative (hardening increasing with time of irradiation)
KBr, KI [4.49]	X-rays	Hardness	Qualitative (hardening increasing with time of irradiation)

of the same crystal, one unirradiated and the other irradiated, an enhancement in hardness by almost a factor of two was clearly established. Hardness was found to increase rapidly initially and then to reach a saturation value. No quantitative correlation could be established between the hardening and the color-centre concentration.

Nadeau [4.45, 4.46] carried out studies of hardness on several alkali halide crystals subjected to γ-irradiation. As a measure of hardness he determined the flow stress. Some of the observations by Nadeau deserve to be noted:

(a) There is an increase in flow stress with γ-ray irradiation though in some cases, a softening was observed at intermediate doses.
(b) The flow stress vs. dosage curve is similar to the F-band vs. dosage curve but not to the M-band vs. dosage curve, thereby indicating that the hardening correlates with F-centre concentration.
(c) The flow stress and F-centre concentration empirically follow the equation:

$$\Delta\tau = K n_F^2 \, , \tag{4.16}$$

where $\Delta\tau$ is the flow stress, K a constant and n_F the (4.15) concentration of F-centres.
(d) Bleaching experiments show that the hardening is not directly due to F-centers but due to interstitials which have the same concentration as F-centers.

As mentioned earlier, Nadeau [4.45, 4.46] chose to study the flow stress which is difficult to measure and is not a property directly related to hardness. Thus, the flow stress does not correlate with other mechanical properties like the elastic moduli [4.46] whereas hardness correlates with the shear modulus [4.50].

Boyarskaya and Zhitaru [4.47] tried to examine the relation between hardening and color-centre concentration but employed X-rays instead of γ-rays. Veeresham et al. [4.48] and Subba Rao et al. [4.49] studied the effect of X-ray irradiation on the hardness of KCl, KBr, KI and their mixed crystals. Radiation hardening was found to be less in mixed crystals than in the pure crystals.

Among other systems studied, mention may be made of study of rochelle salt by Krishnamoorthy and Murthy [4.51] by γ-ray irradiation. The increase in hardness was attributed to an increase in dislocation density caused by irradiation. Ansary et al. [4.52] studied the effect of gamma-irradiation on the hardness of polymers.

Particle Irradiation

Rau and Lacefield [4.53] studied the effect of neutron irradiation on the hardness of BeO. The hardness first increased rapidly and then reached a saturation level. Patel and Desai [4.54] and Patel and Raju [4.55] observed changes in hardness of mineral crystals of fluorite and gypsum respectively on neutron irradiation.

Ascheron et al. [4.56] studied the effect of proton bombardment on the hardness of GaP. They found no effect up to fluences of about 2×10^{16} cm^{-2}. At higher fluences, the hardness increased linearly with the fluence D according to the relation

$$H(D) = H_0 + A \log (D/D_0), \tag{4.17}$$

where H_0 and $H(D)$ are the hardness values before and after bombardment, A is a constant and D_0 the critical fluence where hardening starts.

Mukherjee [4.57] reported an increase in hardness of steel due to plasma implantation of nitrogen ions. Evanov et al. [4.58] studied the effect of plasma irradiation of deuterium on the hardness of vanadium.

4.3.7 Hardness and Chemical Bond

There is an intrinsic relation between hardness and the chemical bond. Powarjonnych [4.59] suggested the following formula connecting hardness and covalency:

$$H_\mathrm{M} = K\, Y\, c/r^2, \tag{4.18}$$

where H_M is the hardness on Mohs scale, Y the sum of the valences, c a constant, r the interionic distance and K is a factor depending on the covalency (1 for 0% covalency and 2 for 100% covalency). Julg [4.60] proposed an empirical formula for crystals with formula A_mB_n

$$H_\mathrm{M} = K\,(1 - \tfrac{2}{3} I^4), \tag{4.19}$$

where K is a constant determined by the number of rows in the periodic table to which atoms A and B belong and I the ionicity. Julg extended the formula to compounds where B is a polyatomic ion. Pillay [4.61] considered a large number of crystals with hardness in the wide range 2–9 (on the Mohs scale) having ionic, covalent and mixed bond character. The ionicities I of these crystals were known. He found that the hardness (on the Mohs scale) and ionicity I empirically fit the quadratic equation

$$H_{\mathrm{M}} = -15.79\, I^2 + 11.33\, I + 7.63. \tag{4.20}$$

Pillay used this equation to estimate the hardness of some crystals with known ionicities and also to estimate the ionicity of some crystals with known hardness.

Gilman [4.62] obtained a linear plot between the hardness H and the shear elastic constant C_{44} for the alkali halides. Later, Chin [4.63] drew such plots for several crystal families. He found that the ratio H_{V}/C_{44} is constant for a family of related crystals and that it has values 0.13, 0.013 and 0.0056 for covalent, ionic and metallic crystals, respectively. This ratio is given the name Gilman–Chin parameter [4.18].

4.3.8 Pressure Variation of Hardness

There is very little work on this aspect. Barbashov and Tkachenko [4.64] determined the microhardness of MgO and a few alkali halide crystals up to a pressure of 15 kbar. The indentations were carried out in a high-pressure chamber with benzene as the pressure-transmitting fluid. The results are shown in Table 4.3.

In all cases, the hardness was more at high pressure. At atmospheric pressure, the hardness H_{V} is related to the Young's modulus E according to the equation:

$$H_{\mathrm{V}} = 0.515\, E^{1.66}. \tag{4.21}$$

Assuming the same relation to hold at high pressures, the hardness at, say, $P = 15$ kbar can be calculated from Young's modulus at the same pressure using data on pressure variation of elastic constants. Such calculated values are

Table 4.3. Hardness H_{V} [kg mm^{-2}] of some crystals with NaCl structure at high pressures

Crystal	H_{V} at $P = 0$	H_{V} at $P = 15$ kbar	H_{V} (calc.)
MgO	799	831	773
LiF	119	128	167
NaCl	21	27	30
KCl	11	20	19
KBr	13	25	16

also given in the table. There is qualitative agreement with the experimental values.

4.3.9 Temperature Variation of Hardness

Atkins and Tabor [4.65] surveyed literature and pointed out that in W, Al, Cu, Pb and Ge, the hardness decreases smoothly as the temperature increases. They quoted similar results for the compounds MgO, WC and SiC. Further, Atkins and Tabor made measurements of hardness at high temperatures on several carbides and found that the indentation hardness falls steeply as the temperature increases.

Merchant et al. [4.66] surveyed high temperature hardness data on a large number of metals. In all the cases, the hardness decreases with increasing temperatures and the temperature variation can be fitted to either of the following two equations:

$$H = A \exp(-bT) \tag{4.22}$$

or

$$H = A' \exp(b'/T). \tag{4.23}$$

Gilman [4.12] examined data on temperature variation of Si crystal. He found that over a temperature range of 0–800°C, the indentation hardness of Si falls from 800 to 80 kg mm^{-2} and follows:

$$(H/H_0) = 1 - (2k_\mathrm{B}\theta/U)\,[\coth(\theta/T) - 1], \tag{4.24}$$

where H_0 is the hardness at 0 K, U an energy barrier and θ the characteristic temperature; $U = 1.48 \times 10^{-13}$ erg and $\theta = 750$ K. Jain and Patel [4.67] measured the hardness of CaF_2 at elevated temperatures. They found that at 350°C, the hardness is 1/3 that at room temperature. Boyarskaya et al. [4.68] measured the microhardness of alkali halide crystals in the range 77–423 K and found that the ratio of hardness at 77 and 293 K was 4.8, 4.5, 2.6 and 2.8 for KCl, NaCl, NaF and LiF, respectively.

4.3.10 Empirical Relations with other Physical Properties

Several empirical relationships have been proposed between the hardness of crystals and their various physical properties. These relations are useful in estimating hardness from these properties or vice versa.

Wooster [4.69] proposed the relation

$$C' = (H_\mathrm{M})^{7/4}. \tag{4.25}$$

Here H_M is the hardness on Mohs scale and C' is an elastic constant. For cubic crystals $C' = C_{11}$ whereas for other symmetries C' is the average of C_{11}, C_{22} and C_{33}. Chin et al. [4.16] found that for alkali halides

$$H_V = 5 \times 10^{-3} E_{111}. \tag{4.26}$$

Here E_{111} is Young's modulus in the <111> direction. While trying to find a theoretical explanation for this relation, Gilman [4.62] was led to the relation

$$H_V = 1.2 \times 10^{-3} C_{44}. \tag{4.27}$$

Gerk [4.50] reported a linear relation between the hardness and the isotropic shear modulus G. Interestingly, there is a single plot between H_V and G for such diverse crystals as diamond, metals and alkali halides though with some scattering. Gerk pointed out that the scattering in the H_V–G plot is reduced by using a modified shear modulus G' given by

$$G' = (C_{44})[(C_{11} - C_{12})/2C_{44}] \;/\; [1 + (C_{11} - C_{12})/2C_{44}]. \tag{4.28}$$

Plendl and Geilisse [4.17] showed the existence of smooth plots between the Mohs hardness and the "volumetric lattice energy" (lattice energy U/molar volume V). A large number of crystals having covalent, ionic and mixed binding were included but not any metals. This relation was used to estimate the lattice energies of some compounds from their scratch hardness.

Gilman [4.4] surveyed work on hardness and discussed several relationships either proposed by others or some observed by him. These are:

1. A linear relation between hardness (H_V) and yield stress (τ). $H_V = 3\tau$ for metals and $H_V = 35\tau$ for alkali halides.
2. Linear relation between hardness and Young's modulus for metals.
3. Linear relationships between hardness of covalent compounds and (a) glide activation energy, (b) energy gap density and (c) bond distance.
4. A linear correlation between hardness and heat of formation for carbides.

4.3.11 Anisotropy of Hardness

Hardness shows directional dependence (anisotropy). The average hardness measured on the face of a crystal differs from face to face. Also, the hardness measured on a given face with a Knoop indenter varies with the orientation of the indenter with reference to a reference direction. This can be seen from Fig. 4.12 which shows the Knoop indentations with <110> and < $\bar{1}00$ > directions on the (001) plane of diamond [4.70]. Figure 4.13 shows the orientation dependence of hardness on the (100), (110) and (111) faces of an iron crystal [4.71]. The hardness has a fourfold, twofold and threefold rotational symmetry on these faces, respectively.

Table 4.4 gives information about the hardness anisotropy on the (001), (110) and (111) planes of several cubic crystals having the NaCl, CaF_2, fcc, diamond and bcc structures.

In a remarkable paper, Brookes et al. [4.72] predicted the hardness anisotropy in Knoop hardness considering the structure and slip system of

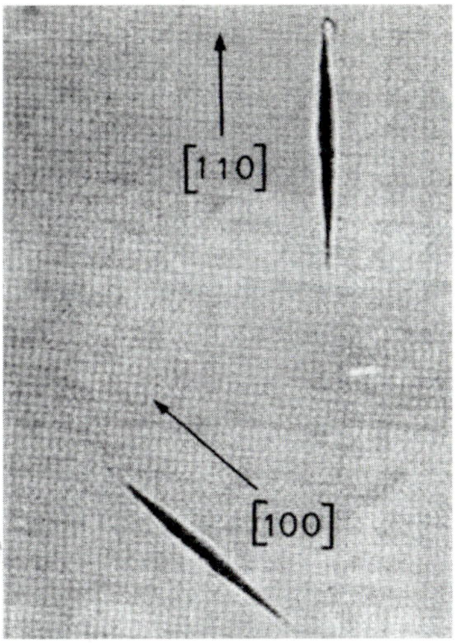

Fig. 4.12. Knoop indentations in the [110] and [$\bar{1}$00] directions on the (001) plane of diamond

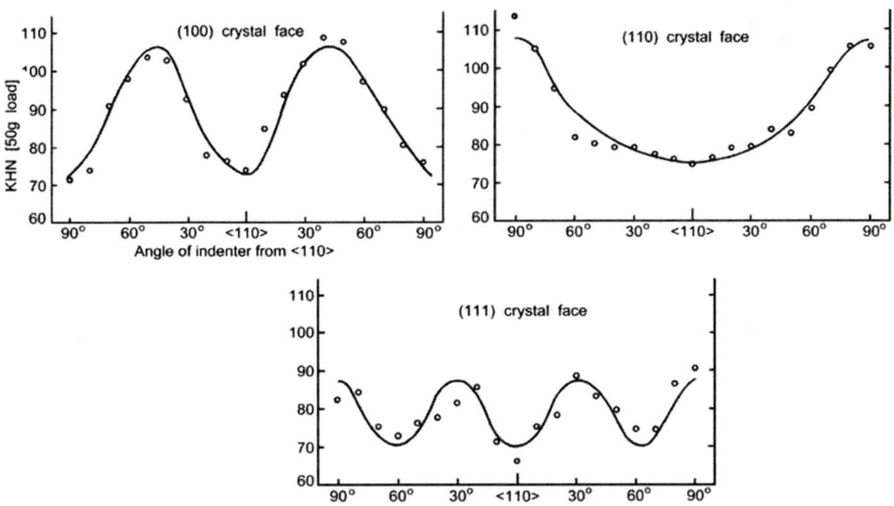

Fig. 4.13. Microhardness anisotropies of (100), (110) and (111) faces of iron crystal

the crystal. These predictions were verified in several cases. In Table 4.5, the hardness anisotropy for some carbides is quoted from [4.70]. In some of these carbides, the anisotropy changes with temperature. Hannink et al. [4.73], attributed this change to a change in the slip system at high temperatures. However, this correlation between slip system and hardness anisotropy was contradicted by Chin et al. [4.74, 4.75] who determined hardness anisotropy of several crystals with NaCl and CaF_2 structures.

Table 4.4. Anisotropy of hardness for several crystalline solids [4.70]

		Knoop hardness [kg mm^{-2}]							
		(001)		(110)			(111)		
Crystal	Structure	$<100>$	$<110>$	$<001>$	$<1\bar{1}1>$	$<1\bar{1}0>$	$<1\bar{1}0>$	$<11\bar{2}>$	Slip system
MnS	Rocksalt	122	142	119	142	142	140	140	$\{110\}<1\bar{1}0>$
LiF	Rocksalt	87	93	87	97	93	93	–	$\{110\}<1\bar{1}0>$
NaCl	Rocksalt	18	20	–	–	–	–	–	$\{110\}<1\bar{1}0>$
MgO	Rocksalt	400	780	420	930	810	–	–	$\{110\}<1\bar{1}0>$
MnO	Rocksalt	252	285	252	287	287	–	–	$\{110\}<1\bar{1}0>$
CaF_2	Fluorspar	178	157	–	–	–	–	–	$\{100\}<011>$
Al	fcc	23	18	23	17	18	–	–	$\{111\}<1\bar{1}0>$
Ni	fcc	105	72	115	84	84	–	–	$\{111\}<1\bar{1}0>$
MnSe	fcc	65	43	73	54	48	–	–	$\{111\}<1\bar{1}0>$
Cu	fcc	47	34	–	–	–	–	–	$\{111\}<1\bar{1}0>$
C	Diamond cubic	9,600	6,900	–	–	–	–	–	$\{111\}<1\bar{1}0>$
W	bcc	445	375	440	360	360	390	408	$\{1\bar{1}0\}<11\bar{2}>$, $\{12\bar{3}\}<111>$
Nb	bcc	81	59	81	63	59	59	–	$\{12\bar{3}\}<111>$
Fe	bcc	229	183	240	196	203	198	214	$\{12\bar{3}\}<111>$
V	bcc	103	79	97	–	78	79	92	$\{12\bar{3}\}<111>$
Cr	bcc	139	108	159	–	115	108	123	$\{12\bar{3}\}<111>$

Table 4.5. Knoop hardness anisotropy of (001) planes of carbides [4.72]

Carbide	Temperature [°C]	Knoop hardness [kg mm^{-2}] <100>	<1$\bar{1}$0>	Slip system[a]
TiC	25	2,020	2,750	{110} <1$\bar{1}$0>
TiC	616	980	880	{111} <110>
$TiC_{0.80}$	25	2,000	2,300	{110} <1$\bar{1}$0>[b]
VC	−196	2,040	2,150	{110} <1$\bar{1}$0>
VC	350	1,390	1,140	{111} <1$\bar{1}$0>
VC	25	2,070	2,600	{110} <1$\bar{1}$0>[b]
HfC	25	1,850	2,500	{110} <1$\bar{1}$0>
ZrC	25	1,980	2,250	{110} <1$\bar{1}$0>
NbC	25	2,300	2,420	{110} <1$\bar{1}$0>[b]
TaC	25	1,650	1,500	{111} <1$\bar{1}$0>

[a] Slip system thought to control indentation process at relevant experimental temperature.
[b] Predominant slip system at relevant experimental temperature with secondary slip system beginning to affect hardness anisotropy.

Offenbacher and Roselman [4.76] measured the hardness on the prismatic and basal planes of ice crystals over the temperature range −5°C to −12°C. They observed that the quantity ($H_{\text{prismatic}}$–H_{basal}) is positive below −8°C and negative at higher temperatures.

Gallaghar et al. [4.77] studied the hardness anisotropy in calcite and sodium nitrate which are isomorphous. Although the rotational symmetry in the anisotropy in hardness in both cases is similar, yet the direction of the maxima is reversed. Gallaghar et al. conclude that the slip systems in the two are different ($\{11\bar{1}\} < 011 >$ in calcite and $\{011\} < 100 >$ in sodium nitrate).

4.3.12 Surface Hardness

There is interest in the hardness in the surface region (depth ~0.001 μm). Information on this aspect is meager and conflicting. Morlin [4.78] observed a decrease in hardness at low depths. Upit and Varchenya [4.79] studied the load variation of hardness at low loads up to 0.3 g. They made measurements on a few alkali halide crystals. Their range of loads covered the depth in the range 5–60 μm. The hardness was found to increase as the surface layer was approached. Upit and Varchenya [4.79] fitted their results to the equation

$$H = 1854\ a\, d^{n-2}, \tag{4.29}$$

where d is the diagonal length and n an index. For the alkali halides, they obtained $n = 1.86$. The coefficient a varied from crystal to crystal and correlated with the surface energy as can be seen from the data in Table 4.6.

Gane [4.80] made ingenuous experiments on gold films employing loads as low as 0.1 mg and incorporating the indenter inside a TEM. He observed a

Table 4.6. Values of coefficient a in (4.29) and the surface energy (σ) of alkali halides

Crystal	a[kg mm^{-2}]	σ [erg cm^{-2}]
LiF	88	250
NaF	58	203
NaCl	15	114
KBr	8	89
KI	7	85

Table 4.7. Values of bulk hardness H_B, index m and surface hardness H_S

Crystal	m	H_B [kg mm^{-2}]	H_S [kg mm^{-2}]
LiF	0.14	100	370
NaCl	0.14	20	60
KBr	0.14	9	31
KI	0.14	7	28
Cu	0.19	30	350
Al	0.25	12	215

substantial increase in hardness but, unfortunately, could not decide whether the effect was genuine or caused by a polymeric film between the sample surface and the indenter.

Upit and Varchenya [4.81] extended their earlier work by including some metals in their study and having a gadget which could measure the depth h of the impression. They found that (a) in all cases the hardness increased with decreasing depth, (b) the variation with depth was given by $H = ch^{-m}$, where c was a constant and m had different values and (c) the surface hardness (H_S) obtained by extrapolation to h = a few lattice constants was much higher than the bulk value (H_B). Their results are summarized in Table 4.7.

4.3.13 Nanohardness of Thin Films

The recently developed technique of nanoindentation has facilitated the measurement of hardness of thin films and coatings. Some recent results on Mo thin films and TiN coatings are given in Table 4.8. The nanohardness changes substantially with the state of the sample. Also, the thin film hardness is much higher than the bulk hardness. The nanohardness is also influenced by the thickness of the coating and by the choice of the substrate [4.84].

4.3.14 Effect of Magnetic Field on Hardness

Smirnov and Urusovskaya [4.85] studied the effect of applied magnetic field on the hardness of LiF:Ni crystals. A static magnetic field of $1.6 \times 10^8\,\mathrm{A\,m^{-1}}$ was applied for 20 s and then switched off. The hardness was measured as a function of time. The hardness was found to increase from 0.82 to 0.91 GPa at 2 s.

Table 4.8. Nanohardness of some thin films and coatings; value of Mo from Yoder et al. [4.82], rest from Rajam and Barshilia [4.83]

Material	Nature of sample	Hardness [kg mm^{-2}]
TiN	Coating	2,400
	Bulk	1,800
TiAlN	Coating	3,600
TiAlN/CrN	Multilayers	3,900
TiN/NbN	Multilayers	4,000
TiN/a-c	Nanocomposites	5,200
Mo	Thin film	600–1,200
	Bulk	156

It then decreased to the original value at 4 s. Thereafter it remained unaltered. The same behaviour was observed in experiments at 200°C with the difference that the maximum occurred at 1 s. The observed variation was attributed to the disturbance of the equilibrium of the crystal–magnetic impurity system and the subsequent relaxation.

Golovin et al. [4.86] studied the effect of magnetic field on the microhardness of NaCl. Microhardness measurements were made on an NaCl crystal to which a magnetic field of 20 T with a gradient was applied for 100 μs. Microhardness measurements were made at different positions along the gradient. The hardness decreased from the zero field value and reached a saturation value which was 20% less than the original value.

4.3.15 Hardness of Organic Crystals

Organic crystals are in general softer than inorganic crystals. There is limited work on the hardness of organic crystals. Sasaki and Iwata [4.87] made microhardness measurements on anthracene crystals on the (001) plane with a Knoop indenter. They found that the hardness has a maximum in the $<110>$ and $<\bar{1}10>$ directions. The mean value on the (001) plane was 4.3 kg mm^{-2}. Joshi and Shah [4.21] made Vickers microhardness measurements on single crystals of anthracene, phenanthrene and benzoic acid and obtained values of 4, 8.2 and 5.2 kg mm^{-2}, respectively. Marwaha and Shah [4.27] made detailed measurements of microhardness as a function of load on benzoic acid crystals. They found that instead of the usual smooth load vs. hardness curve, the P–H_V curve has peaks at 20 and 90 g. These peaks were attributed to activation of slip in the $<010>$ and $<100>$ directions.

4.3.16 Micro-Raman Spectroscopy of Indentations

Many materials undergo structural phase transformations under the application of high pressure. In indentation, although loads of a few grams are applied, the indenter tip area being very small, high stresses are generated.

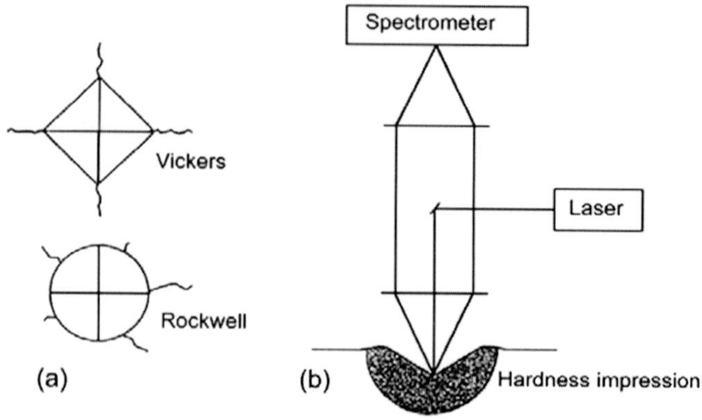

Fig. 4.14. Sketch of the experimental procedure for micro-Raman spectroscopy of indentations. A number of hardness indentations were produced (**a**), which were then examined with a micro-Raman spectrometer

These stresses are sufficient to induce phase transformations which can be probed through the technique of micro-Raman spectroscopy.

The experimental arrangement used by Kailer et al. [4.88] is shown in Fig. 4.14. Indentations were produced by using Vickers or Rockwell indenters and the Raman spectroscopy of the indentation impression was carried out with a LabRam micro-Raman spectrometer using He–Ne or Ar^+ lasers.

Examples of micro-Raman spectra of indented surfaces of a few materials are shown in Fig. 4.15. Figure 4.15a is the spectrum of Si which is known to undergo a phase transition at 10–13 GPa. The spectrum shows, apart from the line for the original diamond structure (Si-I), lines corresponding to the bc8 structure (Si-III) and a rhombohedral phase (Si-XII). In Fig. 4.15b, Ge-I is the line for the original diamond structure. Lines indicated by • and ♦ are assigned to the bc8 and the st12 phases. A hexagonal diamond phase Ge-V is also seen. In Fig. 4.15c, the micro-Raman spectra of indented ZrO_2 are shown; in addition to the tetragonal bands (marked t), those for the monoclinic phase (marked m) are also seen. In the micro-Raman spectra of pristine and indented quartz (Fig. 4.15d), amorphisation of quartz is clearly visible. Similar studies on diamond and SiC have been made by Kailer et al. [4.89].

This is a fast and simple technique for studying phase transformations. It does not need a diamond anvil and requires a small amount of the sample.

4.4 Some of our Results

4.4.1 Load Variation of Hardness

Ideally, the hardness measured by indentation methods should be load-independent. However, in most measurements, the measured values show a

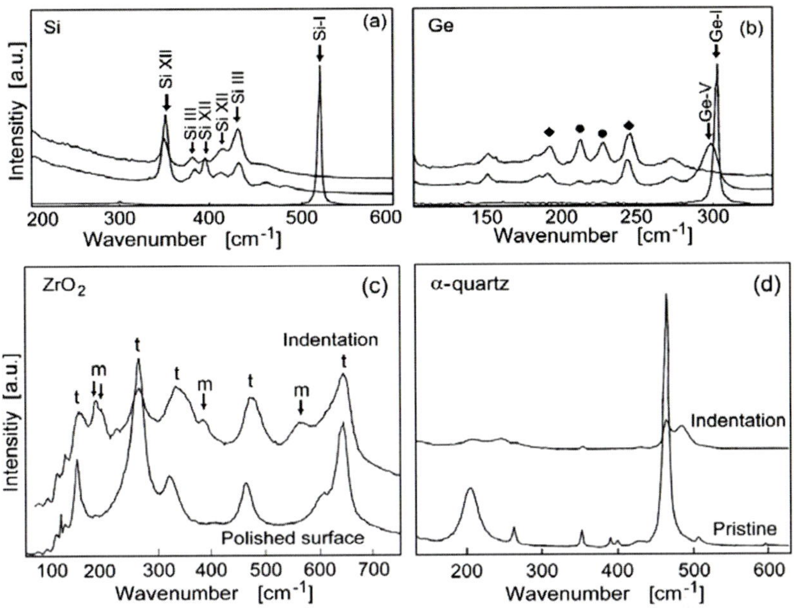

Fig. 4.15. Micro-Raman spectra of: (**a**) Si, (**b**) Ge, (**c**) ZrO_2 and (**d**) quartz before and after indentation

load-dependence. The types of observed load variation and the methods of correction to enable the determination of the true hardness have been discussed in Sect. 4.3.2. Here, we shall discuss some of our own observations.

Type 'c' Variation

Measurement of Vickers hardness of some metals (Zn, Sn, Pb, Bi) was carried out with the help of a Vickers microscope fitted with a diamond pyramidal indenter. The load dependence is shown in Fig. 4.16. This variation corresponds to type 'c' shown in Fig. 4.11. It was also pointed out that for this type of variation, a load correction can be applied by drawing a plot between load (P) and the square of the indentation diagonal (d^2). The intercept of the linear plot on the P-axis gives the correction W. The true hardness is then obtained from (4.12). The values of the true hardness of some metals, thus determined, are given in Table 4.9. The hardness of Zn obtained in these measurements is 33.3 kg mm^{-2} which agrees well with the reported value of 28 kg mm^{-2} [4.5].

Type 'd' Variation

Nearly a hundred crystals and polycrystalline samples have been studied with the help of a miniload Leitz–Wetzlar hardness tester. The load variation, observed in all the cases was of the 'd' type (Sect. 4.3.2) i.e. the hardness is large

Table 4.9. True Vickers hardness H_V of some metals [4.32]

Metal	H_V [kg mm^{-2}]	Metal	H_V [kg mm^{-2}]
Zn	33.3	Bi	11.7
Sn	17.1	Pb	5.1

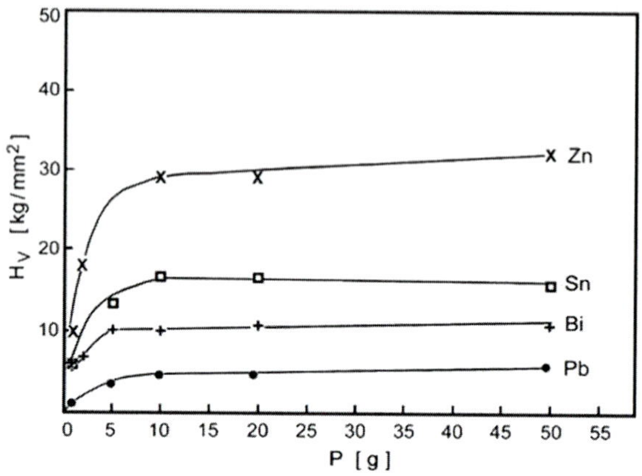

Fig. 4.16. Plot of hardness H_V against load P for some metals

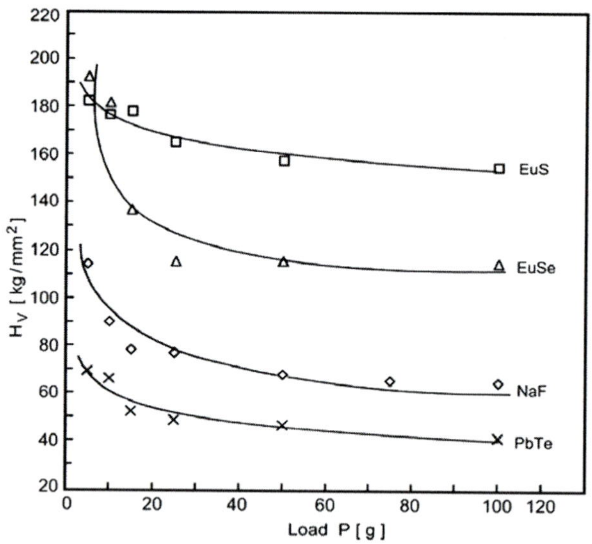

Fig. 4.17. Plot of hardness H_V against load P for some crystals with NaCl structure

at very low loads, decreases steeply with increasing loads and becomes nearly load-independent at loads of 50 gm and above. The load variation observed in some crystals with NaCl structure is shown in Fig. 4.17.

Table 4.10. Values of W and the true hardness of some crystals [4.90]

Crystal	W[g]	H_V (true) [$kg\,mm^{-2}$]	H_V (L-Ind.) [$kg\,mm^{-2}$]
RbCl	2.8	9.3	9.5
RbBr	2.6	7.6	8.0
RbI	2.0	5.9	6.0
NaF	3.7	62.5	63.8
EuS	1.8	152.3	159.5
EuSe	2.4	112.4	116.0
PbTe	3.6	41.0	45.9

As mentioned in Sect. 4.3.2, load variation of this type can be taken into account, again, by drawing a P–d^2 plot. The intercept 'W' on the P-axis in such cases is positive. The true hardness can be obtained from (4.14). The values of W pertaining to some crystals are given in Table 4.10. They are in the range 2–4 g. The true values of H_V are also given. The values of H_V estimated directly from the load-independent part of the P–H_V curves denoted by H_V (L-Ind.) are also given in the table. It is to be noted that these two values agree with the load–variation-corrected true values generally within 5%.

The Case of FeS_2 (Pyrite)

For pyrite, widely differing values have been reported (in units of $kg\,mm^{-2}$): 588–757, 857–1152 [4.5]; 1636–2814 [4.91]; 2951–3618 [4.92]. The values in the lower range are smaller than the microhardness of quartz ($\approx 1,224\,kg\,mm^{-2}$). That these low values for the microhardness of pyrite are in error was shown by indenting crystals of FeS_2 and quartz at identical loads [4.91]; the indentations on FeS_2 were consistently smaller than those on quartz indicating that the hardness of pyrite cannot be less than $1,224\,kg\,mm^{-2}$. Regarding the reported values in the higher range (2,000–3,600 $kg\,mm^{-2}$), it was speculated [4.90] that these differences may be due to different loads employed by different workers but a systematic study at different loads was not made.

Such a study was carried out by Thirmal Rao and Sirdeshmukh [4.93]. The microhardness was found to be load dependent. The values, ranging from $2,000\,kg\,mm^{-2}$ at higher loads to $6,000\,kg\,mm^{-2}$ at lower loads, lie on a smooth curve (Fig. 4.18). It is indeed observed that the several values in the higher range reported earlier are close to the values obtained in the present measurements at different loads, thus supporting the conjecture that the scattering in earlier values may be due to differences in loads employed by different workers.

This is a 'd'-type load variation. A plot of P vs. d^2 gives a straight line with an intercept W and slope ($H_V/1854.4$). For FeS_2 this plot gives a value of 37 g for W. As mentioned earlier, W represents the minimum load to be employed to obtain an indentation. It was verified that no indentation could be observed up to loads of 36 g. A value of $1,743\,kg\,mm^{-2}$ for H_V is obtained from the

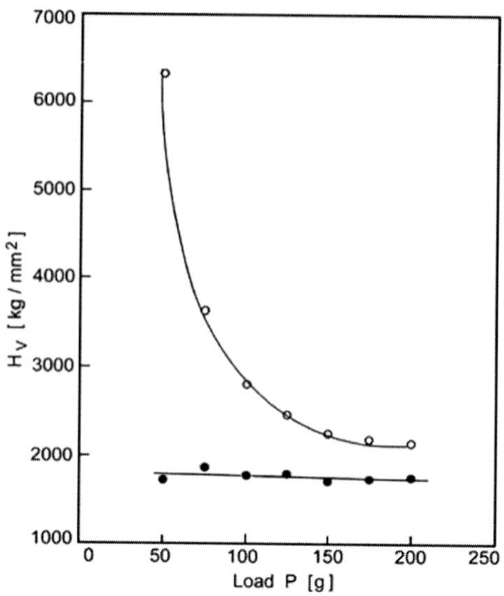

Fig. 4.18. Values of microhardness (H_V) of pyrite (FeS_2) at different loads (P); O observed, • corrected

slope of the plot. As a check, H_V can also be calculated by substituting the value of W in (4.14). These corrected values are shown in Fig. 4.18. It is seen that within experimental error the apparent load variation has disappeared. Further, these values agree with the value obtained from the slope. The true microhardness of pyrite is $1,785 \pm 100\,\mathrm{kg\,mm^{-2}}$ corresponding to a hardness of 8 on the Mohs scale. Mineralogical texts [4.94,4.95] quote the Mohs hardness of pyrite as 6–6.5. Since hardness is used for routine characterisation of minerals, it is necessary to replace the old erroneous values with the new value (8 on the Mohs scale). This study points out that hardness results will be more reliable if they are based on systematic load-variation studies.

The Case of Lithium Niobate ($LiNbO_3$)

Lithium niobate is an optically important crystal which has been studied in detail. There is a large scatter in the reported hardness values of lithium niobate and also the load-variation reported in an earlier study is not normal. Chai [4.96] lists a value of 5 on the Mohs scale for the hardness of $LiNbO_3$which is equivalent to a value of $400\,\mathrm{kg\,mm^{-2}}$ on the Vickers scale. Brown et al. [4.97] made Knoop hardness measurements on the (001) plane at a single load of 50 gm and obtained an average value of $570\,\mathrm{kg\,mm^{-2}}$. It may be mentioned that Knoop and Vickers hardness values generally agree to within 5% [4.5]. Dhanraj et al. [4.98] reported measurements of Vickers hardness on the (001) plane, using crystals grown by them, and obtained a value of about $780\,\mathrm{kg\,mm^{-2}}$ at a load of 50 gm (same load as used in [4.97]).

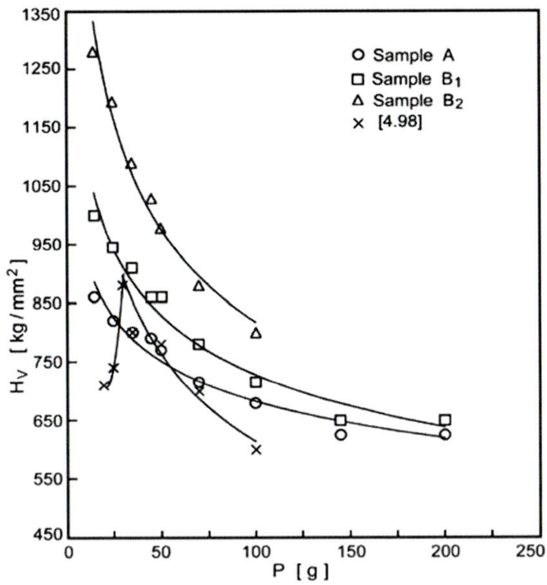

Fig. 4.19. Plots of hardness H_V against load (P) for $LiNbO_3$

Thus, there is a scatter of values from 400 to 800 kg mm^{-2} in the reported microhardness values.

The other discrepancy is related to the load dependence of hardness of $LiNbO_3$. Dhanraj et al. [4.98] observed that the microhardness of $LiNbO_3$ has value of 700 kg mm^{-2} at a low load of 20 g. In the load range 20–30 g, they observed a steep increase in hardness to nearly 900 kg mm^{-2}. At higher loads they observed a decrease in hardness with a value of 600 kg mm^{-2} at 100 g. The increase in hardness in the 20–30 g load range was attributed to work hardening and the decrease at higher loads to plastic flow. This load-dependence (Fig. 4.19) is at variance with several earlier observations. Furthermore, the interpretation of the observed behaviour as due to two different deformation mechanisms in different load regions would have new implications on the indentation process in $LiNbO_3$. In view of these discrepancies, a systematic study of the hardness of $LiNbO_3$ was undertaken by Subhadra et al. [4.99] who used two crystals of $LiNbO_3$ of different origins, one of the crystals being from the same batch as used in [4.98]. A crystal of Fe-doped $LiNbO_3$ was also included in the study.

Pure lithium niobate ($LiNbO_3$) crystals (labeled A) were grown at the Indian Institute of Science, Bangalore, by the Czochralski method. They were oriented, cut and polished to reveal the (001) plane. Crystals of pure $LiNbO_3$ (labeled B_1) and $LiNbO_3$ doped with 0.02 wt% Fe (labeled B_2) were grown by the same method at the Solid State Physics Laboratory, Delhi, and had the same orientation as crystal A. Hardness measurements were made on the (001)

Table 4.11. Load correction (W) and true hardness (H_V) of $LiNbO_3$ [4.99]

Material	Sample code	W [g]	H_V [kg mm^{-2}]
Pure $LiNbO_3$	A	4	630 ± 30
Pure $LiNbO_3$	B_1	7	620 ± 30
0.02% Fe-doped $LiNbO_3$	B_2	6	750 ± 50

plane at several loads in the range 15–200 g. Though hard, $LiNbO_3$ crystals are brittle. As a result, cracks show up in indentations made at loads above 100 g. With some care, measurements could be made up to 200 g on samples A and B_1 whereas in sample B_2 the tendency to crack formation was more and measurements were limited to 100 g. The measured hardness values at a given load are consistent among themselves within 2%.

The variation of hardness with load is shown in Fig. 4.19. The observed load dependence is of type 'd'. The load-independent behaviour is seen to occur beyond 100 g in samples A and B_1; in sample B_2, the load independent region is not reached as measurements were limited to 100 g. These results are analysed following the method discussed earlier for 'd' type variation. The values of the load correction W obtained from the $P - d^2$ plots are given in Table 4.11. The values of the true hardness (H_V) are also given in Table 4.11.

The type of two-regime $P - H_V$ plot, i.e. H_V increasing up to some load and then decreasing, observed in [4.98] has not been observed in these measurements. Instead, the $P - H_V$ plots for all the three crystals are smooth monotonous curves showing the same trend viz. a large hardness value at low loads, followed by a steep fall with increasing loads, with a slower fall at still higher loads.

Based on measurements on two crystals of different origin, the true hardness value of $LiNbO_3$ is estimated as 630 ± 30 kg mm^{-2} on the Vickers scale and 5.6 on the Mohs scale. The value of 570 kg mm^{-2} [4.97] at 50 g load, and the value of 600 kg mm^{-2} [4.98] at 100 g load reconcile with the present values. The extreme values of 400 and 800 kg mm^{-2} referred to earlier, appear to have resulted from some unidentified error.

Though the load vs. hardness plots for crystals A and B_1 nearly converge at high loads, there are large differences in the hardness values measured at low loads (note values of 865 and 1,000 kg mm^{-2} at $P = 15$ g for samples A and B_1 (Fig. 4.19). This re-emphasizes the point that hardness values of different crystals can be compared only when they have been evaluated from a systematic study of load variation.

These studies on load variation of hardness lead to the following conclusions. Variations of type 'c' as well as type 'd' can be corrected by plotting a $P - d^2$ plot and recalculating the hardness by increasing P by a correction W (type c) or decreasing P by a correction W (type d). In both cases, the value obtained is close to the value read off from the flat part of the H_V–P curve (generally 50–100 g). If a flat region is reached the true hardness can be obtained straight away. However, if the flat region is not reached even at high loads (as in FeS_2) or if higher loads cannot be applied due to crack formation (as in Fe-doped $LiNbO_3$), a load correction becomes essential for obtaining the true hardness.

4.4.2 Hardness and Bonding

It is mentioned in Sect. 4.1 that even in a pure (impurity free), well-annealed (nearly dislocation free) crystal, the motion of a dislocation is resisted by the chemical forces. This bestows an intrinsic hardness on the crystal. It follows that the hardness of a crystal should reflect its bonding nature. In our work, the hardness of a large number of crystals has been measured. This hardness data will be analysed to obtain information on the nature of the bonding.

Crystals with the NaCl Structure

Thirmal Rao and Sirdeshmukh [4.100] pointed out that the hardness of the alkali halides with NaCl structure decreases as the value of the lattice constant increases and that a plot of these two parameters (Fig. 4.20) is a smooth curve. Again, Sirdeshmukh et al. [4.18] plotted the values of the hardness of divalent chalcogenides (Fig. 4.20) against the lattice constant and obtained a curve similar to that for the alkali halides. It is well known that a weakening of the interatomic bond-strength increases the lattice constant and decreases the hardness. The H_V–a plots are consistent with this idea but do not give any further insight into the nature of the bonding.

The hardness will now be analysed in terms of the Gilman–Chin parameter. Gilman [4.101] showed that in ionic crystals like alkali halides, the microhardness H_V correlates linearly with the elastic modulus S_{11}^{-1}. Gilman [4.62] showed that a linear relation holds between H_V and the shear elastic constant C_{44} in the alkali halides and, by making several assumptions, obtained the value of the proportionality constant. Chin [4.63] considered a large number

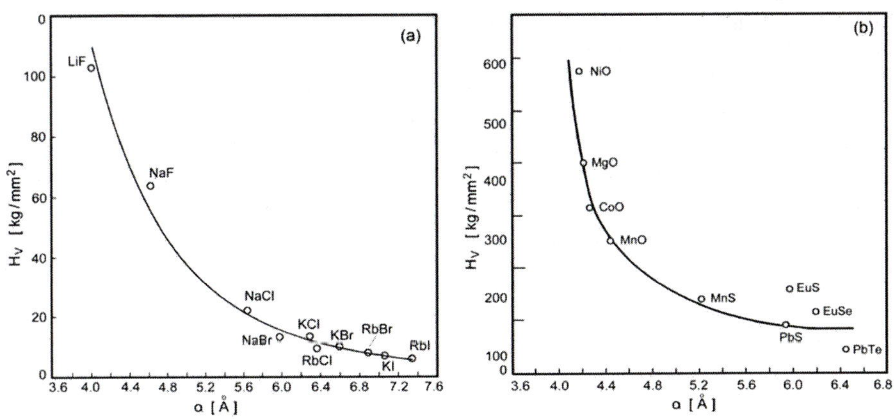

Fig. 4.20. Plots of lattice constant a vs. microhardness H_V for: (**a**) alkali halides, (**b**) divalent chalcogenides

Table 4.12. Values of microhardness H_V, the Gilman–Chin parameter H_V/C_{44} and effective ionic charge per electron (e^*) for some crystals with NaCl structure [4.18]

Crystal	H_V [kg mm^{-2}]	C_{44} [kg mm^{-2}]	H_V/C_{44}	e^*	$(H_V/C_{44})(e^*)^2$
TiC	2,896	1,7845	0.163	0.40	0.031
TaC	1,601	8,056	0.198	0.35	0.024
ZrC	2,701	1,4929	0.181	0.45	0.036
UC	900	6,598	0.136	0.44	0.026
LiF	103	6,404	0.016	0.87	0.012
NaF	64	2,855	0.022	0.93	0.019
NaCl	22	1,290	0.017	0.74	0.009
NaBr	13	1,035	0.013	0.69	0.006
NaI	10	740	0.014	0.71	0.007
KCl	13	642	0.020	0.80	0.013
KBr	10	518	0.019	0.76	0.011
KI	7	377	0.018	0.69	0.009
RbCl	9	484	0.020	0.84	0.014
RbBr	8	392	0.020	0.82	0.013
RbI	6	286	0.021	0.89	0.017
AgCl	9	628	0.015	0.69	0.007
MgO	400	15,439	0.026	0.55	0.008
MnO	252	8,056	0.031	0.54	0.009
NiO	575	10,707	0.054	0.42	0.009
CoO	315	8,158	0.039	0.44	0.007
MnS	140	5,303	0.026	–	–
PbS	92	2,549	0.036	0.75	0.020
PbSe	51	1,621	0.031	0.70	0.015
PbTe	46	1,326	0.035	0.56	0.011
EuS	159	2,784	0.057	0.55	0.017
EuSe	119	2,325	0.051	0.48	0.012

of crystals with different structures and bonding and drew H_V vs. C_{44} plots. He showed that data points for ionic and covalent crystals cluster around two distinctly different straight lines with slopes of 0.01 and 0.12. Data on microhardness and the shear elastic constant for a large number of crystals with NaCl structure are given in Table 4.12 along with values of the Gilman–Chin parameter (H_V/C_{44}). The values of log H_V and log C_{44} are also shown as a plot in Fig. 4.21. Several interesting features can be noted from Table 4.12 and Fig. 4.21.

Although all the crystals have the same structure, the data points do not lie on a single log H_V vs. log C_{44} plot, indicating that there is no unique value for H_V/C_{44} for this suite of crystals. In fact, the Gilman–Chin parameter (H_V/C_{44}) has values ranging from 0.01 to 0.2. It is possible to draw straight lines to represent the data points for three groups of crystals. In Fig. 4.21 line (a) represents the highly ionic alkali halides with NaCl structure and also AgCl. The mean value of H_V/C_{44} for this group is 0.018 which is 40%

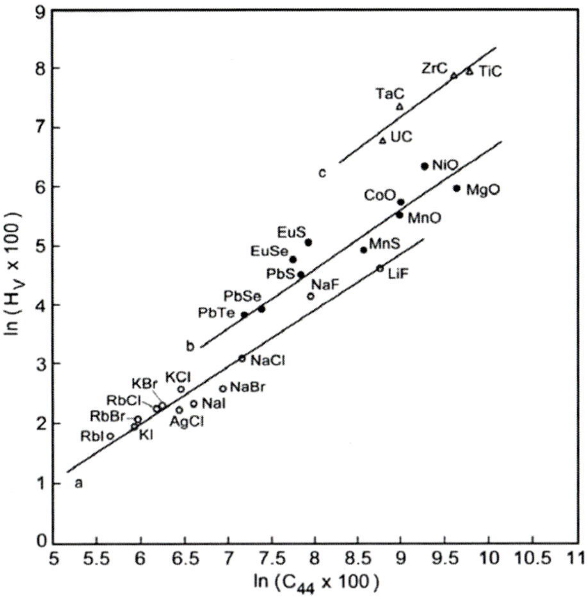

Fig. 4.21. Logarithmic plot of microhardness H_V and shear modulus C_{44} for crystals with NaCl structure: (**a**) alkali halides and AgCl, (**b**) chalcogenides, and (**c**) carbides (H and C_{44} are in GPa)

larger than the value 0.013 given by Chin [4.63] for the alkali halides. The data points cluster around this line with an e.s.d. of 0.003, i.e., about 15%. The line (b) represents the chalcogenides. The mean value of $H_V/C_{44}$for this group is 0.039. The data point for MgO, which deviated from the line for alkali halides in Chin's plot, is now close to this new line. The deviations of the data points from the line are more severe than in the case of line (a) (note that the plot is on log scale). This larger deviation is reflected in an e.s.d. of 0.011, i.e., about 35%. Lastly, the data points for the carbides are clustered along line (c). This group of crystals has a much larger value of 0.169 for the Gilman–Chin parameter. The e.s.d. is again, large with a value of 0.039, i.e., about 25%.

As mentioned earlier, all these crystals have the same structure. The difference in the value of H_V/C_{44} for these three groups cannot be due to the difference in the formal valency, since elementary considerations [4.62], show that the formal valency affects H_V and C_{44} in the same way and thus gets cancelled out in H_V/C_{44}. Further, it is not possible to explain the large deviations in the values of H_V/C_{44} within each group, since each group has the same formal valency.

It is known that the alkali halides are highly (though, not fully) ionic. The chalcogenides are partially ionic and the carbides are highly (though, not fully) covalent. It is conjectured that the difference in the mean values of the Gilman–Chin parameter for these three groups as well as the deviations within each group are due to a continuous change in the bond type from highly ionic to highly covalent as we scan across the whole group of crystals. In other words, the change in the values of the Gilman–Chin parameter over a wide range such as 0.01–0.2 is due to these crystals having different degrees of ionicity.

For this purpose, it was considered worthwhile to seek a correlation between the values of (H_V/C_{44}) and the effective ionic charge e^*. It is known that e^* is close to unity for highly ionic crystals, lesser than unity for partially ionic crystals and still lower for covalent crystals. In Fig. 4.22, the values of the Gilman–Chin parameter are plotted against e^*. A broad correlation is observed between the two parameters. The Gilman–Chin parameter assumes large values for crystals with small values of effective charge i.e., low ionicity and small values for crystals with larger values of e^*, i.e., high ionicity. Further, crystals with intermediate values for the Gilman–Chin parameter have intermediate values of ionicity as indicated by the values of e^*. Empirically, it is observed that the product of H_V/C_{44} and $(e^*)^2$ is a constant with a mean value of 0.020 ± 0.015. The values of $(H_V/C_{44})(e^*)^2$ do show some variation (a factor of about five). This is much less than the variation in the values of the Gilman–Chin parameter (a factor of 20) and also the variation in the hardness values (a factor of 500). Hence, the product $(H_V/C_{44})(e^*)^2$ may be considered to be approximately constant.

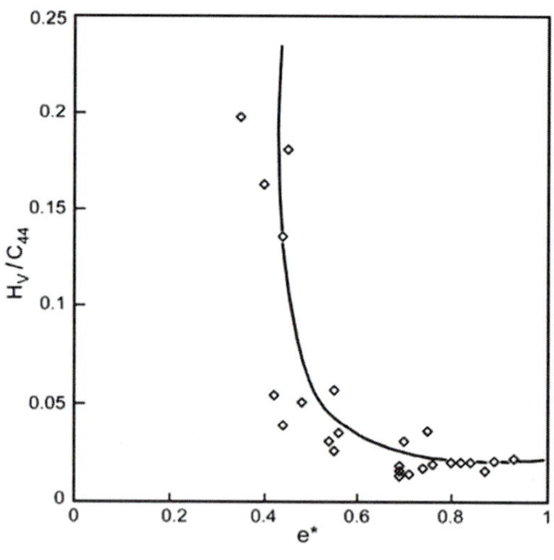

Fig. 4.22. Plot of Gilman–Chin parameter (H_V/C_{44}) against the effective ionic charge e^*

Table 4.13. Values of microhardness (H_V), shear constant (C_{44}) and Gilman–Chin parameter (H_V/C_{44}) for some cubic crystals; hardness values for the nitrates from [4.102] and the rest from [4.103]; C_{44} values from literature

Crystal	Face on which hardness was measured	H_V [kg mm^{-2}]	C_{44} [kg mm^{-2}]	H_V/C_{44}
$Sr(NO_3)_2$	(111)	46	1,460	0.032
$Pb(NO_3)_2$	(111)	35	1,370	0.025
$Ba(NO_3)_2$	(111)	29	1,210	0.024
CaF_2	(111)	185	3,450	0.054
SrF_2	(111)	163	3,100	0.053
BaF_2	(111)	149	2,400	0.062
PbF_2	(111)	133	2,500	0.053
CdF_2	(111)	153	2,100	0.073
EuF_2	(111)	264	3,050	0.087
ThO_2	(111)	750	8,100	0.092
$NaClO_3$	(100)	107	1,179	0.091
$NaBrO_3$	(100)	136	1,536	0.088
$Bi_4(GeO_4)_3$	(100)	561	5,280	0.106
$Bi_4(SiO_4)_3$	(100)	593	4,440	0.134
$Bi_{12}GeO_{20}$(BGO)	(100)	372	2,600	0.143
$Bi_{12}SiO_{20}$(BSO)	(100)	473		

Some Miscellaneous Crystals

Vickers microhardness of 16 crystals with different structures has been measured. The measurements were made only at single loads in the range 25–100 gm. The H_V values are given in Table 4.13. The hardness values are now analysed in terms of the Gilman–Chin parameter. It has been mentioned in the preceding section that the Gilman–Chin parameter (H_V/C_{44}) has values in the range 0.01–0.02 for highly ionic crystals, 0.03–0.05 for partially ionic crystals and 0.1–0.2 for highly covalent crystals.

The values of the Gilman–Chin parameter are given in Table 4.13. For the nitrates, the Gilman–Chin parameter has values close to 0.03 indicating a highly ionic character. This is in agreement with inferences drawn from several other physical properties [4.104]. The Gilman–Chin parameter has values close to 0.06 for CaF_2, SrF_2, BaF_2, CdF_2 and PbF_2, indicating a partially ionic character. This is in agreement with other physical properties like lattice energies [4.105] and thermal expansion [4.106]. In EuF_2, ThO_2, $NaClO_3$ and $NaBrO_3$, the value of the Gilman–Chin parameter is ~0.09 which indicates a partially covalent character. Finally, in $Bi_4(GeO_4)_3$, $Bi_4(SiO_4)_3$ and BGO, the Gilman–Chin parameter is ~0.13 which suggests covalency in these compounds.

Table 4.14. Values of the Vickers hardness H_V, Mohs hardness H_M, lattice constant a, shear elastic constant C_{44}, Gilman–Chin parameter (H_V/C_{44}) and the ionicity I; hardness values from [4.107], C_{44} and a from literature

Sample	Short Symbol	H_V (kg mm^{-2})	H_M	a(Å)	C_{44} (kg mm^{-2})	H_V/C_{44}	I (4.20)
$Gd_3Ga_5O_{12}$	GGG	1,350 ± 70	7.28	12.38	9,220	0.146	0.75
$Gd_3Ga_5O_{12}$	GGG (Nd)	1,300 ± 60	7.23				
$Y_3Al_5O_{12}$	YAG	1,700 ± 120	7.86	12.00	11,730	0.145	0.71
$Y_3Al_5O_{12}$	YAG (Nd)	1,740 ± 120	7.91				
$Gd_3Sc_2Ga_3O_{12}$	GSGG (Nd, Cr)	1,250 ± 60	7.06	12.53			0.77
$Eu_3Ga_5O_{12}$	EuGG	1,150 ± 50	6.91	12.50	7,770	0.142	0.78
$Y_3Fe_5O_{12}$	YIG	1,200 ± 60	7.0	12.38	7,810	0.154	0.77
$Nd_3Ga_5O_{12}$	NdGG	1,300 ± 65	7.19	12.50	8,550	0.152	0.76
$Y_3Ga_5O_{12}$	YGG (Nd)	1,500 ± 90	7.54	12.27	9,740	0.154	0.73
$Tb_3Ga_5O_{12}$	TbGG	1,600 ± 110	7.70	12.35			0.72
$Pr_{2.96}Mn_{1.34}Mg_{0.4}Zr_{0.4}Ga_{2.9}O_{12}$	Pr (Mn, Mg, Zr) GG	1,050 ± 50	6.70				
$Sm_{2.6}Ca_{0.31}Zr_{0.32}Ga_{4.68}O_{12}$	Sm (Ca, Zr) GG	1,250 ± 60	7.10				

Rare Earth Garnet Crystals

Rare earth garnets have a cubic structure [O_h^{10}–Ia3d] with eight molecules per unit cell. They have the general formula $C_3A_2D_3O_{12}$ where C, A and D denote cations and O the oxygen ion. In the study by Sirdeshmukh et al. [4.107], hardness measurements were made on 12 rare earth garnet samples including some doped with 0.5% Nd, measurements were made on (111) planes. The results are given in Table 4.14. The difference in H_V between pure and Nd-doped samples is within error limits. Hence, where only Nd-doped samples were available, the measured hardness is taken to represent the hardness of the pure crystal. The hardness values are analysed to derive information about the binding.

The microhardness of the group of rare earth garnet crystals studied is in the range 1,100–1,700 kg mm^{-2} (7.1–7.8 on the Mohs scale) which by itself is indicative of strong interatomic binding. Within a group of related crystals, the lattice constant is a measure of the strength of binding. Figure 4.23 shows a hardness vs. lattice constant plot for the rare earth garnets. The data points are closely scattered about a line indicating decreasing hardness with increasing lattice constant. Thus the variations of hardness and lattice constant within the group are mutually consistent. The H_V vs. a plot does not yield any further information. The values of the Gilman–Chin parameter (H_V/C_{44}) for the rare earth garnets, for which elastic constant values are available, are given in Table 4.14. The values are in the range 0.14–0.15. This indicates a high degree of covalency in the binding in these garnets.

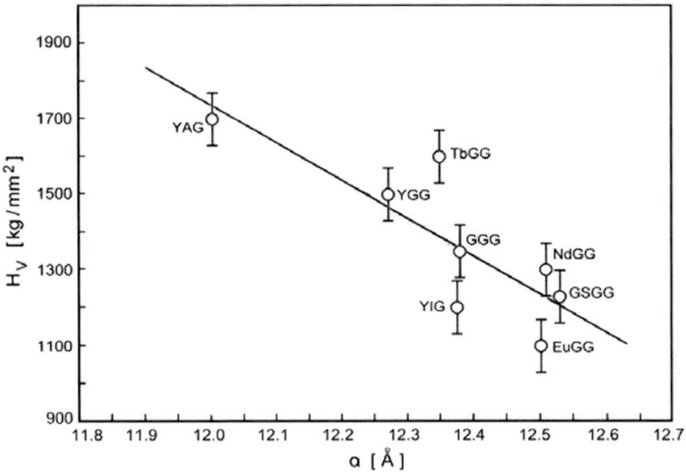

Fig. 4.23. Plot of hardness H_V vs. lattice constant a for rare earth garnets

The Gilman–Chin parameter merely helps to distinguish between different bond types. A different approach makes it possible to estimate the degree of ionicity (I) (or the degree of covalency) from hardness values. Pillay [4.61] considered the hardness of several compounds with known ionicity and obtained the empirical relation

$$H_M = -15.79I^2 + 11.33I + 7.63. \tag{4.20}$$

This equation is valid for a wide range of materials. The ionicities of rare earth garnets estimated using (4.20) are given in Table 4.14. The ionicity values range from 0.71 for YAG to 0.78 for EuGG. These values are considerably less than the ideal value of 1 for purely ionic crystals and are indicative of a high degree of covalency. Further, these values are comparable with the values for several silicates [4.61], which are known to be highly covalent.

Some light can be thrown on what is responsible for the high degree of covalency in these garnets. There are three cation–anion bonds in the garnets corresponding to the tetrahedral (or d) sites, octahedral (or a) sites and dodecahedral (or c) sites. Most of the physical properties of garnets are essentially determined by these bonds. Hofmeister and Campbell [4.108] calculated the compressibility of rare earth garnets from a formula by Brout [4.109] formula which involves the infrared frequencies and the three afore-mentioned bond lengths. Since the calculated values of the compressibility were found to be close to the experimental values, it may be inferred that the three cation–anion bonds play a role in determining the mechanical properties of the rare earth garnets. These bonds should be expected to play a role in determining the hardness also. Chin [4.110] has shown that covalently bonded crystals follow the empirical relation

$$C_{44} = 5.28 \times 10^5 r^{-5}, \quad (4.30)$$

where r is the length of the covalent bond involved. Since the Gilman–Chin parameter (H_V/C_{44}) has a value of 0.15 for YAG, (4.30) may be modified to

$$H_V = 0.15 \times 5.28 \times 10^5 r^{-5}. \quad (4.31)$$

The average cation–anion distance in YAG is 2.18 Å [4.111]. Substituting this value in (4.31), we get a value of 1,500 kg mm^{-2} for the hardness of YAG which may be compared with the experimental value of 1,700 ± 120 kg mm^{-2}. This agreement is fair, considering the approximations involved. Thus the cation–anion bonds in the rare earth garnets appear to be responsible for the large hardness and also, for the high degree of covalency in these crystals.

Crystals with Potassium Dihydrogen Phosphate Structure

Potassium dihydrogen phosphate (KDP) and its isomorphs are tetragonal with space group $I\bar{4}2d$. Microhardness measurements on seven crystals with the KDP structure were made [4.112] and the results are discussed in terms of bonding and structural features. The natural habit of these crystals is a tetragonal prism combined with a tetragonal bipyramid. The prism faces were used for hardness measurements in the a-direction. For measurements in the c-direction, the crystals were cut along the a-direction and the resulting (001) faces were polished. The microhardness measurements were made at loads of 50 and 100 g. Measurements were made on the (100) and (001) faces in the case of KDP, KDA, RbDP, RbDA, CsDP and ADP. The dimensions of the ADA crystal were such that measurements could be made only on the (100) face (i.e. in the <100> direction). The results of the measurements are given in Table 4.15, each value being the mean of several values. The deviations from the mean values are of the order of ±5 kg mm^{-2}.

The microhardness values (H_V) for these crystals are in the range 80–150 kg mm^{-2}. On the Mohs scale, the values are in the range 3–4. These crystals are thus a little harder than the alkali halides and are comparable in hardness with the alkaline earth oxides or the alkaline earth fluorides [4.17].

Haussuhl [4.113] has determined the elastic constants and has calculated the linear compressibilities in the a- and c-directions for five of these crystals. In the case of RbDP, ADP and ADA, the linear compressibility is less in the a-direction than in the c-direction. Khan and Baur [4.114] pointed out that the coefficient of linear expansion is less in the a-direction than in the c-direction in the case of the potassium and rubidium salts. Further, they suggested that the O–H–O bonds which are almost parallel to the basal plane make the structure strong in the a direction. The hardness in the a-direction is, in general, more than that in the c-direction indicating that the structure is stronger in the former than in the latter direction. However, the anisotropy in hardness is not pronounced. In fact, the difference in the hardness in the two directions is only slightly greater than the limits of error. As such, with due hesitation, the larger hardness in the a-direction is attributed to the role of the O–H–O bonds. Khan bonds. Khan and Baur [4.114] also attribute the relatively low thermal expansion of the ammonium compounds to the presence of the N–H—O bonds. But, the value of H_V for the crystals of the ammonium salts is distinctly less than that for the other crystals. Thus the N–H—O bonds not appear to contribute to the hardness. Adhav [4.115] measured the piezoelectric constants of these crystals and observed a systematic difference in the values of the piezoelectric constants of the phosphate crystals and their arsenate counterparts; Adhav [4.115] attributed this difference to the difference in strength of the P–O bond and the As–O bond. However, from the bond distance vs. bond strength relationship proposed by Brown and Shannon [4.116], Khan and Baur [4.114] pointed out that the strength of these two bonds is nearly the same. In any case, we have not observed any systematic difference in the hardness of the phosphate and arsenate crystals.

The KDP structure can be looked upon as a polar structure consisting of the K^+ and $(H_2PO_4)^-$ ions. Considering the interaction between these ions, Hartman [4.117] could explain the growth habit of these crystals. Following Hartman's method, Kishan Rao and Sirdeshmukh [4.112] calculated the lattice energy $U$of these crystals. The values are included in Table 4.15. The lattice energy values are given for unit volume (the 'volumetric lattice energy' [4.17]). It can be seen that there is a correlation between the average hardness (H_V) and lattice energy per unit volume (U/V), H_V being greater for crystals with higher UV. The gross hardness of these crystals thus seems to be determined by the ionic bond.

Table 4.15. Values of microhardness (H_V) and lattice energy per unit volume (U/V) for crystals with KDP structure [4.112]

Crystal	H_V [kg mm^{-2}]			U/V [kcal mol^{-1} cm^{-3}]
	a-Direction	c-Direction	Average	
KDP	145.5	132.0	138.7	2.354
KDA	145.2	145.5	145.5	2.126
RbDP	111.3	97.6	104.4	2.060
RbDA	119.5	93.0	106.2	1.869
ADP	89.0	83.3	86.1	1.978
ADA	75.8	–	75.8	1.802
CsDA	101.9	84.0	92.9	1.608

Table 4.16. Values of k and m (4.33) for some alkali halides [4.90]

Crystal	k	M
RbCl	0.023	1.95
RbBr	0.79	0.35
RbI	0.020	2.00
NaF	2.13	0.95

4.4.3 Radiation Hardening

Some results on effect of irradiation on hardness of crystals have been discussed in Sect. 4.3. We have studied the effect of: (a) γ-ray irradiation on the hardness of some alkali halides and (b) X-ray irradiation on hardness of sodium bromate.

Studies on Alkali Halides

Reference has been made to the effect of irradiation on the hardness of crystals. It was pointed out that hardening has been observed in NaCl and the potassium halides irradiated with X-rays. Several alkali halides have been studied using flow stress as the probe. The rubidium halides have not received any attention.

Thirmal Rao [4.90] undertook a study of the effect of γ-ray irradiation on the hardness of RbCl, RbBr, RbI and NaF. γ-ray irradiation produces colour centers in alkali halides. The concentration n_{F} of color centers was estimated from the Smakula formula [4.118]

$$f n_{\mathrm{F}} = 1.37 \times 10^{17} [n/(n^2+2)^2] \times \Delta \times \alpha_{\max}/\mathrm{cm}^3, \qquad (4.32)$$

where f is the oscillator strength, n the refractive index, Δ the half width of the F-band (in eV) and $\alpha_{\max}$ the absorption coefficient at the F-band peak (in cm^{-1}).

In Fig. 4.24 the increase in hardness (ΔH) is plotted against n_{F}, the concentration of F-centres. It is observed that the hardness increases with the concentration of F-centres in all the four alkali halides. Assuming that the relationship can be described by a general equation of type

$$\Delta H = k\, n_{\mathrm{F}}^{m} \qquad (4.33)$$

the data was processed to obtain the value of the hardening parameter k and the index m. The values of k and m are given in Table 4.16.

The hardening parameter k has values ranging from 0.02 to 2($\mathrm{kg\,mm}^{-2}$). The rubidium halides have low values of k compared to NaF which has the highest value. The index 'm' has values ranging from 0.35 to 2.0. While observing a strong correlation between the flow stress and F-centre concentration, Nadeau [4.46] suggested that other factors like interstitials, colloids, clusters

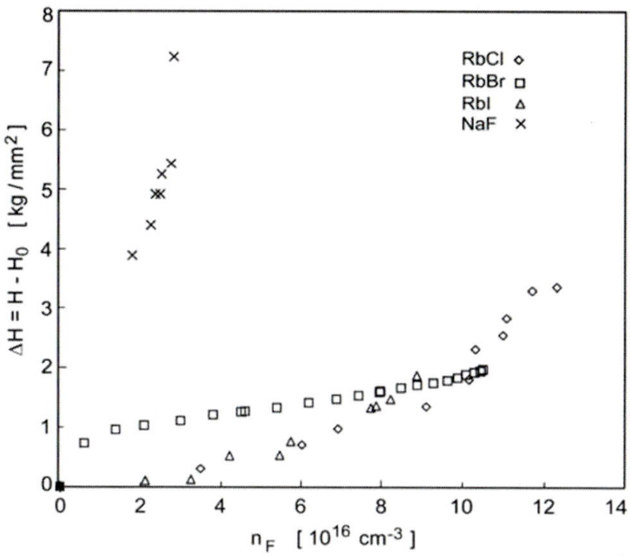

Fig. 4.24. Plot of increase in hardness (ΔH) vs. F-centre concentration n_{F}

may also contribute to the hardening. Boyarskaya and Zhitaru [4.47] also suggested that defects other than F-centres may contribute to the observed hardening. Aerts et al. [4.44] suggested the formation of colloids in the surface layers and their possible effect on the hardness. The deviation of the values of m from crystal to crystal could be due to the role of radiation-induced defects other than F-centres which may differ from crystal to crystal in type and concentration.

Studies on $NaBrO_3$

Sodium bromate ($NaBrO_3$) belongs to the cubic tetrahedral class with point group 23. $NaBrO_3$ is considerably susceptible to X-ray and γ-ray irradiation and is reported to break into fragments on prolonged irradiation [4.119]. The radiation damage in this crystal has been studied using optical absorption, Raman spectroscopy and ESR techniques [4.119–4.121]. There is no report on the effect of irradiation on the hardness. Kishan Rao and Sirdeshmukh [4.122] undertook measurements of microhardness of $NaBrO_3$ crystals subjected to X-ray irradiation. Optical absorption and dislocation density measurements were also made.

Measurements were made on the (111) planes. Hardness measurements were made with a Vickers hardness tester at a load of 50 g. The irradiation was carried out with X-rays from a copper target tube operated at 30 kV and 10 mA.

For the determination of the dislocation density, the crystals were etched with an etchant made of eight parts of glacial acetic acid and two parts of

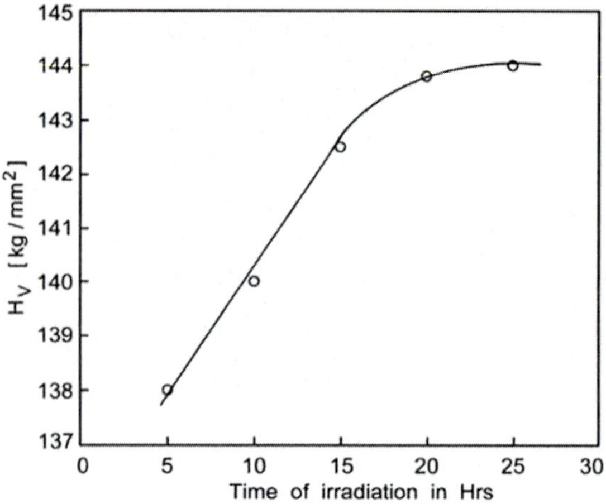

Fig. 4.25. Dependence of hardness (H_V) of $NaBrO_3$ on time (t) of irradiation

formic acid containing 3 mg cm^{-3} of cupric nitrate. The absorption spectra were recorded on a model MPS 5000 Shimadzu UV–visible spectrophotometer.

The variation of microhardness with time of irradiation is shown in Fig. 4.25. The hardness of the unirradiated crystal is 137 ± 2 kg mm^{-2}. It can be seen that the microhardness initially increases with time of irradiation and tends to assume a constant value after about 20 h.

Hardness is a measure of the resistance offered by a crystal to the movement of dislocations. Defects like impurities impede the movement of a dislocation. If the dislocation density is high, the motion of a given mobile dislocation may be affected by the presence of other dislocations. Thus, the observed increase in hardness of $NaBrO_3$ could be attributed either to an increase in dislocation density or to creation of defects which have their origin in radiolysis or to both the mechanisms.

The crystals were etched, as described, before irradiation. An etched surface of an unirradiated crystal is shown in Fig. 4.26a. Triangular etch pits can be observed. The dislocation density (obtained by counting the etch pits in a given area) was about 5×10^4 cm^{-2}. The crystal surface was lightly polished to remove a few layers, the crystal was irradiated for 20 h and etched again. The etched surface of the irradiated crystal is shown in Fig. 4.26b. Most of the pits present in Fig. 4.26a can be seen in Fig. 4.26b and there is no change in the dislocation density. Thus, the observed increase in hardness is not due to any increase in dislocation density.

X-ray irradiation of $NaBrO_3$ causes the breaking up of the BrO_3 ion into several radiolysis products like BrO^-, Br_2 and O_3^-. The presence of O_3^- has been clearly proved from Raman spectra of irradiated $NaBrO_3$ [4.119].The BrO^- fragment gives a band at 330 nm in the absorption spectrum of the X-ray irradiated crystal [4.120]. This is a stable band which is not bleached

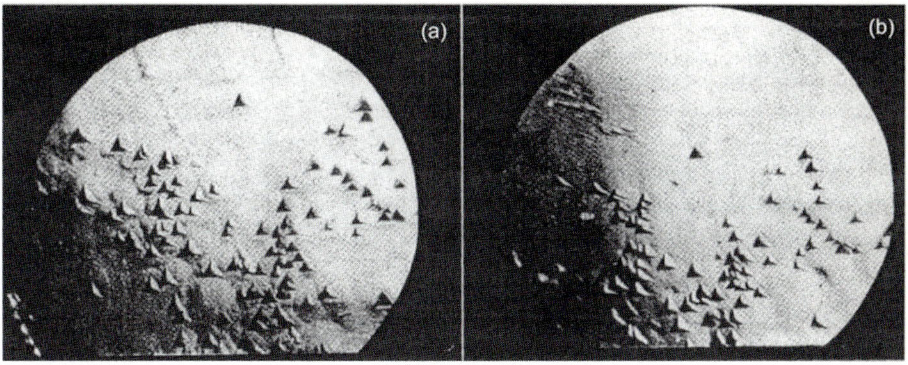

Fig. 4.26. Dislocation etch pits on (111) surface of $NaBrO_3$ (**a**) before and (**b**) after irradiation

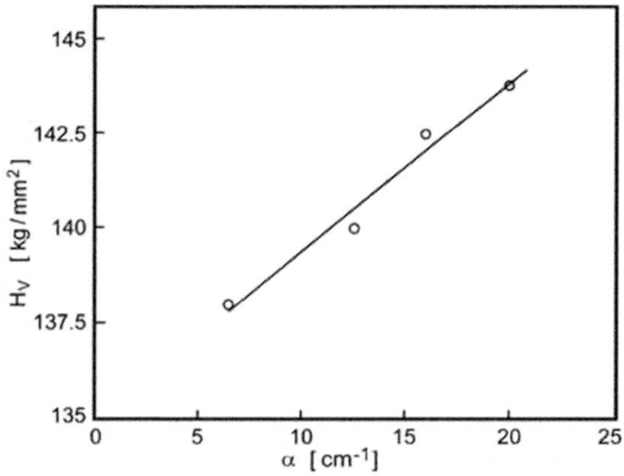

Fig. 4.27. Relation between hardness H_V of $NaBrO_3$ and optical absorption coefficient (α) at 330 nm for different irradiation times

by visible light. Absorption spectra were recorded for unirradiated crystals and for crystals irradiated for different times. The absorption coefficient (α) at 330 nm is taken as a measure of the population of BrO^- in the irradiated crystal. The microhardness and optical absorption measurements are made on a given crystal irradiated for a given time. Such measurements are repeated for increasing irradiation time. The values of the hardness (H_V) and absorption coefficient (α), thus obtained, are plotted in Fig. 4.27. A monotonous linear relationship is seen indicating an increase in both H_V and α as the irradiation increases. A similar correlation might also exist between the hardening and the population of other radiolysis products. Thus, the observed increase in hardness is not due to any increase in the dislocation density but appears to be caused by the radiolysis products which remain dispersed in the crystal and obstruct the motion of dislocations.

4.4.4 Hardness of Doped Crystals

Important observations on the effect of doping on the hardness of host crystals have been referred to in Sect. 4.3; doping, generally, enhances the hardness. The increase in hardness depends on the type and concentration of the dopant as well as on the host crystal. In this section, we present our results on the effect of strontium on the hardness of the rubidium halides.

While there is a good amount of work on the work on the hardness of doped sodium, lithium and potassium halides (see Sect. 4.3), there is no work on the impurity-hardening of rubidium halides. The rubidium halides have useful infrared transmission characteristics and have, recently, been found [4.123] to be efficient X-ray information storage sensors. The utility of alkali halides as device materials is limited by their low hardness. Among the alkali halides for which hardness data is known, the rubidium halides have the lowest hardness. As such, a study of the hardness of rubidium halides doped with different concentrations of Sr^{2+} was undertaken by Sirdeshmukh et al. [4.124].

The results are given in Table 4.17 and are also shown in Fig. 4.28. It is seen that the hardness increases with increase in concentration C of the dopant. The results were fitted to the relation

$$\Delta H_V = k' \, C^{m'}, \tag{4.34}$$

where ΔH_V is the enhancement in hardness and k' and m' are constants; values of k' and m' are given in Table 4.18. On comparing these results with earlier results [4.41] on the hardness of sodium and potassium halides, it is

Table 4.17. Values of Vickers hardness (H_V) for the rubidium halides at different concentrations C of the Sr^{2+} ions

Crystal	C[mol%]	H_V [kg mm^{-2}]
RbCl	0	9.50
	0.27	15.75
	0.40	17.82
	0.59	19.30
	0.83	22.50
RbBr	0	8.10
	0.31	12.40
	0.46	13.61
	0.75	15.52
	1.27	18.30
RbI	0	6.01
	0.13	8.58
	0.21	9.00
	0.35	10.61
	0.58	11.81
	0.85	13.01

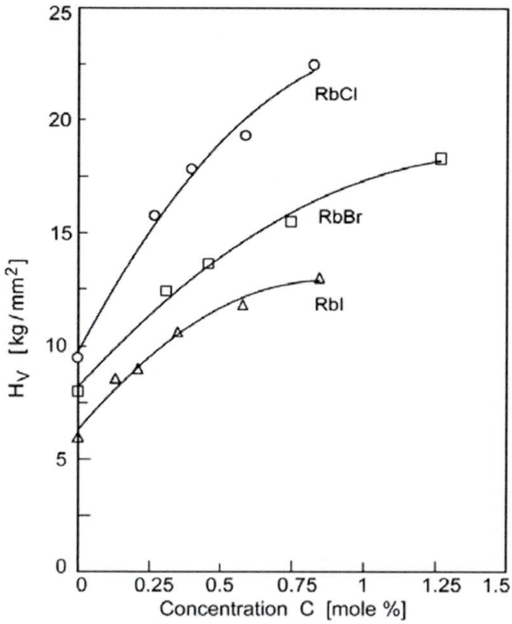

Fig. 4.28. Hardness (H_V) as a function of concentration C of Sr^{2+} ions in rubidium halide crystals

Table 4.18. Values of the constants k' and m' in (4.34)

Crystal	k' [kg mm^{-2}]	m'
RbCl	14.36 ± 1.00	0.63 ± 0.05
RbBr	8.96 ± 1.00	0.60 ± 0.05
RbI	8.15 ± 1.00	0.62 ± 0.05

noted that the observed value of m' ($\sim$0.6) is close to the value of 0.5 obtained by Chin et al. [4.41] and, theoretically, by Gilman [4.42]. On the other hand, the hardening is dependent on the host lattice increasing in the sequence RbI—RbBr–RbCl, unlike in the case of the sodium and potassium halides.

Gilman [4.42] proposed a theory in which hardening is associated with the change in electrostatic energy that occurs when a moving dislocation shears a divalent ion–cation vacancy complex. According to this model, the hardening is proportional to $\kappa\, C^{1/2}$ where the hardening coefficient, κ, is given by

$$\kappa = 4.7\, e^2 (\varepsilon_s\, a^4)^{-1}, \qquad (4.35)$$

here, e is the electron charge, ε_s the static dielectric constant and a the lattice constant of the host crystal and C the concentration of the impurity. As Gilman's immediate interest was in providing an explanation for the results of Chin et al. [4.41], he adopted the unusual procedure of assuming average values for the static dielectric constant and the lattice constants of the four alkali halide crystals (NaCl, NaBr, KCl, KBr) and obtained an average value for κ which was in fair agreement with the experimental value.

Chin et al. [4.41] have concluded that the impurity hardening observed by them in their study of sodium and potassium halides is independent of the host crystal. However, a close look at their diagrams reveals that the data points for each crystal lie on different plots with slightly different slopes which means that the hardening is not altogether independent of the host lattice. Further, calculation of the term $(\varepsilon_s a^4)^{-1}$ for the rubidium halides using individual values for ε_s and a (in Å) from literature [4.19] yields values 1.1, 0.96 and 0.73 (all in 10^{-4}) for RbCl, RbBr and RbI, respectively indicating that the hardening expected from Gilman's theory is host-dependent increasing in the sequence RbI–RbBr–RbCl. Our results are thus consistent with Gilman's theory.

4.4.5 Hardness of Mixed Crystals

The hardness of several mixed crystal systems has been studied as a function of their compositions. Typically, the hardness is a non-linear function of the composition with positive deviations from linearity. In most of the systems, the hardness of mixed crystals in the equimolar region is larger than the hardness of either end members. These results will be discussed in detail in Chap. 7.

4.4.6 Empirical Relations with other Physical Properties

The hardness of crystals correlates with several other physical properties. Some of these have been mentioned in Sect. 4.3. Here we shall consider some new relations.

Sirdeshmukh et al. [4.125] showed the existence of the following relations between hardness H of the alkali halides and their Debye–Waller factors B, the melting point T_m, and the compressibility ψ.

$$H_M - C_1 = C_2(B/a), \tag{4.36}$$

$$H_M - C_3 = C_4(T_m/a^3) \tag{4.37}$$

and

$$H_M - C_5 = C_6\,\psi^{-1}. \tag{4.38}$$

where H_M is the Mohs hardness and a is the lattice constant. The C_i's are constants. The corresponding plots are shown in Figs. 4.29–4.31. Sirdeshmukh et al. [4.125] showed that these relations have a physical basis. For some crystals for which data is not available, the values of H_M were estimated from these relations.

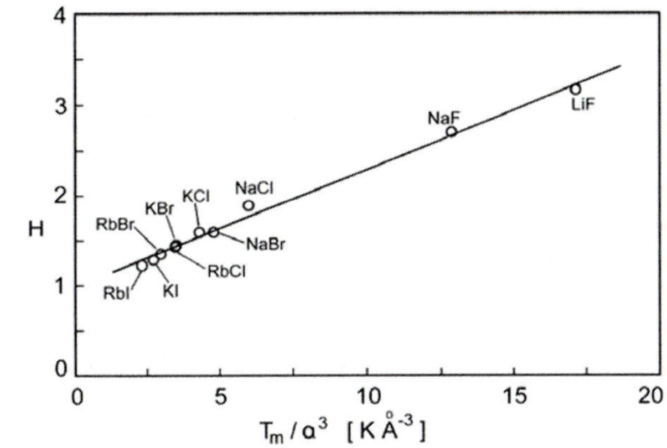

Fig. 4.29. Plot of hardness H_{M} vs. (T_m/a^3) for alkali halides

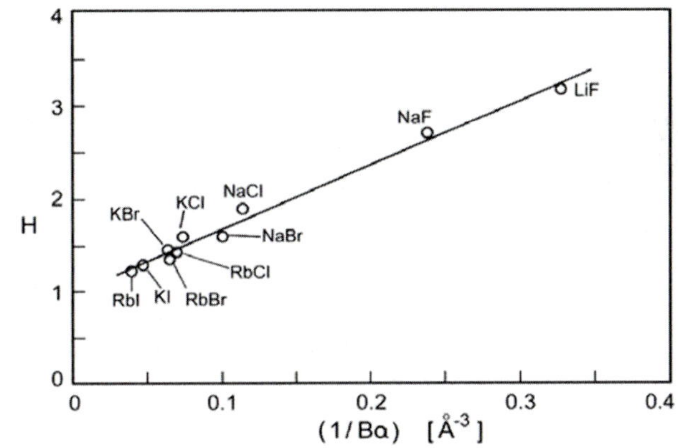

Fig. 4.30. Plot of hardness H_{M} vs. $(1/Ba)$ for alkali halides

The relationship between hardness (H_{V}) and the Debye temperature θ is given by

$$\theta = cH_{\mathrm{V}}^{1/2}\, V^{1/6}\, M^{-1/2}, \tag{4.39}$$

where c is a constant, V the molar volume and M the molar mass. This relation has been used by several authors to estimate H_{V} from θ or θ from H_{V}. The hardness of some rare earth metals was measured with a Leitz indenter and the hardness was estimated from the flat part of the load-hardness curve (see Sect. 4.4.1). The values of the true hardness are given in Table 4.19 along with literature values. As a check on these values, the hardness was calculated from (4.39) using a value of 5×10^{-5} for the constant. The Debye temperatures are taken from Chap. 3. The resulting values are given in Table 4.19. It is seen that there is fair agreement between these values and the experimental values.

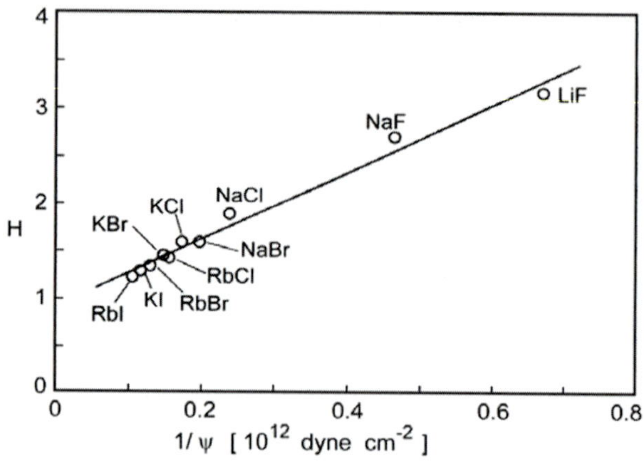

Fig. 4.31. Plot of hardness H_{M} vs. $(1/\psi)$ for alkali halides

Table 4.19. Hardness of some rare earth metals

Metal	Vickers hardness $[\mathrm{kg\,mm^{-2}}]$		
	Experimental [4.126]	Calculated from (4.39)	Others [4.127]
Dy	71	78	39–99
Gd	69	79	33–74
Er	116	104	42–135
Y	88	79	38–130

Also the measured values as well as those calculated from (4.36) are within the range of values quoted by Scott [4.127].

4.4.7 Temperature Variation of Hardness

Using the method of relative hardness, the temperature variation of hardness has been studied in NaCl and KCl by Kishan Rao and Sirdeshmukh [4.128] and in CaF_2, SrF_2 and BaF_2 by Kishan Rao and Sirdeshmukh [4.129]. The hardness is shown as a function of temperature in Fig. 4.32. In all cases, the hardness decreases with increasing temperature, first steeply and then slowly.

For the temperature variation of hardness of metals, Ito [4.130] and Shishokin [4.131] proposed an empirical relation

$$H = A \exp(-bT), \tag{4.22}$$

where A and B are constants. Merchant et al. [4.66] proposed an Arrhenius equation

$$H = A' \exp(b'/T), \tag{4.23}$$

where A' and B' are constants. Gilman [4.12] proposed

$$H/H_0 = 1 - 2k_{\mathrm{B}}\theta/U[\coth(\theta/T) - 1] \tag{4.24}$$

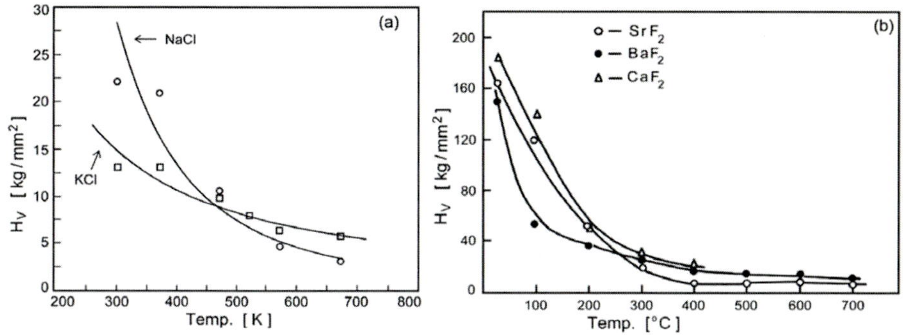

Fig. 4.32. Variation of relative hardness (H) with temperature: (**a**) NaCl and KCl (**b**) CaF_2, SrF_2 and BaF_2

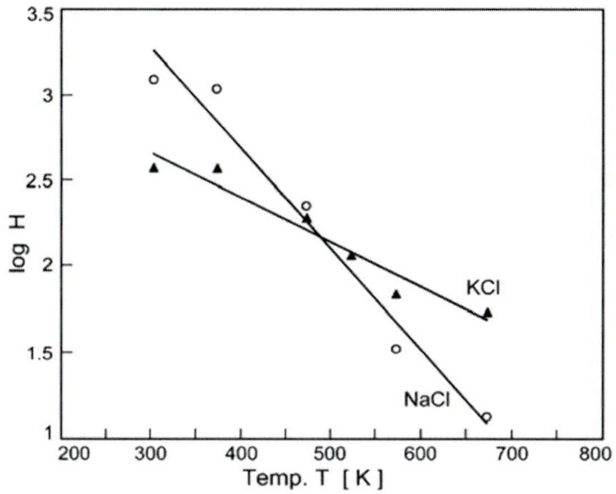

Fig. 4.33. Plot of log H against temperature T for NaCl and KCl

for the temperature variation of microhardness of silicon. Here H_0 is the hardness of the crystal at 0 K, k_B the Boltzmann constant, θ the characteristic temperature and U the energy barrier for plastic flow at 0 K. (4.24) may be written in a linear form:

$$H = A'' - b''\coth\,(\theta/T) \tag{4.40}$$

with $a'' = H_0 + 2k_\mathrm{B}\theta\ H_0/U$ and $b'' = 2k_\mathrm{B}\theta\ H_0/U$.

An attempt is made to assess the applicability of the relations given above to the alkali halides and fluorite type crystals. It can be seen from (4.22), (4.23) and (4.40) that plots of: (a) log H vs. T, (b) log H vs. T^{-1} and (c) H vs. coth (θ/T) should be linear. This has been done for NaCl and KCl. For plotting (c) θ is taken as the Debye temperature and values of 281 and 231 K have been used for NaCl and KCl, respectively [4.118]. The three plots are shown in Figs. 4.33–4.35. The plots are linear with some scattering of data

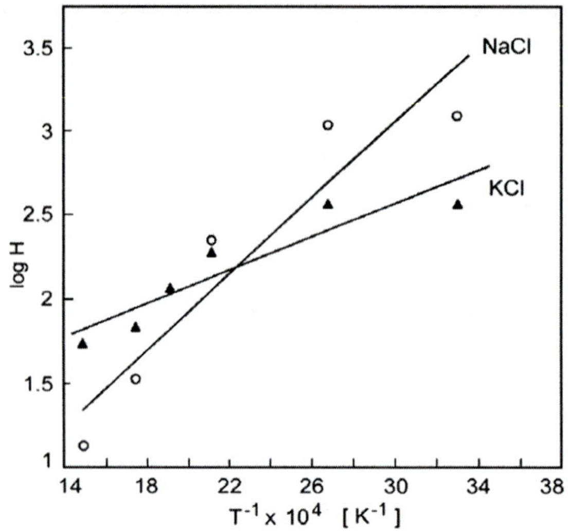

Fig. 4.34. Plot of log H against $(1/T)$ for NaCl and KCl

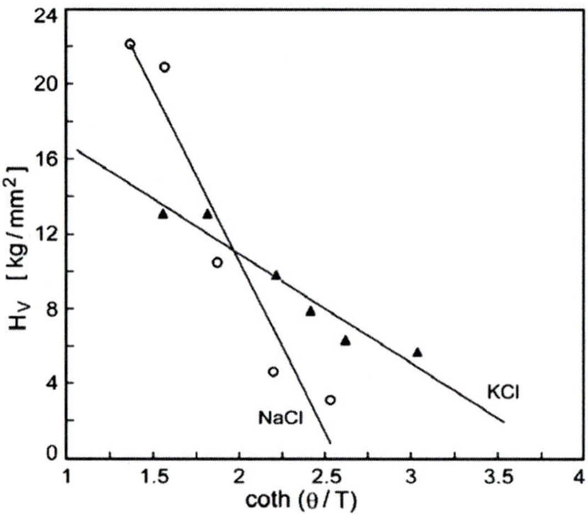

Fig. 4.35. Plot of hardness H against $\coth(\theta/T)$ for NaCl and KCl

points. By least square fitting the values of (A, b), (A', b') and (A'', b'') have been obtained; these are given in Table 4.20 along with their standard errors. It can be seen that for both the crystals the data fit (4.22) with least errors. Next is (4.40). The fit with (4.23) is the poorest. However, it is not possible to draw conclusions about the relative merits of these relations very forcefully in view of the limited accuracy of the present experimental method. Equations (4.22) and (4.40) can therefore be considered to represent empirically the variation of the hardness with temperature for the alkali halides. In the case

Table 4.20. Values of parameters in (4.22, 4.23, 4.39)

Crystal	log H vs. T		log H vs. T^{-1}		H vs. coth (θ/T)	
	A	B	A'	b'	A''	B''
NaCl	5.02 ± 0.29	0.0057 ± 0.0004	0.35 ± 0.55	1,135 ± 235	108.7 ± 13.1	42.76 ± 6.67
KCl	3.43 ± 0.14	0.0025 ± 0.0002	1.08 ± 0.21	1,197 ± 92	54.21 ± 4.17	14.30 ± 1.79
CaF_2	6.58	0.0055				
SrF_2	7.26	0.0073				
BaF_2	5.95	0.0045				

of the crystals of the fluorite type, the data were fitted only to (4.22); the values of A and b are included in Table 4.20.

4.4.8 Surface Hardness of Crystals

There is considerable interest in the hardness in the surface region (depth ~0.001 μm). Information on this aspect is meager and also conflicting. A systematic study of dynamic hardness of crystals in the ultra- and microload range was undertaken by Sirdeshmukh et al. [4.132]. Fifteen crystals belonging to five different structures with their bonding varying from ionic to partially covalent to covalent were studied. The measurements were made by using the Shimadzu dynamic ultrahardness tester DUH 202 described in Sect. 4.2.2.

The results are examined with the following objectives: (a) Is the surface hardness definitely larger than the bulk hardness? (b) Does surface hardness satisfy the correlations with other physical properties observed with microhardness? (c) Does surface hardness show any structural dependence?

Typical results on depth variation are shown in Fig. 4.36. The depth variation is non-linear. In Fig. 4.37 a log H_{DU}–log h plot is shown. This plot is nearly linear with slight non-linearity at low depths. The low-load region can be fitted to an equation of the type

$$\log H_{\mathrm{DU}} = -m \log h + \log c, \tag{4.41}$$

which leads to the following law of depth variation:

$$H_{\mathrm{DU}} = ch^{-m}. \tag{4.42}$$

The values of c and m are given in Table 4.21. Also given in the table are the surface hardness H_{S} obtained by extrapolation of (4.42) to $h = 0.001$ μm (~10 Å), the bulk hardness value H_{B} and the surface hardness coefficient $\delta = H_{\mathrm{S}}/H_{\mathrm{B}}$.

From the data in Table 4.21, the following conclusions can be drawn:

(1) In all cases $H_{\mathrm{S}} \gg H_{\mathrm{B}}$. In fact, H_{S} is larger than H_{B} by an order.
(2) The value of the index m is of the order 0.3–0.5
(3) In four cases δ is in the range 3–20. In eleven crystals, it has a much larger value (20–40)

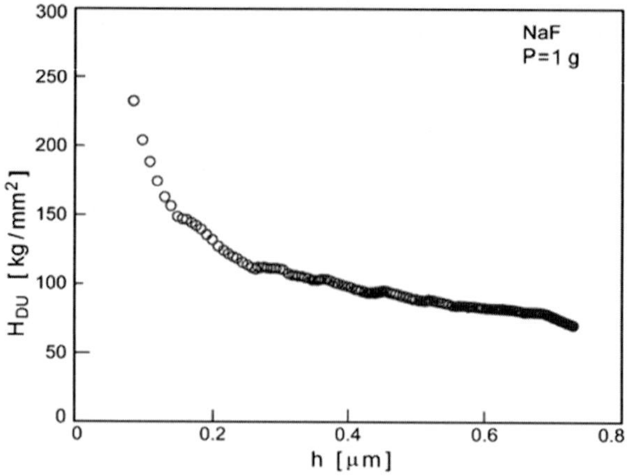

Fig. 4.36. A typical plot of H_{DU} as a function of depth (h)

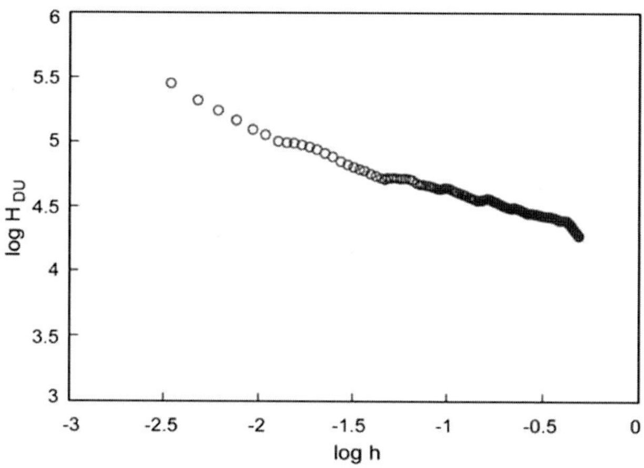

Fig. 4.37. Plot of log H_{DU} vs. log h corresponding to data in Fig. 4.36

In Table 4.22, data on some physical properties needed to obtain correlations are given. These are: the interatomic distance (r), the surface energy and the effective ionic charge (q^*). The effective ionic charge q^* indicates the bonding in the crystals, having values close to unity for highly ionic crystals and having low values for covalent crystals.

The interatomic distance is a useful scaling parameter. Many physical properties like compressibility, elastic constants and Debye temperature form smooth correlations with the interatomic distance [4.19]. The microhardness of a family of crystals correlates with the interatomic distance (Sect. 4.4.2). In Fig. 4.38, the surface hardness H_S is plotted against the interatomic distance. It can be seen that the H_S vs. r plots for each structural family are linear with H_S decreasing as r increases. Since we have only one sample (Si) with

Table 4.21. Values of bulk hardness H_B, coefficients c and m of (4.42), surface hardness H_S, and the surface hardening coefficient δ

Crystal	Plane	c	m	H_S[kg mm^{-2}]	H_B[kg mm^{-2}]	δ
NaCl structure						
NaF	100	67	0.44	1,400	65	21.5
NaCl	100	14	0.55	625	26	24.1
KCl	100	11	0.53	428	11	38.9
KBr	100	15	0.44	313	9	34.8
RbCl	100	14	0.49	413	10	41.3
RbBr	100	20	0.40	317	8	39.6
PbS	100	138	0.42	2,511	78	32.2
CsCl structure						
CsBr	100	27	0.40	425	21	20.2
CsI	100	33	0.36	400	12	33.3
CaF_2 structure						
CaF_2	111	160	0.44	3,342	160	20.9
SrF_2	111	257	0.35	2,883	154	18.7
BaF_2	111	73	0.49	2,145	71	30.3
Zinc blende and diamond structure						
CdTe	111	83	0.45	1,858	54	34.4
GaAs	111	452	0.32	4,122	724	5.7
Si	100	678	0.28	4,690	1228	3.8

diamond structure, we have associated the data point with the plot for the closely related family of cubic ZnS-type crystals. The PbS data point deviates from the plot for the crystals with NaCl structure. This may be because PbS is a divalent crystal unlike the rest of the NaCl-type crystals which are monovalent.

Upit and Varchenya [4.79] suggested a correlation between the microhardness and surface energy of alkali halides; very few crystals were considered due to lack of data on surface energy. Intuitively, it is felt that the surface energy should be correlated with the surface hardness rather than the bulk hardness. In Fig. 4.39, a plot of surface energy (σ) vs. the surface hardness (H_S) is shown. In the absence of experimental values of surface energy, we have considered the theoretically calculated values. The plot is non-linear with H_S increasing as σ increases.

To examine the factors on which the surface hardening coefficient ($\delta = H_\mathrm{S}/H_\mathrm{B}$) depends, the values of δ are plotted against log H_B in Fig. 4.40. In spite of large scattering, a linear trend can be seen with δ decreasing as H_B increases. Here again, the data point for PbS deviates from the plot for crystals with NaCl structure. Similarly, a plot between the effective ionic charge q^* and log δ (Fig. 4.41) indicates reduction in δ as q^* decreases. Fig. 4.40 shows that surface hardening is more for soft crystals than for hard crystals while Fig. 4.47 shows that δ is more for ionic crystals and less for partially ionic and covalent crystals.

Table 4.22. Data for correlation

Crystal	$\langle r \rangle$ [Å] [4.133]	q^* [4.19, 4.134, 4.135]	σ [erg cm^{-2}] [4.136, 4.137]
NaCl structure			
NaF	2.310	0.83	221
NaCl	2.814	0.77	158
KCl	3.146	0.81	145
KBr	3.300	0.78	130
RbCl	3.291	0.83	140
RbBr	3.427	0.80	125
RbI	3.671	0.77	110
PbS	2.968		
CsCl structure			
CsBr	3.720	0.82	
CsI	3.955	0.78	
CaF_2 structure			
CaF_2	2.364	0.833	543
SrF_2	2.511	0.875	437
BaF_2	2.672	0.896	393
Zinc blende and diamond structure			
CdTe	2.80	0.74	
GaAs	2.44	0.51	
Si	2.34	0.0	

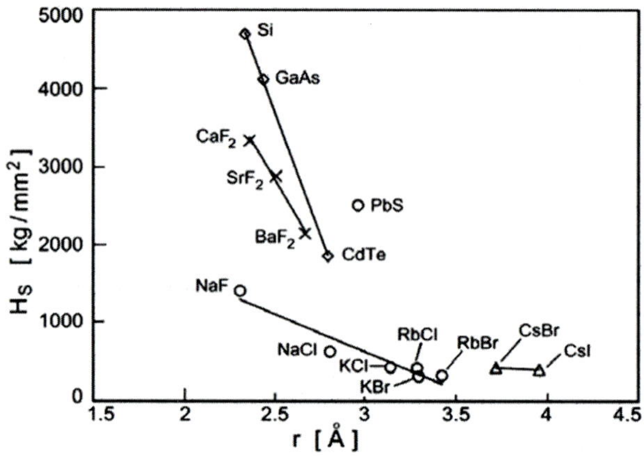

Fig. 4.38. Plot of r vs. H_S

Different structural families appear to have different ranges of H_S values. There is some overlapping between the ranges of different groups. Thus, crystals with CsCl structure have H_S values of the order of 400 kg mm^{-2}, whereas crystals with NaCl structure have values ranging from 300 to –2,500 kg mm^{-2}. Crystals with CaF_2 structure have H_S varying from 2,000 to 3,000 kg mm^{-2}.

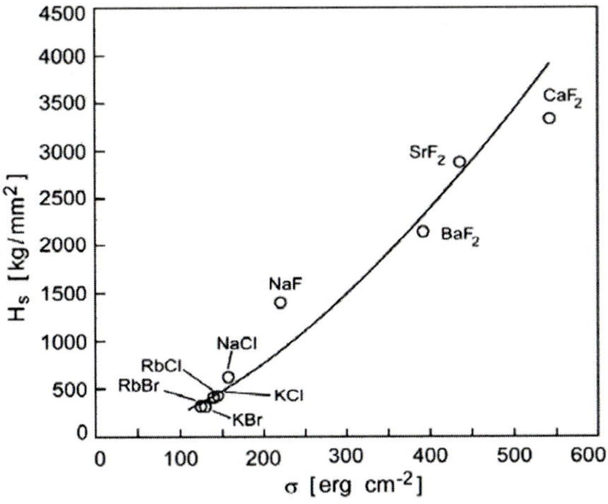

Fig. 4.39. Plot of σ vs. H_S

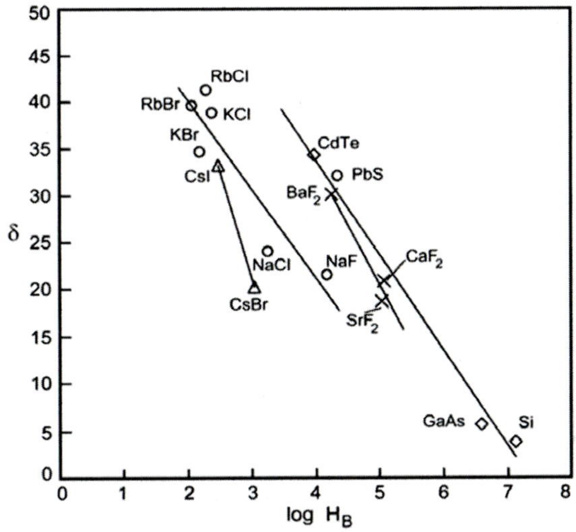

Fig. 4.40. Plot of log H_B vs. δ

Crystals with the ZnS and diamond structure have the highest H_S values ranging from 2,000 to 4,700 kg mm^{-2}.

With regard to the observed correlations, the H_S vs. r plots (Fig. 4.38) resolve into separate plots for each structure. The same applies to the log H_B vs. δ plot (Fig. 4.40). On the other hand, there is no structure-wise resolution in the H_S vs. σ lot (Fig. 4.39) and the q^* vs. log δ plot (Fig. 4.41).

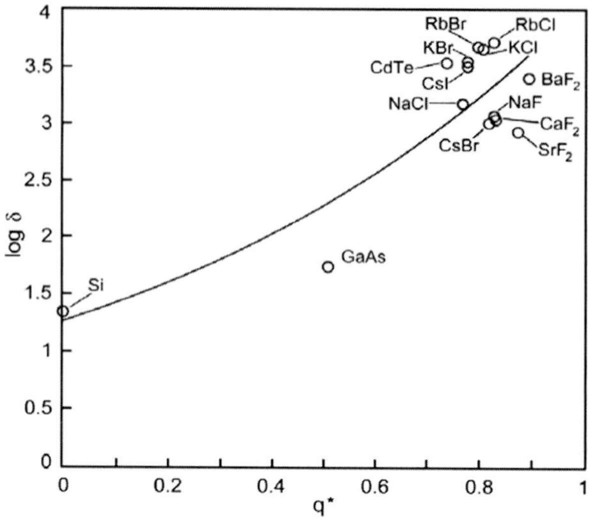

Fig. 4.41. Plot of q^* vs. log δ

4.4.9 Anisotropy of Hardness

Some observations on the anisotropy of hardness have been referred to in Sect. 4.3. In this section, some of our results on anisotropy will be discussed. In the case of crystals with KDP structure the anisotropy was studied by measuring Vickers hardness on different faces. In all other cases, the anisotropy was studied by measuring Knoop hardness with the long axis of the Knoop indenter set at different angles (ϕ) with reference to the <100> direction on the (100) plane.

Hardness Anisotropy vis-à-vis Point Group Symmetry

The variation of hardness with the orientation of the indenter on the (100) plane for several crystals has been studied [4.100, 4.138]. The variation is shown in Figs. 4.42 and 4.43. In the case of the three rubidium halides and PbS which have NaCl structure and in NH_4Cl which has the CsCl structure, the polar diagram shows a maximum at $\phi = 0°, 90°$ and 180° i.e. the feature repeats after 90°. Thus, the hardness shows a fourfold symmetry. The first four crystals as well as NH_4Cl have m3m point group in which the (100) face has a fourfold symmetry. On the other hand, in $NaClO_3$, $NaBrO_3$ and $Ba(NO_3)_2$, the maximum in hardness shows up at $\phi = 0°$ and 180°. Thus, the hardness anisotropy has twofold symmetry. $NaClO_3$ and $NaBrO_3$ have 23 point group and $Ba(NO_3)_2$ has m3 symmetry. In both these point groups, the (100) plane has two fold symmetry. An interesting observation was made by Kishan Rao and Sirdeshmukh [4.139] in the case of $NaBrO_3$. Measuring the Vickers hardness on the (111) and $(\bar{1}\bar{1}\bar{1})$ faces, they found that the hardness

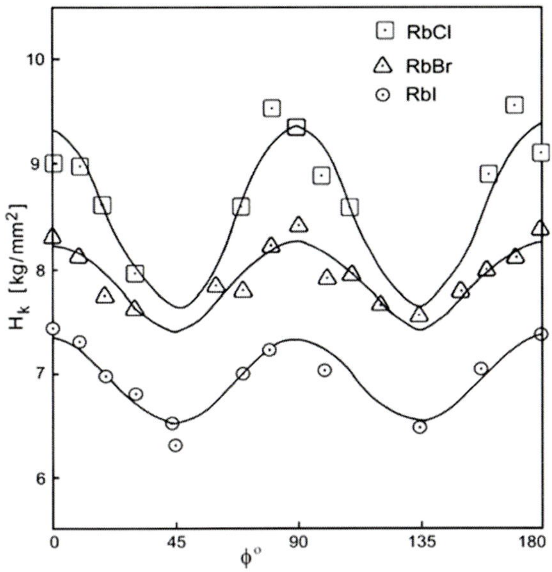

Fig. 4.42. Plot of Knoop hardness H_K against orientation ϕ

on these faces was different (137 kg mm^{-2} and 160 kg mm^{-2}, respectively). In this crystal which has 23 point group, there is no center of symmetry and the two faces are not equivalent. Thus, the anisotropy of hardness is consistent with the point group symmetry of the respective crystals.

Orientation Dependence of Hardness in NaCl Type Crystals

Thirmal Rao and Sirdeshmukh [4.100] measured the hardness of rubidium halides corresponding to different orientations of the Knoop indenter and found that the angular variation in hardness on the (100) plane of these crystals can be represented by:

$$H(\phi) = A + b\cos^2(2\phi), \tag{4.43}$$

where A and b are constants and ϕ is the angle between the direction of measurement (direction of long axis of Knoop indenter) and the <100> direction. The values of A and b for the rubidium halides are given in Table 4.23.

Hardness Anisotropy vis-à-vis Young's Modulus Anisotropy

In Table 4.24 the experimental data on the hardness anisotropy (H_{100}/H_{110}) are given for a number of cubic ionic crystals together with some related information. H_{100} and H_{110} are the hardness values measured on the (100) plane with the long axis of a Knoop indenter parallel to the <100> and <110>

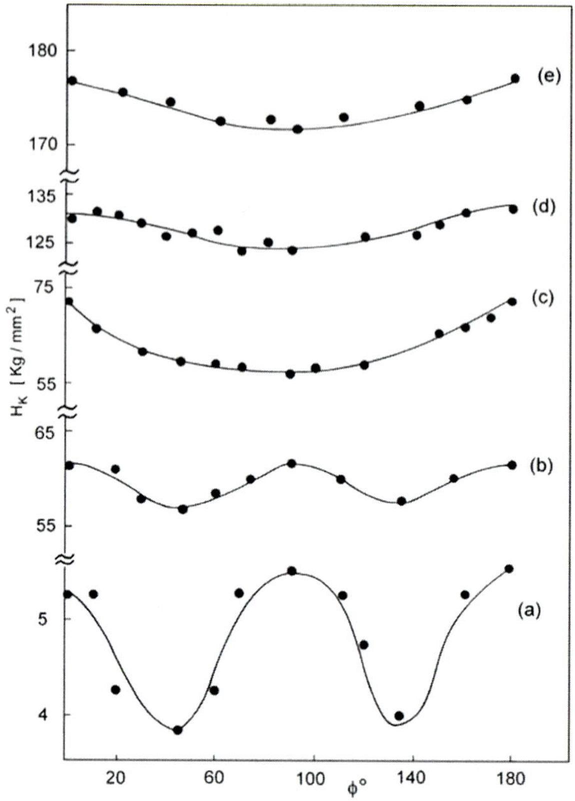

Fig. 4.43. Knoop hardness H_K as a function of indenter orientation ϕ on (100) faces of: (**a**) NH_4Cl, (**b**) PbS, (**c**) $Ba(NO_3)_2$, (**d**) $NaClO_3$ and (**e**) $NaBrO_3$

Table 4.23. Values of A and b (in 4.43) for the rubidium halides

Crystal	A[kg mm^{-2}]	b[kg mm^{-2}]
RbCl	7.6	1.7
RbBr	7.4	0.9
RbI	6.6	0.7

directions, respectively. The crystals include those studied by us as well as those studied by other workers. This appears to be all the information available on hardness anisotropy in cubic ionic crystals. For most of the crystals listed in the table H_{100}/H_{110} is >1. However, for four crystals, LiF, MgO, MnO and MnS, H_{100}/H_{110} is <1.

Brookes et al. [4.72] concluded that materials having the same crystal structure and common slip systems possess similar hardness anisotropy and, on this basis, Hannink et al. [4.73] proceeded to predict slip systems from the observed anisotropy in hardness. Chin et al. [4.75] pointed out that

Table 4.24. Hardness anisotropy ratio (H_{100}/H_{110}), Young's modulus anisotropy ratio E_{100}/E_{110} and related data for some cubic ionic crystals [4.90, 4.140]

S. No.	Crystal	Structure	Slip system	H_{100} [kg mm^{-2}]	H_{100}/H_{110}	E_{100}/E_{110}
1	NaCl	NaCl	$\{110\} <\bar{1}10>$	17.0	1.11	1.20
2	NaBr	NaCl	$\{110\} <\bar{1}10>$	–	1.17	1.28
3	KCl	NaCl	$\{110\} <\bar{1}10>$	9.5	1.24	1.90
4	KBr	NaCl	$\{110\} <\bar{1}10>$	8.1	1.28	2.05
5	KI	NaCl	$\{110\} <\bar{1}10>$	9.5	1.14	1.96
6	AgCl	NaCl	$\{110\} <\bar{1}10>$	6.3	1.23	1.63
7	MnS	NaCl	$\{110\} <\bar{1}10>$	122.0	0.86	–
8	LiF	NaCl	$\{110\} <\bar{1}10>$	96.0	0.93	0.78
9	MgO	NaCl	$\{110\} <\bar{1}10>$	400.0	0.51	0.80
10	MnO	NaCl	$\{110\} <\bar{1}10>$	252.0	0.88	0.77
11	CaF_2	CaF_2	$\{001\} <110>$	178.0	1.11	1.55
12	SrF_2	CaF_2	$\{001\} <110>$	154.0	1.10	1.36
13	BaF_2	CaF_2	$\{001\} <110>$	87.0	1.14	1.00
14	RbCl	NaCl	$\{001\} <\bar{1}10>$	9.3	1.23	2.20
15	RbBr	NaCl	$\{001\} <\bar{1}10>$	8.2	1.11	2.40
16	RbI	NaCl	$\{001\} <\bar{1}10>$	7.3	1.13	2.60
17	NH_4Cl	CsCl	$\{110\} <001>$	9.2	1.12	1.77

there are several crystals with different slip systems having similar hardness and anisotropy. This is also borne out from the data in Table 4.24. KCl and CaF_2 have different structures and different slip systems but similar hardness anisotropy. On the other hand, KCl and LiF have same structure and slip systems but different hardness anisotropy. Boyarskaya et al. [4.141] suggested that hardness anisotropy depends on the magnitude of average hardness. For soft crystals such as KCl and NaCl (average hardness ~10–20 kg mm^{-2}) H_{100}/H_{110} is >1, whereas for harder crystals such as LiF and MgO (average hardness ~100–800 kg mm^{-2}) H_{100}/H_{110} is <1. This suggestion does not hold in the case of the crystals with fluorite structure. The three crystals with this structure which are included in Table 4.24 have average hardness value comparable with LiF and MgO but have H_{100}/H_{110} is <1.

Several correlations between the average hardness and elastic properties of crystals have been suggested (Sect 4.3.10). While these relationships differ among themselves, a common feature amongst them is that they connect hardness with elastic properties of crystals. In view of this, it appears worthwhile to examine whether a correlation also exists between the anisotropy of hardness and the anisotropy of elastic modulus. For this purpose, the value of the Young's moduli in the <100> and <110> directions (E_{100} and E_{110}, respectively) and their ratio were calculated from literature data on single crystal elastic constants; the values of the ratio E_{100}/E_{110} are given in Table 4.24. As in the case of the ratio H_{100}/H_{110}, the ratio E_{100}/E_{110} also has values >1 for some crystals and <1 for others. A striking feature is that the crystals for which H_{100}/H_{110} is >1 have E_{100}/E_{110} >1 and vice versa. Thus, a

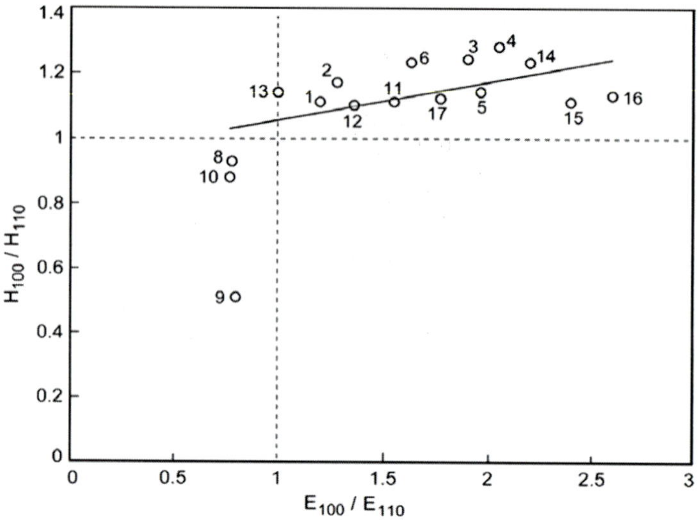

Fig. 4.44. Plot of (H_{100}/H_{110}) against (E_{100}/E_{110}) (numbers as in Table 4.24)

correlation exists between the two ratios. In order to examine further the nature of the correlation, the values of the two ratios are plotted in Fig. 4.44; a near-linear relationship between H_{100}/H_{110} and E_{100}/E_{110} is revealed. Only the data point for magnesium oxide deviates significantly from the linear plot although it satisfies the correlation qualitatively inasmuch as the values of H_{100}/H_{110} and E_{100}/E_{110} are both <1.

Thus, the anisotropy in Young's modulus seems to be an important factor in determining the hardness anisotropy of a crystal. It is interesting to note that the correlation between hardness anisotropy and elastic anisotropy is applicable to the entire group of crystals listed in Table 4.24, unlike the correlations with slip mode and magnitude of hardness proposed by earlier workers [4.72, 4.73]. In addition, the earlier correlations were purely qualitative whereas the present correlation is quantitative as a linear relationship is clearly indicated. However, the present correlation is limited to ionic cubic crystals; it does not, for instance, account for the hardness anisotropy in the refractory carbides which have the NaCl structure but a complex-bonding scheme.

Anisotropy of Hardness in Potassium Dihydrogen Phosphate and its Isomorphs

Potassium dihydrogen phosphate (KDP) is tetragonal. Kishan Rao and Sirdeshmukh [4.112] have measured the Vickers hardness on the (100) and (001) faces of KDP and six of its isomorphs at a load of 100 gm. The hardness on these faces and also the average hardness are given in Table 4.15. In all cases, the hardness in the a direction (100) is larger than in the c-direction (001) although the relative difference differs from crystal to crystal. A discussion of these results in terms of the bonding in these crystals is given in Sect. 4.4.2.

5

Dielectric and Electrical Properties of Solids

5.1 Introduction

The dielectric constant is a solid state property which is technically important and is also helpful in understanding basic crystal physics. Combined with other information like the refractive index and the absorption frequency it throws light on the bonding in crystals. In theoretical studies of lattice dynamics, the dielectric constant forms one of the input parameters.

Measurement of dielectric constant and loss as a function of frequency and temperature helps in understanding the polarization mechanism, process of conduction, influence of impurities and phase transition. AC conductivity obtained from the dielectric properties combined with the data on DC conductivity yields useful information on defect formation and nature of conduction.

Theory of dielectric behaviour of solids is discussed in literature [5.1–5.4]. However, for completeness, the basic aspects are discussed in brief.

5.1.1 Dielectric Polarization

Polarization is the result of ordering of the electrically charged particles under the action of external field. Macroscopically, it shows up as an increase in the capacitance of a condenser in the presence of a dielectric. Microscopically, it is described in terms of the induced electric moment.

The electric moment acquired by an atom or a molecule under the influence of an electric field is proportional to the applied field. The proportionality constant is known as the polarizability (α). It is the measure of the ability of the material to respond to the field. The polarization P is defined as the electric moment per unit volume of the dielectric and is given by

$$P = N\alpha E, \tag{5.1}$$

where N is the number of particles (atoms, ions or molecules) in unit volume and E the applied field.

In general, the electric moment may be induced by different mechanisms. In a most general case the total polarizability α may be expressed as

$$\alpha = \alpha_{\mathrm{e}} + \alpha_{\mathrm{a}} + \alpha_{\mathrm{d}} + \alpha_{\mathrm{s}}, \tag{5.2}$$

where α_{e}, the electronic polarizability, arises due to the deformation of the spherical distribution of negative charge around the nucleus. In addition to the deformation of the electronic cloud, the positively and negatively charged ions are displaced. This leads to the atomic polarizability α_{a}. If the material consists of permanent dipoles, they align along the direction of the field and an additional factor α_{d} will be added. The fourth component α_{s} is the polarizability arising out of the space charge.

The electronic polarization P_{e} is related to the high frequency dielectric constant through the well-known Clausius–Mossotti relation [5.5]:

$$P_{\mathrm{e}} = \frac{\varepsilon_{\infty} - 1}{\varepsilon_{\infty} + 2}\frac{M}{\rho} = \frac{4}{3}\pi N_{\mathrm{A}}\alpha_{\mathrm{e}}, \tag{5.3}$$

where, ε_{∞} is the high frequency (or optical) dielectric constant, M the molecular weight, ρ the density and N_{A} is the Avogadro number. The electronic polarization P_{e} is also known as molar refraction and is related to the refractive index n and α_{e} through the Lorentz–Lorenz formula

$$P_{\mathrm{e}} = \frac{n^2 - 1}{n^2 + 2}\frac{M}{\rho} = \frac{4}{3}\pi N_{\mathrm{A}}\alpha_{\mathrm{e}}. \tag{5.4}$$

For an ionic crystal

$$P_{\mathrm{e}} + P_{\mathrm{a}} = \frac{\varepsilon_{\mathrm{s}} - 1}{\varepsilon_{\mathrm{s}} + 2}\frac{M}{\rho} = \frac{4}{3}\pi N_{\mathrm{A}}(\alpha_{\mathrm{e}} + \alpha_{\mathrm{a}}), \tag{5.5}$$

where ε_{s} is the static dielectric constant. For a material with permanent dipoles, the modified Clausius-Mossotti–Debye equation relates the dielectric constant and the total polarization

$$P_{\mathrm{e}} + P_{\mathrm{a}} + P_{\mathrm{d}} = \frac{\varepsilon_{\mathrm{s}} - 1}{\varepsilon_{\mathrm{s}} + 2}\frac{M}{\rho} = \frac{4}{3}\pi N_{\mathrm{A}}(\alpha_{\mathrm{e}} + \alpha_{\mathrm{a}} + \alpha_{\mathrm{d}}). \tag{5.6}$$

For an ideal ionic crystal, (5.5) holds good. But in natural samples or crystals grown in the laboratory, defects and impurities are present. Very often they result in charges which are not locally bound to the crystal lattice. Under the influence of the field such charges are capable of migrating through some distance inside the dielectric and produce distortion of the field. In addition, trapped electrons and holes, defects like grain boundaries, voids and dislocations too lead to accumulation of charges on the surface of the material and cause space charge polarization which is observed as additional polarizability (α_{s}). The frequency variation of the total polarizability (α) is shown in Fig. 5.1. The space charge polarization results in the enhancement of the capacitance and hence an increase in the measured dielectric constant.

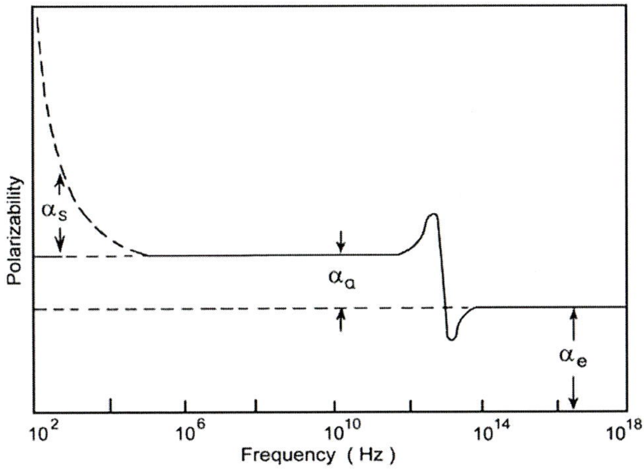

Fig. 5.1. Variation of polarizability with frequency in an ionic crystal

5.1.2 Dielectric Dispersion and Dielectric Loss

The variation of the dielectric constant with frequency of an ionic crystal is similar to the variation of polarizability and polarization. At low frequencies of the order of a few Hz, the dielectric constant is made up of contributions from electronic, atomic and space charge polarization. When measurements are carried out as a function of frequency, the space charge polarization ceases after a certain frequency and the dielectric constant becomes frequency independent. The frequency beyond which the variation ceases may fall in the range of a few kHz to MHz [5.6]. The frequency-independent value is taken as the true static dielectric constant.

Hence, generally, the dielectric constant is measured as a function of frequency to obtain the true static dielectric constant. The dielectric constant measured in the frequency independent region is taken as static or low frequency dielectric constant ε_s (sometimes referred to as infrared dielectric constant ε_{ir}). As the frequency is increased further, the value remains unaffected till the strong resonance absorption frequency is approached in the infrared region. Beyond the resonant frequency, since the ions cannot follow the field, the polarization due to electronic contribution alone persists. Hence the dielectric constant in this region is termed 'optical' (ε_{opt}) or high frequency dielectric constant (ε_∞).

Under the influence of static field, the dielectric constant is treated as a real number. The system is assumed to get polarized instantaneously on the application of the field. When the dielectric is subjected to alternating field, the displacement cannot follow the field due to inertial effects and spatially-oriented defects. The dielectric constant is then treated as a complex quantity $\bar{\varepsilon}(\omega)$. The variation of real and imaginary parts of the complex dielectric constant with frequency ω is given by the Debye equations [5.7]

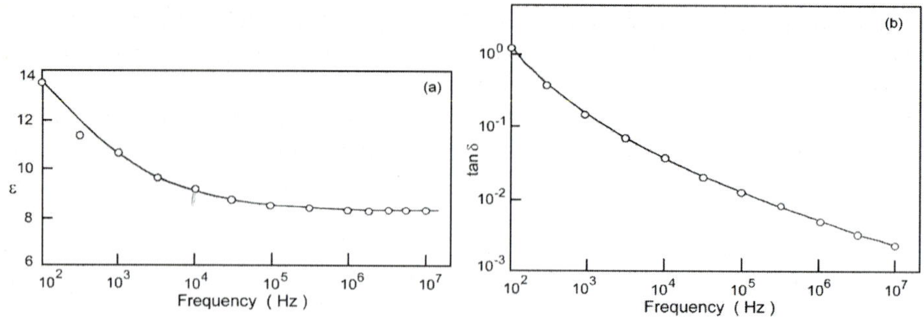

Fig. 5.2. Variation of (**a**) ε and (**b**) $\tan\delta$ with frequency for an ionic crystal

$$\bar{\varepsilon}(\omega) = \varepsilon'(\omega) - \mathrm{i}\varepsilon''(\omega), \tag{5.7}$$

where

$$\varepsilon'(\omega) = \varepsilon_\infty + \frac{\varepsilon_\mathrm{s} - \varepsilon_\infty}{1 + \omega^2\tau^2} \tag{5.8}$$

and

$$\varepsilon''(\omega) = (\varepsilon_\mathrm{s} - \varepsilon_\infty)\frac{\omega\tau}{1 + \omega^2\tau^2}, \tag{5.9}$$

where τ is the relaxation time. ε' is identified with the measured dielectric constant ε and ε'' is a measure of the average power loss in the system. Very often, the loss is expressed in terms of the phase angle δ as

$$\tan\delta = \varepsilon''/\varepsilon'. \tag{5.10}$$

The frequency variation of dielectric constant and loss for a typical ionic crystal is shown in Fig. 5.2.

5.1.3 Dielectric Loss and Conduction

The loss ($\tan\delta$) in general consists of two contributions; one, due to conduction and the other due to relaxation effects. The loss due to conduction of free carriers is expressed in terms of conductivity σ [5.5]:

$$\tan\delta = 4\pi\sigma/\omega\varepsilon'. \tag{5.11}$$

A plot of log ($\tan\delta$) against log ω will give a straight line. If ε'' (i.e. $\varepsilon'\tan\delta$) is proportional to inverse frequency, the conduction is frequency independent and is equivalent to DC energy loss.

The dipolar impurities or dipoles created by defects show relaxation effects. The loss due to relaxation effects is obtained by combining (5.8) and (5.9) as

$$\tan\delta = \varepsilon''(\omega)/\varepsilon'(\omega) = \frac{(\varepsilon_\mathrm{s} - \varepsilon_\infty)\omega\tau}{\varepsilon_\mathrm{s} + \varepsilon_\infty\omega^2\tau^2}. \tag{5.12}$$

The frequency dependence of the loss due to relaxation effects differs from that of conduction loss. Unlike the conduction loss, which shows a linear log ($\tan\delta$) versus frequency plot, the loss due to relaxation effects shows a maximum at a certain frequency. The net effect on the appearance of the curve when both the effects are present will be a peak superimposed on a straight line. The departure from a straight line depends on the features of the peak like breadth, sharpness, etc.

The total contribution to dielectric loss ($\tan\delta$) is the sum of the conduction loss given by (5.11) due to free vacancies and the dipolar Debye loss (5.12). Thus

$$\tan\delta = \frac{4\pi\sigma}{\omega\varepsilon'} + \frac{(\varepsilon_{\mathrm{s}} - \varepsilon_{\infty})\omega\tau}{\varepsilon_{\mathrm{s}} + \varepsilon_{\infty}\omega^2\tau^2}. \tag{5.13}$$

At normal temperatures and low frequencies up to a few kHz the second term is negligible. The first term becomes predominant and conductivity may be obtained from the dielectric loss from the equation

$$\sigma = \varepsilon_0\, \varepsilon'\omega \tan\delta, \tag{5.14}$$

where ε_0 is the vacuum dielectric constant. Unlike the conduction loss, the Debye loss is not explicitly defined. The loss may be due to rotating polar entities, defects such as impurity-vacancy, impurity-interstitial pairs, electrode polarization or even due to the presence of air gaps [5.8]. It may be seen from (5.14) that the conductivity can be estimated using data on dielectric constant and loss.

5.1.4 Temperature Variation of Dielectric Constant and Loss

Ionic crystals with high purity show a frequency independent slow temperature variation of dielectric constant up to a certain temperature and a frequency dependent larger variation at higher temperatures (Fig. 5.3). The temperature beyond which the second region commences depends on the individual sample. The variation in the first region is attributed to crystal expansion and the electronic and ionic polarization. At higher temperatures the increase is mainly attributed to the thermally generated charge carriers and impurity dipoles. The data in the first region can be fitted to a linear equation, while the second requires higher order terms.

Depending on the purity of the sample the loss may be negligible up to a certain temperature. If loss is present, a frequency dependent variation of loss is observed as in the case of dielectric constant. The variation for a typical ionic crystal is shown in Fig. 5.4. The smooth variation in $\tan\delta$ is the result of conduction loss. The presence of Debye type loss shows a peak at a certain temperature superimposed on the plot due to conduction loss. Further, a shift in the peak is observed on change of frequency. From the magnitude of the shift the relaxation frequency is obtained. The variation of loss ($\tan\delta$) when both types of losses are present is shown in Fig. 5.5 for a typical case.

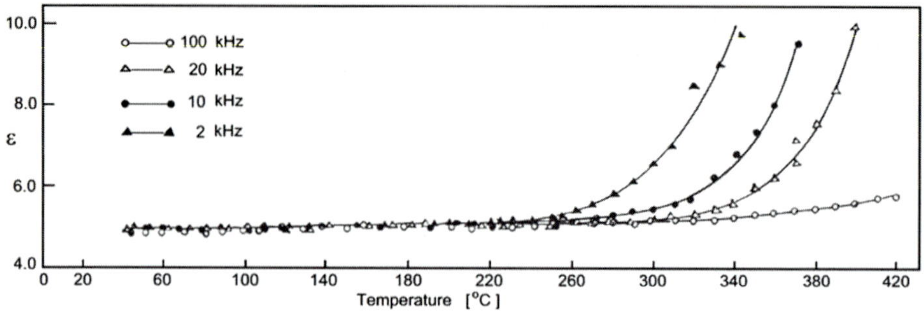

Fig. 5.3. Variation of ε with temperature at different frequencies for a typical ionic crystal (KBr)

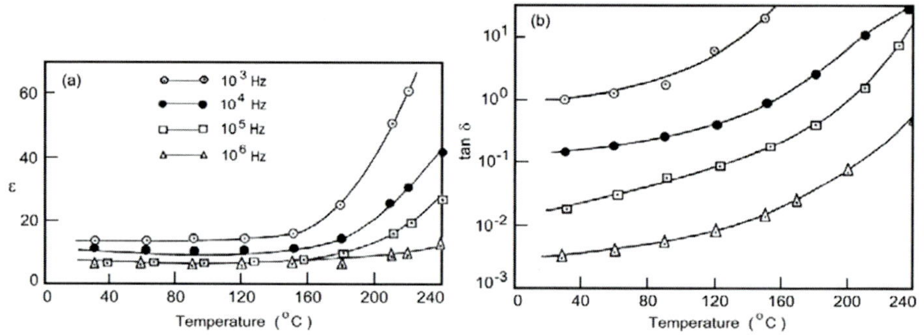

Fig. 5.4. Variation of (**a**) ε and (**b**) tan δ with temperature at different frequencies for a crystal with conduction loss ($NaIO_4$)

5.2 Experimental

5.2.1 Measuring Instruments

TF 329G and TF 1245 Marconi Circuit Magnification Meter

These instruments are based on the principle of resonance circuit [5.2, 5.9]. The capacitor with the dielectric medium is made part of an LC circuit which is loosely coupled to an oscillator. The circuit is made to resonate by means of a variable capacitor with and without the sample. The resonance is detected by noting the changes in the voltage developed across the test circuit. The dielectric loss is obtained from the width of the resonance curve for the empty cell and the cell filled with the sample. By changing the inductors, the desired frequency is obtained.

GR 716C Capacitance Bridge

This is a modified form of conventional AC Scherring bridge. The capacitance and loss ($\tan \delta$) are directly measured in the frequency range 10^2–10^7 Hz.

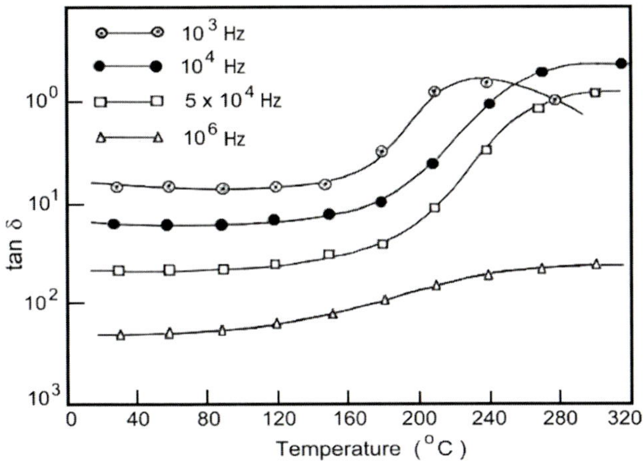

Fig. 5.5. Variation of $\tan\delta$ with temperature at different frequencies for a crystal with Debye-type loss (Sm-doped EuF_2)

GR 1620A Capacitance Measuring Assembly

The GR 1620A capacitance measuring assembly consists of an oscillator (1311A), a transformer ratio arms capacitance bridge (1615A) and a tuned amplifier and null detector (1232A). Since the inbuilt oscillator is only up to 10 kHz, an external oscillator has to be used for frequencies beyond 10 kHz.

The dielectric constant is determined from the measured value of the capacitance with and without the sample. A schematic diagram of the basic capacitance bridge and the modified bridge incorporating the transformer ratio arms is shown in Fig. 5.6. Use of transformer ratio arms increases the accuracy. The null balance can be obtained either in terms of a dissipation factor (for low loss materials) or conductance. The capacitance may be measured from 10^{-5} pF to a maximum of 1 μF. The dielectric loss ($\tan\delta$) is measured in terms of the dissipation factor up to a value of 0.01. For higher ranges, measurements are done in terms of conductance. The balance controls are operated with lever type switches. The read-out is digital and decimal points are automatically positioned.

The bridge may be used in two-terminal configuration or three-terminal configuration. The principle of three-terminal measurement is shown in Fig. 5.7a. The direct capacitance between H and L represents the capacitance to be measured and G is the ground. In the two-terminal configuration G and L are connected together to the ground. For three-terminal measurement G is connected to a shield which completely surrounds the terminals H and L and their connecting wires as shown in Fig. 5.7b. The changes in the environment and losses in the external part of the circuit can change the terminal capacitance C_{HG} and C_{LG} but the direct capacitance C_{HL} is determined only by the internal geometry. By such an arrangement the direct capacitance as low as a fraction of a pF can be measured.

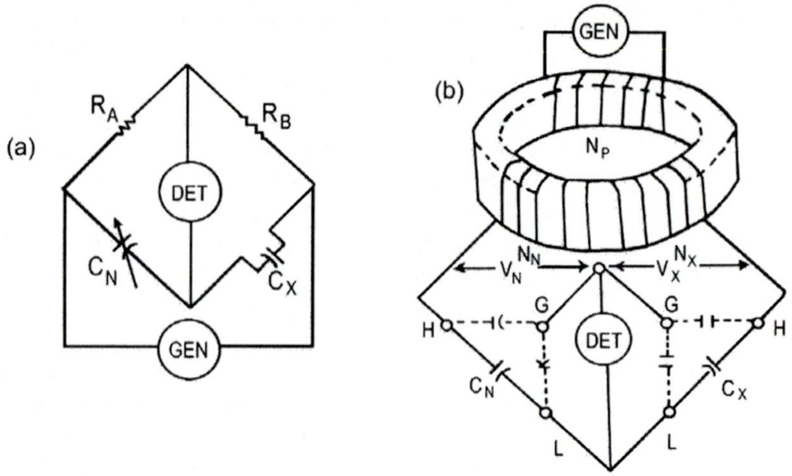

Fig. 5.6. (**a**) Basic ratio bridge [R_A: variable ratio arm; C_N: calibrated variable standard capacitor; C_X: unknown capacitance]; (**b**) Capacitance bridge with transformer ratio arms

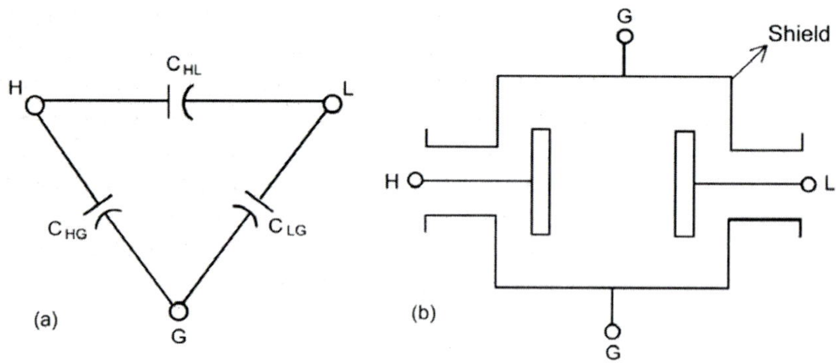

Fig. 5.7. Three terminal capacitor: (**a**) principle and (**b**) shielded cell

Others

For measurements at higher frequencies >1 MHz and low temperatures, the following instruments are used:

(a) HP 4275A multifrequency LCR meter
(b) CGA 83 transformer ratio arm three terminal bridge
(c) HP 4192A impedance analyzer
(d) Keithley 610C electrometer is commonly employed for DC conductivity measurements

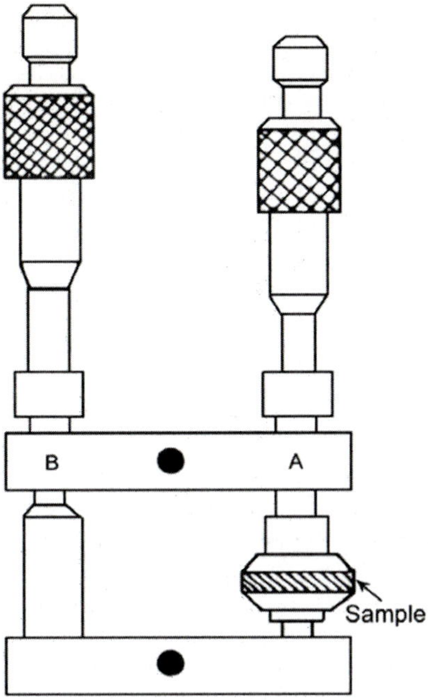

Fig. 5.8. Dielectric test jig TJ155C/1

5.2.2 Cell Designs

Test Jig TJ 155C/I

It has a built-in micrometer arrangement with two linear capacitors in parallel as shown in Fig. 5.8. One of the micrometers, which is a linear low capacitor, is used for band-width determination. The other is a plate capacitor of 1 in. diameter to hold the specimen under investigation. This test jig supplied with Marconi magnification meter can be used only at room temperature.

Sample Holders used with GR 716C Capacitance Bridge

The sample holder for high temperature measurement (Fig. 5.9) consists of a brass hollow cylinder. The flattened end of the cylinder serves as one of the electrodes. The second electrode is fixed to the brass rod held by a spring. The other components are shown in the figure with description. The current variation is done using a variac and the thermocouple close to the electrode registers the temperature of the sample.

For measurements at low temperatures, the cell shown in Fig. 5.10 can be used. The cell with the sample in place is inserted into a liquid nitrogen bath (−180°C). The sample temperature could be varied from −180°C to room temperature by raising the cell upward.

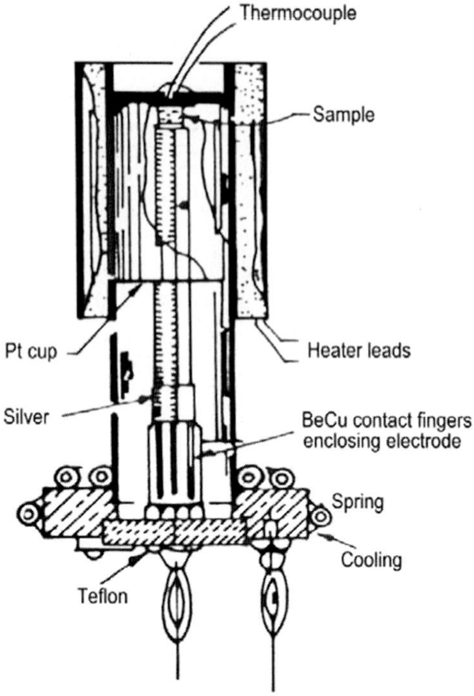

Fig. 5.9. Sample holder for high temperature measurements

Laboratory Fabricated Cell for use with GR 1620 Capacitance Assembly

The cell fabricated for use up to 400°C is shown in Fig. 5.11. The sample holder has a lower electrode (E_1) which is a brass disc of 2 in. diameter fixed on the top of a ceramic rod (C.R) which is held tightly in a brass holder (B). The upper electrode (E_2) which is a cylindrical brass rod of diameter 1.5 cm is movable and is spring-loaded. The movement of the upper electrode is controlled by the micrometer screw. The assembly with the upper electrode is fixed on to an ebonite rod (Eb.R). The spacing between the upper and lower electrodes can be varied with the upper assembly sliding along the ebonite rod. Fine adjustments with the sample are made with the help of a micrometer screw. The temperature of the sample can be varied with the help of a heater surrounding the two electrodes. The heater is constructed with super kanthal wire wound over a muffle to withstand temperatures up to ≈500°C. The heater is placed in a metallic enclosure (Sh) held in position with the help of pillars (P). The heater with the metallic enclosure is capable of sliding up and down as required. The heater is fed through a stabilizer and the temperature is steady within 1°C. The temperature is measured using a copper constantan thermocouple and a digital panel meter. Well-shielded coaxial leads are used

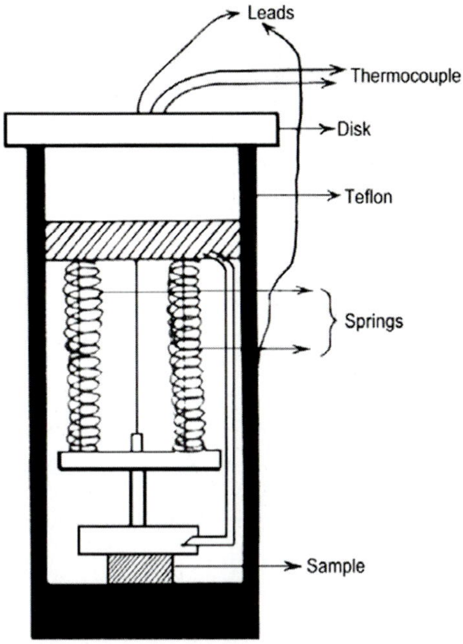

Fig. 5.10. Sample holder for low temperature measurements

to connect the two electrodes to the H and L terminals of the bridge. The metal enclosure acts as a shield to form the ground terminal G of the bridge. This arrangement completely eliminates external effects.

The special features of the cell are the replaceable arrangement for the two electrodes and the moveable heater assembly. The sample-in and sample-out readings are taken with the rigid electrode at room temperature and for high temperature measurements, the rigid electrode is replaced by an electrode with a light spring to allow for sample expansion. The movable heater assembly serves a dual purpose. First, it allows the sample to be properly placed by lowering the heater assembly and second, when lifted to cover the sample, it serves as a shield to form the third terminal.

High Temperature Cell for Measurement of Dielectric Constant Above 400°C

Various parts of the cell designed for use above 400°C are indicated in Fig. 5.12. As shown, the sample holder is placed in a stainless steel cylinder C. The circular disc with the sample holder is fixed firmly at the top of the cylinder with the help of 'O' ring R. The cell is provided with water circulation facility at the top, so that the connectors and the Teflon bushes do not get heated. Valve (V) is provided to allow for evacuation or to fill with inert gas. Well-shielded co-axial leads are used for connecting the two electrodes

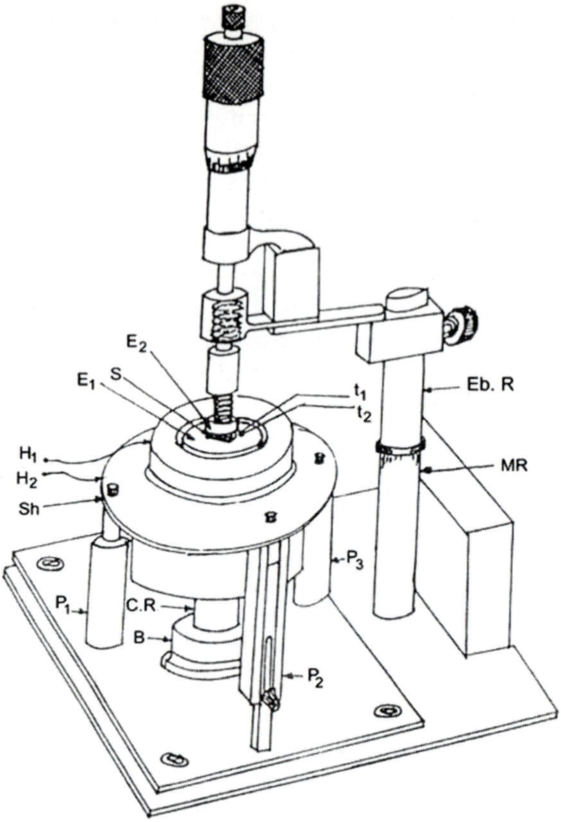

Fig. 5.11. Sample holder for high temperature dielectric measurements (E_1 – lower electrode; E_2 – upper electrode; S – sample; t_1, t_2 – thermocouple terminals; H_1, H_2 – heater terminals; Sh. – shield; P_1, P_2, P_3 – pillars; C.R. – ceramic rod; B – brass holder; Eb.R – ebonite rod; M.R – metallic rod)

to the instrument. All the coaxial shields and the outer cylinder which also acts as a shield are connected to the ground terminal of the three terminal configurations. The temperature of the sample is varied by keeping the cell in a furnace. The furnace is designed to provide automatic preset temperatures up to 800°C. The cell is also useful for DC conductivity measurements.

5.2.3 Procedural Details

General

The procedure for dielectric constant and dielectric loss measurement is described in this section. The procedure may appear specific to our work but is of general relevance. As a check on the experimental set-up, the cell capacitance was initially determined as a function of frequency and temperature. The capacitance was found to be less than 1 pF throughout the range of measurements. This indicates a good shielding of the sample holder and low loss in the insulating materials used in the cell.

For good electrical contact and to avoid air gaps between the surface of the sample and the electrodes, several materials were tested such as carbon, graphite, silver paint, aluminium foil and aluminium foil fastened on silver paint. A pure KCl crystal supplied by Harshaw Chemical Company was used as a standard. Measurements were carried out as a function of temperature at 100 kHz. For the electrode combination of silver paint and aluminium foil, the dielectric constant was found to show identical values both for heating and cooling runs. The loss was found to be minimum when compared to other electrodes. The values were found to be consistent and reproducible. Hence, all further measurements were done with silver paint and aluminium foil.

For measurements using the sample holder shown in Fig. 5.12, the capacitance with and without the sample are obtained at room temperature initially.

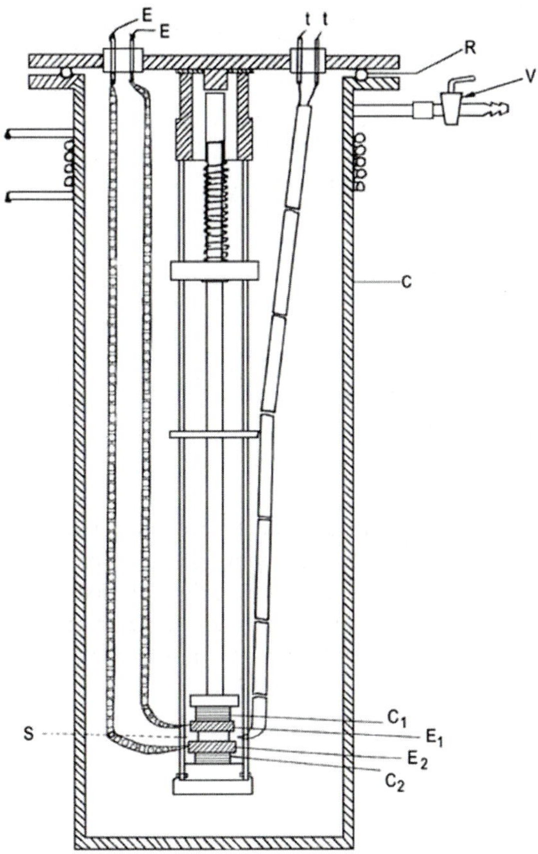

Fig. 5.12. Cell for high temperature dielectric measurements above 400°C (S–sample; E_1, E_2–the electrodes; C_1, C_2 – ceramic blocks; t_1, t_2–thermocouple connectors; E_1, E_2–electrode connectors; C–stainless steel cylindrical enclosure; R – 'O' ring; V – valve)

The upper rigid electrode is then replaced by the spring electrode for temperature variation measurements. The capacitance is adjusted with the help of the micrometer screw of the upper electrode to read the same value as with the rigid electrode. The heater is now brought into position and the capacitance and loss ($\tan\delta$) are measured as a function of frequency and temperature.

Single Crystals

For alkali halides, plates with parallel cleavage planes were used. Other crystals were cut along the required direction to obtain samples of 1–2 mm thickness. They were polished with emery powder or used as supplied if the samples were already polished. The area of the samples was generally less than that of the electrodes. Accurate measurement of area was required to ensure an accurate value of dielectric constant. For samples of irregular shape the area was obtained by graphical method. Alternatively, the area was determined from the mass of the sample obtained with a single pan balance with an accuracy of 10^{-6} g combined with the density and thickness. The samples were annealed at 100°C for a couple of hours before taking measurements.

The dielectric constant is calculated using the formula

$$\varepsilon = 1 + \frac{\Delta C}{C_{\mathrm{g}}}, \tag{5.15}$$

where ΔC is the difference of the capacitance with the sample (C_{s}) and the capacitance without the sample (C_0). C_{g} is the geometrical capacitance of the sample. The geometrical capacitance is $\varepsilon_0(A/t)$; ε_0 is the vacuum dielectric constant, A the area and t the thickness of the sample.

Powder Samples

Powders of required particle size were obtained by using a mortar and were sieved through a mesh. The powders were dried for a few hours at 100°C to remove moisture. Die-pressed samples were prepared using pressures of the order of 20,000 lbs in^{-1}. The densities of the pellets were determined from the mass obtained using a balance with an accuracy of 10^{-6} g. Pellets of diameter equal to that of Marconi test jig and thickness 1–2 mm were used. For smaller quantities of samples, pellets were obtained with a suitable die with diameter less than that of the electrode. The dielectric constant is obtained by using a suitable relation in the two cases.

Several correlation equations are available in literature to obtain dielectric constant ε of the solid having the crystal density from the value measured with powder samples [5.10–5.13]. Two such equations are:

$$\delta_{\mathrm{p}} = \frac{\varepsilon_{\mathrm{pr}}^{1/3} - 1}{\varepsilon^{1/3} - 1}, \tag{5.16}$$

$$\varepsilon = \frac{1}{\delta_{\mathrm{P}}{}^2}[(\varepsilon_{\mathrm{pr}} - b\}^{1/2} - (1 - \delta_{\mathrm{P}})(1 - b)^{1/2}]^2 + b, \tag{5.17}$$

where δ_{p} is the packing fraction, $\varepsilon_{\mathrm{pr}}$ the dielectric constant of the powder sample and 'b' a constant with a value of 0.5. Equation (5.16) by Looyenga [5.12] is found suitable for materials of dielectric constant <12 and (5.17) by Dube and Prasad [5.13] for those with dielectric constant >12.

Accuracy of Measurement

A wide range of capacitance can be measured with GR 1620 bridge from a fraction of a pF to a maximum of 1 μF. The loss component is read as dissipation factor up to a maximum of 0.01. For higher values, measurements are done in terms of the conductance. The frequency can be varied continuously from 50 Hz to 10 kHz. Above this, the frequency of measurement is fixed at 20, 50 and 100 kHz using external oscillator. Accuracy in the measurement of capacitance is 10^{-5} pF and in loss 10^{-4}. The overall accuracy in the measurement of dielectric constant and tan δ is of the order of 1% and 3%, respectively.

5.2.4 Measurement in the Microwave Region

For measurements in the microwave region, resonant cavity methods are used in the frequency range 10^8–10^9 Hz [5.9] while transmission line methods are employed at frequencies 10^9–10^{11} Hz [5.2]. Special oscillators, detectors and frequency-measuring cavity wave-meters are employed. A typical experimental set-up is shown in Fig. 5.13. The measurement of the dielectric constant consists essentially of the measurement of wavelength of the microwave radiation in the waveguide with and without the sample. Chandra [5.14] used this set-up to measure the dielectric constants of several alkali halides up to temperatures close to the melting point. This technique is particularly suited for studying samples which cannot be studied by the conventional AC methods because of high conduction at the lower frequencies.

5.2.5 Dielectric Constants from IR Reflectivity

The IR reflectivity curves, generally recorded over the 100–1,000 cm^{-1} range, provide a useful method for determination of the dielectric constants. The working equations involved in this method [5.15] are:

$$\varepsilon' = n^2 - k^2 = \varepsilon_\infty + \sum_j \frac{4\pi\rho_j \nu_j^2(\nu_j^2 - \nu^2)}{(\nu_j^2 - \nu^2)^2 + (\gamma_j \nu_j \nu)^2}, \tag{5.18}$$

$$\varepsilon'' = 2nk = \sum_j \frac{4\pi\rho_j \nu_j^2(\gamma_j \nu_j \nu)^2}{(\nu_j^2 - \nu^2)^2 + (\gamma_j \nu_j \nu)^2} \tag{5.19}$$

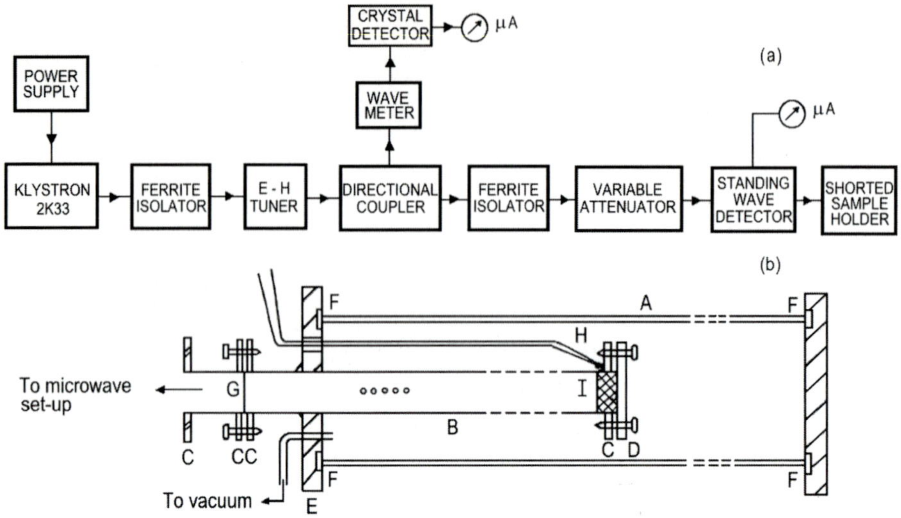

Fig. 5.13. (**a**) Apparatus for microwave dielectric measurements; (**b**) waveguide sample holder: (A) Vycor tube placed in a three zone furnace, (B) stainless steel K-band waveguide, (C) waveguide flanges, (D) stainless steel short, (E) end seals, (F) gasket, (G) mica window, (H) Pt–Pt 10% Rh thermocouple, (I) sample

and

$$R = [(n-1)^2 + k^2]/[(n+1)^2 + k^2], \tag{5.20}$$

where n is the refractive index, k the extinction coefficient; ν_j, γ_j and $4\pi\rho_j$ are, respectively, the resonant frequency, the damping factor and strength of the jth resonance. The procedure is to assume a set of values for ν_j, γ_j, ρ_j and evaluate ε' and ε'' for a given value of ν. From these values of ε' and ε'', the values of n, k and the reflectivity R are calculated. The best fit is obtained by minimizing the quantity:

$$S = \sum_i [(R_i)_{\text{expt}} - (R_i)_{\text{calc}}]^2 \tag{5.21}$$

With the best values of these parameters, ε_∞ and $\varepsilon_s (= \varepsilon')$ are calculated from (5.18, 5.19). As an example, the analysis of the reflectivity curve for MnO is shown in Fig. 5.14. Plendl et al. [5.16] assumed two resonances for MnO and obtained the values: $\nu_1 = 262\,\text{cm}^{-1}, \nu_2 = 445\,\text{cm}^{-1}, 4\pi\rho_1 = 17.5, 4\pi\rho_2 = 0.05, \gamma_1 = 0.092, \gamma_2 = 0.088, \varepsilon_\infty = 4.95$ and $\varepsilon_s = 22.5$.

This is a versatile method which yields not only the values of ε_∞ and ε_s but also the resonance frequencies. It is useful in handling materials like semiconductors which cannot be studied otherwise due to the conduction effects. Thus, the dielectric constants of CdO and mixed crystals GaSb–InSb were determined by this method by Finkenrath and Uhle [5.17] and Brodsky et al. [5.18], respectively.

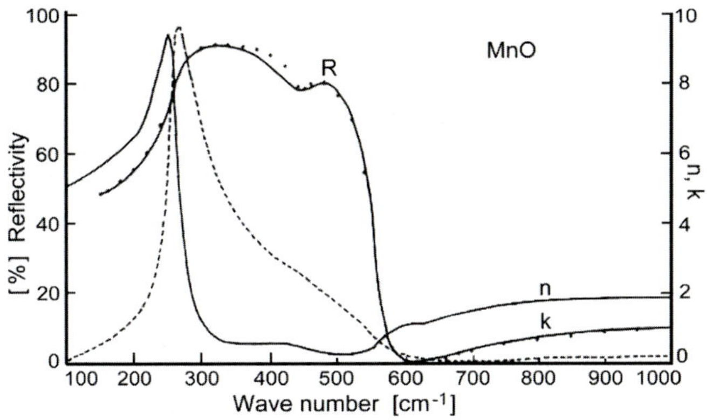

Fig. 5.14. IR reflectivity curve of MnO. *Dots* are calculated points using a two-resonance damped oscillator

The analysis of IR reflectivity curves discussed above is based on the classical damped oscillator model. Alternatively, the analysis can also be carried out by the Kramers–Kronig method [5.19].

5.2.6 Impedance Spectroscopy

The Cole–Cole plot [5.20] of ε' and ε'' for a lossy dielectric with a single relaxation time is a semi-circle with a radius $(\varepsilon_s - \varepsilon_\infty)/2$ and with the centre at $(\varepsilon_s - \varepsilon_\infty)/2$. The data points are obtained over a wide range of frequencies.

Complex impedance analysis is a powerful method to analyse multiple relaxation processes particularly in polycrystalline materials. The complex impedance is described by the Cole–Cole diagram at different temperatures to obtain relaxation times arising due to different mechanisms.

In recent years sophisticated automatic impedance measuring instruments are available. They are versatile and allow the measurements of several parameters like impedance, admittance, phase angle, capacitance and dissipation factor. They have a frequency range of 1 Hz to 20 MHz. Hewlett-Packard HP 4192A analyzer covers the range from 5 Hz to 13 MHz.

5.2.7 Comparison of Methods

Several methods for the determination of dielectric constant have been discussed in Sect. 5.2.1–5.2.6. The results obtained for KCl and KBr by these methods are given in Table 5.1.

The results given in [5.21–5.26] are consistent among themselves within 2%. The technique and results of Fontanella et al. [5.26] are by far the best (accuracy ~0.1%). The IR reflectivity method, though useful in some situations, gives less reliable values. Thus the value for KCl obtained from classical oscillator analysis is the least in the set and that for KBr is the largest. More

Table 5.1. Static dielectric constant (ε_s) of KCl and KBr

Method	Frequency	ε_s		Accuracy	Ref.
		KCl	KBr		
Immersion	1 MHz	4.68	4.78	–	[5.21]
Microwave	23 GHz	4.85	4.89	–	[5.22]
Scherring bridge		4.68	4.75	–	[5.23]
3-Terminal		4.72	4.76	1 %	[5.24]
Immersion		4.80	4.87	0.5%	[5.25]
3-Terminal		4.8126	4.8762	0.1%	[5.26]
IR reflectivity					
Classical oscillator		4.6	4.9		[5.27]
Kramers—Kronig		5.3	6.5		[5.27]

seriously, the values obtained from Kramers–Kronig analysis are clearly out of range.

5.3 An Overview

There is extensive literature on the studies of various aspects of dielectric behaviour of solids. This overview is thus bound to be limited. Further, in this overview ferroelectric materials are not included as it is a field by itself.

A good treatment of the basics of dielectric behaviour is given in [5.2, 5.4, 5.5, 5.28–5.30]. Compilations of dielectric data on inorganic materials are given in [5.31, 5.32]. Compilations of dielectric properties of biological systems are made by Stefan [5.33] and Ahmed et al. [5.34]. A review by Pethig [5.35] deals with protein-water interactions.

5.3.1 Some Important Experimental Results

While reference has been made to data compilations, mention is made here of some significant contributions. Being the simplest of ionic crystals, the alkali halides have received much attention. Lowndes and Martin [5.36, 5.37] made measurements of the static dielectric constants as well as the optical dielectric constants of several alkali halides at very low temperatures. Rao [5.38] and Chandra [5.14, 5.39] made measurements of the dielectric constants of NaCl, the potassium halides and the rubidium halides at high temperatures up to the melting point. The temperature variation of the static dielectric constants of rubidium halides [5.39] is shown in Fig. 5.15. The pressure variation of dielectric constant of alkali halides was studied by Jones [5.24], Lowndes and Martin [5.37] and Fontanella et al. [5.26] who gave the first as well as second pressure derivatives of dielectric constant. The pressure variation of dielectric constants of NaCl and RbCl was studied by Samara [5.40, 5.41]. His results

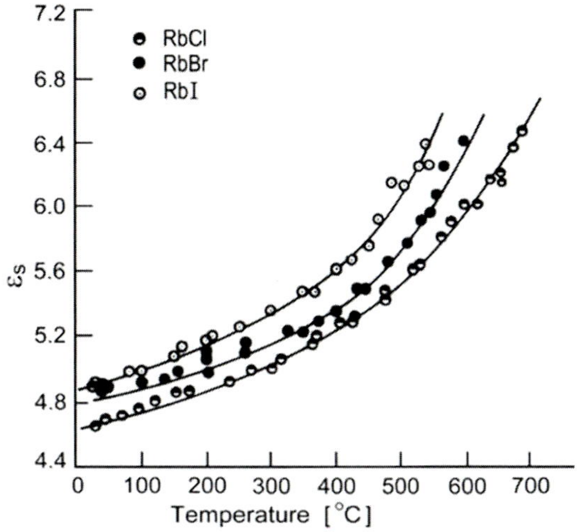

Fig. 5.15. Static dielectric constant (ε_s) for rubidium halides as a function of temperature

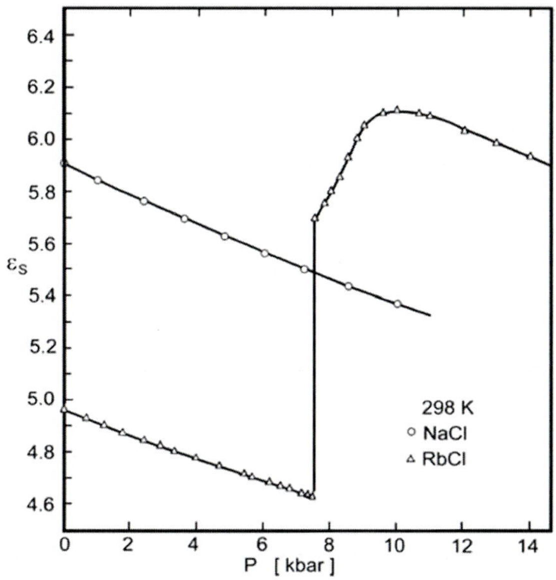

Fig. 5.16. Pressure dependence of the static dielectric constants of NaCl and RbCl. Changes in ε_s associated with the pressure-induced NaCl–CsCl phase transition in RbCl are shown

are shown in Fig. 5.16. In the case of RbCl the high pressure transition is observed at 7.6 kbar. An integrated data compilation on dielectric properties of alkali halides is given by Sirdeshmukh et al. [5.42].

The temperature and pressure dependence of dielectric constants of thallium halides was studied by Samara [5.43]. The temperature variation of the dielectric properties of the transition metal oxides has been reported in [5.44, 5.45].

A method for the accurate measurement of dielectric constant by substitution was developed by Andeen et al. [5.46] and was used by them to determine the dielectric constants of the alkaline earth fluorides [5.47]. The pressure and temperature dependence of the dielecteric constants of the alkaline earth fluorides was studied by Andeen et al. [5.48], of CdF_2 by Young and Frederikse [5.49] and of PbF_2 by Samara [5.50]. The work on dielectric behaviour of fluorite type crystals has been reviewed by Hayes [5.51].

The dielectric constants at room temperature for several alums were reported by Haussuhl [5.52]. Rama Rao et al. [5.53] reported the temperature variation of the dielectric constants of six alums. Dielectric constants of cubic nitrates were determined by Badr and Kamel [5.54]. Conductivity studies on the nitrates were made by Badr and Kamel [5.55] and El-Kabbany et al. [5.56].

The rare earth garnets are important optoelectronic materials. The dielectric and electrical properties of some rare earth garnets were reported in [5.57–5.60]. The dielectric constants of several rare earth garnets were measured at room temperature by Shannon et al. [5.61].

The low temperature dielectric constants of some anisotropic crystals were studied by Shelby et al. [5.62]. The anisotropic crystals included MgF_2 which is tetragonal with the rutile structure. Other rutile type crystals for which dielectric constants have been reported are GeO_2 [5.15] and TiO_2 [5.63]. The dielectric constants of several tetragonal crystals with scheelite structure have been reported by Fang and Brower [5.64] and Brower and Fang [5.65, 5.66].

Lal et al. [5.67] reported the temperature variation of dielectric constant for $EuWO_4$. They observed a sharp increase in dielectric constant which was attributed to formation of large polarons. The dielectric constant of several rare earth sesquioxides have been reported by Lal et al. [5.68].

The dielectric properties of hexagonal PbI_2 were investigated at different frequencies and temperatures by Dugan and Henisch [5.69]. Fernandez and Srivastava [5.70] studied the dielectric loss spectra of several crystals of CdI_2 and correlated their observations with polytypic structures.

Although it was stated that ferroelectric crystals will be excluded from this overview, one aspect will be mentioned as it involves a novel experimental study. This is the particle size dependence of dielectric behaviour. The changes in dielectric behaviour with change of particle size have been studied by Mansingh and Bawa [5.71] for KDP and Rochelle salt powders and by Rao and Bhanumati [5.72] for $RbHSO_4$. Figure 5.17 shows results of Mansingh and Bawa [5.71] on KDP. The dielectric constant decreases with particle size but there is no change in the transition temperature. The most dramatic observation, however, was made by Goswami [5.73] who observed that ferroelectricity in $BaTiO_3$ vanishes at a particle size of 100 μm.

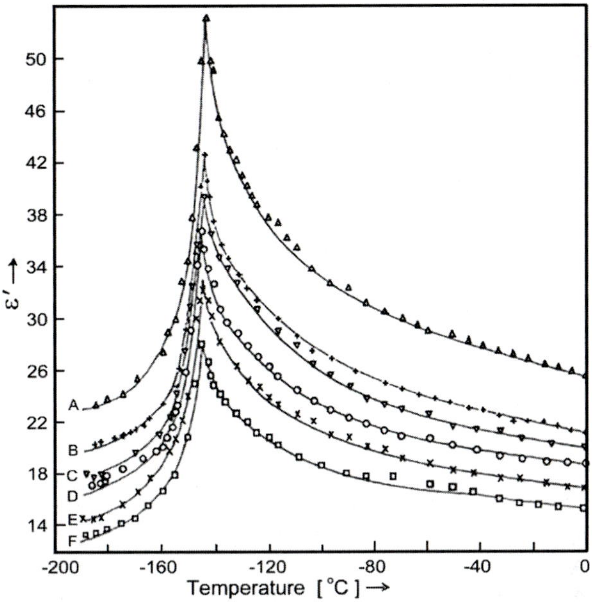

Fig. 5.17. Dielectric constants of KDP for powders of different particle size; A: 300–600 μm, B: 150–300 μm, C: 125–150 μm, D: 100–125 μm, E: 53–125 μm, F: 20–53 μm

A compilation of dielectric properties of biological systems by Stefan [5.33] covers work on several biomaterials such as DNA, cells and membranes, enzymes, botanical and animal components. A dielectrospectrometer has been designed by Sing et al. [5.74] for an in vivo study to detect benign and cancerous tissues. Agarwal and Singh [5.75] reported the electrical and dielectric properties of kidney stones. Saraswati [5.76] investigated the dielectric properties of natural tooth which essentially consists of hydroxy apatite.

Studies of proteins reveal that they may be compared to semiconductors. Dry proteins behave like perfect insulators, with a large energy gap (>4 eV). Conduction is induced either by complexing with a donor to form an electronic conductor or by hydration to induce protonic conduction [5.35, 5.77]. Measurements of dielectric constant and loss as a function of water content and temperature have yielded useful information. Ahmed et al. [5.34] observed in their studies on temperature variation that the dielectric constant shows sharp variations corresponding to change in the crystal structure or denaturing.

5.3.2 Temperature Variation of Dielectric Constant

Theoretical evaluation of temperature dependence of dielectric constant has been attempted by several investigators [5.78–5.84].

Owens' Relation

By differentiation of the equation given by Szigeti [5.85, 5.86], Owens [5.81] obtained the relation

$$(\varepsilon_{\mathrm{s}} - \varepsilon_\infty)_T = (\varepsilon_{\mathrm{s}} - \varepsilon_\infty)_0[1 - \beta(2\gamma_t - 1)\Delta T]^{-1} \tag{5.22}$$

to predict the temperature variation of the dielectric constant. Here β is the volume expansion coefficient and γ_{t} the mode Gruneisen constant defined by

$$\gamma_{\mathrm{t}} = -[d(\log \omega_t)/d(\log V)], \tag{5.23}$$

ω_{t} being the transverse optical frequency.

Method of Varotsos

Varotsos [5.82] has proposed a simple formula which permits the evaluation of dielectric constant at any temperature from data at room temperature and other parameters like thermal variation of bulk modulus and lattice parameter. This approach has a limited applicability due to lack of input data over a wide range of temperature.

Method of Havinga and Bosman

Havinga [5.83] and Havinga and Bosman [5.84] have shown that the temperature variation of dielectric constant could be traced to the temperature variation of polarizability. They have adopted a macroscopic model to evaluate three contributions known as A, B, C, terms. Differentiating the Clausius–Mossotti equation

$$\frac{\varepsilon_{\mathrm{s}} - 1}{\varepsilon_{\mathrm{s}} + 2} = \frac{4}{3}\frac{\pi\alpha}{V} \tag{5.24}$$

(where α is the polarizability of a small sphere of volume V), they obtained

$$\frac{1}{(\varepsilon_{\mathrm{s}} - 1)(\varepsilon_{\mathrm{s}} + 2)}\left(\frac{\partial\varepsilon_{\mathrm{s}}}{\partial T}\right)_P = -\frac{\beta}{3} + \frac{\beta}{3}\left\{\frac{V}{\alpha}\left(\frac{\partial\alpha}{\partial V}\right)_T\right\} + \frac{1}{3\alpha}\left(\frac{\partial\alpha}{\partial T}\right)_V = A + B + C, \tag{5.25}$$

$$\frac{1}{(\varepsilon_{\mathrm{s}} - 1)(\varepsilon_{\mathrm{s}} + 2)}\left(\frac{\partial\varepsilon_{\mathrm{s}}}{\partial P}\right)_T = \frac{\psi}{3} - \frac{\psi}{3}\frac{V}{\alpha}\left(\frac{\partial\alpha}{\partial V}\right)_T, \tag{5.26}$$

where β is the volume thermal expansion coefficient and ψ the isothermal compressibility.

The term A is obtained from the thermal expansion β and represents the decrease in the number of polarizable particles per unit volume with increase in temperature. The term B is the contribution due to the increase of polarizability of a fixed number of particles with increase of available volume as the temperature increases. B is evaluated from the data on pressure variation

Table 5.2. Values of the quantities $(\varepsilon_s - 1)^{-1}(\varepsilon_s + 2)^{-1}(\partial\varepsilon_s/\partial T)_P$, A, B and C [5.84]

Crystal	$(\varepsilon_s - 1)^{-1}(\varepsilon_s + 2)^{-1}$ $(\partial\varepsilon_s/\partial T)_P$ $[10^{-5}\ \mathrm{K}^{-1}]$	$A[10^{-5}\ \mathrm{K}^{-1}]$	$B[10^{-5}\ \mathrm{K}^{-1}]$	$C[10^{-5}\ \mathrm{K}^{-1}]$
LiF	3.7	−3.4	6.4	0.7
NaCl	5.2	−4.0	8.2	1.0
KCl	5.6	−3.8	7.7	1.7
KBr	5.9	−4.0	7.4	2.5

of dielectric constant and coefficient of expansion. In the absence of experimental data on pressure variation of ε_s an alternate method to evaluate B was suggested by Havinga and Bosman [5.84] based on a simple ionic model. The term C is obtained as the difference between the value of $(A + B + C)$ obtained experimentally from temperature variation of dielectric constant and $(A + B)$. The C contribution which is related to the temperature variation of IR polarizability plays a dominant role in determining the temperature variation of dielectric constant. C reflects the anharmonic effects since it is the pure temperature derivative of α at constant volume.

This method of interpretation of the temperature variation of the dielectric constant was applied to the alkali halides by Havinga and Bosman [5.84]. Their results are given in Table 5.2. This method has also been applied to alkaline earth fluorides [5.50], CdF_2 [5.49] and TiO_2 [5.87].

5.3.3 Szigeti's Theory (Effective Ionic Charge and Anharmonicity)

According to Born and Mayer [5.88] the lattice contribution to dielectric constant for a diatomic crystal is given by

$$\varepsilon_s - \varepsilon_\infty = \frac{2\pi e^2}{r^3\omega_t^2}\frac{1}{\mu}, \qquad (5.27)$$

where ε_s is the static (low frequency) dielectric constant, ε_∞ the optical (high frequency) dielectric constant, r the nearest neighbour distance, ω_t the transverse optical mode frequency and μ the reduced mass.

In an ideal crystal consisting of deformable ions which do not overlap, the individual ions carry their formal charge ze. But in real crystals, the ions overlap. The net charge on the ions will differ from their formal charge. Szigeti [5.85, 5.86] was the first to take into account the polarization effects due to this overlap by introducing the concept of effective ionic charge. The effective ionic charge (q^*) is expressed as the effective ionic charge per electron (ze*/ze). Szigeti's theory has also been discussed in [5.89, 5.90]. The main result of Szigeti's theory is the relation

$$\varepsilon_s - \varepsilon_\infty = \frac{4\pi z^2 e^2 q *^2\, (\varepsilon_\infty + 2)^2}{9V_i\omega_t^2\mu}, \qquad (5.28)$$

Table 5.3. Values of q^* calculated from (5.28)

Crystal	q^*	Crystal	q^*	Crystal	q^*
C	0	GaP	0.58	LiF	0.87
Si	0	GaAs	0.51	NaF	0.93
Ge	0	GaSb	0.33	NaCl	0.74
		InP	0.68	KCl	0.80
		InAs	0.56	RbCl	0.84
		InSb	0.42	CsCl	0.84

Values for alkali halides from [5.85] and others from [5.93]

where z is the valency, e the formal electron charge, V_i the volume per ion pair. q^*, the ratio of effective charge to formal charge (ze*/ze), has come to be known as the Szigeti charge.

Values of q^* obtained from (5.28) for several crystals are given in Table 5.3. It can be seen that the effective ionic charge varies from zero for covalent crystals diamond, germanium and silicon to a value nearly unity for alkali halides. Partially covalent crystals have intermediate values. Szigeti [5.85] and Lowndes and Martin [5.36] attributed the deviation from unity to greater polarizability whereas Denham et al. [5.19], Gielisse et al. [5.91] and Bansigir [5.92] treat the value of effective ionic charge as a measure of ionicity.

Equation (5.28) represents the lattice contribution to dielectric constant modified to include overlap effects. Under harmonic approximation, the parameters on the right-hand side of (5.28) are unique functions of volume. Hence $(\partial\varepsilon_s/\partial T)_V$ should be equal to zero. Experimental results on dielectric constant by Fontanella et al. [5.26], Havinga and Bosman [5.84] and Bartels and Smith [5.94] indicate that this is not true. $(\partial\varepsilon_s/\partial T)_V$ has been shown to reflect the anharmonic contribution. It is found to be negative or positive depending on the crystal structure, magnitude of dielectric constant and temperature. These aspects are discussed by Havinga and Bosman [5.84], Rao and Smakula [5.95] and Samara [5.43]. Szigeti [5.96, 5.97] discussed the effect of anharmonicity on lattice contribution to dielectric constant and modified (5.28) to

$$\begin{aligned} \varepsilon_s - \varepsilon_\infty &= \frac{4\pi z^2 e^2 q^{*2}(\varepsilon_\infty + 2)^2}{9V_i\omega_t^2}\frac{1}{\mu} + G \\ &= \eta + G . \end{aligned} \tag{5.29}$$

The first term η (which is the right-hand side of (5.28)) is the harmonic contribution whereas the second term G represents the anharmonic contribution. Further, for temperatures $T > \theta_D$ where θ_D is the Debye temperature, Szigeti [5.97] has shown that

$$G = AT, \tag{5.30}$$

where A is a constant independent of temperature. Hence

$$\left[\frac{\partial(\varepsilon_s - \varepsilon_\infty)}{\partial T}\right]_V = \left(\frac{\partial \eta}{\partial T}\right)_V + \left(\frac{\partial G}{\partial T}\right)_V . \tag{5.31}$$

Since $(\partial\eta/\partial T)_V = 0$ under harmonic assumption

$$\left(\frac{\partial G}{\partial T}\right)_V = \left[\frac{\partial(\varepsilon_s - \varepsilon_\infty)}{\partial T}\right]_V = A \tag{5.32}$$

and, therefore, neglecting the variation of ε_∞ with T

$$G = T\left[\frac{\partial(\varepsilon_s - \varepsilon_\infty)}{\partial T}\right]_V \cong T\left(\frac{\partial \varepsilon_s}{\partial T}\right)_V . \tag{5.33}$$

The magnitude and sign of G can be obtained from experimental data on temperature dependence of dielectric constant.

Starting with the Clausius–Mossotti relation, the temperature derivative of dielectric constant at constant pressure is obtained in terms of A, B, C components (5.25). By differentiation of (5.24) with respect to temperature, the temperature derivative at constant volume is given by

$$\left(\frac{\partial \varepsilon_s}{\partial T}\right)_V = (\varepsilon_s - 1)(\varepsilon_s + 2)\frac{1}{3\alpha}\left[\frac{\partial \alpha}{\partial T}\right]_V = (\varepsilon_s - 1)(\varepsilon_s + 2)C. \tag{5.34}$$

Hence, the anharmonic contribution G is obtained in terms of C as

$$G = T(\varepsilon_s - 1)(\varepsilon_s + 2)C. \tag{5.35}$$

Typical values of η and G for some alkali halides are given in Table 5.4.

The results show that anharmonic effects account for about 30–40% of the lattice contribution to dielectric constant of the thallous halides as compared to less than 5% in the case of alkali halides. Further for thallous halides and alkali halides having CsCl structure, the anharmonicities act to reduce the harmonic contribution. On the other hand, for alkali halides having the NaCl structure, the anharmonicities enhance the harmonic contribution.

Table 5.4. Values of η and G expressed as a fraction of $(\varepsilon_s - \varepsilon_\infty)$ [5.43]

Crystal	$\eta/(\varepsilon_s - \varepsilon_\infty)$	$G/(\varepsilon_s - \varepsilon_\infty)$
NaCl	0.969	0.031
KCl	0.950	0.050
CsCl	1.047	−0.047
CsBr	1.041	−0.041
TlCl	1.371	−0.371
TlBr	1.338	−0.338

5.3.4 Spectroscopic Aspects

Self-energy

When a photon interacts with a diatomic lattice, two mechanisms are possible. A photon can be absorbed in the creation of a single long wavelength transverse optical phonon (TO) or it can create several phonons simultaneously through its interaction with the non-linear dipole moment of the crystal. The former decays into two or more finite wavelength phonons giving rise to width. The second effect is very small and is generally neglected [5.37,5.98]. The width of the long wavelength optic phonon is temperature dependent. In addition a shift is observed with temperature. Since the changes in the width and shift with temperature are related to fundamental infrared absorption bands, the data on phonon width and shift provides useful information regarding the dynamics of the lattice.

The relation between dielectric phenomena and phonon processes has been investigated by many workers [5.98–5.102]. The temperature dependence of the frequency shift and the phonon width has been interpreted in terms of anharmonic effects. Chang and Mitra [5.103] have studied the temperature variation of shift for NaF and LiF from IR reflectivity curves. Lowndes and Martin [5.37] measured the low frequency dielectric constants and their temperature and pressure dependence for a large number of alkali halides and used the data to interpret their phonon spectra.

The measured shift in the phonon frequency $\Delta\omega(\text{obs})$ consists of two contributions: one, the shift due to volume change $\Delta\omega(\text{vol})$ and the other, the shift due to anharmonic effect $\Delta\omega(\text{anh})$. $\Delta\omega(\text{vol})$ is the shift due to the pure volume change that the crystal will undergo if only the expansion effect is present. This is obtained from measurement of pressure variation of frequency. The shift due to anharmonic effect (self-energy shift) is obtained by eliminating the pure volume effect from the observed shift $\Delta\omega(\text{obs})$. Thus

$$\Delta\omega(\text{anh}) = \Delta\omega(\text{obs}) - \Delta\omega(\text{vol}). \tag{5.36}$$

From dielectric measurements Lowndes and Martin [5.37] obtained the anharmonic contribution $(\Delta\varepsilon)_V(\text{anh})$ to the change in the dielectric constant with temperature using the relation:

$$(\Delta\varepsilon_V)(\text{anh}) = (\Delta\varepsilon_T)(\text{obs}) - (\Delta\varepsilon_P)(\text{vol}), \tag{5.37}$$

where $(\Delta\varepsilon)_P$ and $(\Delta\varepsilon)_T$ are the measured quantities. $(\Delta\varepsilon)_P$ is the change in ε on raising the temperature from 0 to T K at constant pressure. $(\Delta\varepsilon)_T$ is the change in ε on raising the pressure at constant temperature to restore the volume to its original value. Lowndes and Martin [5.37] obtained the individual values of $(\Delta\varepsilon)_P$ and $(\Delta\varepsilon)_T$ to estimate $(\Delta\varepsilon)_V$ for a large number of crystals. Further, they have shown the following correlation between the anharmonic contribution to change in the dielectric constant with temperature and the anharmonic frequency shift $\Delta\omega(\text{anh})$:

$$\frac{(\Delta\varepsilon)_V}{\varepsilon(0)} = -2\left[\frac{\Delta(T)}{\omega_{\rm qh}} - \frac{\Delta(0)}{\omega_{\rm qh}}\right], \quad (5.38)$$

where $\Delta\omega(\text{anh})$ is expressed as the temperature dependent self-energy shift Δ. The observed variations in self-energy shift Δ are interpreted in terms of anharmonic contributions from higher order terms in potential energy based on the work of Szigeti [5.97] and Cowley [5.99]. The variations are the result of the sum of two contributions, the cubic and quartic terms in the crystal potential. The shifts due to the two terms are in opposite directions. The cubic term results in negative shift and the quartic term in a positive shift to $\Delta(T)$.

Lowndes and Martin [5.37] have studied the influence of anharmonicity on the longitudinal optical modes of several alkali halides in terms of (5.38). Their results for some alkali halides are shown in Fig. 5.18. They find that the self-energy contribution $(\Delta\varepsilon)_V$ is positive for those salts crystallizing in the NaCl structure and negative for the crystals having CsCl structure. This has been interpreted in terms of third-order and fourth-order anharmonic contribution to the potential energy. Enhanced anharmonic contribution in CsCl arising out of larger contribution from the fourth order term leads to a net negative contribution. Bosman and Havinga [5.23] suggested that the cubic term is

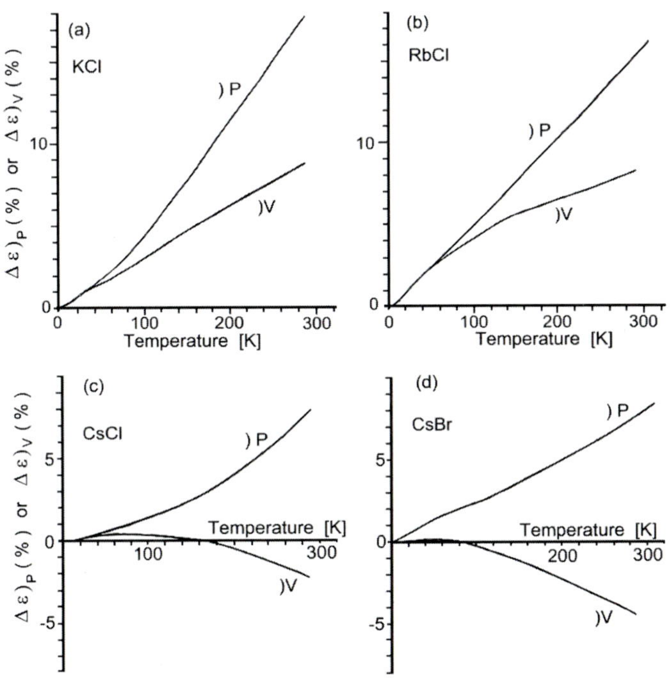

Fig. 5.18. Variation of $(\Delta\varepsilon)_P$ and $(\Delta\varepsilon)_V$ with temperature for (**a**) KCl (**b**) RbCl (**c**) CsCl and CsBr; $(\Delta\varepsilon)_V$ is a measure of self-energy shift (5.38)

of lesser significance in CsCl structure because of the higher coordination number which reduces the fluctuations among the pairs of bonds formed by an ion with its neighbours.

Evaluation of self-energy is a difficult task both theoretically and experimentally. Theoretically, it involves rigorous computational analysis using many body Green's function approach. Experimentally, it requires accurate measurement of temperature and pressure dependence of transverse optic mode frequency.

Sano [5.104] adopted a semi-empirical method to obtain the self-energy of the TO phonon. By logarithmic differentiation of the Szigeti equation (5.28) with respect to temperature and with some approximations he obtained

$$(\Delta\varepsilon)_V = (\Delta\varepsilon)_P - \frac{\omega_t^2(\varepsilon_s - \varepsilon_\infty)}{\omega_{qh}^2}\left[(2C-1) + 2\gamma_t\right]\int_0^T \beta\, dT, \qquad (5.39)$$

where $C = \frac{\partial \log q^*}{\partial \log V}$ and γ_t is the transverse mode Gruneisen parameter. Sano estimated the self-energy values, $\Delta(T)$ of the TO phonon as a function of temperature from $(\Delta\varepsilon)_V$ for KCl. The values were found to be positive in the high temperature range.

Damping Constant

For ionic crystals under the damped oscillator model [5.103], the dielectric constant $\varepsilon(\omega)$ is given by

$$\varepsilon(\omega) = \varepsilon_\infty + \frac{(\varepsilon_s - \varepsilon_\infty)}{1 - (\omega/\omega_t)^2 - i(\omega/\omega_t)(\gamma_d/\omega_t)}, \qquad (5.40)$$

where (γ_d/ω_t) is the damping constant. In terms of the refractive index n and the extinction coefficient k

$$\varepsilon(\omega) = n^2 - k^2 - 2ink. \qquad (5.41)$$

The imaginary part of ε is

$$2nk = \frac{(\varepsilon_s - \varepsilon_\infty)(\gamma_d/\omega_t)(\omega/\omega_t)}{[1 - (\omega/\omega_t)^2]^2 + (\omega/\omega_t)^2(\gamma_d/\omega_t)^2}, \qquad (5.42)$$

$2nk$ has maximum value at $\omega = \omega_t$. At this frequency

$$(\gamma_d/\omega_t) = \frac{(\varepsilon_s - \varepsilon_\infty)\omega_t}{(2nk\omega)_{max}} = \frac{\varepsilon_s - \varepsilon_\infty}{(2nk)_{max}}. \qquad (5.43)$$

The damping constant can also be expressed in terms of the half-width of the resonance ω_t as

$$(\gamma_d/\omega_t) = \frac{2(\omega_t - \omega_{1/2})}{\omega_t}. \qquad (5.44)$$

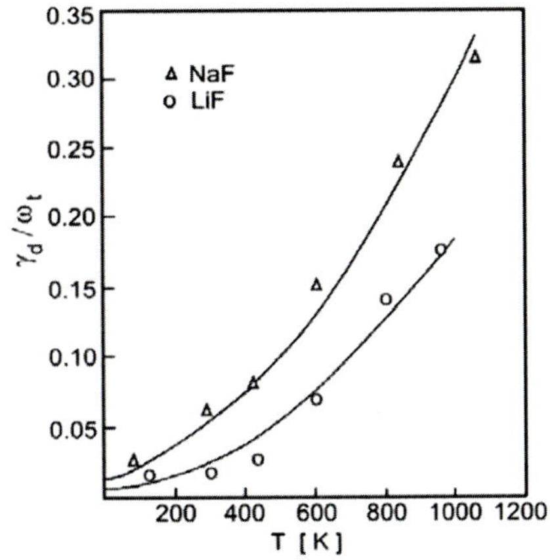

Fig. 5.19. Temperature dependence of the damping constant $(\gamma_{\mathrm{d}}/\omega_{\mathrm{t}})$ of NaF and LiF

Chang and Mitra [5.103] have obtained the temperature variation of $(\gamma_{\mathrm{d}}/\omega_{\mathrm{t}})$ from the half-width of the resonance curves. Their data for LiF and NaF are shown in Fig. 5.19. The temperature dependence of $(\gamma_{\mathrm{d}}/\omega_{\mathrm{t}})$ is found to be of the form $AT + BT^2$.

The 'LST' Relation

Lyddane et al. [5.105] derived the relation:

$$(\varepsilon_{\mathrm{s}}/\varepsilon_{\infty}) = (\omega_{\mathrm{l}}/\omega_{\mathrm{t}})^2. \tag{5.45}$$

Now known as the 'LST ' relation, it relates the static and high frequency dielectric constants (ε_{s}, ε_{∞}) to the transverse optical and longitudinal optical frequencies ($\omega_{\mathrm{t}}, \omega_{\mathrm{l}}$). The equation holds good for cubic diatomic crystals; these crystals have an LO mode and a doubly degenerate TO mode. For a more general case, Cochran [5.106] derived the following modified form:

$$(\varepsilon_{\mathrm{s}}/\varepsilon_{\infty}) = \prod_i (\omega_{\mathrm{l}})_i^2/(\omega_{\mathrm{t}})_i^2. \tag{5.46}$$

The product Π is over all the optic modes.

Some of the applications of the 'LST' relation are:

(a) ω_{l} is difficult to be determined experimentally. In many cases, only ω_{t} is available. In such cases, ω_{l} can be estimated from the LST relation.
(b) If both ω_{l} and ω_{t} are known from experiment and if diverse values of ε_{s} are available for a crystal, the appropriate value of ε_{s} may be identified with the help of the LST relation.
(c) Ferroelectric crystals follow the Curie–Weiss law; $\varepsilon_{\mathrm{s}} \to \infty$ (or a large value) as $T \to T_{\mathrm{C}}$, the Curie point. In these crystals, one of the lattice frequencies shows anomalous variation near the Curie temperature assuming

a very low value. According to the generalized LST relation, $\varepsilon_s \to \infty$ if $\omega_t \to 0$. This is the 'soft mode' interpretation of ferroelectricity.

5.3.5 Conductivity of Ionic Crystals

In ionic crystals, conduction is mainly due to the motion of defects. Depending on the nature of the crystal, the defects may be Schottky pairs (+ve and −ve ion vacancies) or Frenkel pairs (vacancies and ions in interstitial positions). The theory of ionic conduction is discussed in [5.7, 5.29]. Energy (E_f) is required to form a defect and, once a defect is created, energy E_M is required for its migration. The conductivity (σ) is given by

$$\sigma = ne\mu, \tag{5.47}$$

where n is the concentration of defects, e the charge and μ the mobility of the defect. Substitution for n from the theory of ionic conductivity [5.29] leads to:

$$\sigma = A_1 \exp\left(-\frac{E_M}{k_B T}\right) + A_2 \exp\left(-\frac{E_M + (E_f/2)}{k_B T}\right), \tag{5.48}$$

where, A_1 and A_2 are constants for a given crystal and k_B the Boltzmann constant. At low and moderate temperatures, only the first term is operative whereas at high temperatures the second term dominates. Hence, a $\log \sigma$ vs. $(1/T)$ plot will have different slopes at low and high temperatures. These two regions are called "extrinsic" and "intrinsic", respectively. The energies of migration and formation of defects can be calculated from these slopes which are, respectively, equal to $[E_M/k_B]$ and $[E_M + (E_f/2)]/k_B$.

A typical $\log \sigma$ vs. $(1/T)$ plot obtained from DC measurement of σ is shown in Fig. 5.20. As expected, it consists of two linear regions. As mentioned in Sect. 5.1.4, σ can be obtained from measurement of dielectric loss. A typical plot of $\log \sigma$ vs. $(1/T)$ from AC measurements is shown in Fig. 5.21. Here

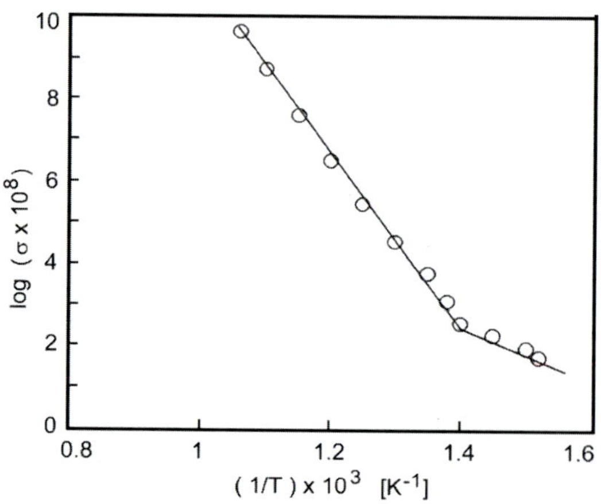

Fig. 5.20. Plot of $\log \sigma$ vs. $(1/T)$ for KCl (DC measurement)

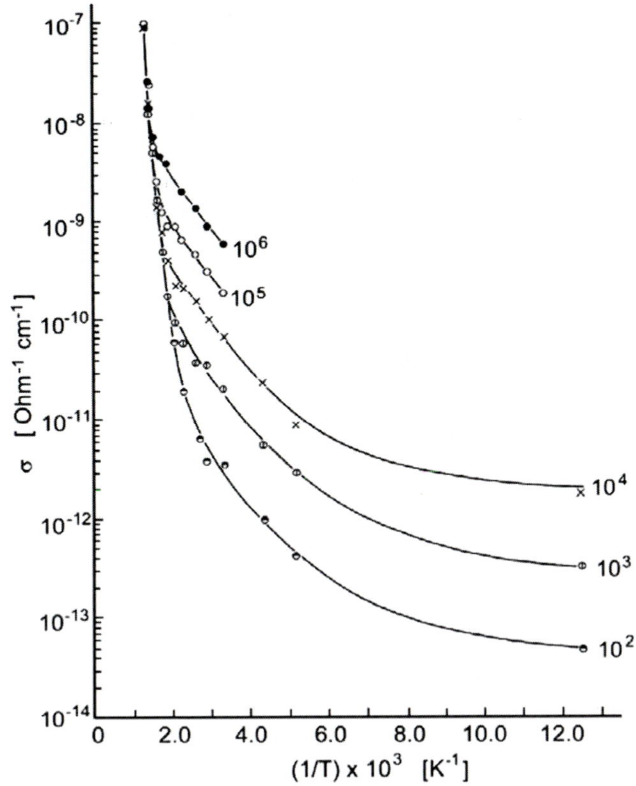

Fig. 5.21. Plot of log σ vs. $(1/T)$ for calcite (a-axis) at different frequencies (Hz)

Table 5.5. Values of E_M and E_f for some crystals

Crystal	Method	E_M [eV]	E_f [eV]	Ref.
KCl	DC	0.51	2.04	[5.107]
$CaCO_3$ a-axis	AC	0.11	1.88	[5.108]

it is seen that the plot corresponding to the extrinsic region is frequency dependent. However, the slopes at different frequencies are nearly the same. In Table 5.5, the E_M and E_f values obtained from DC and AC conductivities are given. E_f values for several alkali halides are given in [5.42].

5.3.6 Dielectric Constant and Polaron Conduction

Theoretical work by several workers over the years has provided some understanding of conduction in oxides and transition metal compounds [5.109, 5.110]. For these materials the interaction between electrons and optical phonons is strong and the conduction is explained on the basis of 'small polarons'. A 'small polaron' is a slow moving electron in an ionic lattice. These conduction electrons polarize the surrounding lattice in an ionic crystal. The electrons together with the associated lattice deformation polarization are termed small polarons. The 'small polarons' conduct in a band-like manner at low temperatures and by hopping mechanism at high temperatures ($T > \theta_{\mathrm{D}}/2$ where θ_{D} is the Debye temperature) [5.111, 5.112].

A brief sketch of the theory of polaron conduction based on the treatment by Mott and Davis [5.113] follows. The potential energy of an electron at a distance r from another electron is given by $e^2/\varepsilon_{\infty}r$ for an immobile lattice and $e^2/\varepsilon_{\mathrm{s}}r$ for a mobile lattice. Introducing a distance r_{p} from the electron beyond which the medium is fully polarized and ε_{p} as the effective dielectric constant

$$\frac{1}{\varepsilon_{\mathrm{p}}} = \frac{1}{\varepsilon_{\infty}} - \frac{1}{\varepsilon_{\mathrm{s}}}. \tag{5.49}$$

The energy required to polarize the medium is $(1/2)(e^2/\varepsilon_{\mathrm{p}}r_{\mathrm{p}})$ and the kinetic energy of the electron is $(\pi^2\hbar^2/2m^*{r_{\mathrm{p}}}^2)$ where m^* is the effective electron mass. The total energy defined as polaron energy W_{p} is given by

$$W_{\mathrm{p}} = \frac{\pi^2\hbar^2}{2m^*{r_{\mathrm{p}}}^2} - \frac{e^2}{2\varepsilon_{\mathrm{p}}r_{\mathrm{p}}}. \tag{5.50}$$

Minimising the above equation r_{p} is obtained as

$$r_{\mathrm{p}} = \frac{2\pi^2\hbar^2}{m^*e^2}\varepsilon_{\mathrm{p}}. \tag{5.51}$$

Neglecting the kinetic energy

$$W_{\mathrm{p}} = \frac{e^2}{2\varepsilon_{\mathrm{p}}r_{\mathrm{p}}}. \tag{5.52}$$

Frohlich [5.114] and Allcock [5.115] by an independent self-consistent calculation of potentials obtained W_{p} in terms of the electron–phonon coupling constant α_{e}. α_{e} is given by

$$\alpha_{\mathrm{e}} = (e^2/\varepsilon_{\mathrm{p}})(m^*/2\hbar^3\omega_{\mathrm{D}})^{1/2}, \tag{5.53}$$

where $\omega_{\mathrm{D}} = k_{\mathrm{B}}\theta_{\mathrm{D}}/\hbar$ and θ_{D} is the Debye temperature. They also obtained a relation for polaron mass m_{p}

$$m_{\mathrm{p}} = 0.02m^*{\alpha_{\mathrm{e}}}^4. \tag{5.54}$$

The treatment given by Bogomolov et al. [5.116] gives r_{p} as

$$r_{\mathrm{p}} = 1/2(\pi/6N)^{1/3}, \tag{5.55}$$

where N is the number of available sites per unit volume.

For $\alpha_{\mathrm{e}} > 5$ and r_{p} less than the distance between the lattice sites, the polarons are "small polarons". The polaron parameters may be calculated from the dielectric constant and other crystal parameters. A knowledge of the magnitude of the coupling constant α_e and the polaron radius r_p helps in understanding the nature of polaron conduction.

From AC conductivity studies on materials exhibiting small polaron conduction, it is observed that below 300 K the conductivity obeys the relation [5.117]

$$[\sigma(\omega) - \sigma(0)] = \text{constant}\left[\frac{\omega^2\tau}{1+\omega^2\tau^2}\right]. \tag{5.56}$$

This frequency dependent conductivity is independent of temperature below 300 K. At higher temperatures, the conductivity is temperature dependent and is expected to be frequency independent.

The hopping conduction by small polarons is not observed in DC measurements but it is observed only when the measurements are carried out in AC field at different frequencies. Since τ is of the order of 10^{-10} s [5.118], at low frequencies ($\omega < 10^5$ Hz), $\omega^2\tau^2 << 1$. Hence, $\omega^2\tau^2$ can be neglected in (5.56). A plot of log $[\sigma(\omega)–\sigma(0)]$ vs. log ω^2 should therefore yield a straight line.

5.3.7 Dielectric Constant and Additivity of Polarizability

The concept of additivity of polarizabilities implies that the molecular polarizability of a complex substance can be expressed as the sum of polarizabilities of simpler molecular entities constituting the molecule. Roberts [5.119] applied this rule to simple ionic crystals and barium titanate. Tessman et al. [5.120] have demonstrated the validity of the additivity rule for electronic polarizabilities of alkali halides and alkaline earth chalcogenides. Shannon et al. [5.61] and Shannon [5.121] have applied the additivity rule of molecular polarizabilities to oxides having a variety of structures including some garnets. The additivity rule of molecular polarizabilities may be illustrated as follows:

$$\alpha_D(M_2M'X_4) = 2\alpha_D(MX) + \alpha_D(M'X_2), \tag{5.57}$$

where α_D is the dielectric polarizability for a molecular unit, $M_2M'X_4$ may stand for Mg_2SiO_4, Be_2SiO_4, etc. The basic equation is the Clausius–Mossotti equation which relates the polarizability with the measured dielectric constant. Other input data are the density and molecular weight. Good agreement between calculated and observed polarizabilities demonstrate that the additivity rule is useful in predicting dielectric constants of new materials whose dielectric constant has not been measured. Shannon [5.122] has given data on polarizabilities of a large number of ions derived from the dielectric constants of several oxides and fluorides.

5.3.8 Dielectric Behaviour of Proteins Dielectric Properties and Protein Hydration

In dealing with biological systems like proteins, the role of water is vital in understanding their behaviour. This aspect has received the attention of many workers [5.77, 5.123, 5.124]. Kuntz and Kauzmann [5.124] reviewed the hydration of proteins and polypeptides which reveal the existence of water in many forms, about 0.1% as strongly bound, 10% as less strongly bound and a major part as free or bulk water.

The dielectric behaviour can be understood by considering three different stages of protein-water interactions. Initially, when the dielectric properties of dry protein are measured in vacuum no dispersion will be observed. The dielectric constant and loss show frequency independent behaviour. The contribution to polarization is by electronic and ionic displacements. As the water content increases, a large increase in dielectric constant is observed around the water content of about 1 wt%. The water content at which large increase occurs is termed 'critical' hydration (h_C). The critical hydration is identified as the amount of water required to cover the protein molecules with a single layer of water. Beyond critical hydration water exists as adsorbed water (w_{ads}) and is tightly bound to the protein molecule. The polarization contribution increases to $(P_e + P_a) + P_{ads}$. At the third stage where the hydration level is further increased, some of the water molecules (w_d) exist as loosely bound molecules and induce a dipolar contribution P_d to the polarization. Hence at this level of hydration, a large increase in dielectric constant is observed. The total polarization P_{total} is expressed as

$$P_{total} = (P_e + P_a) + P_{ads} + P_d. \tag{5.58}$$

The molar polarization P, polarizability α and the dielectric constant are related through Clausius–Mosotti relation. Hence, the dielectric constant ε changes with the water content due to the increase in polarization.

Dielectric Relaxation and Dispersion

Protein samples in solid state exhibit three regions of dielectric dispersion. The first dispersion denoted as Ω occurs at very low frequencies and is attributed to the electrode effects resulting from the accumulation of ions at the sample electrode interface.

The second dispersion denoted as α-dispersion occurs beyond a few hundred Hz and is associated with the hopping of charge carriers between the localized sites. The dispersion peaks are found to move to higher frequencies with the increase of hydration. For ovalbumin containing 11 wt% water Takashima and Schwan [5.125] observed a broad dielectric dispersion around 1 kHz. Lawton et al. [5.126] found this dispersion around 500 Hz for urease containing 13.6 wt% water at room temperature (Fig. 5.22).

Dielectric measurements in the microwave range of 10 MHz to 25 GHz show dispersion associated with the orientational relaxations of water bound to protein molecules. Harvey and Hoekstra [5.127] have observed this effect for lysozyme around room temperature. Bone and Pethig [5.128] estimated the

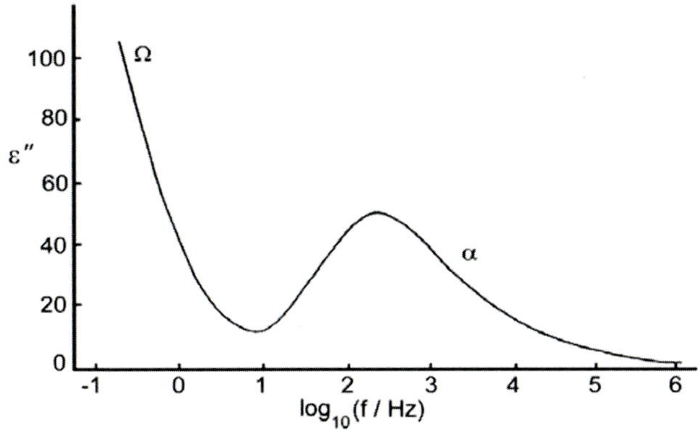

Fig. 5.22. Variation of the dielectric loss parameter $\varepsilon''(\omega)$ for urease containing 13.6 wt% water at 294 K, to show the forms of the Ω and α-dispersion commonly exhibited by solid-state protein samples

number of bound water molecules irrotationally bound to protein molecules. They carried out a detailed study on a number of proteins to understand the role of bound water in facilitating protein flexibility.

Conduction Mechanism

Pure proteins are insulators at normal temperatures with energy gaps in the range 4–12 eV between the valence and conduction bands. Electron delocalization may be induced within the valence bands and they show solid state physical properties similar to those of semiconductors. For instance, electronic conduction is induced when the proteins are complexed with methylglyoxal as a result of charge transfer interaction [5.129]. From steady-state measurements on several proteins Bone [5.130] attributed the dispersion observed in the low frequency range up to 100 Hz to electronic activity by creation of mobile electrons or holes in the protein complexes.

For hydrated proteins the α-dispersion discussed earlier is associated with the charge carriers hopping between localized sites. The relaxation time is related to the steady state conductivity due to proton transport. The mobile protons originate from ionizable protein molecule and the mobility is induced due to hydrogen bond network formed by the water molecules. At higher level of hydration where free hydration or structural flexibility is made possible, one more component due to orientation of permanent dipoles will be added to dielectric loss. Pethig [5.35] stated that in the medium frequency range (100 Hz to few kHz) measurements on dried protein powders overcome complications associated with the rotation of the protein molecules so that a more direct estimate of adsorbed water can be made.

5.3.9 Irradiation Effects

Rao [5.131] reported the dielectric properties of X-irradiated NaCl and LiF crystals. The relative change in the dielectric loss was found to be more than the relative change in the dielectric constant. Subrahmanyam and Badarinath [5.132] observed an increase in dielectric constant of RbCl on γ-irradiation. Subrahmanyam [5.133] measured the dielectric constant on quenched and X-irradiated NaCl crystal. He observed an increase of dielectric constant at low frequencies by more than ten times but on prolonged irradiation the original value was restored. At higher frequencies no change was observed.

Changes in optical and dielectric properties of calcite crystals are reported by Rao [5.134]. He observed a correlation between the change in the dielectric loss and thermoluminescence glow. On heating the samples to 350°C and cooling to room temperature the dielectric constant was found to recover its original value. Govinda and Rao [5.135] observed a decrease in dielectric constant of TGS on γ-irradiation.

An interesting effect in dielectric properties of γ-irradiated fused silica was reported by Fontanella et al. [5.136]. Their results shown in Fig. 5.23 show large changes in the relaxation peaks on irradiation. The unirradiated sample has relaxation peaks at 40 and 240 K. The 240 K peak was found to disappear with creation of an additional loss peak at about 60 K. No recovery of original peak was noted even after annealing the sample at room temperature for a couple of months. A corresponding growth of optical absorption bands in the visible region was observed. The relaxation was attributed to an aluminium–alkali centre and the new relaxation at 60 K to an aluminium–oxygen-hole centre.

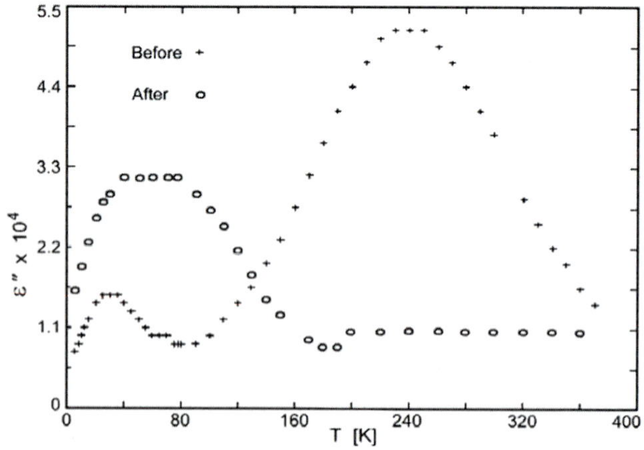

Fig. 5.23. ε'' vs. temperature for a sample of Infrasil 2 before and after irradiation with about 1.5×10^8 rad of gamma rays from a ^{60}Co source

5.4 Some of our Results

Using the experimental techniques and procedures discussed in Sect. 5.2, measurements of dielectric constant, dielectric loss and conductivities (AC and DC) have been carried out for a variety of materials at various temperatures and frequencies. The frequencies ranged generally from 100 Hz to 100 kHz and in some cases up to 10 MHz and the temperatures from liquid nitrogen temperature to 700°C. The exact range of temperature differed from crystal to crystal.

The materials studied included inorganic substances of different structures, minerals, organic compounds, biomaterials and a few polymers. Some materials were studied in the powder form and others as single crystals. In a few cases, studies have been made both on powders and single crystals.

The experimental results have been analysed and interpreted in terms of relevant theories to obtain information like effective charges, activation energies, spectroscopic parameters and polaron parameters.

5.4.1 Dielectric Properties – Data Generation

While detailed information regarding the frequency and temperature variation of dielectric properties are available in our publications, the main results are given in Tables 5.6–5.9. These are the static dielectric constant (ε_s), the loss ($\tan\delta$) at room temperature and the temperature coefficient of dielectric constant $(1/\varepsilon_s)(d\varepsilon_s/dT)$ for the range RT–100°C.

5.4.2 Analysis of Temperature Variation of Dielectric Constant

The theory proposed by Havinga and Bosman (1965) [5.84] has been discussed in Sect. 5.3.2. It was shown that

$$\frac{1}{(\varepsilon_s-1)(\varepsilon_s+2)}\left(\frac{\partial\varepsilon_s}{\partial T}\right)_P=-\frac{\beta}{3}+\frac{\beta}{3}\left\{\frac{V}{\alpha}\left(\frac{\partial\alpha}{\partial V}\right)_T\right\}+\frac{1}{3\alpha}\left(\frac{\partial\alpha}{\partial T}\right)_V=A+B+C. \tag{5.25}$$

The macroscopic polarizability α consists of two components, one the optical polarizability α_{op} due to the displacement of electrons and the other, α_{ir} due to the displacement of ions. Thus

$$\alpha=\alpha_{op}+\alpha_{ir}. \tag{5.59}$$

Substituting for α in the B term in (5.25)

$$\begin{aligned}\frac{1}{(\varepsilon_s-1)(\varepsilon_s+2)}\left(\frac{\partial\varepsilon_s}{\partial T}\right)_P&=A+B+C\\&=-\frac{\beta}{3}+\frac{\beta}{3}\left[\frac{\alpha_{op}}{\alpha}\left\{\frac{V}{\alpha_{op}}\left(\frac{\partial\alpha_{op}}{\partial V}\right)_T\right\}\right.\\&\qquad\left.+\frac{\alpha_{ir}}{\alpha}\left\{\frac{V}{\alpha_{ir}}\left(\frac{\partial\alpha_{ir}}{\partial V}\right)_T\right\}\right]+C.\end{aligned} \tag{5.60}$$

Table 5.6. Static dielectric constant (ε_s) and loss ($\tan\delta$) at room temperature for some polycrystalline materials

Material	ε_s	$\tan\delta$ [10^{-2}]	Ref.
NaCl structure			
CaS	7.11	2	[5.137]
SrS	8.50	1.2	[5.138]
BaS	9.10		[5.138]
$Sr(NO_3)_2$ structure			
$Sr(NO_3)_2$	5.17	–	[5.139]
$Ba(NO_3)_2$	4.59	–	[5.139]
Langbeinite structure			
$K_2Zn_2(SO_4)_3$	7.63	0.3	[5.140]
$(NH_4)_2Mg_2(SO_4)_3$	4.80	0.05	[5.140]
Scheelite structure			
KIO_4	5.09	0.004	[5.141]
$NaIO_4$	6.30	0.01	[5.141]
Baryte structure			
$CaSO_4$	8.50	4.0	[5.142]
$SrSO_4$	8.93	1.8	[5.143]
$BaSO_4$	9.60	0.04	[5.143]
$PbSO_4$	19.43	7.2	[5.142]
Miscellaneous compounds			
$3MgCO_3.Mg(OH)_2.3H_2O$	5.13	1.0	[5.144]
Didymium oxide (R_2O_3)	55.0	42.0	[5.145]

In the above equation, α is obtained from the experimental values of static dielectric constant using Clausius–Mossotti relation. α_{op} is obtained from data on the refractive index n. α_{ir} is the difference between α and α_{op}. Following Havinga and Bosman [5.84], the IR polarizability derivative is obtained from the relation

$$\frac{V}{\alpha_{\text{ir}}}\left(\frac{\partial\alpha_{\text{ir}}}{\partial V}\right)_T = \frac{(r/\rho)^2 - 2(r/\rho) + 2}{3[(r/\rho) - 2)]}, \tag{5.61}$$

where (r/ρ) is the stiffness constant occurring in the Born repulsion term in the potential function for an ionic crystal.

In the absence of experimental values of strain optical polarizability constant, it is evaluated using the following equation:

$$\frac{V}{\alpha_{\text{op}}}\left(\frac{\partial\alpha_{\text{op}}}{\partial V}\right)_T = \frac{1}{3\alpha}\left[(r/\rho) - \frac{2}{(r/\rho) - 2}\right] + (2\alpha_{+-} - \alpha_{++})(1 - e^*/e) \tag{5.62}$$

proposed by Shankar et al. [5.172] where α_{+-} and α_{++} are the free ion polarisabilities of cation and anion, respectively, and $e^*/e = q^*$ is the effective ionic charge. Substituting the values of $\left(\frac{V}{\alpha_{\text{ir}}}\right)\left(\frac{\partial\alpha_{\text{ir}}}{\partial V}\right)_T$ and $\left(\frac{V}{\alpha_{\text{op}}}\right)\left(\frac{\partial\alpha_{\text{op}}}{\partial V}\right)_T$ in (5.60), the B term can be calculated and, in turn, C is obtained as the difference of $(A + B + C)$ and $(A + B)$.

Table 5.7. Dielectric constant (ε_s), loss ($\tan\delta$) at room temperature and temperature coefficient of dielectric constant $[(1/\varepsilon_s)(d\varepsilon_s/dT)]$ of some cubic single crystals

Material	ε_s	$\tan\delta$ $[10^{-2}]$	$(1/\varepsilon_s)(d\varepsilon_s/dT)$ $[10^{-4}\,K^{-1}]$	Ref.
NaCl structure				
KCl	4.81	–	2.85	[5.146]
KBr	4.87	–	2.73	[5.146]
RbCl	4.91	–	3.22	[5.146]
RbBr	4.88	–	2.87	[5.146]
LiF	9.30	1.0	2.82	[5.146]
NaF	5.10	–	3.05	[5.146]
CsCl structure				
CsI	5.70	0.2	1.55	[5.146]
Fluorite structure				
PbF_2	29	0.01	9.65	[5.147]
EuF_2	8.30	0.06	–	[5.148]
ThO_2	17.16	1.0	0.30	[5.149]
$NaClO_3$ structure				
$NaClO_3$	5.80	1.0	5.20	[5.150]
$NaBrO_3$	5.96	0.07	3.50	[5.150]
Eulytite structure				
$Bi_4(SiO_4)_3$	13.7	0.003	5.80	[5.151]
$Bi_4(GeO_4)_3$	16.4	0.001	3.24	[5.151]
Selenite structure				
$Bi_{12}SiO_{20}$	45.5	0.008	0.58	[5.152]
$Bi_{12}GeO_{20}$	40	0.5	0.6	[5.153]
Langbeinite structure				
$K_2Mn_2(SO_4)_3$	8.70	–	1.32	[5.154]
$(NH_4)_2Mn_2(SO_4)_3$	16.52	1.0	0.84	[5.154]
$Sr(NO_3)_2$ structure				
$Sr(NO_3)_2$	5.18	–	0.29	[5.155]
$Ba(NO_3)_2$	4.85	–	0.26	[5.156]
$Pb(NO_3)_2$	15.52	–	–	[5.157]
Garnet structure				
Alamandine–pyrope	11.07	0.80	0.80	[5.158]
$Gd_3Ga_5O_{12}$	12.08	0.30	0.30	[5.159]
$Nd_3Ga_5O_{12}$	12.19	0.47	0.47	[5.160]
$Tb_3Ga_5O_{12}$	12.48	0.15	0.15	[5.161]
$Y_3Fe_5O_{12}$	12.5	0.6	0.6	[5.162]
$Y_3Al_5O_{12}$	12.01	0.2	0.2	[5.163]
$Eu_3Ga_5O_{12}$	12.61	0.2	0.2	[5.164]
$Gd_3Sc_2Ga_5O_{12}$	12.15	0.45	0.45	[5.164]
$Y_{3.1}Tb_{.003}Ga_{4.9}O_{12}$	12.60	0.04	0.04	[5.164]
$Sm_{2.69}Ca_{0.31}Ga_{4.68}Zr_{0.32}O_{12}$	13.00	0.10	0.10	[5.164]
$Pr_{2.96}In_{1.34}Mg_{0.4}Zr_{0.4}Ga_{2.9}O_{12}$	12.61	0.05	0.05	[5.164]

Table 5.8. Dielectric constant (ε_s), loss ($\tan\delta$) and temperature coefficient of dielectric constant $[(1/\varepsilon_s)(d\varepsilon_s/dT))]$ at room temperature of some anisotropic single crystals

Material	Crystal class	Orientation	ε_s	$\tan\delta$ $[10^{-3}]$	$(1/\varepsilon_s)(d\varepsilon_s/dT)$ $[10^{-3}\,K^{-1}]$	Ref.
$BaSO_4$	Orthorhombic	100	7.56	6.0	1.60	[5.149]
		010	12.52	3.9	1.72	[5.149]
		001	8.52	6.0	0.26	[5.149]
$SrSO_4$	Orthorhombic	100	7.51	1.0	1.49	[5.149]
		010	18.46	2.5	0.05	[5.149]
		001	8.26	0.3	3.12	[5.149]
$CaWO_4$	Tetragonal	Along *c*-axis	9.10	10	–	[5.164]
$CaMoO_4$	Tetragonal	Along *c*-axis	7.02	–	–	[5.164]
$PbWO_4$	Tetragonal	Along *c*-axis	32.00	40	–	[5.164]
$LiNbO_3$	Trigonal	Along *c*-axis	77.10	1	–	[5.164]
$LaGaO_3$	Trigonal	Along *c*-axis	25.75	4.5	–	[5.164]
Corundum	Trigonal	Along optic axis	10.44	0.40	–	[5.165]
Topaz	Orthorhombic	Along optic axis	6.40	0.20	–	[5.165]
Talc	Monoclinic	Along optic axis	8.90	0.63	–	[5.165]
Tourmaline	Trigonal	Along optic axis	6.84	1.51	–	[5.159]
Apatite	Hexagonal	Along optic axis	9.45	5.15	–	[5.166]

The theory is applied to crystals with fluorite structure and to some cubic nitrates to evaluate the three terms. For PbF_2, EuF_2, ThO_2 and the nitrates, our experimental data was used. For alkaline earth fluorides and CdF_2 experimental data of earlier workers was taken from literature. The input data required for the evaluation of the three terms are given in [5.147,5.149,5.173]. The calculated values of strain polarizability constants and also the values of A, B and C are given in Table 5.10.

The C values in Table 5.10 show that it is negative for PbF_2 and $Pb(NO_3)_2$ while for all the other crystals it has a positive value. The experimental $(A+B+C)$ value is also negative at room temperature. For better understanding of this result, A, B, C values for PbF_2 and EuF_2 obtained at various temperatures from the corresponding input data at these temperatures are shown in Table 5.11. First, the results given in Table 5.11 show that for EuF_2, the temperature coefficient of dielectric constant is positive and the C value is also positive throughout the temperature range. However, in PbF_2, the temperature coefficient of dielectric constant is negative at low temperatures but becomes positive at elevated temperatures. Correspondingly, the partameter

Table 5.9. Dielectric constant (ε_s) and loss ($\tan\delta$) of some organic compounds, biomaterials and inorganic coordinated polymers

Material	ε_s	$\tan\delta[10^{-3}]$	Ref.
Organic compounds			
Urea	4.09	0.5	[5.167]
N-Methyl Urea	4.18	0.6	[5.167]
NN'-dimethyl urea	3.47	0.6	[5.167]
NN'-diethyl urea	2.89	0.6	[5.167]
Thiourea	10.2	0.6	[5.167]
NN'-dimethyl thiourea	4.04	1.0	[5.167]
NN'-diethyl thiourea	3.2	–	[5.167]
Oxamide	3.9	0.05	[5.167]
Dithiooxamide	4.49	0.05	[5.167]
Biomaterials			
Urinary stones – major organic components	3–5	0.01	[5.168]
Urinary stones – major inorganic components	5–7	0.05	[5.168]
Dental materials and apatites			
Zinc oxide	3.44	–	[5.168]
Zinc phosphate	7.43	–	[5.168]
Polycarboxylate	6.49	0.05	[5.168]
Glass ionomer	16.50	0.1	[5.168]
Composite resin	3.50	0.1	[5.168]
Detry-*RR*	2.36	0.1	[5.168]
Human tooth (adult)	7.46	0.040	[5.166]
Human tooth (milky)	10.92	0.093	[5.166]
Mineral hydroxyapatite	6.33	0.020	[5.166]
Synthetic hydroxyapatite	7.47	0.060	[5.166]
Proteins			
Lisozyme	3.00	–	[5.168]
Bovine serum albumin (BSA)	2.60	–	[5.168]
Casein	2.15	–	[5.168]
Gelatin	2.75	–	[5.168]
Typsin	2.98	–	[5.168]
Pepsin	5.00	–	[5.168]
Papain	5.25	–	[5.168]
Egg albumin	5.50	–	[5.168]
Inorganic coordinated polymers			
Dihydroxybenzoquinone (DHBQ)	2.89	–	[5.169]
Fe-DHBQ	12.54	0.2	[5.169]
Co-DHBQ	5.10	0.10	[5.169]
Ni-DHBQ	22.27	0.40	[5.169]
Cu-DHBQ	4.03	–	[5.169]
Zn-DHBQ	3.39	–	[5.169]
Dihydroxynapthaquinone (DHNQ)	3.02	0.1	[5.170]
Fe-DHNQ	213	2.2	[5.170]
Co-DHNQ	17.8	0.47	[5.170]
Ni-DHNQ	124	1.0	[5.170]
Cu-DHNQ	19.3	0.58	[5.170]
Zn-DHNQ	4.01	0.07	[5.170]

Table 5.9. Continued

Material	ε_s	$\tan\delta[10^{-3}]$	Ref.
Tetrahydroxy benzoquinone (THBQ)	3.76	–	[5.171]
Fe-THBQ	17.76	0.28	[5.171]
Co-THBQ	9.65	0.2	[5.171]
Ni-THBQ	6.79	0.19	[5.171]
Cu-THBQ	10.9	0.09	[5.171]
Zn-THBQ	12.9	0.19	[5.171]
Dihydroxy anthraquinone DHAQ	3.08	–	[5.171]
Zn-DHAQ	4.06	0.01	[5.171]

Table 5.10. Strain polarizability constants and calculated values of A, B, C for some fluorite type crystals and some nitrates at room temperature

Material	$\frac{V}{\alpha_{op}}\left(\frac{\partial\alpha_{op}}{\partial V}\right)_T$	$\frac{V}{\alpha_{ir}}\left(\frac{\partial\alpha_{ir}}{\partial V}\right)_T$	$(A+B+C)^a$ $[10^{-5}\,K^{-1}]$	A $[10^{-5}$ $K^{-1}])$	B $[10^{-5}$ $K^{-1}]$	C $[10^{-5}$ $K^{-1}]$	Ref.
CaF_2	0.40	3.00	3.29	−1.85	3.68	1.46	[5.147]
SrF_2	0.21	3.07	2.92	−1.93	3.68	1.17	[5.147]
BaF_2	0.08	3.14	2.68	−1.98	3.72	0.94	[5.147]
CdF_2	0.26	3.23	3.76	−2.16	3.89	2.03	[5.147]
PbF_2	−0.04	2.88	−0.79	−2.45	3.92	−2.26	[5.147]
EuF_2	0.14	2.90	1.75	−1.54	2.56	0.73	[5.147]
ThO_2	0.59	2.18	0.97	−0.86	1.22	0.63	[5.149]
$Sr(NO_3)_2$	0.51	4.21	5.10	−2.88	5.95	2.03	[5.173]
$Ba(NO_3)_2$	0.57	4.05	4.67	−1.82	3.67	2.82	[5.173]
$Pb(NO_3)_2$	0.33	4.15	0	−2.48	5.49	−3.01	[5.173]

[a] $[1/(\varepsilon_s - 1)(\varepsilon_s + 2)](\partial\varepsilon_s/\partial T)_P$ obtained from observed temperature variation of ε_s

C also is negative at low temperatures and positive at elevated temperatures. It is to be noted that A and B parameters are of the same sign in the entire suite of crystals and over the entire range of temperatures. Thus the contribution C plays a dominant role in determining the temperature variation of dielectric constant. It is to be noted that the C contribution reflects the anharmonic effects. Second, it is seen that in the absence of data on pressure variation of dielectric constant, reasonable values of A, B and C terms can be obtained by a semi-empirical approach based on a simple ionic model.

Based on the analysis of the data on dielectric constant for PbF_2, a similar procedure was adopted for strontium nitrate, barium nitrate and lead nitrate. It was observed that for the Sr and Ba nitrates the values of $(A+B+C)$ and C are negative while for Pb nitrate, as no change in the dielectric constant was found at room temperature, the C term is again negative.

Table 5.11. A, B, C values at various temperatures for PbF_2 and EuF_2 [5.147]

Material	Temperature [K]	$(A+B+C)$ [$10^{-5}\,K^{-1}$]	$A[10^{-5}$ $K^{-1}]$	$B[10^{-5}$ $K^{-1}]$	$C[10^{-5}$ $K^{-1}]$
PbF_2	180	−0.95	−2.41	3.86	−2.40
	295	−0.79	−2.45	3.92	−2.26
	395	5.30	−3.40	5.44	3.26
EuF_2	295	1.75	−1.54	2.56	0.73
	395	5.43	−1.89	3.14	4.80

Table 5.12. Effective ionic charge q^* calculated from (5.28)

Material	q^*	Ref.
MgO	0.60	[5.174]
CaO	0.64	[5.174]
SrO	0.78	[5.174]
BaO	0.66	[5.174]
CaS	0.41	[5.137]
SrS	0.58	[5.138]
BaS	0.49	[5.138]
PbF_2	0.93	[5.142]
EuF_2	0.86	[5.142]
$Sr(NO_3)_2$	0.84	[5.139]
$Ba(NO_3)_2$	0.71	[5.139]
$Pb(NO_3)_2$	1.04	[5.139]
$NaClO_3$	0.92	[5.142]
$NaBrO_3$	1.05	[5.142]
$CaSO_4$	1.10	[5.143]
$SrSO_4$	1.07	[5.143]
$BaSO_4$	0.94	[5.143]
$CaWO_4$	0.92	[5.164]
$CaMO_4$	0.78	[5.164]
$PbMO_4$	0.95	[5.164]

5.4.3 Application of Szigeti's Theory

Effective Ionic Charge

Employing the values of the static dielectric constants at room temperature in conjunction with other data, the effective ionic charges q^* have been calculated from (5.28). The static dielectric constants obtained in our work have been used in the calculation except for the alkaline earth oxides for which data from literature have been employed. Although Szigeti derived the equations specifically for the alkali halides, he extended it to triatomic crystals like CaF_2 and to non-cubic crystals like TiO_2. In the same spirit, we have extended the calculations to polyatomic and non-cubic crystals. The values are shown in Table 5.12.

It can be seen that except for the sulphides and to some extent the alkaline earth oxides, the values of q^* for other materials are close to unity indicating an ionic character. The values of q^* for oxides are in the range 0.6–0.8 showing lesser ionicity compared to alkali halides and alkaline earth nitrates and sulphates. For the alkaline earth sulphides q^* values are in the range 0.4–0.6 which is very much lower than unity. These low values indicate strong deviation from ionic nature.

Generally, the q^* values increase with the increase in cation size. Thus the q^* value for the Sr compounds is more than that for the Ca compounds. However, there is a reversal of this trend when we compare the q^* values of Sr and Ba compounds, the latter being systematically less. Srinivasan [5.175] considered a number of properties of the cubic nitrates and observed that the bonding in the Ba compound is less ionic than that in the Sr compound. The reversal of q^* values in Ba compounds vis-à-vis the Sr compounds may be related to this aspect.

Anharmonic Contribution to Dielectric Constant for some Fluorides and Nitrates

Using (5.35), the anharmonic parameter $G/(\varepsilon_s - \varepsilon_\infty)$ was evaluated for the alkaline earth fluorides, CdF_2, PbF_2 and EuF_2 and also for the nitrates, $Sr(NO_3)_2$, $Ba(NO_3)_2$ and $Pb(NO_3)_2$. The values of C are taken from Table 5.10. The results are given in Table 5.13. The values agree well with those obtained by Samara [5.50] for the fluorides. Amongst the nitrates, only $Pb(NO_3)_2$ has a negative value as in the case of PbF_2.

5.4.4 Spectroscopic Aspects

Self-energy

The relation between dielectric phenomenon and phonon processes was discussed briefly in Sect. 5.3.4. Using (5.38) and (5.39), the self-energy values

Table 5.13. Values of the anharmonicity parameter $G/(\varepsilon_s\text{-}\varepsilon_\infty)$

	$G/(\varepsilon_s - \varepsilon_\infty)$		
Crystal	Expt. [5.50]	Calculated value (5.37)	Ref.
CaF_2	0.040	0.046	[5.147]
SrF_2	0.032	0.035	[5.147]
BaF_2	0.036	0.033	[5.147]
CdF_2	–	0.070	[5.147]
PbF_2	−0.197	−0.225	[5.147]
EuF_2	–	0.028	[5.147]
$Sr(NO_3)_2$	–	0.06	[5.173]
$Ba(NO_3)_2$	–	0.09	[5.173]
$Pb(NO_3)_2$	–	−0.18	[5.173]

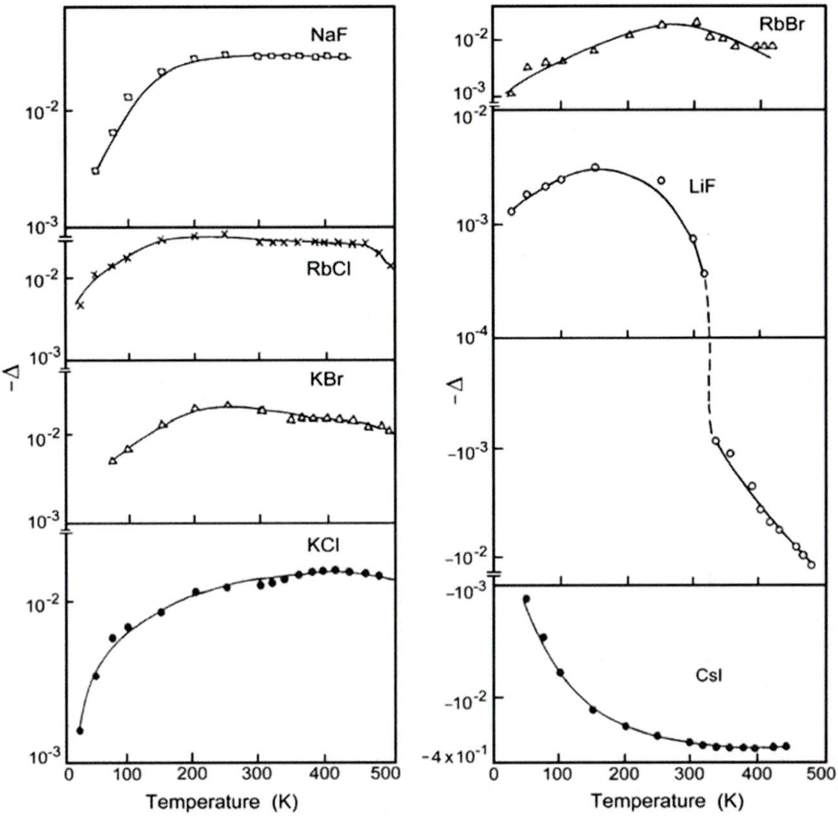

Fig. 5.24. Plots of self-energy ($-\Delta$) against temperature for some alkali halides

have been estimated for seven crystals (KCl, KBr, RbCl, RbBr, LiF, NaF and CsI). The input parameters required are: the static dielectric constant (ε_s), optical dielectric constant (ε_∞), dielectric constant at absolute zero $\varepsilon(0)$, Gruneisen parameter (γ_t), the parameter C and the volume thermal expansion coefficient β.

The values of $\varepsilon_\infty, \varepsilon(0)$ and ε_s up to 300 K are taken from Lowndes and Martin [5.37]. Beyond 300 K our values of ε_s are used. γ_t, C and β are from [5.176–5.178]. The self-energy values ($-\Delta$) are plotted as a function of temperature in Fig. 5.24. It can be seen that for KCl, KBr, RbCl and NaF they are positive and show only a slight variation with temperature beyond 300 K. For RbBr the values are positive with a decrease beyond 300 K. For LiF the values are positive up to 300 K and beyond 300 K they are negative with a larger variation at higher temperatures. For CsI ($-\Delta$) is negative for the entire range.

The negative anharmonic contributions obtained for CsI and LiF at higher temperatures indicate that in these compounds the fourth-order contribution to the lattice potential is quite large. For the rest of the crystals, third-order contribution dominates. These results agree qualitatively with the results obtained by Lowndes and Rastogi [5.176] purely from spectroscopic data.

Damping and Dielectric Constant

The values of the damping constant (γ_d/ω_t) were calculated for LiF, KBr and NaI from dielectric data using (5.43):

$$(\gamma_d/\omega_t) = \frac{(\varepsilon_s - \varepsilon_\infty)\omega_t}{(2nk\omega)_{max}} = \frac{\varepsilon_s - \varepsilon_\infty}{(2nk)_{max}}. \quad (5.43)$$

Values at different temperatures were calculated using the input parameters at those temperatures. Dielectric constant data for LiF and KBr were from [5.46] and for NaI from [5.37].

Figure 5.25 shows the variation of γ_d/ω_t and ε_s as a function of temperature. It can be seen that for KBr and NaI there is a good correlation between the values of γ_d/ω_t and ε_s. For LiF there is a clear departure in γ_d/ω_t and ε_s values from linearity. The damping is expected to show a variation with temperature as $AT + BT^2$. The first term is due to the third-order harmonicity contribution and the second term represents the fourth-order contribution. Departure from linearity is observed at higher temperatures in the case of LiF.

The self-energy values obtained from our dielectric data as input yielded reasonably good results. The temperature dependence of damping is explained well in terms of the temperature dependence of dielectric constant.

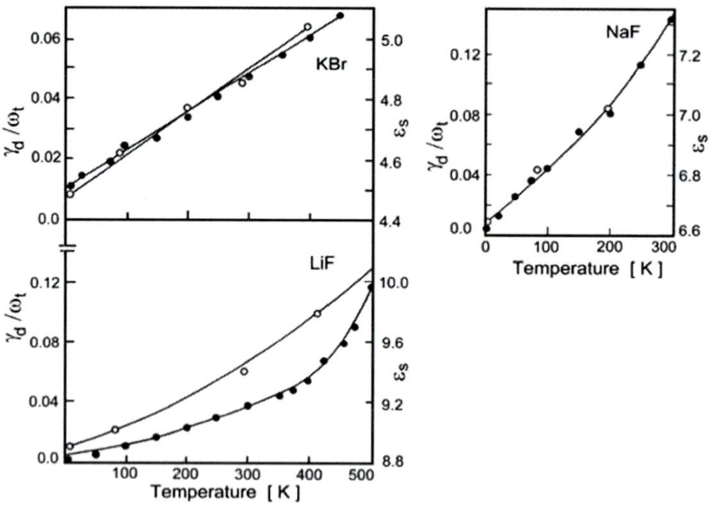

Fig. 5.25. Plots of γ_d/ω_t (open circles) and ε_s (full circles) against temperature for KBr, LiF and NaF crystals

5.4.5 Polaron Conduction in Garnets

A brief account of polaron conduction was discussed earlier in Sect. 5.3.6. The conductivity data obtained for the garnets is examined in the light of polaron conduction. Polaron parameters have been calculated for some rare earth garnets using the dielectric data obtained by us together with other required parameters from literature.

The input data for evaluating polaron parameters are given in Table 5.14. The calculated values of the polaron parameters for three garnets are shown in Table 5.15. From the table, it is seen that the coupling constant α_e is >5 for all the garnets and that the polaron radius r_p is less than the interionic distance. These are the required conditions for small polaron conduction mechanism to be effective in a given lattice.

Apart from these calculations, further supporting evidence is obtained from AC conductivity (σ_{AC}) measurements carried out by us at low temperatures. From these measurements, it is observed that the plots of log (σ_{AC}) vs. ω^2 for YGG and YIG at 200 K and NdGG at 300 K are linear. A typical plot is shown in Fig. 5.26. As mentioned in Sect. 5.3.6, this is to be expected for small polaron conduction. Petrov et al. [5.182] interpreted high temperature conduction in rare earth iron garnets in terms of small polaron conduction. Detailed measurement of AC conductivity of YIG and its interpretation in terms of small polaron theory was carried out by Sirdeshmukh et al. [5.162]. Thus calculations of polaron parameters and measurements of AC conductivity at low as well as high temperature clearly indicate the role of small polarons in the conduction in rare earth garnets.

Table 5.14. Input parameters for evaluating polaron parameters for conduction in garnets; lattice constants (a) and Debye temperatures (θ_D) from [5.179, 5.180], ε_∞ from [5.181] and ε_s from [5.160, 5.162, 5.164]

Crystal	a [Å]	ε_∞	ε_s	θ_D [K]
$Y_3Ga_5O_{12}$	12.27	3.72	12.60	584.5
$Y_3Fe_5O_{12}$	12.37	4.84	12.55	562.4
$Nd_3Ga_5O_{12}$	12.54	3.72	12.19	513.4

Table 5.15. Polaron parameters for garnets

Polaron parameter	YGG	YIG	NdGG
$r_p = 1/2(\pi/6N)^{1/3}$ [Å]	0.90	0.918	0.930
$\varepsilon_p = \varepsilon_\infty\varepsilon_s/(\varepsilon_s - \varepsilon_\infty)$	5.39	7.90	5.28
$W_p = e^2/4\pi\varepsilon_0\varepsilon_p r_p$ [eV]	1.47	0.99	1.46
$\alpha_e = (e^2/\varepsilon_p)(m^*/2\hbar^3\omega_D)^{1/2}$	17.07	13.85	18.15
$m_p = 0.02m^*{\alpha_e}^4$ [$10^5 m_e$]	3.75	4.33	4.63
$\omega_D = k_B\theta_D/\hbar$ [Hz]	7.64	7.35	7.35

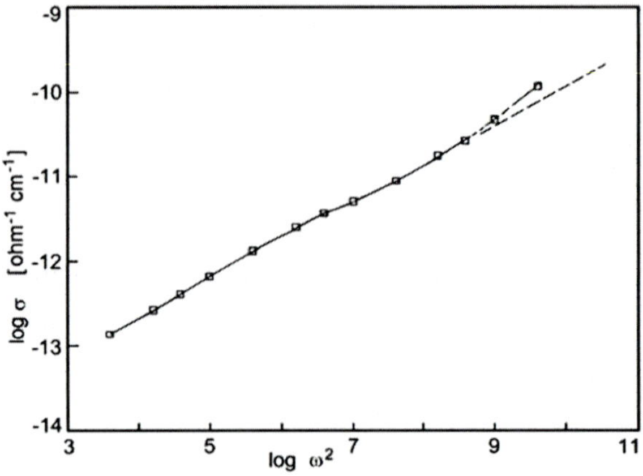

Fig. 5.26. Plot of $\log \sigma$ vs. $\log \omega^2$ for YIG at 200 K

5.4.6 Dielectric Constant and Additivity of Polarizability

The rule of additivity of polarizability discussed earlier in Sect. 5.3.7 was employed to evaluate the dielectric constants of mineral garnets and some rare earth garnets.

Mineral garnets are represented by the formula $X_3Al_2Si_3O_{12}$, where X = Fe for almandine, Mg for pyrope and Mn for spessartine. The molecular polarizability α_D is expressed in terms of the polarizabilities of three simple molecules as

$$\alpha_D[X_3Al_2Si_3O_{12}) = 3\alpha_D(SiO_2) + 3\alpha_D(XO) + \alpha_D(Al_2O_3). \quad (5.63)$$

The molecular polarizabilities of the compounds on the right hand side of the above equation are evaluated from Clausius–Mossotti relation in conjunction with the data on density and dielectric constant [5.158]. From the sum of the polarizabilities, the dielectric constants of the three pure mineral garnets are obtained. The mineral garnet for which the dielectric constant was measured was a mixed crystal of almandine (Alm), pyrope (Pyr) and spessertine (Spe). The composition of the mineral sample was determined by chemical analysis. For the calculation of dielectric constant, the composition was taken into account to obtain the polarizability.

For the rare earth garnets the equation for the polarizabilities is given by

$$2\alpha_D(X_3Y_5O_{12}) = 3\alpha_D(X_2O_3) + 5\alpha_D(Y_2O_3), \quad (5.64)$$

where X may be Ga, Y, Nd, etc. and Y represents Ga, Fe or Al. Literature values of dielectric constants of rare earth sesquioxides and their densities were used to evaluate the polarizabilities of oxides on the right-hand side. The value for Ga_2O_3 was taken from Shannon et al. [5.61].

The calculated values of dielectric constant along with the measured values are given in Table 5.16. It can be seen that there is reasonably good agreement

Table 5.16. Measured and calculated dielectric constants of some garnets

Material	ε Calculated	Measured	Ref.
Mineral Garnets			
$Fe_3Al_2Si_3O_{12}$ (Almandine)	12.52	–	[5.158]
$Mg_3Al_2Si_3O_{12}$ (Pyrope)	10.42	–	[5.158]
$Mn_3Al_2Si_3O_{12}$ (Spessertine)	13.16	–	[5.158]
Alm—Pyr–Spe (mineral sample)	11.73	11.70	[5.158]
Rare earth garnets			
$Gd_3Ga_5O_{12}$	12.43	12.08	[5.159]
$Tb_3Ga_5O_{12}$	12.08	12.45	[5.159]
$Nd_3Ga_5O_{12}$	13.16	12.19	[5.159]
$Y_3Fe_5O_{12}$	11.22	12.55	[5.159]
$Eu_3Ga_5O_{12}$	12.12	12.61	[5.164]

in all cases. Thus, the oxide additivity rule is useful for predicting the dielectric constant of a complex material if the molecular polarization of the simple molecular components is known.

5.4.7 Ferroelectric Behaviour in $NaClO_3$ and $NaBrO_3$

Mason [5.183] has measured the dielectric constant of $NaClO_3$ at 1 kHz in the temperature range −100°C to 200°C. He represented the results in the form of Curie–Weiss type equations. The equations predict a ferroelectric transition at 320°C and 414°C for $NaClO_3$ and $NaBrO_3$ respectively. These temperatures are above the melting points. The ferroelectricity was attributed to a change in the orientation of the chlorate (or bromate) ion or to apex reversal of the halogens. Prasad Rao et al. [5.184] and Dawson [5.185] carried out Raman studies on these crystals but failed to observe any soft mode behaviour.

In order to reexamine the suggestion of ferroelectricity in these crystals we have measured the dielectric constant as a function of temperature from −180°C to 240°C. We find that the data could be fitted to two types of equations. One, in the form of a polynomial and the other, a Curie–Weiss type equation. These equations are:

Curie–Weiss form:

$$\varepsilon = 5.4 + \frac{106}{273 - t} \quad \text{for sodium chlorate,} \tag{5.65}$$

$$\varepsilon = 5.4 + \frac{143}{281 - t} \quad \text{for sodium bromate.} \tag{5.66}$$

Polynomials:

$$\varepsilon = 5.69 + 3.06 \times 10^{-3} t \quad \text{at} -180°\text{C} < t < 150°\text{C}. \tag{5.67}$$

$$\varepsilon = 7.51 - 3.55 \times 10^{-2} t + 1.64 \times 10^{-4} t^2 \quad \text{at } t > 170°\text{C} \tag{5.68}$$

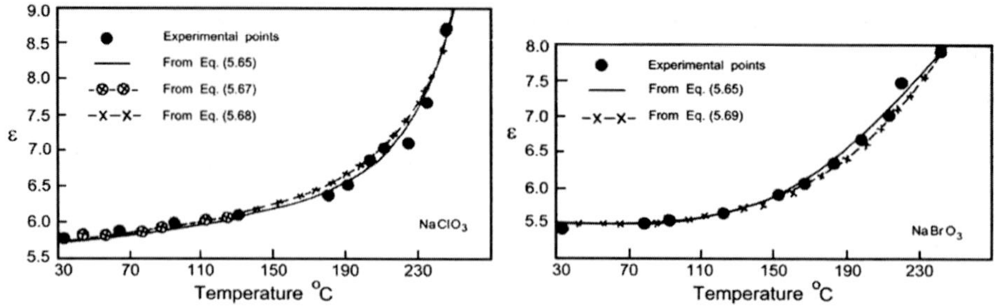

Fig. 5.27. Plots of ε at 10^6 Hz against temperature for $NaClO_3$ and $NaBrO_3$

for sodium chlorate, and

$$\varepsilon = 6.45 - 1.34 \times 10^{-2}t + 9 \times 10^{-5}t^2 \quad \text{at } 30°\text{C} < t < 170°\text{C}. \tag{5.69}$$

for sodium bromate.

In Fig. 5.27 the values of ε calculated from these equations are plotted along with the experimental values. It can be seen that the two sets of equations represent experimental data equally well within the limits of errors. Thus, the data obtained by Mason could as well have been represented by a power series equation which does not suggest ferroelectricity. Our analysis indicates that it is hazardous to predict ferroelectricity in a crystal merely on the basis of an empirical Curie–Weiss type of equation.

5.4.8 Analysis of Conductivity Data

Activation Energy and Defect Formation Energy

From the data on dielectric constant and loss as a function of temperature, conductivity σ is obtained. Activation energies (E_{ex} and E_{in}) are obtained from the slopes of the log σ vs. $1/T$ plot in the extrinsic and intrinsic regions respectively (Sect. 5.3.5). The results are shown in Tables 5.17–5.19. E_{ex} is equal to E_{M} the energy for migration of the defect. On the other hand, E_{in} is the sum of E_{M} and half the defect formation energy E_{f}. The values of E_{f} are also given in the table.

The value of 0.92 for E_{f} in the case of PbF_2 is lower than the rest of the materials in the table and alkaline earth fluorides which are of the order 2–3 eV [5.186]. The smaller value for PbF_2 has been attributed by Samara [5.50] to the large dielectric constant ε, existence of soft mode and strong anharmonic effects. The value for EuF_2 is of the same order as those of alkaline earth fluorides. The results indicate that the predominant defect in the crystals with fluorite structure is Frenkel anion pair. The mobile charge carrier is F^- ion. On the basis of polarizability values, the conduction in $NaClO_3$ and $NaBrO_3$ can be attributed to the migration of the Na^+ ions. The formation energy is highest for $Bi_{12}SiO_{20}$ amongst the materials studied.

For the nitrates, the values are from DC conductivity measurements. The values of energy of motion and formation indicate that while the formation of

Table 5.17. Activation energies (E_{ex} and E_{in}) and defect formation energy (E_f)

Material	E_{ex} (eV)	E_{in} (eV)	Temp. range (°C)	E_f (eV)	Ref.
PbF_2	0.32	0.78	130–230	0.92	[5.142]
EuF_2	0.16	1.05	200–300	1.78	[5.148]
ThO_2	0.19	0.79	300–500	1.20	[5.149]
$NaClO_3$	0.18	2.78	200–240	5.22	[5.150]
$NaBrO_3$	0.22	1.60	180–240	2.76	[5.150]
$Bi_4(SiO_4)_3$	0.07	0.95	230–400	1.76	[5.151]
$Bi_4(GeO_4)_3$	0.06	1.20	230–400	2.28	[5.151]
$Bi_{12}SiO_{20}$	0.05	5.33	300–360	10.56	[5.152]
$Bi_{12}GeO_{20}$	0.49	1.12	300–480	1.26	[5.153]
$Sr(NO_3)_2$	0.28	4.15	180–230	7.74	[5.155]
$Ba(NO_3)_2$	0.34	2.03	210–320	3.38	[5.156]
$Pb(NO_3)_2$	0.71	1.86	150–210	2.30	[5.157]

Table 5.18. Activation energies in the extrinsic (E_{ex}) and intrinsic region (E_{in})

Material	Direction	E_{ex} (eV)	E_{in} (eV)	Temp. range (°C)	Ref.
$BaSO_4$	100	0.62	0.78	130–230	[5.149]
	010	0.49	1.25	280–480	[5.149]
	001	0.58	0.99	270–460	[5.149]
$SrSO_4$	100	0.52	0.79	210–490	[5.149]
	010	0.51	1.23	180–490	[5.149]
	001	0.58	0.83	220–490	[5.149]
$K_2Zn_2(SO_4)_3$	–	0.47	0.67	280–460	[5.140]
$(NH_4)_2Mg_2(SO_4)_3$	–	0.90	1.98	310–460	[5.140]
$K_2Mn_2(SO_4)_3$	100	0.58	1.04	300–480	[5.154]
$(NH_4)_2Mn_2(SO_4)_3$	100	0.71	1.45	300–500	[5.154]

the defect is difficult, the motion is much easier than in fluorite type crystals. This is perhaps due to the large space of about 40 Å^3 available in the lattice. It is observed that in ionic solids the size of the ions and the polarizability influence the process of conduction.

For some materials for which the defect mechanism is not clear; only the values of E_{ex} and E_{in} are given (Table 5.18). In the case of $SrSO_4$ and $BaSO_4$, measurements were made in three directions. In both cases, the activation energies are higher in the <010> direction than in the other two directions.

For the minerals and garnets, the log σ vs. $1/T$ plot is linear at higher temperatures but highly non-linear at moderate temperatures. In these cases (Table 5.19) only the values of activation energies are given.

Relaxation Effects

The loss ($\tan\delta$) versus temperature curves at different frequencies for $PbF_2, EuF_2, Bi_4(SiO_4)_3, Bi_4(GeO_4)_3$ and YIG show broad peaks at certain temperatures. A shift in the peaks is observed for a change in the frequency. The peaks shift towards higher frequencies. The relaxation peaks observed in YIG are shown in Fig. 5.28. Such peaks have been observed by Agrawal [5.187]

Table 5.19. Conductivity (σ) and activation energy for garnets and mineral samples

Material	σ_{AC} at 300°C $[(10^{-9}\Omega\ \mathrm{cm})^{-1}]$	σ_{DC} at 300°C $[(10^{-9}\Omega\ \mathrm{cm})^{-1}]$	Activation energy [eV]		Ref.
			<200°C	>200°C	
Garnets	7.01	6.80	0.18	0.57	[5.158]
GGG	8.09	0.21	0.15	0.84	[5.159]
NdGG	23.60	0.78	0.10	0.37	[5.160]
TbGG	0.45	0.04	0.47	0.78	[5.161]
YIG	6.0×10^4	1.8×10^3	0.42	0.87	[5.162]
Minerals					
Corundum	93	17.0	0.08	0.31	[5.165]
Topaz	11.20	9.00	0.07	0.65	[5.165]
Opal	184	41.00	0.06	0.23	[5.165]
Talc	1.4×10^3	740	0.06	0.11	[5.165]
Tourmaline	456	400	0.19	0.46	[5.159]
Apatite	0.63	0.52	–	0.88	[5.159]

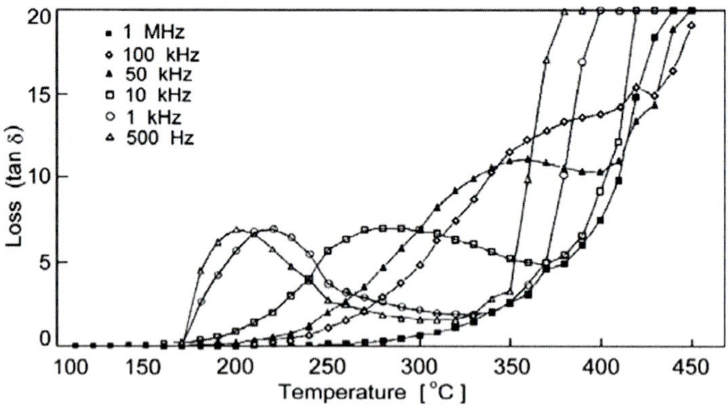

Fig. 5.28. Variation of $\tan\delta$ with temperature at different frequencies showing relaxation peaks

in doped CaF_2 crystals and Young and Frederikse [5.49] in the case of CdF_2 due to dipolar impurity. The EuF_2 crystal contained Sm^{3+} impurity and the YIG crystal contained Si^{4+} impurity. The observed relaxations are attributed to the dipoles formed by these impurities. The relaxation peaks in the other crystals may be due to unidentified impurities. The frequencies and temperatures at which the peaks are observed are given in Table 5.20.

The activation energy for the rotation E_R of the dipoles in the crystals has been calculated using the relation

$$f = f_0 e^{-E_R/k_B T}, \tag{5.70}$$

Table 5.20. Temperatures at which relaxation peaks are observed for PbF_2, EuF_2, $Bi_4(SiO_4)_3$ and $Bi_4(GeO_4)_3$

Crystal	10^3 Hz	10^4 Hz	10^5 Hz	10^6 Hz	E_R [eV]
PbF_2	–	95°C	120°C	220°C	0.06
EuF_2	230°C	285°C	–	–	0.82
$Bi_4(SiO_4)_3$	270°C	340°C	–	–	0.25
$Bi_4(GeO_4)_3$	250°C	300°C	360°C	–	0.23
YIG	230°C	270°C	340°C	360°C	0.75

where f is the frequency of the relaxation peak at temperature T, f_0 a constant and E_R the activation energy for the rotation of the dipoles. The values of the activation energy are also given in Table 5.20. The peaks are rather broad when compared to the peaks generally observed. As such the values obtained for activation energies are to be considered as approximate.

5.4.9 γ-Irradiation Studies

Measurements of dielectric constant and loss ($\tan\delta$) have been carried out on γ-irradiated crystals of some alkali halides [5.188], cubic nitrates [5.173] and some minerals [5.159]. The crystals were irradiated with γ-radiation at a dose rate of 112 krad h^{-1} from a Co^{60} source.

For alkali halides, the results are shown in Figs. 5.29 and 5.30. The curves show large variations in dielectric constant. Peaks are observed (Fig. 5.29) for 25 h of irradiation and again for 70 h; in KCl a peak is observed only at 25 h of irradiation. The percentage increase in dielectric constant is more for

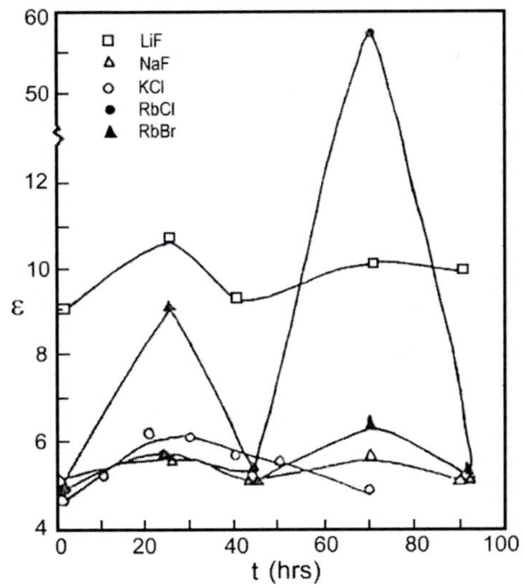

Fig. 5.29. Plots of ε vs. time of γ-irradiation at 100 kHz for some alkali halides

Table 5.21. Dielectric constant (ε) and tan δ at 100 kHz of some minerals

Sample	Before irradiation		After irradiation (72 hours)		After annealing	
	ε	tan δ	ε	tan δ	ε	tan δ
Garnet	11.75	–	12.53	–	11.88	–
Corundum	10.54	0.001	11.13	0.005	10.60	0.002
Topaz	6.42	–	6.55	–	6.47	–
Tourmaline	8.85	–	9.35	–	8.90	–
Apatite	9.67	–	10.57	–	9.76	–
Talc	16.48	0.4	22.05	0.50	17.63	0.42

RbBr and RbCl than for NaF and LiF. Similar trend is observed in loss also. (Fig. 5.30). It may be mentioned that in the study on γ-irradiation effects on F-centre growth curves and microhardness, Thirmal Rao [5.189] found that the percentage increase was more in rubidium halides when compared to those of LiF and NaF. Our observation agrees with that of Thirmal Rao.

In an earlier study, Subrahmanyam [5.190] observed that in X-irradiated NaCl crystal the dielectric constant increases considerably up to a certain duration, but on prolonged irradiation, the original value is restored. It is believed that the observed initial increase is due to creation of F-centres. The charged vacancies accumulate leading to space charge till equilibrium is reached after which the dielectric constant reaches the original value. Further irradiation results in the repeat of the process.

The dielectric constants of the nitrates were determined and their optical absorption recorded. The results for $Sr(NO_3)_2, Ba(NO_3)_2$ and $Pb(NO_3)_2$ are shown in Figs. 5.31 and 5.32. It can be seen that peaks are observed in the ε vs. irradiation time curves (Fig. 5.31) as in the case of alkali halides. Broad peaks were also observed in optical spectra recorded after 40 h of irradiation. The bands are the F-centre bands. Tagaya [5.191] reported F-centre formation in $Sr(NO_3)_2$. In these crystals, the presence of free NO_3^- ion products NO_2^- and O_2 is reported by Tagaya [5.192] and Bhatta et al. [5.193]. On further irradiation beyond 80 h, random recombination takes place [5.193]. Prolonged irradiation results in recovery of original ions and initial dielectric constant.

The results for the mineral samples are shown in Table 5.21. An increase in dielectric constant was observed after 72 h of irradiation. The effect was stable and no change in dielectric constant was observed even after 15 days. In order to examine the effect of annealing, the samples were kept at 300°C for about 10 h and the measurements were repeated for all the samples at room temperature. The results show that for all the mineral samples the dielectric constant assumes values close to those for unirradiated samples.

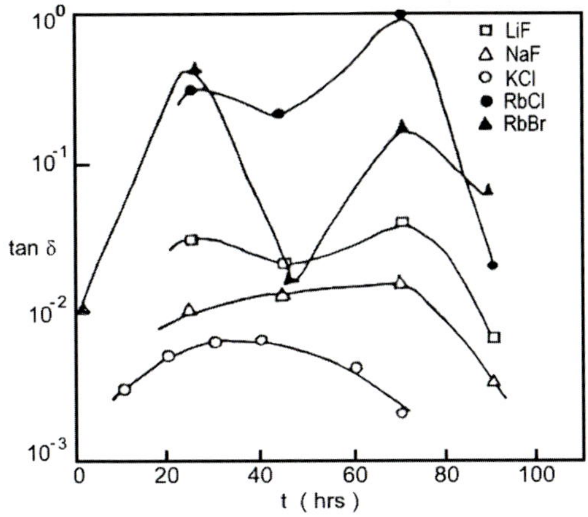

Fig. 5.30. Plots of $\tan\delta$ vs. time of γ-irradiation at 100 kHz for some alkali halides

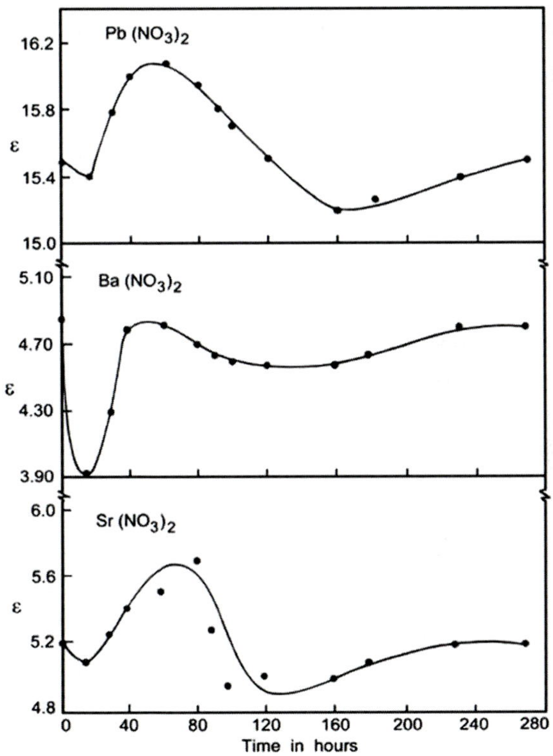

Fig. 5.31. Variation of ε with time of γ-irradiation for $Sr(NO_3)_2$, $Ba(NO_3)_2$ and $Pb(NO_3)_2$

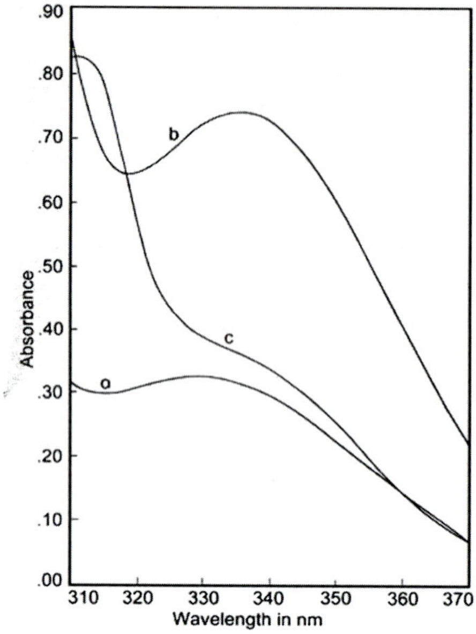

Fig. 5.32. Optical absorption spectra for (**a**) $Sr(NO_3)_2$, (**b**) $Ba(NO_3)_2$ and (**c**) $Pb(NO_3)_2$

5.4.10 Dielectric Properties and Protein Hydration

Krishna Murthy [5.168] carried out dielectric measurements on the proteins lysozyme, bovine serum albumin (BSA), casein, gelatin, trypsin, pepsin, papain and egg albumin. Hydration effect was studied by measurements at various levels of hydration. Though the samples were dehydrated and preserved in vacuum at room temperature, the samples were found to retain certain amount of water [5.35]. Hence, the dielectric constant for dry protein is obtained by extrapolating the curves of dielectric constant vs. wt% of water to zero hydration. The values thus obtained are shown in Table 5.22.

Figure 5.33 shows the plots of $\varepsilon''(\varepsilon \tan \delta)$ for dry and wet (more than 20 wt% of water) samples of lysozyme and gelatin. As the frequency range

Table 5.22. Hydration data on proteins

Sample	ε of dry protein	h_C	w_{ads} [wt % of H_2O]	w_{dip} [wt% of H_2O]
Lysozyme	3.00	0.91	4.1	7.8
BSA	2.60	1.02	3.5	6.5
Casein	2.15	1.08	6.5	12.5
Gelatin	2.75	1.10	5.5	11.5
Trypsin	2.98	0.85	3.5	7.25
Pepsin	5.00	0.95	3.8	6.0
Papain	5.25	0.90	2.0	4.8
Egg albumin	5.50	0.87	2.5	4.5

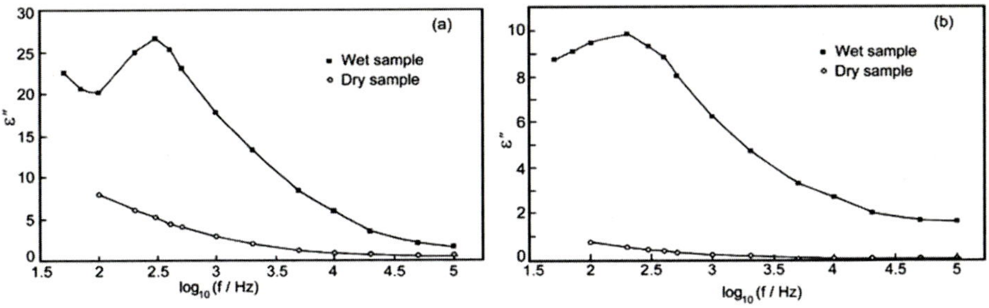

Fig. 5.33. Variation of ε'' with frequency for (**a**) lysozyme and (**b**) gelatin

was limited to 100 kHz, only α dispersion was observed. The α dispersion peaks are clearly seen for both the hydrated samples. From the peak values, the relaxation times are obtained as 5.3×10^{-4} and 7.8×10^{-4} s for lysozyme and gelatin, respectively. These observations are similar to α-dispersion for urease shown in Fig. 5.22.

6

Theoretical Evaluation of Some Crystal Properties

6.1 Introduction

Most of the work described in the earlier chapters relates to the experimental determination of properties of lattice dynamical interest. Such measurements can be complemented by theoretical evaluation of physical properties of crystals. Such evaluations facilitate a check, on the one hand, on the assumed models and, on the other, on the reliability of the measured properties.

With this approach, an attempt has been made to evaluate elastic constants, coefficients of thermal expansion, Debye temperatures and Gruneisen parameters. One approach is to calculate crystal properties from the interatomic potentials. This approach is employed to evaluate the elastic constants, the thermal expansion and the Gruneisen parameter. Another approach is to evaluate some crystal properties from other measured properties. Thus the Debye temperature can be calculated from the elastic constants and the Gruneisen constant from the pressure derivatives of elastic properties. The Debye temperatures and the Gruneisen parameter can also be calculated from properties other than the elastic properties. The results of such calculations are presented.

6.2 Elastic Constants of Ionic Crystals

Krishnan and Roy [6.1, 6.2] made a detailed analysis of lattice vibrations in alkali halides. One of the results of their study is the derivation of expressions for the elastic constants of ionic crystals with NaCl and CsCl structure in terms of the interionic potential. They assumed these crystals to be ionic with a lattice potential ϕ given by

$$\phi = -(Az^2e^2/r) + b\exp(-r/\rho), \tag{6.1}$$

where A is the Madelung constant, z the valency, e the electron charge, r the interionic distance, b a constant and r/ρ the repulsion parameter. The

Table 6.1. Elastic constants of some alkali halides; calculated values from [6.2] and observed values from [6.3]

Crystal		C_{11}	C_{12}	C_{44}
			$[10^{11}\,\text{dynes cm}^{-2}]$	
LiF	Calc.	10.7	4.9	4.9
	Obs.	11.8	4.33	6.28
NaCl	Calc.	5.0	1.3	1.3
	Obs.	4.87	1.24	1.26
KBr	Calc.	3.5	0.7	0.7
	Obs.	3.46	0.58	0.50
KI	Calc.	3.1	0.5	0.5
	Obs.	2.67	0.43	0.42

parameter r/ρ can be obtained from the compressibility ψ using the relation:

$$\psi = 18r^4/\{Az^2e^2[(r/\rho) - 2]\}. \tag{6.2}$$

The expressions for the elastic constants C_{ij} derived by Krishnan and Roy [6.2] are:

$$C_{11} = \{2A[(r/\rho) + 1)] - 6X\}(z^2e^2/12r^4), \tag{6.3}$$

$$C_{12} = C_{44} = (-3A + 3X)(z^2e^2/12r^4). \tag{6.4}$$

Here X is a lattice sum with a value of 3.14 for the NaCl lattice. Similar expressions were obtained for the CsCl lattice. The equality of C_{12} and C_{44} is the consequence of the assumption of central forces.

Krishnan and Roy calculated the elastic constants of several alkali halides. Their values for some alkali halides are given in Table 6.1. These calculated values were compared by Krishnan [6.3] with experimental values. The agreement is fair indicating that the Krishnan–Roy theory is applicable for ionic crystals.

The transition metal oxides have the NaCl structure. Their effective ionic charges calculated by Kinney and O'Keefe [6.4] from Szigeti's theory indicate a high degree of ionicity. Sirdeshmukh and Subhadra [6.5] applied the Krishnan–Roy theory to this group of crystals. The results are given in Table 6.2 along with experimental data. The agreement of calculated values of C_{11} and C_{12} with experimental values is as good as in the case of alkali halides. The agreement is not so good in the case of C_{44}. As mentioned earlier, any central force model would predict $C_{12} = C_{44}$.

Subhadra and Sirdeshmukh [6.7] also applied the Krishnan and Roy theory to LiH, an ionic crystal with NaCl structure. Here, several different compressibility values were available which yielded different r/ρ values and hence, different values of C_{ij}s (Table 6.3). There was considerable difference in these different sets. The strategy adopted to sift the best set and hence the best value of compressibility is discussed in Chap. 8.

Table 6.2. Elastic constants of transition metal oxides; calculated values from [6.5] and observed values from [6.6]

Crystal		C_{11}	C_{12}	C_{44}
			$[10^{11}\ \mathrm{dyne\,cm^{-2}}]$	
CoO	Calc.	26.5	15.6	15.6
	Obs.	25.6	14.4	8.0
NiO	Calc.	26.3	16.8	16.8
	Obs.	27.0	12.5	10.5
MnO	Calc.	16.7	13.1	13.1
	Obs.	22.2	11.0	7.8

Table 6.3. Elastic constants of LiH; calculated values from [6.7]; sources of compressibility as given in Chap. 8

$\psi[10^{-12}\ \mathrm{cm^2\,dyne^{-1}}]$	r/ρ	C_{11}	$C_{12} = C_{44}$
		$[10^{11}\ \mathrm{dyne\,cm^{-2}}]$	
2.0	5.88	5.8	4.6
2.83	4.74	1.4	4.6
3.70	4.10	−1.1	4.6
4.38	3.77	−2.4	4.6

6.3 Coefficient of Thermal Expansion from Interatomic Potentials

6.3.1 Thermal Expansion Coefficient of Crystals with Fluorite Structure

Smyth [6.8] has developed a simple method to estimate the coefficient of thermal expansion α of a crystal from its interatomic potential ϕ. According to this method, α is given by

$$\alpha = -(C_V/2r)(\mathrm{d}^3\phi/\mathrm{d}r^3)/[(\mathrm{d}^2\phi/\mathrm{d}r^2)]^2. \tag{6.5}$$

This method has been used by Sharma and Madan [6.9] and Kachhava and Saxena [6.10] to estimate the thermal expansion coefficient of various crystals.

The interatomic potential ϕ for an ionic crystal [6.11] is given by:

$$\phi = -\phi_\mathrm{e} - \phi_\mathrm{d}(6) - \phi_\mathrm{d}(8) + \phi_\mathrm{r}, \tag{6.6}$$

where the terms, respectively, represent the Coulomb, dipole–dipole, dipole–quadrupole and repulsion interactions. Benson and Dempsey [6.12] carried out a term-by-term evaluation of the lattice potential and lattice energy of fluorite-type crystals. In their work, the repulsion term was, further, composed of four terms $\phi_\mathrm{r}(1)$, $\phi_\mathrm{r}(2)$, $\phi_\mathrm{r}(3)$, and $\phi_\mathrm{r}(4)$ which corresponded to different ion pairs. Subjecting the potential in (6.6) to (6.5), Sirdeshmukh [6.13] obtained

Table 6.4. Values of coefficients of thermal expansion (α) of some fluorite-type crystals

Crystal	$\alpha[10^{-6}\ \text{per}^\circ\text{C}]$	
	Calculated [6.13]	Observed [Chap. 2]
CaF_2	19.0	18.3
SrF_2	19.6	18.0
BaF_2	21.0	19.0

$$\begin{aligned}\alpha = -\left(\frac{C_V}{2}\right) & [6\phi_e + 336\phi_d(6) + 720\phi_d(8) - (r/\rho)^3\phi_r(1) - (k_2 r/\rho)^3\phi_r(2) \\ & - (k_1 r/\rho)^3\phi_r(3) - (k_1 r/\rho)^3\phi_r(4)]/ \\ & [-2\phi_e - 42\phi_d(6) - 72\phi_d(8) + (r/\rho)^2\phi_r(1) + (k_2 r/\rho)^2\phi_r(2) \\ & + (k_1 r/\rho)^2\phi_r(3) + (k_1 r/\rho)^2\phi_r(4)]^2,\end{aligned} \tag{6.7}$$

where k_1 and k_2 are constants in the repulsion terms. Room temperature values of C_V were calculated from known values of the Debye temperature. Other parameters were taken from Benson and Dempsey [6.12].

Values of the coefficients of thermal expansion calculated from (6.7) are given in Table 6.4 along with experimental values. It can be seen that the agreement is fairly good. These calculations provide additional evidence that these crystals are essentially ionic in character.

6.3.2 Thermal Expansion Coefficients of Some Anisotropic Elements

The interatomic potential in metals is described by the Morse potential:

$$\phi = L\{\exp[-2a(r - r_0)] - 2\exp[-a(r - r_0)]\}, \tag{6.8}$$

where L is heat of vaporization, a the Morse potential parameter, r the interatomic distance and r_0 its equilibrium value. Applying (6.5), Joshi and Mitra [6.14] obtained the following expression for α:

$$\alpha = (C_V/2)(\psi/2VL)^{1/2}, \tag{6.9}$$

where V is the molar volume. Joshi and Mitra [6.14] used (6.9) to estimate α for some cubic elements and found good agreement with experimental values.

Sirdeshmukh [6.15, 6.16] assumed the Morse potential to be applicable to anisotropic systems like the hcp and trigonal elements and used (6.9) to estimate the average (isotropic) coefficient of thermal expansion of some hcp and trigonal elements. Values of the quantities in (6.9) were taken from Gschneidner [6.17]. The results are given in Table 6.5 The agreement between the calculated values is fair, particularly in the case of hcp metals.

Table 6.5. Average (isotropic) coefficient of expansion α of some hcp and trigonal elements; calculated values from [6.15,6.16], observed values from [6.17]

Crystal	α [10^{-6}per°C]	
	Calc.	Obs.
hcp		
Be	13.0	11.5
Cd	32.9	30.6
Er	17.9	12.3
Ho	14.5	10.7
Lu	15.3	8.1
Mg	30.7	25.7
Os	5.2	4.7
Re	5.4	6.6
Ru	7.8	9.4
Tc	6.2	8.1
Ti	10.9	8.3
Tm	17.5	13.3
Y	15.4	12.0
Zn	31.7	29.7
Trigonal		
Te	29.6	16.7
Se	50.1	36.9

6.4 Debye Temperatures from Elastic Constants

6.4.1 General

The Debye characteristic temperature θ for an isotropic solid is given by the well-known relation:

$$\theta = (h/k_{\mathrm{B}})(9pN_{\mathrm{A}}/4\pi V)^{1/3}(1/C_{\mathrm{l}}^3 + 3/C_{\mathrm{t}}^3)^{-1/3}, \tag{6.10}$$

where h is Planck's constant, k_{B} the Boltzmann constant, N_{A} the Avogadro number, p the number of vibrating units in the molecule, V the molar volume and C_{l} and C_{t} the velocities of longitudinal and transverse waves. Since the condition of isotropy is not usually satisfied, the velocities have to be averaged over all directions and the Debye temperature is given by:

$$\theta = (h/k_{\mathrm{B}})(9pN_A/4\pi V)^{1/3}C_{\mathrm{m}}, \tag{6.11}$$

where the mean sound velocity C_m is defined by

$$3C_{\mathrm{m}}^{-3} = \int_0^{4\pi} \sum_{i=1}^{3} C_i^{-3}\,\mathrm{d}\Omega/4\pi. \tag{6.12}$$

$\mathrm{d}\Omega$ denotes an element of solid angle and C_i the three sound velocities which depend on direction and are obtained from Christoffel's equations of elasticity. The mean sound velocity must be evaluated from the integral of (6.12) to obtain the Debye temperature. Equation (6.12) is not integrable analytically and hence numerical methods, exact or approximate, are employed to obtain values of C_{m} from (6.12). Alers [6.18] has reviewed the methods of calculating Debye temperature from single crystal elastic constants for all the crystal systems. Anderson [6.19] has discussed the methods of obtaining the Debye temperature from polycrystalline elastic data. These methods are briefly reviewed. The discussion is limited to crystals of cubic and tetragonal symmetries.

6.4.2 Debye Temperatures from Single Crystal Elastic Constants

Direct Numerical Integration

For cubic crystals, the integrand in (6.12) simplifies considerably as there are only three elastic constants, viz. C_{11}, C_{12} and C_{44}. Hence it is enough to consider direction cosines confined to 1/48 of a sphere. In this discussion, the values of θ obtained by this method will be designated as θ_{exact}.

de Launay's Method (DLM)

de Launay [6.20] developed the lattice dynamics of cubic metals using a model consisting of ion points embedded in an electron gas. The ion point lattice is of Born–Von Karman type with nearest neighbour interactions. The electron gas is assumed to possess bulk modulus but no shear modulus. With this model, de Launay obtained expressions for Debye temperatures of cubic solids in terms of single crystal elastic constants with and without the electron gas contribution.

In a subsequent paper de Launay [6.21] showed that the cumbersome expression given earlier reduces to the simpler form:

$$\theta = (h/k_{\mathrm{B}})(9pN_{\mathrm{A}}/4\pi V)^{1/3}(C_{44}/\rho)^{1/2}[9f_{\mathrm{E}}/(18+\sqrt{3})]^{1/3}, \tag{6.13}$$

where f_{E} is a function of elastic constants. The values of this function are provided in the form of tables. de Launay's method has been very popular and has been found to give highly reliable results.

Graphical Method (GM)

Marcus [6.22] has given a graphical method to calculate the Debye temperature in terms of the ratios of elastic constants. The equation is

$$\theta = (h/k_{\mathrm{B}})(9N_{\mathrm{A}}/4\pi V)^{1/3}(C_{11}/\rho)^{1/2}g, \tag{6.14}$$

where the parameter g is obtained from the curves between $(C_{11}-C_{12})/2C_{11}$ and C_{44}/C_{11}. Alers [6.18] observed that this method gives values in fair agreement with those from numerical integration.

Power Series Expansion (PSEM)

Hopf and Lechner [6.23] proposed a polynomial in the place of the integrand in (6.12). An improvement of this method by Quimby and Sutton [6.24] and Sutton [6.25] gives the relation:

$$\theta = (h/k_{\mathrm{B}})(9N_{\mathrm{A}}/4\pi V)^{1/3}(C_{11} - C_{44}/\rho)^{1/2}Y_{\mathrm{R}}^{-1/3}(1+\Delta)^{-1/3}, \qquad (6.15)$$

where Y_{R} is a function of elastic constants and Δ a correction term. Leibfried [6.26] has given a graphical method to calculate the terms Y_{R} and Δ.

Harmonic Series Expansion (HSEM)

This method of solving (6.12) involves the substitution of a series of cubic harmonics in the place of the integration. Houston [6.27] originally developed this method and Bhatia and Tauber [6.28] have applied it to calculate the Debye temperature with a three term expansion. Betts et al. [6.29] gave a six term expansion. The values of θ thus calculated are referred to as θ_{series}. The equation is given by

$$\theta = (h/k_{\mathrm{B}})(9pN_{\mathrm{A}}/4\pi V)^{1/3}\rho^{-1/2}J^{-1/3}. \qquad (6.16)$$

For cubic crystals, Betts et al. [6.29] showed that

$$\begin{aligned} J = {} & 0.10878I_1 + 0.070803I_2 + 0.016184I_3 + 0.352656I_4 + 0.287712I_5 \\ & + 0.163861I_6 \end{aligned} \qquad (6.17)$$

and

$$\begin{aligned}
I_1 &= 2[C_{44}]^{-3/2} + [C_{11}]^{-3/2}, \\
I_2 &= [C_{44}]^{-3/2} + [(C_{11} - C_{12})/2]^{-3/2} + [(C_{11} + C_{12} + 2C_{44})/2]^{-3/2}, \\
I_3 &= 2[(C_{44} + C_{11} - C_{12})/3]^{-3/2} + [(C_{11} + 2C_{12} + 4C_{44})/3]^{-3/2}, \\
I_4 &= [C_{44}]^{-3/2} + [C_{44} + (\alpha/2) + \{(9\alpha^2 + 16\beta^2)^{1/2}/10\}]^{-3/2} \\
&\quad + [C_{44} + (\alpha/2) - \{(9\alpha^2 + 16\beta^2)^{1/2}/10\}]^{-3/2}, \\
I_5 &= [C_{44} + (\alpha - \beta)/6]^{-3/2} + [C_{44} + \{(5\alpha + \beta)/12\} \\
&\quad + \{(9\alpha^2 + 33\beta^2 - 6\alpha\beta)^{1/2}/12\}]^{-3/2} + [C_{44} + \{(5\alpha + \beta)/12\} \\
&\quad - \{(9\alpha^2 + 33\beta^2 - 6\alpha\beta)^{1/2}/12\}]^{-3/2}, \\
I_6 &= [C_{44} + 4(\alpha - \beta)/9]^{-3/2} + [C_{44} + \{(5\alpha + 4\beta)/18\} \\
&\quad + \{(9\alpha^2 + 48\beta^2 + 24\alpha\beta)^{1/2}/18\}]^{-3/2} + [C_{44} + \{(5\alpha + 4\beta)/18\} \\
&\quad - \{(9\alpha^2 + 48\beta^2 + 24\alpha\beta)^{1/2}/18\}]^{-3/2}.
\end{aligned}$$

In these equations

$$\alpha = (C_{11} - C_{44}) \qquad (6.18)$$

and

$$\beta = (C_{12} + C_{44}). \tag{6.19}$$

Betts [6.30] extended the expansion to 9 and 15 terms. Konti and Varshni [6.31, 6.32] used these expressions to calculate the Debye temperatures of several crystals.

Similarly for tetragonal symmetry, Betts et al. [6.33] gave

$$1,290J = 288I_1 + 144I_2 + 512I_3 - 125I_4 + 96I_5 + 375I_6, \tag{6.20}$$

where

$$I_1 = [C_{44}]^{-3/2} + [C_{66}]^{-3/2} + [C_{11}]^{-3/2},$$
$$I_2 = 2[C_{44}]^{-3/2} + [C_{33}]^{-3/2},$$

$$\begin{aligned} I_3 = {} & [(C_{66} + C_{44})/2]^{-3/2} + \Big[\frac{1}{4}(C_{11} + 2C_{44} + C_{33}) \\ & + \frac{1}{4}\left\{(C_{11} - C_{33})^2 + 4(C_{13} + C_{44})^2\right\}^{1/2}\Big]^{-3/2} \\ & + \Big[\frac{1}{4}(C_{11} + 2C_{44} + C_{33}) - \frac{1}{4}\left\{(C_{11} - C_{33})^2 + 4(C_{13} + C_{44})^2\right\}^{1/2}\Big]^{-3/2}, \\ I_4 = {} & \Big[\frac{1}{5}(4C_{66} + C_{44})\Big]^{-3/2} + \Big[\frac{1}{10}(4C_{11} + 5C_{44} + C_{33}) \\ & + \frac{1}{10}\left\{(4C_{11} - 3C_{44} - C_{33})^2 + 16\,(C_{13} + C_{44})^2\right\}^{1/2}\Big]^{-3/2} \\ & + \Big[\frac{1}{10}(4C_{11} + 5C_{44} + C_{33}) - \frac{1}{10}\left\{(4C_{11} - 3C_{44} - C_{33})^2 \right. \\ & \left. + 16(C_{13} + C_{44})^2\right\}^{1/2}\Big]^{-3/2}, \\ I_5 = {} & [C_{44}]^{-3/2} + \Big[\frac{1}{2}(C_{11} - C_{12})\Big]^{-3/2} + \Big[\frac{1}{2}(C_{11} + 2C_{66} + C_{12})\Big]^{-3/2} \\ I_6 = {} & \Big[\frac{1}{5}(2C_{11} - 2C_{12} + C_{44})\Big]^{-3/2} + \Big[\frac{1}{10}(2C_{11} + 2C_{12} + 4C_{66} + 5C_{44} + C_{33}) \\ & + \frac{1}{10}\left\{(2C_{11} + 2C_{12} + 4C_{66} - 3C_{44} - C_{33})^2 + 16(C_{13} + C_{44})^2\right\}^{1/2}\Big]^{-3/2} \\ & + \Big[\frac{1}{10}(2C_{11} + 2C_{12} + 4C_{66} + 5C_{44} + C_{33}) \\ & - \frac{1}{10}\left\{(2C_{11} + 2C_{12} + 4C_{66} - 3C_{44} - C_{33})^2 + 16(C_{13} + C_{44})^2\right\}^{1/2}\Big]^{-3/2}. \end{aligned}$$

Fedorov's Method (FM)

Fedorov [6.34, 6.35, 6.37] presented a general theory for the propagation of elastic waves in homogeneous crystalline solids. Fedorov and Bystrova [6.38] developed two approximations to calculate the Debye temperatures based on Fedorov's theory. The expression for the Debye temperature based on the second approximation, which is found to be more accurate is given by

$$\theta = (h/k_{\mathrm{B}})(9pN_{\mathrm{A}}/4\pi V)^{1/3} I^{-1/3}, \tag{6.21}$$

where I is a function of the C_{ij}s.

6.4.3 θ from Polycrystalline Elastic Data

The problem of averaging single crystal elastic constants to obtain effective values of polycrystalline elastic constants is important in metallurgy and geophysics [6.39, 6.40] and in other fields where mechanical properties of aggregates are of interest.

Voigt [6.41] showed that the elastic moduli of a polycrystal can be calculated from the single crystal elastic moduli assuming the strain to be uniform in all grains. Reuss [6.42] assumed a uniform stress over all grains and obtained the relations for the elastic moduli of crystals. The general expressions for obtaining the average moduli from the single crystal elastic constants are given below. The suffixes V and R relate to the two assumptions discussed.

$$\left.\begin{array}{ll} \textit{Voigt} & \textit{Reuss} \\ K_{\mathrm{V}} = (A + 2B)/3 & K_{\mathrm{R}} = 1/(3a + 6b) \\ G_{\mathrm{V}} = (A - B + 3C)/5 & G_{\mathrm{R}} = 5/(4a - 4b + 3c) \end{array}\right\} \tag{6.22}$$

where

$$\begin{array}{ll} 3A = C_{11} + C_{22} + C_{33} & 3a = S_{11} + S_{22} + S_{33} \\ 3B = C_{23} + C_{31} + C_{12} & 3b = S_{23} + S_{31} + S_{12} \\ 3C = C_{44} + C_{55} + C_{66} & 3c = S_{44} + S_{55} + S_{66} \end{array}$$

Here C_{ij}s are the elastic stiffness constants and S_{ij}s the compliances. Hill [6.43] showed that the Voigt and Reuss averages are the upper and lower bounds for the elastic moduli, the actual value lying in between. He suggested the arithmetic mean as a better choice. The geometric mean has also been recommended. Several other averages of the Voigt and Reuss values have been proposed [6.44–6.47].

Binnie [6.48] and, independently, Gilvarry [6.49] were the first to suggest that the Debye temperature could be calculated from the averaged elastic moduli. Anderson [6.50] recommended the use of the HAM (Hill arithmetic mean) moduli rather than the V or R moduli. Anderson's working equations are:

$$\begin{aligned} C_{\mathrm{l}} &= [\{K_{\mathrm{HAM}} + (4/3)G_{\mathrm{HAM}}\}/\rho]^{1/2} \\ C_{\mathrm{t}} &= (G_{\mathrm{HAM}})/\rho]^{1/2} \end{aligned} \tag{6.23}$$

The equations for C_m and θ are as already given (see (6.10) and (6.11)).

Verma and Aggarwal [6.51] empirically proposed an average (VAA) for the shear modulus given by

$$G_{VAA} = [(G_V + 2G_R)/3]. \tag{6.24}$$

Verma and Aggarwal [6.51,6.52] calculated the Debye temperatures of a number of crystals and found that the Debye temperatures obtained from this averaging scheme (θ_{VAA}) are closer to θ_{exact} than the values from other averaging schemes. Recently, Verma et al. [6.53] and Basu and Verma [6.54] extended this averaging scheme to the orthorhombic, hexagonal and tetragonal systems and again observed that the θ_{VAA} values are superior to those from other methods.

Reddy [6.55] made an interesting suggestion that the average velocity C_m in (6.12) can be identified with the sound velocity values obtained from polycrystalline samples. Substituting these experimental values in (6.11), he obtained the Debye temperatures of a number of solids.

6.4.4 Brief Review of Earlier Work

There is a large body of data on the values of Debye temperatures of crystals calculated from elastic constants. These reports are scattered in literature. We discuss here some of the reports, limiting the choice to papers in which a large amount of data are given or where the results have led to important observations.

Betts et al. [6.29] employed the method developed by them to evaluate the Debye temperatures of nine cubic crystals. Betts [6.30] extended the calculations to a number of crystals with NaCl structure (other than alkali halides) and some alkaline earth nitrates. Betts et al. [6.33] used the expressions developed by them for lower symmetries to obtain the Debye temperatures of a number of tetragonal, trigonal and hexagonal crystals. Bolef and Menes [6.56] who measured the elastic constants of several alkali bromides and iodides calculated the Debye temperatures using them and compared the values with the results from specific heats. Joshi and Mitra [6.57, 6.58] used the method developed by Betts et al. [6.29–6.33] to calculate the Debye temperatures of a large number of crystals for which data on elastic constants had become available. The crystals included metals like Be and In, semiconductors like GaAs and GaSb, compound crystals like $CaCO_3$ and $NaNO_3$ and organic crystals like $C_6H_{12}N_4$ and $C_{14}H_{10}O_2$.

Robie and Edwards [6.59] used the numerical integration method to calculate the mean sound velocity and Debye temperatures of 14 solids, many of them of geological importance. The interesting features of their work are the construction of the velocity surfaces for calcite and a study of the effect of limiting the angular range in the integration on the values of the mean velocity.

Using the most up-to-date data which were then available on the elastic constants of the alkali halides at liquid helium temperatures, Lewis et al. [6.60] calculated the Debye temperatures by DLM for 13 alkali halides and compared them with the values from specific heats; agreement was observed within limits of errors. An interesting observation made by Lewis et al. was that when the Debye temperatures were tabulated ion-wise, equality of Debye temperatures was noted in a diagonal block. Thus the Debye temperatures are roughly equal for RbCl, KBr and NaI ($\theta \approx 170\,\mathrm{K}$) or for RbBr and KI ($\theta \approx 135\,\mathrm{K}$).

Konti and Varshni [6.31] calculated the Debye temperatures of 24 cubic elements from the single crystal elastic constants using PSEM, FM and the HSEM. They observed good agreement between the values obtained by these different methods and also with the values from specific heats. Konti and Varshni [6.32] calculated the Debye temperatures of all the alkali halides except LiI and CsF using low temperature elastic constants. Good agreement was observed with values from specific heats. Konti and Varshni looked for regularities in the Debye temperature values. They observed linear relationships between the Debye temperature and reduced mass for a given alkali halide group (common alkali ion or common halogen ion). From these relationships a value of 210 K was estimated for LiI for which elastic constant data was not available.

Michard et al. [6.61] determined the elastic constants of nitrates of Sr, Ba and Pb, which are cubic, and calculated their Debye temperatures. The values of the Debye temperature were found to decrease from strontium to lead and thus depend on the size of the metal ion. Large discrepancies exist in the values of the Debye temperatures of these crystals as reported by Michard et al. [6.61], Anderson [6.19] and Betts [6.30]. Seshagiri Rao et al. [6.62] determined the elastic moduli of an important class of solids, the ferrites, and calculated the Debye temperatures. Alers [6.18] in his review gave data on Debye temperatures for nearly 30 solids. Anderson [6.19] reviewed the methods of evaluation of elastic moduli for polycrystalline aggregates. He applied these methods for calculating the Debye temperatures of over a hundred solids belonging to cubic, tetragonal, trigonal, hexagonal and orthorhombic symmetries.

Extensive calculations of Debye temperatures have been made by Verma and co-workers. Aggarwal and Verma [6.63] calculated the Debye temperatures of about a dozen II–VI and III–V compounds using the arithmetic mean and geometric mean values of Voigt and Reuss averages of elastic moduli and recommended the use of GMA values. Aggarwal et al. [6.64] calculated the Debye temperatures of about two dozen elements using the de Launay's method and the VRH method. Basu and Verma [6.65] calculated the Debye temperatures of a number of crystals with the perovskite structure and found that the values correlate with the volume. Verma et al. [6.53] and Basu and Verma [6.54] extended the VAA averaging scheme for the calculation of Debye temperatures of orthorhombic, tetragonal and hexagonal crystals. Ledbetter [6.66] calculated the Debye temperatures for six elemental solids using nine approximations and found that the geometrical mean approximation gave best results. Padial et al. [6.67] extended the calculations to 81 crystals.

6.4.5 Some of Our Results

Using the methods discussed in the preceding sections, the Debye temperatures of several systems have been calculated. These results are now discussed system wise.

NaCl-Type Crystals

Subhadra and Sirdeshmukh [6.68] calculated the Debye temperatures of eight crystals with NaCl structure using elastic constants. The VAA method was used. The results are given in Table 6.6.

For KCN, the specific heat calculated from these Debye temperatures will give only the contribution of the relative vibrations of the K^+ and $(CN)^-$ ions and the contribution due to any rotation of the $(CN)^-$ ion will have to be estimated separately.

Rare Earth Chalcogenides

Gopi Krishna et al. [6.69] evaluated the Debye temperatures of six rare earth chalcogenides with NaCl structure using VAAM. The elastic constants and lattice constants used in these calculations and the results are given in Table 6.7.

Baldwin and Tompson [6.70] pointed out that the Debye temperatures of related compounds correlate linearly with the reciprocal of the lattice constant. The values of θ for the rare earth chalcogenides are plotted in Fig. 6.1 against a^{-1}. The Debye temperatures of some rare earth pnictides reported by Mullen et al. [6.71] and some europium chalcogenides reported by Shapira and

Table 6.6. Debye temperatures (θ) of some crystals with NaCl structure

Crystal	C_{11}	C_{12}	C_{44}	ρ	θ [K]	
	$[10^{11}$ dyne cm$^{-2}]$			[g cm^{-3}]	[6.68]	By other methods
NH_4I	2.448	0.428	0.240	2.514	119	First report
CsF	4.406	1.380	0.703	4.627	157	First report
MnO	22.2	11.0	7.8	5.365	533	First report
CoO	25.6	14.4	8.6	6.438	514	First report
MnS	14.3	7.4	5.2	3.99	420	First report
BaO	11.2	3.6	3.4	5.992	291	232–370 (sp. heats)
NiO	27.0	12.5	10.5	6.809	578	595 (sp. heats)
KCN	1.936	1.164	0.1545	1.548	123	

Table 6.7. Debye temperatures (θ) of some rare earth compounds with NaCl structure

Crystal	C_{11}	C_{12}	C_{44}	θ [K]	Lattice constant
	[10^{11} dynes cm^{-2}]			(VAAM)	a[Å]
SmS	1.2	0.11	0.25	268	5.97
YS	2.5	0.2	0.3	387	5.50
$Sm_{0.75}Y_{0.25}S$	1.35	−0.5	0.3	307	5.69
GdS	3.1	0.3	0.3	325	5.56
TmSe	1.85	−0.65	0.26	258	5.68
TmTe	1.02	0.06	0.186	193	6.35
TmS				(323)	

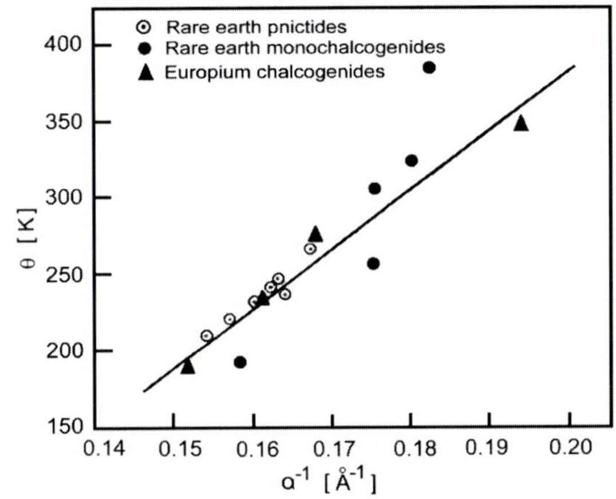

Fig. 6.1. Plot of Debye temperature (θ) against reciprocal of lattice constant (a^{-1})

Reed [6.72] are also included in the plot. The plot is linear. The data point for YS shows maximum deviation from the linear plot. The Debye temperature of TmS is not known. Using the value of the lattice constant, the value of θ is estimated from the plot. This value is given in Table 6.7 in parenthesis.

Rare Earth Garnets

Nagaiah et al. [6.73] evaluated the Debye temperatures of seven rare earth garnets using the elastic constants reported by Haussuhl and Mateika [6.74] and Haussuhl et al. [6.75]. HSEM was used for the calculation. The resulting values are given in Table 6.8.

As in the case of the rare earth chalcogenides, a plot of θ vs. a^{-1} is drawn (Fig. 6.2). It is linear. For three rare earth garnets, the Debye temperatures are not known. These estimated values are also given in Table 6.8.

Table 6.8. Debye temperatures (θ) of some rare earth garnets

Crystal	C_{11} [10^{11} dynes cm^{-2}]	C_{12}	C_{44}	ρ[g cm^{-3}]	Lattice constant a[Å]	θ[K]	Estimated from $\theta - a^{-1}$ correlation
$Y_3Al_5O_{12}$	33.35	11.1	11.5	4.559	12.003	747	
$Y_3Ga_5O_{12}$	29.0	11.7	9.55	5.799	12.273	585	
$Y_3Fe_5O_{12}$	26.85	10.95	7.65	5.169	12.376	562	
$Eu_3Fe_5O_{12}$	25.1	10.7	7.62	6.307	12.498	499	
$Gd_3Ga_5O_{12}$	28.70	11.60	9.04	7.087	12.379	515	
$Nd_3Ga_5O_{12}$	27.78	11.15	8.38	6.614	12.506	513	
$Sm_3Ga_5O_{12}$	28.076	11.35	8.60	6.857	12.437	512	
$Sm_3Fe_5O_{12}$					12.530		463
$Ho_3Fe_5O_{12}$					12.380		545
$Lu_3Fe_5O_{12}$					12.277		584

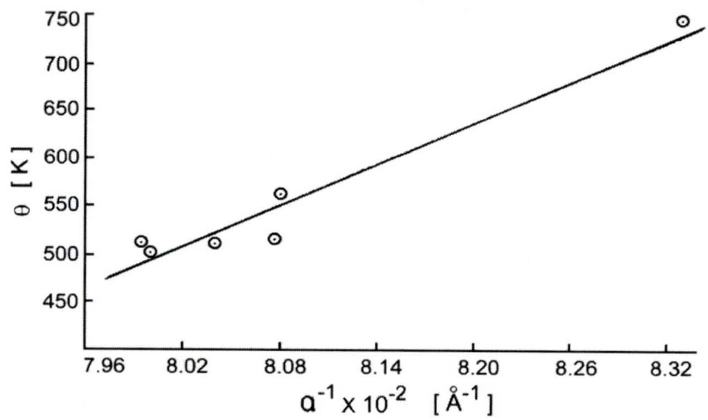

Fig. 6.2. Debye characteristic temperature (θ) vs. reciprocal of the lattice constant (a^{-1}) for rare earth garnets

Bismuth Orthosilicate and Orthogermanate

Bismuth orthosilicate [$Bi_4(SiO_4)_3$] and [$Bi_4(GeO_4)_3$] orthogermanate are cubic crystals with several applications. Rao et al. [6.76] calculated the Debye temperatures of these two crystals by the VAA method using the elastic constants reported by Schweppe [6.77]. The input data and the resulting values are given in Table 6.9. Nedyukha and Chernyi [6.78] suggested that the parameter $m\theta^2$ (m being the mass) is a measure of the interatomic bond strength. The values of this parameter are also given in Table 6.9. m$\theta^2$is larger for the silicate than the germanate indicating that the bonding is stronger in the silicate than in the germanate.

Rutile-Type Crystals

Sirdeshmukh and Rao [6.79] calculated the Debye temperatures of MgF_2 and MnF_2 using the elastic constants reported by Haussuhl [6.80]. These

Table 6.9. Debye temperatures (θ) of $Bi_4(SiO_4)_3$ and $Bi_4(_4)_3$

Crystal	C_{11}	C_{12}	C_{44}	V	θ [K] VAAM	mθ^2 [10^6]
	[10^{11} dyne cm^{-2}]			[cm^3]		
$Bi_4(SiO_4)_3$	11.58	2.70	4.36	175.5	747	9.91
$Bi_4(GeO_4)_3$	13.57	2.27	5.18	163.5	271	11.67

Table 6.10. Debye temperatures and mean amplitudes of vibration ($\sqrt{\langle u^2\rangle}$) of some rutile-type crystals

Crystal	θ [K]	$\sqrt{\langle u^2\rangle}$ [Å]
MgF_2	602	0.139
MnF_2	367	0.180
TiO_2	726	0.076

are tetragonal crystals and HSEM was used for the calculations. The results are given in Table 6.10. The value of θ for TiO_2 is also quoted from Anderson [6.19]. The r.m.s. amplitude of vibration $\sqrt{\langle u^2\rangle}$ is calculated from θ; these values are included in Table 6.10. Nedyukha and Chernyi [6.78] pointed out that the values of $\sqrt{\langle u^2\rangle}$ are a measure of the strength of interatomic binding, being smaller for stronger binding. It is seen that the bonding in TiO_2 is much stronger than that in MgF_2 and MnF_2. Since covalent bonding is stronger than ionic bonding, TiO_2 may be partially covalent.

Zircon ($ZrSiO_4$)

Zircon is a tetragonal crystal. There are three reports on its elastic constants. They are given in Table 6.11. It can be seen that the values of C_{ij}s in sets ii and iii differ from those in set i by an order. Sirdeshmukh and Subhadra [6.81] calculated the Debye temperature by Anderson's arithmetic averaging method. These values are given in Table 6.11 along with a range of values from specific heats. It is clear that the values of θ from sets ii and iii are closer to the value from specific heat than the value from set i. That sets ii and iii are physically more acceptable has been shown by these results as well as by several other considerations [6.81].

Crystals with Scheelite Structure

Crystals with scheelite structure are useful as laser hosts. These crystals are tetragonal. The elastic constants and derived Debye temperatures are available only for some of them. For several other crystals in this family, neither specific heat data nor the elastic constants were available. Sirdeshmukh and Rao [6.85] estimated the Debye temperatures from Lindemann's [6.86] melting theory. According to this theory, the Debye temperature θ is given by

$$\theta = C[T_m/MV^{2/3}]^{1/2}, \tag{6.25}$$

Table 6.11. C_{ij}s and Debye temperature (θ) of zircon

Set no.	C_{11}	C_{33}	C_{44}	C_{66}	C_{12}	C_{13}	Ref.	θ [K]	
	[10^{11} dynes cm^{-2}]							From C_{ij}	From sp. heats
i	5.7	4.6	1.4	3.2	2.5	0.5	[6.82]	300	
ii	33.01	38.05	7.33	3.97	10.75	15.42	[6.83]	601	
iii	42.37	49.00	11.36	4.85	7.03	14.95	[6.84]	725	
									600–900

Table 6.12. Debye temperatures (θ) for some scheelite-type crystals

Crystal	T_m [K]	θ [K]	
		From elastic constants [6.87]	From T_m [6.85]
$SrMoO_4$	1738	252	
$CaWO_4$	1853	241	
$CaMoO_4$	1753	300	
$SrWO_4$	1808		219
$BaWO_4$	1748		178
$PbWO_4$	1401		164
$BaMoO_4$	1673		212
$PbMoO_4$	1341		181
KIO_4	855		172
$KReO_4$	823		154

where C is a constant for a family of related crystals, T_m the melting temperature, M the mass and V the molar volume. Substituting known values of θ of $CaWO_4$, $CaMoO_4$ and $SrMoO_4$ [6.87] in (6.25), a value of 355.4 was obtained for C. Using this value of C and known values of T_m, Debye temperatures for several crystals were calculated; these are given in Table 6.12. It may be mentioned that the Debye temperature for $PbMoO_4$ was estimated for the first time by this method. The calculation of θ from elastic constants for $PbMoO_4$ was done later by Suryanarayana [6.88] and Basu and Verma [6.89]; they reported values of 190 K and 54 K, respectively. The value reported by Suryanarayana [6.88] agrees with the value 181 K obtained from (6.25).

6.5 Gruneisen Parameter

Some aspects of Gruneisen's theory of thermal expansion have been discussed in Chap. 2. It was mentioned there that an important outcome of Gruneisen's theory is the emergence of the parameter γ, subsequently called the "Gruneisen constant" or "the Gruneisen parameter". Gruneisen showed that thermodynamically

$$\gamma_{th} = 3\alpha V/\psi C_V, \tag{6.26}$$

where α is the coefficient of linear expansion, V the molar volume, ψ the compressibility and C_V the molar specific heat. Here, γ carries the suffix "th" to indicate its thermodynamic origin.

The microscopic definition of the Gruneisen parameter is

$$\gamma = -(\mathrm{d}\ \log\ \nu/\mathrm{d}\ \log\ V), \tag{6.27}$$

where ν is a lattice frequency. γ, thus, represents the volume dependence of lattice frequencies. In the Debye model, it would become

$$\gamma_{\mathrm{Debye}} = -(\mathrm{d}\ \log\ \nu_{\mathrm{D}}/\mathrm{d}\ \log\ V) = (\mathrm{d}\ \log\ \theta_{\mathrm{D}}/\mathrm{d}\ \log\ V), \tag{6.28}$$

where ν_{D} and θ_{D} are the Debye frequency and the Debye temperature, respectively.

With further development of lattice dynamics it was realized that a Gruneisen parameter has to be associated with each mode of vibration. Thus, the mode Gruneisen parameter for the ith mode is given by

$$\gamma_i = -(\mathrm{d}\ \log\ \nu_i/\mathrm{d}\ \log\ V). \tag{6.29}$$

Some of the modes of vibration can be observed in IR or Raman or neutron scattering. The corresponding γ_is can be experimentally determined. Others can be calculated lattice dynamically. The mean $\bar{\gamma}$ of these mode gammas is given by

$$\bar{\gamma} = \sum C_i\gamma_i / \sum C_i, \tag{6.30}$$

where C_i is the contribution of the ith mode to the specific heat. $\bar{\gamma}$ should be close to γ_{th}.

The Gruneisen parameter is no longer a parameter just related to thermal expansion. It is a link between various aspects of crystal physics like the elastic properties, their pressure derivatives, the thermal conductivity, the Debye temperature, the lattice dynamical models and, as will be shown, the dielectric properties.

Results of our calculations of the Gruneisen parameter from the interatomic potentials, from the pressure variation of Debye temperatures, from the pressure variation of elastic properties and from dielectric properties are discussed in the following sections.

6.5.1 Gruneisen Parameter from Interatomic Potentials

The evaluation of Gruneisen parameter from interatomic potentials has been discussed by Slater [6.90], Das et al. [6.91] and Kachhava and Saxena [6.92]. If ϕ is the interatomic potential, the Gruneisen constant γ is given by

$$\gamma = -(r/6)[(\mathrm{d}^3\phi/\mathrm{d}r^3)/(\mathrm{d}^2\phi/\mathrm{d}r^2)]. \tag{6.31}$$

This method has been used for the evaluation of γ of some ionic crystals and for some elemental solids.

γ of Some Ionic Crystals

For ionic crystals, a commonly employed potential is

$$\phi = -(AZ^2e^2/r) + Be^{-r/\rho}, \tag{6.32}$$

where the first term represents Coulomb attraction and the second term the overlap repulsion. A is the Madelung constant, Z the valence of the ions, e the electron charge, r the interionic distance and B and r/ρ the repulsion constants. The procedure to evaluate r/ρ from the compressibility has been discussed in Sect. 6.2.

From (6.31) and (6.32), we get

$$\gamma = -[(r/\rho)^2 - 6)]/[6(r/\rho) - 2]. \tag{6.33}$$

As an example of the application of this method, values of γ for some alkali halides calculated from (6.33) by Kachhava and Saxena [6.92] are given in Table 6.13 along with values of γ_{th}. It can be seen that the agreement between $\gamma_{\mathrm{calc.}}$ and γ_{th} is fair, the difference being about 20%.

Sirdeshmukh and Rao [6.93] used the same potential to evaluate γ of the alkaline earth oxides which also have the NaCl structure. The values of $\gamma_{\mathrm{calc.}}$ and γ_{th} for this group of crystals are also given in Table 6.13 . The alkaline earth oxides are believed to be slightly less ionic than the alkali halides. Yet, the difference between $\gamma_{\mathrm{calc.}}$ and γ_{th} is of the same order as in the alkali halides. Sirdeshmukh and Rao [6.94, 6.95] also evaluated γ of another group of ionic crystals, the fluorite compounds, using the same potential with appropriate values for A and Z. The values of $\gamma_{\mathrm{calc.}}$ and γ_{th} for some fluorides and oxides with CaF_2 structure are given in Table 6.14. Here, it is observed that the agreement between $\gamma_{\mathrm{calc.}}$ and γ_{th} for the fluorides is of the same order as for the alkali halides. However, the difference between $\gamma_{\mathrm{calc.}}$ and γ_{th} is clearly larger in the case of the oxides suggesting partial covalency in these compounds.

Table 6.13. Values of $\gamma_{\mathrm{calc.}}$ from (6.33) and γ_{th} from (6.26) for some ionic crystals with NaCl structure [6.92, 6.93]

Crystal	$\gamma_{\mathrm{calc.}}$	γ_{th}
LiF	1.46	1.64
NaCl	1.95	1.64
KCl	1.95	1.49
KBr	2.00	1.46
RbI	1.92	1.51
MgO	1.17	1.54
CaO	1.30	1.76
SrO	1.32	1.68

Table 6.14. Values of $\gamma_{calc.}$ from (6.33) and γ_{th} from (6.26) for some ionic crystals with CaF_2 structure [6.94, 6.95]

Crystal	$\gamma_{calc.}$	γ_{th}
EuF_2	1.64	1.17
CaF_2	1.68	1.68
SrF_2	1.77	1.62
BaF_2	1.85	1.57
CdF_2	1.85	2.2
UO_2	1.7	3.2
ThO_2	1.7	3.0

γ of Some Elements

Sirdeshmukh and Rao [6.96] evaluated γ of some group-V and group-VI elements. For this purpose, they assumed that the interatomic interactions for these solids can be described by the Morse potential given by

$$\phi = L\{\exp[-2a(r - r_0)] - 2\exp[-a(r - r_0)]\}. \tag{6.8}$$

Subjecting this potential to (6.31), we get

$$\gamma = ar/2. \tag{6.34}$$

The constant a can be expressed as

$$a = 3(N_A cr/2\psi L), \tag{6.35}$$

where N_A is the Avogadro number, ψ the compressibility, L the heat of sublimation and c a constant in the expression for the volume V:

$$V = N_A cr^3. \tag{6.36}$$

Thus,

$$\gamma = (9V/8\psi L)^{1/2}. \tag{6.37}$$

For evaluation of γ, values of V,ψ and L were taken from Gschneidner [6.17]. The values of $\gamma_{calc.}$ and γ_{th} are given in Table 6.15. γ_{th} values have been calculated using more recent data. The comparison of $\gamma_{calc.}$ and γ_{th} shows that the Morse potential is capable of estimating the Gruneisen parameters of metals.

6.5.2 γ from Pressure Variation of Debye Temperature

According to (6.29)

$$\gamma = -(\mathrm{d}\log\theta_D/\mathrm{d}\log V). \tag{6.28}$$

Table 6.15. Values of $\gamma_{\text{calc.}}$ and γ_{th} for some group-V and group-VI elements [6.96]

Crystal	$\gamma_{\text{calc.}}$	γ_{th}
As	2.58	1.44
Sb	1.74	1.18
Bi	1.92	1.11
Te	1.66	1.28
Se	1.25	1.59

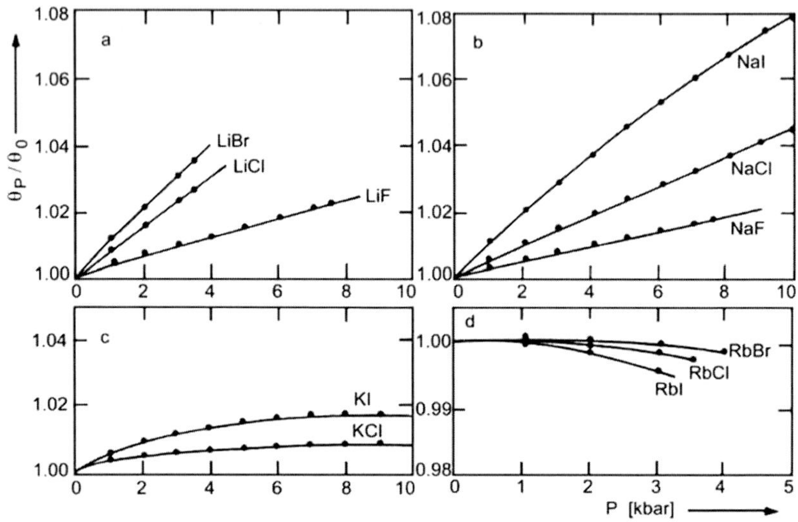

Fig. 6.3. Pressure dependence of Debye temperature for alkali halides

This can be written as:

$$\gamma = (1/\theta_{\mathrm{D}}\psi)(\mathrm{d}\theta_{\mathrm{D}}/\mathrm{d}P). \tag{6.38}$$

Sirdeshmukh and Subhadra [6.97] calculated the Debye temperatures of some alkali halides with NaCl structure at different pressures using the elastic constant data at different pressures. The Debye temperatures were calculated by the methods discussed in Sect. 6.4. The pressure variation of θ_{D} is shown in Fig. 6.3. In most cases the Debye temperature increases with pressure; in the rubidium halides it shows a decrease with pressure.

The values of the Debye temperatures at normal pressure and at 3 kbar are given in Table 6.16. From these values, the Gruneisen parameter can be calculated using (6.38). These values are also given in Table 6.16 For comparison, γ_{th} values at room temperature and at low temperature are quoted from [6.98]. It is noticed that (6.38) yields values of γ of the correct order. The γcalc. values for the rubidium halides are negative. It may be noted that while γ_{th} for the rubidium halides is +ve at room temperature (RT), it is

Table 6.16. Debye temperatures (θ_D) of some alkali halides at $P = 0$ and 3 kbar and the Gruneisen constants [6.97]

Crystal	θ_D [K]		$\gamma_{calc.}$	γ_{th}	
	$P = 0$	$P = 3$ kbar	(6.38)	RT	LT
LiF	697.4	703.3	1.89	1.64	1.70
LiCl	390.5	399.3	2.23	1.69	0.90
LiBr	246.8	254.0	2.31	1.88	–
NaF	474.5	478.0	1.14	1.83	–
NaCl	306.1	310.4	1.12	1.64	0.90
NaI	156.1	160.5	1.41	1.66	–
KCl	224.0	225.0	0.26	1.49	0.32
KI	125.4	126.7	0.43	1.45	0.28
RbCl	161.5	161.4	−0.03	1.57	0.00
RbBr	128.5	128.3	−0.07	1.43	−0.03
RbI	100.8	100.4	−0.14	1.51	−0.18

Table 6.17. Pressure variation of Debye temperatures of some fluorite-type crystals and values of γ [6.99]

Crystal	$d\theta_D/dP$[K bar^{-1}]	$\gamma_{calc.}$ (6.38)	γ_{th}
CaF_2	0.695	1.16	1.68
SrF_2	0.400	0.76	1.62
BaF_2	0.175	0.36	1.57
CdF_2	0.688	2.32	2.2
PbF_2	0.307	0.84	2.08

negative at low temperature (LT). The $\gamma_{calc.}$ values correspond better with the low temperature values of $\gamma_{th.}$

Hussain [6.99] calculated the Debye temperatures of some crystals with fluorite structure using literature data on the pressure variation of elastic constants. From the values of $(d\theta_D/dP)$, values of γ were calculated using (6.38). These results are given in Table 6.17. Once again, an order-of-magnitude agreement is observed between $\gamma_{calc.}$ and γ_{th}.

6.5.3 Evaluation of γ from Pressure Derivatives of Elastic Moduli

The evaluation of γ of crystals with NaCl structure by Slater's method and Knopoff and Shapiro's method are considered in this section.

Slater's Formulation

Slater [6.90] starts with the definition of γ given by

$$\gamma = -(d \log \theta_D / d \log V). \tag{6.28}$$

The characteristic frequency ν_D is given by the expression:

$$\nu_D = (9N_A/4\pi V)^{1/3}(1/C_l^3 + 2/C_t^3)^{-1/3}, \tag{6.39}$$

where C_l and C_t are the velocities of longitudinal and transverse waves, respectively. It is to be noted from the above expression that the Debye characteristic frequency is related to the velocities of the longitudinal as well as transverse waves. However, Slater [6.90], considering only the velocity of the longitudinal waves, assumed that the Poisson's ratio of the solid is independent of volume and obtained from (6.28) and (6.39)

$$\gamma_{SL} = -(1/6) + (1/2)(\mathrm{d}B/\mathrm{d}P), \tag{6.40}$$

where $\mathrm{d}B/\mathrm{d}P$ is the pressure derivative of bulk modulus. The values obtained from Slater's equation for the Gruneisen parameter are generally higher than the thermal values.

The Contribution of Shear Waves: Formulation of Knopoff and Shapiro

Knopoff and Shapiro [6.100] deduced an equation for the Gruneisen parameter by taking into consideration the dependence of the Debye characteristic frequency ν_D on the velocities of transverse as well as longitudinal waves. They have also taken into account the variation of Poisson's ratio with volume.

Starting with (6.28) and (6.39) we have for the Gruneisen parameter

$$\gamma = (1/3)[\mathrm{d}\log \rho\{(1/C_l^3) + (2/C_t^3)^{-1/3}\}^{-1}]/\mathrm{d}\log \rho, \tag{6.41}$$

where ρ is the density. The result of the differentiation is

$$\gamma = -(1/6) + (1/2)(\mathrm{d}B/\mathrm{d}P) - g(1-2\sigma)(\mathrm{d}B/\mathrm{d}P) + (2g/3)(1+\sigma)(\mathrm{d}G/\mathrm{d}P), \tag{6.42}$$

where $\mathrm{d}B/\mathrm{d}P$ is the pressure derivative of bulk modulus, $\mathrm{d}G/\mathrm{d}P$ the pressure derivative of shear modulus and g a function of Poisson's ratio (σ) given by

$$g(\sigma) = [\{1/(3-3\sigma)^{5/2}\} + \{3 \times 2^{3/2}/(3-6\sigma)^{5/2}\}]\Big/[\{1/(3-3\sigma)^{3/2}\} + \{2^{5/2}/(3-6\sigma)^{3/2}\}]. \tag{6.43}$$

Under the assumption $\mathrm{d}\sigma/\mathrm{d}P = 0$, (6.42) reduces to Slater's formula. For a special case of $\sigma = 0.25$, (6.42) turns out to be

$$\gamma_{KS} = -(1/6) + 0.024(\mathrm{d}B/\mathrm{d}P) + 0.793(\mathrm{d}G/\mathrm{d}P). \tag{6.44}$$

It may be seen from (6.44) that for the evaluation of the Gruneisen parameter, the pressure derivative of shear modulus is more important than the pressure derivative of bulk modulus.

Knopoff and Shapiro [6.100] calculated γ from (6.40) and (6.44) and compared these values with γ_{th}. They found that there was no systematic behaviour and concluded that the results were not very encouraging. However, they used the method only for eight substances belonging to different structures. For a better appreciation of the method of Knopoff and Shapiro, it is necessary to apply the method to a larger number of crystals with a common structure. Rao [6.101] calculated γ_{SL} and γ_{KS} for a number of crystals with NaCl structure.

In carrying out the calculations, a problem that is encountered is the lack of experimental data for the pressure derivative of shear modulus for polycrystalline aggregates. These values have to be calculated from single crystal elastic constants and their pressure derivatives using the procedures suggested by Voigt [6.41] and Reuss [6.42].

For crystals of cubic symmetry, the elastic moduli are given by the following expressions:

$$B_{\mathrm{V}} = B_{\mathrm{R}} = (C_{11} + 2C_{12})/3, \tag{6.45}$$

$$G_{\mathrm{V}} = (C_{\mathrm{a}} + 3C_{44})/5 \tag{6.46}$$

and

$$G_{\mathrm{R}} = 5C_{\mathrm{a}}C_{44}/(3C_{\mathrm{a}} + 4C_{44}), \tag{6.47}$$

where

$$C_{\mathrm{a}} = C_{11} - C_{12}. \tag{6.48}$$

In these expressions, the subscripts V and R stand for Voigt and Reuss values. The Voigt and Reuss values are the same for the bulk modulus but different for the shear modulus.

Chung and Buessem [6.102] defined a dimensionless quantity A^* for elastic anisotropy in terms of the Voigt and Reuss values of shear modulus. The expression is

$$A^* = (G_{\mathrm{V}} - G_{\mathrm{R}})/(G_{\mathrm{V}} + G_{\mathrm{R}}). \tag{6.49}$$

A^* is always a positive quantity for anisotropic crystals and is a measure of the relative magnitude of the elastic anisotropy. A^* is zero for an isotropic crystal.

The pressure derivatives of shear modulus for polycrystalline aggregates from single crystal data on the pressure variation of the elastic stiffness constants are given by the following expressions:

$$\mathrm{d}G_{\mathrm{V}}/\mathrm{d}P = (1/5)(\mathrm{d}C_{\mathrm{a}}/\mathrm{d}P) + (3/5)(\mathrm{d}C_{44}/\mathrm{d}P) \tag{6.50}$$

and

$$\mathrm{d}G_{\mathrm{R}}/\mathrm{d}P = (4/5)(G_{\mathrm{R}}/C_{\mathrm{a}})^2(\mathrm{d}C_{\mathrm{a}}/\mathrm{d}P) + (3/5)(G_{\mathrm{R}}/C_{44})^2(\mathrm{d}C_{44}/\mathrm{d}P) \tag{6.51}$$

and

$$\mathrm{d}C_{\mathrm{a}}/\mathrm{d}P = (\mathrm{d}C_{11}/\mathrm{d}P) - (\mathrm{d}C_{12}/\mathrm{d}P). \tag{6.52}$$

The Gruneisen parameter from Slater's formula is calculated using the pressure variation of bulk modulus from (6.40). Using the method of Knopoff and Shapiro, the Gruneisen parameter is calculated from (6.44). For this, the Voigt as well as the Reuss values of pressure derivatives of shear modulus obtained from (6.50) and (6.51), respectively, are used. Equation (6.44) involves Poisson's ratio which is calculated from the formula:

$$\sigma = (3B - 2G)/2(G + 3B). \tag{6.53}$$

Here, two values of the Poisson's ratio, σ_V and σ_R are obtained corresponding to the Voigt and Reuss values of shear modulus. The procedure suggested by Hill [6.43] to find the average values of the elastic moduli between the Voigt and Reuss limiting values has also been attempted. The average shear modulus, G^*, the average pressure derivatives of shear modulus, $\mathrm{d}G^*/\mathrm{d}P$ and the average Poisson's ratio using G^*, viz. σ^* are calculated. Using these average values, the average Gruneisen parameter, γ^* is calculated.

Results

The results of the calculations are given in Table 6.18. The table gives the values of the Gruneisen parameter obtained from thermal data (γ_{th}), Slater's equation (γ_{SL}), Knopoff and Shapiro's equation using the Voigt approximation for the pressure derivatives of shear modulus ($\gamma_{\mathrm{KS-V}}$), Knopoff and Shapiro's equation using the Voigt approximation for the pressure derivatives of shear modulus ($\gamma_{\mathrm{KS-R}}$), Knopoff and Shapiro's equation using the Hill average for the pressure derivatives of shear modulus (γ^*) and the anisotropy factor A^*. For these calculations, the input data are the elastic constants and their pressure derivatives. These are taken from [6.93, 6.103–6.105].

The Gruneisen parameters $\gamma_{\mathrm{KS-V}}, \gamma_{\mathrm{KS-R}}$ and γ_{th} are plotted against the anisotropy factor A^* in Fig. 6.4. From an examination of the data given in Table 6.18 and Fig. 6.4 the following conclusions can be drawn:

(a) Slater's formula for the Gruneisen parameter always gives a higher value than the thermal gamma. In some cases, the values from Slater's formula are higher by a factor of nearly 2. It has been pointed out by earlier workers that this difference is due to Slater's neglect of the volume dependence of Poisson's ratio in the derivation of his formula.

(b) The modification introduced by Knopoff and Shapiro [6.100], which allows for the volume dependence of Poisson's ratio through the inclusion of the pressure derivative of the shear modulus, in general, gives values of the Gruneisen parameters which are in better agreement with the thermal gammas. The present results support the remark made by Knopoff and Shapiro that the pressure derivative of shear modulus is of greater significance than the pressure derivative of bulk modulus for the evaluation of the Gruneisen parameter.

Table 6.18. Gruneisen parameters from pressure derivatives of elastic moduli for crystals with NaCl structure

Crystal	γ_{th}	γ_{SL}	γ_{KS-V}	γ_{KS-R}	γ^*	$A^* \times 100$
LiF	1.63	2.94	1.46	2.56	1.89	5.2
LiCl	1.81	2.61	1.82	2.72	2.27	4.5
LiBr	1.94	2.53	1.95	2.99	2.44	4.5
LiI	2.19	2.73	1.95	3.29	2.59	4.6
NaF	1.51	2.42	1.46	1.23	1.24	0.8
NaCl	1.62	2.47	1.97	1.41	1.69	1.5
NaBr	1.65	2.48	1.86	1.31	1.60	1.6
NaI	1.71	2.53	2.01	1.41	1.70	1.9
KF	1.52	2.47	1.22	0.41	0.88	5.5
KCl	1.49	2.51	1.59	0.11	0.89	11.1
KBr	1.50	2.53	1.66	0.18	0.93	12.6
KI	1.54	2.56	1.91	0.20	1.18	14.1
RbF	1.40	2.62	1.49	0.20	1.01	7.0
RbCl	1.39	2.57	1.60	−0.22	0.76	15.3
RbBr	1.42	2.56	1.75	−0.30	0.85	17.7
RbI	1.56	2.55	1.89	−0.74	0.70	20.8
MgO	1.54	2.08	1.21	1.66	1.43	2.30
CaO	1.76	2.83	1.12	1.12	1.12	0.06
SrO	1.68	2.83	1.03	0.45	0.98	0.25

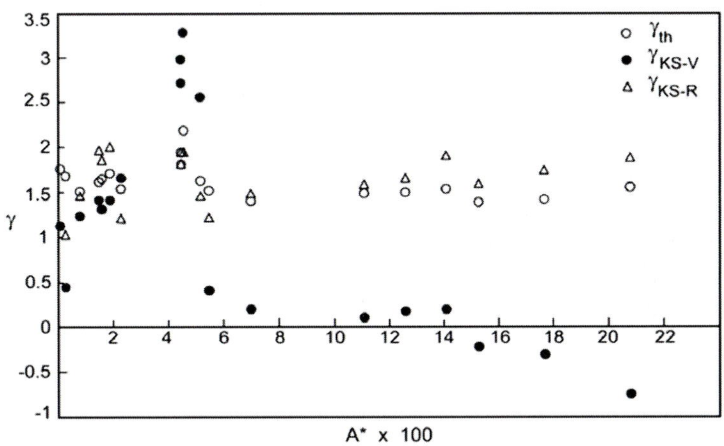

Fig. 6.4. Plot of Gruneisen parameter (γ) against A^* for crystals with NaCl structure

(c) Considerable difference is observed in the values of γ_{KS-V} and γ_{KS-R}. As shown in Fig. 6.4, this difference seems to be related to the anisotropy factor A^*, increasing as A^* increases.

(d) A careful comparison between the values of thermal gamma and the gamma values calculated from the elastic moduli shows that, in most of the cases, there is closer agreement between thermal gamma and $\gamma_{\mathrm{KS-V}}$ than between thermal gamma and $\gamma_{\mathrm{KS-R}}$. It implies that, as far as pressure derivatives of elastic moduli are concerned, the Voigt assumption is better than the Reuss assumption.

6.5.4 Mode Gruneisen Parameters of Fluorite-Type Crystals

The concept and definition of mode Gruneisen parameter was discussed earlier. A mode Gruneisen parameter is given by

$$\gamma_i = -(\mathrm{d}\log \nu_i/\mathrm{d}\log V). \tag{6.30}$$

In the vibration spectrum of crystals with the fluorite structure, there are six optic branches. Out of the six long-wave phonons one (ω_{t1}) is infrared active and another ω_{t2}) is Raman active and is designated (ω_{R}). ω_{l1} is difficult to observe and the others are degenerate either with ω_{t1} or ω_{t2}. Studies of pressure variation of these phonon frequencies are available and these have led to the evaluation of the mode Gruneisen parameters γ_{t} and γ_{R}.

The experimental values of γ_{t} and γ_{R} for CaF_2, SrF_2 and BaF_2 are summarized in Table 6.19. It can be seen that there is a large difference (of 100–200%) between the values of γ_{t} reported by Ferraro et al. [6.106] and those by Lowndes [6.107]. On the other hand there is very good consistency among the values of γ_{R} reported by Ferraro et al. [6.106], Ruppin [6.108] and Kessler et al. [6.109].

For the theoretical evaluation of γ_{t}, we may start with the equations given by Szigeti [6.110, 6.111]:

$$\frac{1}{\psi} = \frac{\mu\omega_{\mathrm{t}}^2}{6r}\left(\frac{\varepsilon_{\mathrm{s}}+2}{\varepsilon_\infty+2}\right) \tag{6.54}$$

and

$$ze^* = \left(\frac{\varepsilon_{\mathrm{s}}-\varepsilon_\infty}{4\pi}\right)^{1/2}\left(\frac{3}{\varepsilon_\infty+2}\right)(\mu V_{\mathrm{i}})^{1/2}\,\omega_{\mathrm{t}}, \tag{6.55}$$

where ψ is the compressibility, r the interionic distance, z the valency, e the formal electron charge, V_{i} the volume per ion pair, γ_{t} the transverse optical

Table 6.19. Experimental values of γ_t and γ_R

Crystal	γ_{t}			γ_{R}		
	Ferraro et al. [6.106]	Lowndes [6.107]	Andeen et al. [6.113]	Ferraro et al. [6.106]	Ruppin [6.108]	Kessler et al. [6.109]
CaF_2	1.8	3.2	2.85	1.9	1.83	1.85
SrF_2	1.2	3.1	2.61	–	–	1.60
BaF_2	0.8	2.4	2.61	1.8	2.03	2.00

mode frequency and μ the reduced mass. Differentiation of these equations leads to

$$\gamma_t = \frac{1}{2\psi}\left[\frac{1}{\psi}\left(\frac{\partial\psi}{\partial P}\right) + \frac{\psi}{3} + \frac{1}{\varepsilon_s + 2}\left(\frac{\partial\varepsilon_s}{\partial P}\right) - \frac{1}{\varepsilon_\infty + 2}\left(\frac{\partial\varepsilon_\infty}{\partial P}\right)\right], \quad (6.56)$$

$$\frac{V}{e^*}\left(\frac{\partial e^*}{\partial V}\right) = -\frac{1}{2\psi}\left[\frac{1}{(\varepsilon_s - \varepsilon_\infty)}\left\{\left(\frac{\partial\varepsilon_s}{\partial P}\right) - \left(\frac{\partial\varepsilon_\infty}{\partial P}\right)\right\} - \left(\frac{2}{\varepsilon_\infty + 2}\right)\left(\frac{\partial\varepsilon_\infty}{\partial P}\right) - (1 + 2\gamma_t)\right]. \quad (6.57)$$

Equation (6.56) enables the calculation of γ_t from the compressibility and pressure derivatives of compressibility and dielectric constants. Values of γ_t for some alkali halides have been calculated from (6.56) by Jones [6.112]. Equation (6.57) can be used in two ways. If values of the strain derivative of the effective ionic charge are known, the equation can be used to calculate γ_t. On the other hand, if γ_t is known, (6.57) yields the value of $(V/e^*)(\mathrm{d}e^*/\mathrm{d}V)$ which is another important and interesting entity.

Accurate data on the pressure variation of ε_s and ε_∞ are available for the alkaline earth fluorides [6.113]. Using these values, Andeen et al. tried to evaluate the parameter $(V/e^*)(\mathrm{d}e^*/\mathrm{d}V)$. However, for this calculation γ_t is also needed as an input parameter. Andeen et al. found that the two sets of γ_t values (given in Table 6.19) lead to widely different values for $(V/e^*)(\mathrm{d}e^*/\mathrm{d}V)$. Andeen et al. therefore stated that "no conclusion should be drawn with respect to the sign of $(V/e^*)(\mathrm{d}e^*/\mathrm{d}V)$ on the basis of (their data)". On the other hand, Andeen et al. equated $(V/e^*)(\mathrm{d}e^*/\mathrm{d}V)$ to zero and used (6.57) to estimate the maximum value of γ_t. The values of γ_t thus obtained are quoted in Table 6.19. An attempt is made to sort out the best value of γ_t from the several available values and then to estimate the sign and value of $(V/e^*)(\mathrm{d}e^*/\mathrm{d}V)$.

Axe [6.114] applied the dipole shell model to the lattice dynamics of crystals with the fluorite structure and obtained the following expressions:

$$\frac{1}{\psi} = \frac{1}{12r}\left(\frac{2(\varepsilon_s + 2)\mu\omega_t^2}{(\varepsilon_\infty + 2)} + \mu_R\omega_R^2\right), \quad (6.58)$$

$$ze* = \left(\frac{\varepsilon_s - \varepsilon_\infty}{4\pi}\right)^{1/2}\left(\frac{3}{\varepsilon_\infty + 2}\right)(\mu V_i)^{1/2}\omega_t. \quad (6.59)$$

Here, $\mu = (m_1 m_2)/(m_1 + 2m_2)$ and $\mu_R = m_2$ and ω_R is the Raman mode frequency. Equation (6.59) is identical to (6.55). Equation (6.58) is similar to (6.53) but it contains an additional term involving the Raman mode frequency. Differentiation of (6.58) with respect to volume (and conversion of some volume derivatives into pressure derivatives) leads to the following equation for the Raman mode Gruneisen parameter for the fluorite-type structure compounds:

Table 6.20. Values of γ_t, γ_R and $(V/e^*)(\mathrm{d}e^*/\mathrm{d}V)$

Crystal	γ_t (exp.)	γ_R Calc. (6.60)	γ_R Exp. [6.109]	$(V/e^*)(\mathrm{d}e^*/\mathrm{d}V)$ Calc. (6.57)
CaF_2	1.8	1.41	1.85	0.86
	3.2	−0.65		
	2.85	0.16		
SrF_2	1.2	3.1	1.6	1.42
	3.1	−0.94		
	2.61	−0.05		
BaF_2	0.8	2.4	2.0	1.82
	2.40	0.03		
	2.61	−0.48		

$$\gamma_R = -\left(\frac{2r}{\mu_R \psi \omega_R^2}\right)\left[1 + \frac{3}{\psi^2}(\partial\psi/\partial P)_T\right] - \left(\frac{\mu}{\mu_R}\right)\left(\frac{\omega_t^2}{\omega_R^2}\right)\left(\frac{\varepsilon_s + 2}{\varepsilon_\infty + 2}\right)\left(\frac{1}{\psi}\right)$$
$$\times \left[4\psi\gamma_t - \left(\frac{1}{\varepsilon_s + 2}\right)\left(\frac{\partial \varepsilon_s}{\partial P}\right)_T + \left(\frac{1}{\varepsilon_\infty + 2}\right)\left(\frac{\partial \varepsilon_\infty}{\partial P}\right)_T\right]. \quad (6.60)$$

This is the first attempt to express the Raman mode Gruneisen parameter in terms of dielectric data. The input parameters for (6.60) are the compressibility, γ_t and the pressure derivatives of ψ, ε_s and ε_∞. If accurate values of γ_t were available, (6.60) could have been used to estimate γ_R from dielectric data and to compare these values with γ_R obtained from direct experiments. But, as pointed out earlier, there is a very large scatter in the experimental values of γ_t. Since experimental values of γ_R are dependable, (6.60) can be used to calculate γ_R with all available values of γ_t and by comparing these calculated γ_R values with the experimental ones, a choice of γ_t could be made. Prameela Devi [6.115] made such calculations using the γ_t values given in Table 6.19. The results are given in Table 6.20. When the γ_t values reported by Lowndes [6.107] and those by Andeen et al. [6.113] are employed, they lead to γ_R values which are either negative or too small. On the other hand, the γ_t values given by Ferraro et al. [6.106] lead to values which agree in sign and magnitude with the experimental values of γ_R. Thus, on the basis of the comparison between the calculated and experimental values of γ_R, it is possible to sift out the most acceptable values of γ_t from the several available values, viz. the set of γ_t given by Ferraro et al. [6.106].

It is now possible to consider the parameter $(V/e^*)(\mathrm{d}e^*/\mathrm{d}V)$, the sign of which could not be uniquely determined by Andeen et al. [6.113] due to the uncertainty in the γ_t values. The value of $(V/e^*)(\mathrm{d}e^*/\mathrm{d}V)$ obtained by substituting the "most acceptable" values of γ_t in (6.57) are given in Table 6.20. The values of this parameter for all the three alkaline earth fluorides are positive. The deviation of e^* from unity is interpreted as due to ionic distortion caused by neighbouring ions. Barron and Batana [6.116] pointed out that since a reduction in volume increases this distortion, the parameter $(V/e^*)(\mathrm{d}e^*/\mathrm{d}V)$ should be positive. The positive sign obtained in the present calculation is thus consistent with the interpretation of e^*.

7

The Physics of Mixed Crystals

7.1 Introduction

7.1.1 General

When a physical mixture of two compounds is subjected to a crystal growth process (melting-and-cooling or dissolving-and-evaporation), if a single-phase crystalline compound results having the same structure as the two pure compounds, such a crystal is called a mixed crystal or a solid solution.

For mixed crystal formation, the two components should have (1) same structure, (2) same valency, and (3) reasonably close values of lattice dimensions (differing by not more than 6–7%). If these conditions are fulfilled, mixing takes place over the entire range of compositions. If there are deviations from the above conditions, mixing may not take place or may take place over a limited composition range.

If the two pure components have the same structure, the mixed crystals also will have the same structure for all compositions. Thus the KCl_xBr_{1-x} mixed crystals have the NaCl structure for all values of x. If the two components have different structures, the mixed crystal will have the structure of the dominant member. Thus in $(NH_4)_xRb_{1-x}Cl$ mixed crystals, the mixed crystals have the NaCl structure in the RbCl-rich region and the CsCl structure in the NH_4Cl-rich region. The atoms or ions of one component may occupy substitutional or interstitial positions in the lattice of the other. The distribution of the solute ions may be random (disordered mixed crystal) or preferential (ordered mixed crystal).

What makes the mixed crystals interesting is that their properties are intermediate between those of the mixing components. Thus, mixed crystals of two compounds provide crystals of the same structure but with physical properties which change continuously from one end of the system to the other. This creates the possibility of tailor-made crystals.

In the study of mixed crystals, the knowledge of the exact composition is necessary. This is obtained by the use of the techniques of titration potentiometry, polarography or atomic absorption spectroscopy.

7.1.2 Earlier Reviews on Mixed Crystals

Mohanlal [7.1] reviewed the properties of the KCl–KBr and $NaClO_3-NaBrO_3$ mixed crystal systems. The properties considered by him were (1) the lattice constants of the two systems, (2) Debye–Waller factors of the KCl–KBr system, and (3) the optical rotation in the $NaClO_3-NaBrO_3$ system. Kitaigorodski's [7.2] treatise considers a variety of mixed crystals with an emphasis on thermodynamics and organic crystals. Hari Babu and Subba Rao [7.3] reviewed the properties of alkali halide mixed crystals studied by the techniques of ionic conductivity, dielectric loss, microhardness, radiation hardening, colour centres and thermoluminescence.

Sirdeshmukh and Srinivas [7.4] reviewed the composition dependence of several physical properties of alkali halide mixed crystals. These properties include the bulk modulus, the elastic constants, the Debye temperatures, the Debye–Waller factors and spectroscopic properties. Sirdeshmukh and Srinivas [7.4] pointed out that on the basis of the trends in composition dependence, the properties of mixed crystals can be grouped in four categories as follows:

Type A: Properties which show a linear composition dependence.
Type B: Properties which show a slightly non-linear composition dependence
Type C: Properties which show a highly non-linear composition dependence with values for the mixed crystals exceeding those for end members.
Type D: New phenomena (properties displayed by mixed crystals but not displayed by the pure crystals).

This grouping of the properties is shown in Table 7.1 Some properties not discussed in [7.4] but discussed in [7.3] are also included in the table.

Most of the properties show either a linear composition dependence (Type A) or a slightly non-linear composition dependence (Type B). This trend in the composition dependence is consistent with the facts that (1) these properties are by and large, determined by the interatomic forces, (2) the nature of the interatomic forces is the same for the mixed crystals in a given series, and (3) the magnitude of the interatomic interactions varies smoothly from one end member to the other.

The properties which show a highly non-linear composition dependence (Type C) are influenced by additional factors not present in the pure crystals. Thus, the presence of a second atom at lattice points normally occupied by one atom results in a static displacement which enhances the Debye–Waller factor. This displacement (or strain) is also responsible for a higher dislocation density which in turn increases the microhardness. Although, on an average the alkali halide mixed crystals are thoroughly disordered, the ions around a solute ion assume configuration of minimum energy. The observed heats of formation can be satisfactorily accounted for only by taking into consideration these site energies. The substitution of a second ion (say bromine) in a host crystal (say KCl) affects the symmetry, at least locally. This gives rise to new

Table 7.1. Categorization of some physical properties of mixed crystals on the basis of their composition dependence (Type A, linear; Type B, slightly non-linear; Type C, highly non-linear; Type D, new phenomenon)

Property	Type			
	A	B	C	D
Lattice constant	*****************			
Thermal expansion	********			
Bulk modulus		********		
Elastic constants	********			
Debye temperature	******************			
Refractive index	********			
Dielectric constant		******************		
Debye–Waller factor			********	
Dislocation density			********	
Microhardness			********	
Heat of formation			********	
IR spectra				
Single mode	******************			
Two modes				********
Raman spectra (first order)				********

features (Type D) in symmetry-dependent properties like the infrared and Raman spectra of mixed crystals.

7.1.3 Theoretical Models

In much of the earlier work on ionic mixed crystals, the Born–Mayer model of the following type was used

$$-U = \phi = -(\alpha e^2/r) - (C/r^6) - (D/r^8) + b\,\exp(-r/\rho). \qquad (7.1)$$

Here U is the lattice energy, ϕ the potential and the terms on the RHS, respectively, represent the Coulomb, dipole–dipole, dipole–quadrupole and repulsive interactions. Such a model was used by Wallace [7.5] to estimate the heats of formation (ΔH_f) of alkali halide mixed crystals from the relation

$$\Delta H_\mathrm{f} = U - [xU_1 + (1-x)U_2], \qquad (7.2)$$

where U, U_1 and U_2 are the lattice energies of the mixed crystal, and of the two pure components.

Krishnamurthy and Murti [7.6, 7.7] used the same potential as in (7.1) but made the model more elaborate by considering several distributions of the solute ions and finding the associated configurational entropy. They also assumed a single frequency. By this procedure they were able to determine the phase diagram of the mixed crystal system, the lattice constants, bulk modulus and formation energy of Schottky defects.

Chang and Mitra [7.8] were specifically interested in a spectroscopic problem related to mixed crystals. In some mixed crystal systems like KCl_xBr_{1-x}, the IR frequency has a single value changing smoothly from the value for KCl to that for KBr. This is called 'one-mode behaviour'. On the other hand, in some systems like InP_xAs_{1-x}, two IR frequencies are observed with values close to those of the end members. This is called 'two-mode behaviour'. Chang and Mitra [7.8] modified the real-element-isodisplacement (REI) model originally introduced by Chen et al. [7.9] and set up equations of motion for the ions A, B and C in a mixed crystal AB_xC_{1-x} in terms of force constants F_{AB}, F_{AC} and F_{BC} and the masses m_A, m_B and m_C. The force constants were expressed in terms of the dielectric constants and frequencies of the end-point compounds. Solving the equation of motion, Chang and Mitra [7.8] obtained the following simple conditions for one-mode and two-mode behaviour:

$$\begin{aligned} m_B &> \mu_{AC} \quad \text{one-mode behaviour,} \\ m_B &< \mu_{AC} \quad \text{two-mode behaviour.} \end{aligned}$$

Here μ_{AC} is the reduced mass of crystal AC.

7.2 An Overview

Several properties of mixed crystals have been discussed in the reviews mentioned in Sect. 7.1.2. In this section we shall discuss some select properties in detail. This overview differs from the earlier reviews in two respects. Firstly, the earlier reviews are confined mostly to the alkali halide mixed crystals but here other systems are also included. Secondly, in earlier reviews a large body of numerical results are included but here the emphasis is on the physics of the property.

7.2.1 Molar Volume and Lattice Parameters

Retgers [7.10] proposed that the molar volume of mixed crystals is additive. If V_1 and V_2 and V are the molar volumes of crystal 1, 2 and the mixed crystal with composition x and $(1-x)$ of the two compounds, Retgers' law of volumes states that

$$V = xV_1 + (1 - x)V_2. \tag{7.3}$$

The densities of mixed crystal systems $KCl_{1-x}Br_x$, $NaCl_{1-x}Br_x$ and $Na_xK_{1-x}Cl$ have been reported by Slagle and Mckinstry [7.11], Wollam and Wallace [7.12] and Barrett and Wallace [7.13]. The volumes calculated from these densities are plotted against x in Fig. 7.1. A linear behaviour is observed in agreement with (7.3).

Expressing the volume in terms of the lattice constant a, (7.3) can be written as

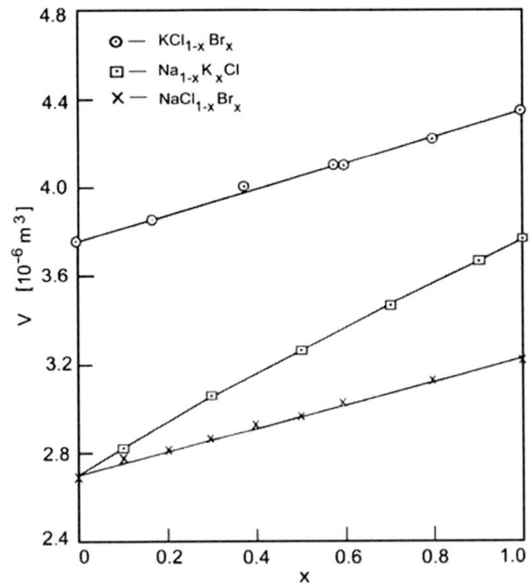

Fig. 7.1. Composition dependence of the molar volume of some mixed crystals

$$a^3 = xa_1^3 + (1-x)a_2^3, \tag{7.4}$$

where a_1 and a_2 and a are the lattice constants of crystal 1, 2 and the mixed crystal, respectively. This is known as Retgers' law of lattice constants.

A linear law for the composition dependence of lattice constants was proposed by Vegard [7.14]. This law is given by

$$a = xa_1 + (1-x)a_2. \tag{7.5}$$

Grimm and Herzfeld [7.15] derived the following law on the basis of the theory of ionic crystals,

$$a^8 = xa_1^8 + (1-x)a_2^8. \tag{7.6}$$

The available data on lattice constants of mixed crystals [7.14, 7.16] were not sufficiently accurate to permit a critical evaluation of (7.4)–(7.6). The first systematic attempt to critically compare the three laws vis-à-vis experimental data was made by Slagle and Mckinstry [7.17]. They determined the lattice constants of the KCl–KBr system with an accuracy of 0.005% and calculated the lattice constant a from the general equation

$$a^n = xa_1^n + (1-x)a_2^n, \tag{7.7}$$

by assigning values 1, 3, 4, 6 and 8 to the index n. The deviation ($a_{\text{exp}} - a_{\text{calc}}$) was calculated and plotted against x. The plot is shown in Fig. 7.2. It was found that the best fit was obtained with $n = 3$. The value $n = 8$ was clearly rejected.

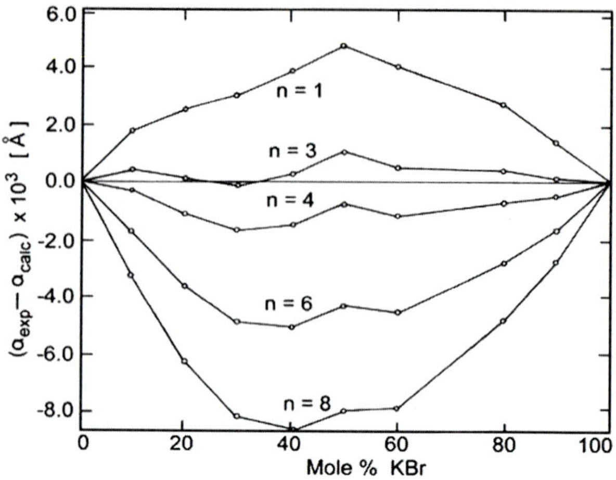

Fig. 7.2. Plot of $(a_{\mathrm{exp}} - a_{\mathrm{calc}})$ vs. x for the $KCl_{1-x}Br_x$ system for different values of n

Ahtee [7.18] compared the measured values of the lattice constants of seven alkali halide mixed crystal systems and examined the deviations from Vegard's law and Retgers' law. He concluded that the deviations from Retgers' law were less than those from Vegard's law and yet recommended the use of Vegard's law in view of its simplicity. Gielisse et al. [7.19] found that the lattice constants of the NiO–CoO system have a linear composition dependence. Zimmerman [7.20] determined the lattice constants of the $Li(H)_x(D)_{1-x}$ system and found that Vegard's law was followed.

While in most mixed crystal systems, the lattice constants follow either Vegard's law or Retgers' law, strong deviations from these laws have been observed in some cases. These cases are now discussed.

(1) Metal alloys

Gschneidner and Vineyard [7.21] analysed lattice constant data on 44 binary alloy systems. They observed positive deviations in every case. Several explanations were offered to explain these deviations like the compressibilities, short range order and electron gas effect. Finally, a model developed in [7.21] based on second-order elasticity was found to predict both the sign and magnitude of the deviation from Vegard's law in 29 out of the 44 alloy systems.

(2) $NH_4Cl_{1-x}Br_x$

Anselmo and Smith [7.22] studied the $NH_4Cl_{1-x}Br_x$ system. The lattice constants were determined from X-ray photographs with an accuracy of 0.005 Å. The values of a are shown in Table 7.2 along with deviations from Vegard's law. It is found that Vegard's law is followed except in the high chloride region where negative deviations much larger than the experimental error are observed.

Table 7.2. Lattice constants (a) of $NH_4Cl_{1-X}Br_x$ solid solutions

x	a [Å]	Deviations from Vegard's law [10^{-3} Å]	x	a [Å]	Deviations from Vegard's law [10^{-3} Å]
0.00	3.875	–	0.47	3.963	+2
0.04	3.875	−7	0.70	4.009	+5
0.08	3.882	−8	0.99	4.060	+3
0.11	3.883	−12	1.00	4.059	–
0.23	3.907	−10			

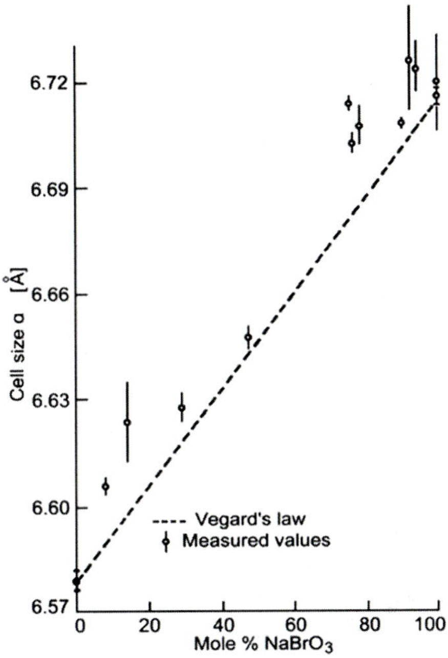

Fig. 7.3. Lattice constant as a function of composition for the $Na(ClO_3)_x(BrO_3)_{1-x}$ system

(3) $Na(ClO_3)_{1-x}(BrO_3)_x$

This system was studied by Mohanlal [7.1]. This is a rare case where Weissenberg photographs were used to determine lattice constants since powder photographs showed considerable broadening. The data obtained by Mohanlal are shown in Fig. 7.3. Severe positive deviations from Vegard's law are observed. Mohanlal calculated the lattice constants for values of $n = 1$, 2 and 3 in (7.7) but found that the positive deviations still persisted.

(4) $KBr_{1-x}I_x$

Nair and Walker [7.23] studied the $KBr_{1-x}I_x$ system with the help of X-ray powder photographs. Their results are shown in Fig. 7.4. They observed overall Vegard's law behaviour with large positive deviations in the KBr-rich region.

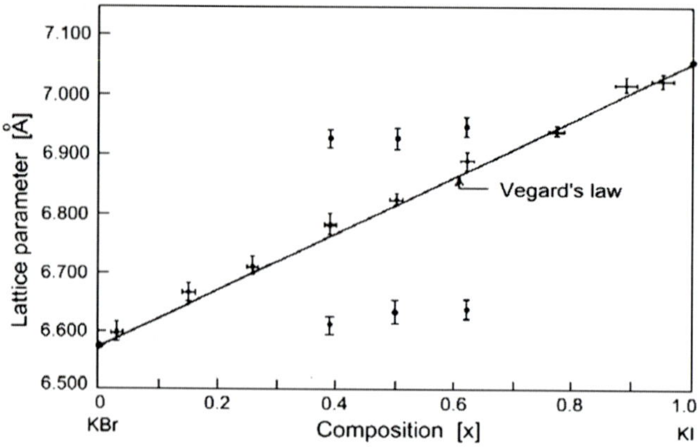

Fig. 7.4. Lattice parameter variation of $KBr_{1-x}I_x$ with composition x

Table 7.3. Values of α, β and γ (7.8) for the $CuInSe_{2(1-x)}Te_{2x}$ system

Parameter [Å]	α [Å]	β [Å]	γ [Å]
a	5.796	0.339	0.057
c	11.635	0.516	0.242

Nair and Walker [7.23] mention that the Debye–Scherrer lines in this region were "rather broad". Also, in the region $0.3 < x < 0.7$, the crystal samples showed existence of three phases, one corresponding to the experimentally determined composition and the other two having compositions $x = 0.08$ and $x = 0.75$ as indicated by their lattice constants.

(5) $CuInSe_{2(1-x)}Te_{2x}$

Sridevi and Reddy [7.24] prepared mixed crystals of $CuInSe_{2(1-x)}Te_{2x}$ which are tetragonal. The lattice parameters a and c were determined from X-ray powder diffractograms. Vegard's law requires the lattice parameters to show a linear composition dependence but in this case a significantly different parabolic composition dependence was observed. The a and c parameters were fitted to equations of the type

$$Y = \alpha + \beta x + \gamma x^2, \tag{7.8}$$

where Y is a or c. The values of α, β and γ obtained by least squares fitting are given in Table 7.3.

Thus, although most mixed crystal systems follow either Vegard's law or Retgers' law of lattice constants, cases do occur which show deviations. Hence, whenever a new mixed system is prepared, it is routine practice to examine the composition dependence of its lattice constants.

7.2.2 Debye–Waller Factors

Atomic thermal vibrations in crystals affect the intensities of diffracted X-rays (or neutrons). If $I(T)$ is the intensity at temperature T and I_0 that at 0 K, their relation is given by

$$I(T) = I_0 e^{-2B(\sin\theta/\lambda)^2}. \tag{7.9}$$

Here, B is the Debye–Waller factor, θ the Bragg angle and λ the wavelength of X-rays. The Debye–Waller factor is an important parameter which is related to the amplitude of vibration, the Debye temperature and, in general, to the vibration spectrum. Several aspects of the Debye–Waller factor are discussed in Chap. 3.

The Debye–Waller factors of $KCl_{0.5}Br_{0.5}$ were determined by Wasastjerna [7.25] and Ahtee et al. [7.26] from X-ray intensities. Mohanlal et al. [7.27] determined the Debye–Waller factors for two compositions in the KCl–KBr system from neutron diffraction intensities. These studies indicate that the Debye–Waller factors of mixed crystals are larger than those predicted by additivity. In fact, the Debye–Waller factors in the equimolar region are larger than those of the end members.

A lattice distortion is present in mixed crystals which contribute to the measured Debye–Waller factor. This can best be understood with reference to a KCl–KBr mixed crystal. Each K^+ ion in a KCl (or KBr) lattice is surrounded by a combination of Cl^- and Br^- ions. The number of Cl^- and Br^- ions surrounding a K^+ ion varies from ion to ion as the mixed crystal is disordered. This results in a distortion of the field around a K^+ ion which in turn causes a static displacement of the K^+ ion and enhances the Debye–Waller factor B in a mixed crystal. Thus, we have

$$B_{\text{measured}} = B_{\text{thermal}} + B_{\text{static}}. \tag{7.10}$$

Several models [7.28–7.31] have been proposed to estimate B_{static}. The Weiss model [7.30] leads to the expression

$$B_{\text{static}} = 24x(1-x)(a_1 - a_2)^2, \tag{7.11}$$

where a_1 and a_2 are the lattice constants of the pure compounds. The values of B_{static} obtained from experiment and from (7.11) for the KCl_xBr_{1-x} system are given in Table 7.4.

Table 7.4. Static contribution (B_{static}) to the Debye–Waller factor for KCl_xBr_{1-x} mixed crystals

x	B_{static} [$Å^2$]		
	Experimental	Reference	Calculated (7.11)
0.2	0.34	[7.27]	0.362
0.4	0.47	[7.27]	0.543
0.5	0.51	[7.25]	0.566

7.2.3 Debye Temperatures

Debye [7.32] proposed a vibration spectrum for a crystal lattice which has a cut-off frequency ν_{max} or ν_{D}. Further, he developed an expression for the specific heat in terms of a parameter $\theta_{\mathrm{D}} = h\nu_{\mathrm{D}}/k$. This parameter which has dimensions of temperature is called the Debye temperature. The Debye temperature is derivable from the specific heats, elastic constants and X-ray (or neutron) diffraction intensities. Various methods of determination of Debye temperatures have been reviewed [7.33–7.36]. Details of the determination of θ from X-ray diffraction intensities are given in Chap. 3.

Several relations have been proposed either empirically or semi-theoretically to describe the composition dependence of the Debye temperatures of mixed crystals. Thus, by assuming the additivity of specific heats and using the low-temperature expression from Debye's theory for the specific heats, the following equation is obtained

$$\theta^{-3} = x\theta_1^{-3} + (1-x)\theta_2^{-3}, \tag{7.12}$$

where θ, θ_1 and θ_2 are the Debye temperatures of the mixed crystals and the two end members. This is known in literature as the Kopp–Neumann relation. From empirical considerations Karlsson [7.37] proposed the relation:

$$\theta^{-2} = x\theta_1^{-2} + (1-x)\theta_2^{-2}. \tag{7.13}$$

Giri and Mitra [7.38] proposed the relation

$$M\theta^2 = xM_1\theta_1^2 + (1-x)M_2\theta_2^2, \tag{7.14}$$

where M, M_1 and M_2 are the masses of the mixed crystal and those of the end members; M is given by $[xM_1 + (1-x)M_2]$.

A number of reports on the Debye temperatures of mixed crystals are available. The results in these reports are summarized in Table 7.5. From these results, it may be concluded that the composition dependence of Debye temperatures of mixed crystals is slightly non-linear with negative deviations from linearity.

7.2.4 Hardness of Mixed Crystals

Various aspects of hardness of crystals have been discussed in Chap. 4. There are several studies on the composition dependence of hardness of mixed crystals. Plendl et al. [7.46] quote references on the microhardness of the following systems: Ag–Au, Ag–Cu, K_2SO_4–$(NH_4)_2SO_4$, KCl–KBr, InP–GaP, CaF_2–SrF_2, SrF_2–BaF_2 and NiO–CoO. The microhardness of alkali halide mixed crystal systems has been studied extensively [7.47–7.49]. Sarma and Suryanarayana [7.50] carried out a detailed study of the CaF_2–SrF_2 system. Bodnar et al. [7.51] made measurements of the hardness of mixed crystal

Table 7.5. Summary of reports on Debye temperatures of mixed crystals

System	Reference	Method	Remarks
KCl–KBr	[7.37]	Specific heats	Equation (7.13) suitable
	[7.38]	Elastic constants	Equation (7.14) suitable
	[7.39]	Elastic constants	Negative deviations from additivity
KCl–RbCl	[7.40]	Elastic constants	Almost linear
KCl–NaCl	[7.41]	Elastic constants	Large deviation from linearity attributed to low stability
KBr–KI	[7.42]	Specific heats	Equation (7.13) suitable
	[7.43]	Elastic constants	Negative deviations from additivity
NaCl–NaBr	[7.44]	Elastic constants	Equation (7.12) suitable
	[7.38]	Elastic constants	Equation (7.14) suitable
KI–RbI	[7.45]	Neutron diffraction	Equation (7.12) suitable
AgCl–AgBr	[7.38]	Elastic constants	Equation (7.14) suitable

systems $CuGaS_{2x}Se_{2(1-x)}$, $CuInS_{2x}Se_{2(1-x)}$, $CuGa_xIn_{1-x}S_2$, $CuGa_xIn_{1-x}Se_2$ and $AgGa_xIn_{1-x}S_2$.

The trends observed in all these studies are: (1) the microhardness shows a severe non-linear composition dependence with positive deviations from linearity, and (2) the microhardness around the equimolar region ($x = 0.4$–0.6) is often larger than the hardness of either end member. Some approaches to explain these observations will now be discussed.

Plendl and Gielisse [7.52] gave an atomistic interpretation for hardness and obtained the expression:

$$H = (1/2)(zm/V)[2\pi\nu_{\mathrm{t}}r/\psi(\mathrm{Anh})]^2. \tag{7.15}$$

Here H is the hardness, z the valency, m the mass, r the interatomic distance, ν_{t} the transverse optical frequency and $\psi(\mathrm{Anh})$ the anharmonicity factor. Considering the structure and expressing V in terms of r, (7.15) takes the form

$$H = \pi^2 zm\nu_{\mathrm{t}}^2/[r\psi^2(\mathrm{Anh})] \tag{7.16}$$

for NaCl structure and

$$H = 4\pi^2 zm\nu_{\mathrm{t}}^2/[r\psi^2(\mathrm{Anh})] \tag{7.17}$$

Table 7.6. Values of $r, \nu_t, \psi(\text{Anh})$, H_{calc} and H_{obs} for the CaF_2–SrF_2 system

CaF_2/SrF_2	$r[10^{-8}\,\text{cm}]$	$\nu_t[\text{cm}^{-1}]$	$\psi(\text{Anh})$	$H_{\text{calc}}\,[\text{kg mm}^{-2}]$	$H_{\text{obs}}\,[\text{kg mm}^{-2}]$
100:0	2.36	291	0.87	170	166
75:25	2.39	290	0.825	195	199
50:50	2.43	281	0.82	195	208
25:75	2.46	268	0.85	170	180
0:100	2.50	252	0.84	155	137

for CaF_2 structure. Values of H calculated from these expressions agree well with experimentally determined values for several crystals with NaCl and CaF_2 structures. Plendl et al. [7.46] assumed that these expressions apply to mixed crystals also. Each of the component factors r, ν_t and $\psi(\text{Anh})$ have their own composition dependence and their combined composition dependence accounts for the observed composition dependence of H. The relevant data for the CaF_2-SrF_2 system given in Table 7.6 show fair agreement between H_{calc} and H_{obs}. It may be mentioned that the anharmonicity factor $\psi(\text{Anh})$ is determined from infrared reflectivity curves. Such data are limited to the CaF_2-SrF_2 system and some compositions in the NiO–CoO system. Hence, (7.16) and (7.17) have not been applied to other systems.

The above discussion is along empirical lines. A semi-theoretical explanation for the composition dependence of hardness of alkali halide mixed crystals has been given by Shrivastava [7.53] based on the Kataoka–Yamada model [7.54]. Kataoka and Yamada [7.54] considered the problem of hardening in mixed crystals in terms of the critical resolved shear stress (CRSS). Considering elastic interaction due to size misfit between the edge dislocations and the solute atoms, Kataoka and Yamada derived an expression for the enhancement τ_0 in the value of the CRSS. The expression obtained by them is

$$\tau_0 = \frac{4\sqrt{2}}{\pi}\alpha\left\{\frac{\alpha}{\alpha'}\right\}^{1/2}\frac{\mu^2 b^2}{\Gamma}\left[\frac{\Delta V}{b^3}\right]^2 x(1-x), \tag{7.18}$$

where

$$\alpha = \sum_{n=0}^{\infty}\int_0^{\infty}\int_0^{\infty}\frac{x^2(2n+1/4)^2}{[x^2+(2n+1/4)^2+z^2+R_0^2]^5}\mathrm{d}x\mathrm{d}z,$$

$$\alpha' = \sum_{n=0}^{\infty}\int_0^{\infty}\int_0^{\infty}\frac{25x^2(2n+1/4)^2}{[x^2+(2n+1/4)^2+z^2+R_0^2]^7}\mathrm{d}x\mathrm{d}z,$$

$$\mu = (1/2)(C_{11}-C_{12}),$$

$$\Gamma = (1/2)(C_{11}+C_{12})\left[\frac{C_{44}(C_{11}-C_{12})}{C_{11}(C_{11}+C_{12}+2C_{44})}\right]^{1/2} b^2.$$

Here, C_{11}, C_{12} and C_{44} are the elastic constants and ΔV is the difference in anion volumes of the two components. In the NaCl structure, the Burgers vector b is equal to the interionic distance r. Kataoka and Yamada observed a good agreement between the measured values of τ_0 and values calculated from (7.18).

Since hardness of a pure crystal is linearly related to the CRSS of a pure crystal [7.55,7.56], Shrivastava [7.53] proposed that the enhancement of CRSS in a mixed crystal should result in corresponding enhancement in hardness. Thus, he proposed for the hardness of a mixed crystal

$$H = xH_{\mathrm{A}} + (1 - x)H_{\mathrm{B}} + A(\tau_0), \tag{7.19}$$

where A is a constant. Hence,

$$\Delta H = A[\tau_0] = A\left[\frac{4\sqrt{2}}{\pi}\alpha\left\{\frac{\alpha}{\alpha'}\right\}^{1/2}\frac{\mu^2 b^2}{\Gamma}\left\{\frac{\Delta V}{b^3}\right\}^2\right]x(1 - x). \tag{7.20}$$

For mixed crystals, the values of μ, b and Γ are taken from linear interpolation of the values for the end members whereas values of α and α' are the same for all the alkali halides and ΔV is a constant independent of composition for a given system. The constant A is obtained by using the experimental value of H for one of the compositions as an input to (7.20). Shrivastava [7.53] applied (7.19) and (7.20) to the KCl–KBr mixed crystal system and observed good agreement with experimental results (Table 7.7).

Subba Rao and Hari Babu [7.48] considered various factors that affect the hardness of mixed crystals and pointed out that there are mainly two contributions. The first is the lattice contribution, i.e. the hardness due to interatomic forces in the lattice. The second is the defect contribution, i.e. the effect of various defects like vacancies, dislocations and grain boundaries which act as obstacles to dislocation motion. Subba Rao and Hari Babu [7.48] proposed the following equation

$$H = Cr^{-n} + kx(1 - x), \tag{7.21}$$

where n is a constant whose value depends on a particular system under consideration and k is the coefficient of hardening. Here, the two terms represent the lattice and defect contributions, respectively, C is a constant and r is

Table 7.7. Hardness (H) values for $\mathrm{KCl}_x\mathrm{Br}_{1-x}$ system

x	H [kg mm^{-2}]	
	Calc. (7.19), (7.20)	Obs. [7.48]
0.94	13.77	14.6
0.87	17.55	19.4
0.62	24.62	24.7
0.39	22.96	23.3
0.29	20.29	20.9
0.15	15.00	16.3

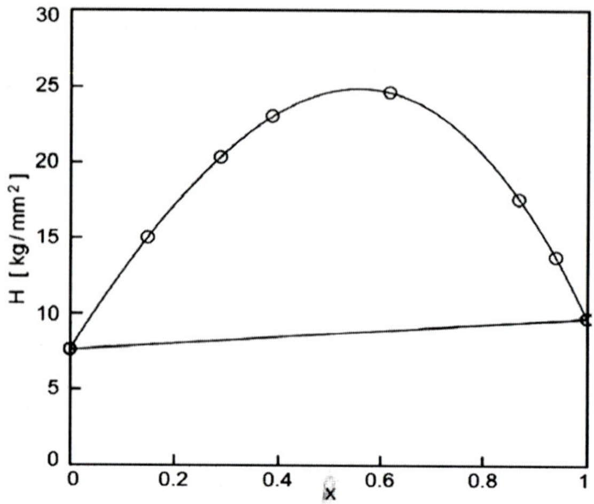

Fig. 7.5. Composition dependence of hardness of KCl_xBr_{1-x} mixed crystal system

the interionic distance. Using the values of r for mixed crystals of KCl–KBr, Subba Rao and Hari Babu [7.48] found that the first term shows an effectively linear variation with x. Thus we may write

$$\Delta H = kx(1 - x), \tag{7.22}$$

where ΔH now is the difference between the observed value and the value obtained from linear interpolation; k is a constant. The values of H for the KCl_xBr_{1-x} system obtained from additivity and from (7.22) are shown in Fig. 7.5.

7.2.5 Dielectric Constant

There is limited work on the static dielectric constant of mixed crystals. Gielisse et al. [7.19] determined the static dielectric constant of the NiO–CoO system from infrared reflectivity curves and found that its composition dependence is linear. Fertel and Perry [7.57] measured the dielectric constant of the KCl–KBr system from infrared reflectivity curves and found that the variation with composition was not smooth. Kamiyoshi and Nigara [7.58] determined the dielectric constant of the NaCl–NaBr, KCl–KBr, KCl–RbCl, RbCl–RbBr and KI–RbI systems at 1 MHz using the "immersion method". Their results for some of the systems are shown in Fig. 7.6. It is seen that the composition dependence is non-linear with positive deviations from linearity.

Combining the Clausius–Mosotti equation, the Lorentz–Lorenz equation and Vegard's law variation for the lattice constant and the infrared absorption wavelengths, Kamiyoshi and Nigara [7.58] obtained the following equation for the static dielectric constant ε_s of a binary ionic mixed crystal like AB_xC_{1-x}

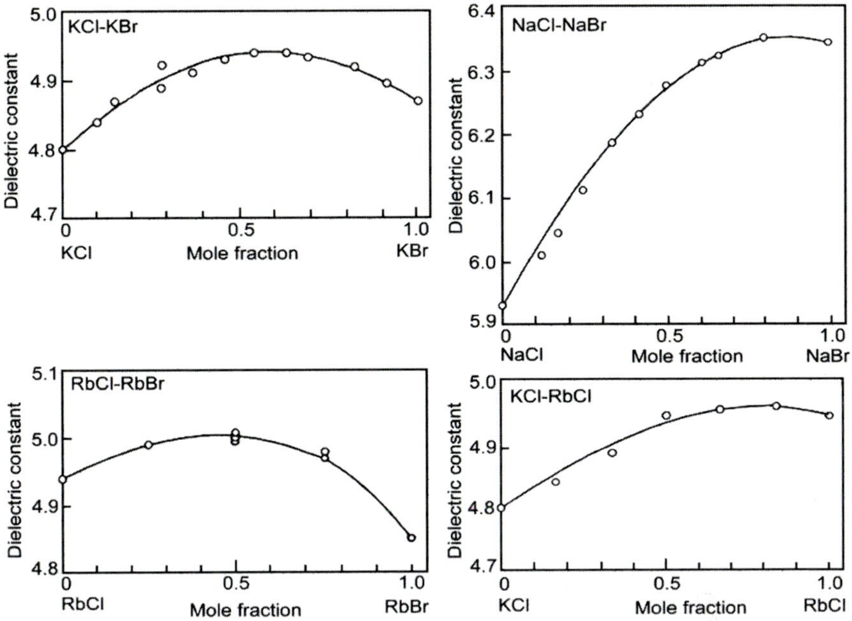

Fig. 7.6. Static dielectric constant of alkali halide mixed crystals; continuous curve calculated from (7.23)

$$
\begin{aligned}
(\varepsilon_s - 1)/(\varepsilon_s + 2) = {} & x(\lambda/\lambda_1)^2 (a_1/a)^6 (\varepsilon_1 - 1)/(\varepsilon_1 + 2) + (1 - x)(\lambda/\lambda_2)^2 \\
& (a_2/a)^6 (\varepsilon_2 - 1)/(\varepsilon_2 + 1) + x(a_1/a)^3 \{1 - (\lambda/\lambda_1)^2 \\
& (a_1/a)^3\}(n_1^2 - 1)/(n_1^2 + 2) + (1 - x)(a_2/a)^3 \\
& \{1 - (\lambda/\lambda_2)^2 (a_2/a)^3\}(n_2^2 - 1)/(n_2^2 + 2), \qquad (7.23)
\end{aligned}
$$

where (a, λ), (a_1, λ_1) and (a_2, λ_2) are the lattice constants and infrared absorption wavelengths of the mixed crystal and two end members and (ε_1, n_1), (ε_2, n_2) are static dielectric constant and refractive index of the end members, respectively. The values obtained from this equation are shown in Fig. 7.6 by the solid line. Excellent agreement is observed with experimental data.

Varotsos [7.59] modified (7.23) by bringing in the polarisability and bulk modulus. Sinha and Shanker [7.60] further modified (7.23) and calculated the dielectric constants of alkali halide mixed crystals using the equations of Kamiyoshi and Nigara [7.58], Varotsos [7.59] and their own equation. They found that, by and large, Kamiyoshi and Nigara's equation gives the best agreement with experimental data.

7.2.6 Effective Ionic Charge in Mixed Crystals

Szigeti [7.61, 7.62] proposed a theory for dielectric crystals which led to the equation

$$(q^*)^2 = (ze^*/ze)^2 = (9V_\mathrm{i}/4\pi^2)(1/ze)^2(\varepsilon_\mathrm{s} - \varepsilon_\infty)\mu\omega_\mathrm{t}^2/(\varepsilon_\infty + 2)^2, \qquad (7.24)$$

where q^* is called the effective ionic charge, z the valency, e the electron charge, e^* its effective value, V_i the volume of an ion pair, ε_s the static dielectric constant, ε_∞ the optical dielectric constant, μ the reduced mass and ω_t the transverse optical frequency. Calculations of q^* for several crystals are given in Chap. 5. The values of q^* range from 0 for purely covalent crystals to nearly unity for purely ionic crystals.

The composition dependence of q^* for NiO–CoO mixed crystals was studied by Gielisse et al. [7.19]. Using experimental values for V_i, ε_s, ε_∞ and ω_t for the mixed crystals, they found that q^* varies linearly with composition. Shanker and Jain [7.63] and Sinha and Shanker [7.60] calculated q^* for NaCl–NaBr, KCl–KBr and KBr–KI systems and the KCl–RbCl system, respectively. For their calculations they used input data obtained by linear interpolation of data for end members. They found that q^* varies nearly linearly with composition.

Starting with Szigeti's [7.61, 7.62] dielectric function

$$\varepsilon(\omega) = \varepsilon_\infty + (e^{*2}/\varepsilon_0 V_\mathrm{i}\mu)[(\varepsilon_\infty + 2)/3]^2[1/(\omega_\mathrm{t}^2 - \omega_\mathrm{l}^2)] \qquad (7.25)$$

and putting $\varepsilon(\omega) = 0$, we get

$$\mu(\omega_\mathrm{l}^2 - \omega_\mathrm{t}^2) = (e^{*2}/\varepsilon_0\varepsilon_\infty V_\mathrm{i})[(\varepsilon_\infty + 2)/3]^2. \qquad (7.26)$$

Here ε_0 is the vacuum dielectric constant. Scott [7.64] introduced a reduced charge (Z^*e) defined by

$$(Z^*e) = e^{*2}(\varepsilon_\infty + 2)^2/9\varepsilon_\infty \qquad (7.27)$$

which leads to

$$(\omega_\mathrm{l}^2 - \omega_\mathrm{t}^2) = (Z^*e)^2/(\mu\varepsilon_0 V_\mathrm{i}). \qquad (7.28)$$

Gervais [7.65] calculated (Z^*e) for several binary and ternary compounds using (7.28).

Wakamura and Arai [7.66] applied (7.28) to calculate (Z^*e) as a function of the composition for several mixed crystal systems with NaCl and ZnS structures and found the variation of (Z^*e) with composition to be slightly non-linear. For a binary mixed crystal denoted by $AB_{1-x}C_x$, (Z^*/Z_A) follows the equation

$$(Z^*)/(Z_\mathrm{A}) = a + bx + cx(1 - x). \qquad (7.29)$$

The values of a, b and c for a few systems are given in Table 7.8.

7.2.7 Colour Centres in Alkali Halide Mixed Crystals

There is a vast amount of literature on the colour centres in alkali halide crystals. This includes books [7.67, 7.68], review articles [7.69, 7.70] and a

Table 7.8. Values of (Z^*/Z_A) and parameters a, b, c for some mixed crystal systems

System	(Z^*/Z_A)		Parameters		
	$x = 0$	$x = 1$	a	b	c
$KCl_{1-x}Br_x$	0.747	0.736	0.750	−0.016	−0.015
$K_{1-x}Rb_xCl$	0.751	0.839	0.756	0.080	0.036
$AgCl_{1-x}Br_x$	0.791	0.672	0.791	−0.114	0.019
$Ni_{1-x}Co_xO$	0.448	0.470	0.450	0.020	0.000
$ZnS_{1-x}Se_x$	0.463	0.419	0.455	0.036	0.132

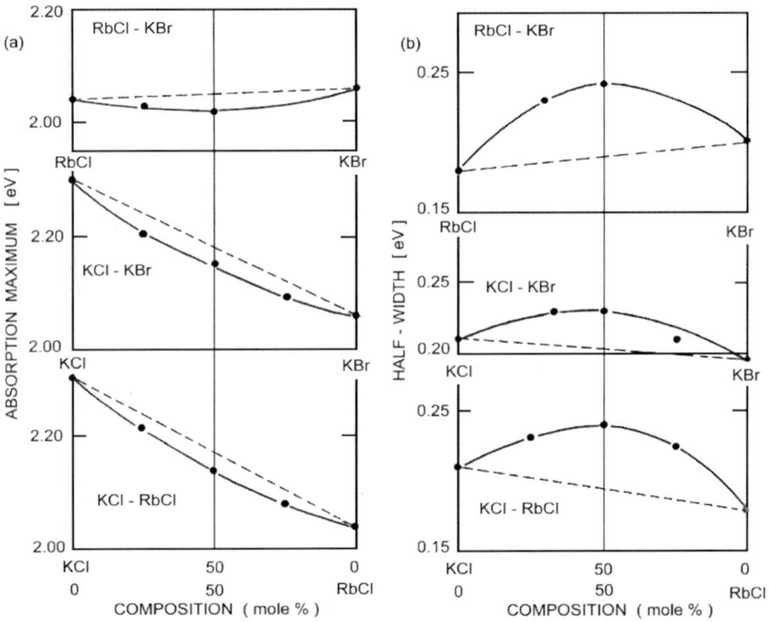

Fig. 7.7. Influence of composition on (**a**) spectral position of the F band and (**b**) the half-width of the F band at −190°C

comprehensive compilation of data [7.71]. Suffice it to state that irradiation of alkali halides by X-rays or γ-rays results in several bands in their absorption spectrum, the most prominent being the F band.

The composition dependence of F band parameters has been studied in several alkali halide mixed crystals. Smakula et al. [7.72] studied the shift in F band position and composition dependence of half-width in KCl–KBr, KCl–RbCl and RbCl–KBr mixed crystals. Their results are shown in Fig. 7.7. Both the absorption maximum and the half-width show non-linear composition dependence. Thyagarajan [7.73] recorded the F bands in the NaCl–NaBr system and found that the F band wavelength and the half-width show a non-linear composition dependence. Similar behaviour was observed in the KBr–KI and KI–RbI systems [7.74, 7.75].

Table 7.9. Values of C and s (7.30) for absorption bands in KCl–KBr mixed crystals

Absorption band	$C[\text{Å}^{1-s}]$		s	
	92 [K]	300 [K]	92 [K]	300 [K]
F	58	80	2.46	2.31
M	98	100	2.39	2.39
R_1	117	–	2.19	–
R_2	196	–	1.97	–
N_1	152	–	2.26	–

The wavelengths of colour centres in alkali halides correlate with the lattice constant/nearest neighbour distance through equations of the type

$$\lambda = Cr^s, \tag{7.30}$$

where C and s are constants for a family of related crystals. Such relations are known as the Mollwo–Ivey relations. The index s generally has values $\sim$1–2. Values of s for some prominent colour centres in alkali halides with NaCl and CsCl structures are quoted by Sirdeshmukh et al. [7.71].

Murti and Prasad [7.76] found that the Mollwo–Ivey relations are followed in the mixed crystals also. The values of C and s for the F, M, R_1, R_2 and N_1 centres in the KCl–KBr system are given in Table 7.9. Hovi [7.75] found that there are strong deviations from the Mollwo–Ivey relation in the F centre data in the KI–RbI system.

7.2.8 Defects in Mixed Crystals

As discussed in Chap. 1, the number of vacancies in a crystal can be estimated by comparing the measured density with the density calculated from the lattice constant (X-ray density). Such density studies on the KCl–KBr and NaCl–NaBr systems [7.12, 7.77] have shown that the numbers of vacancies in the mixed crystals are a few per cent more than in the pure crystals.

Vacancies give rise to ionic conductivity. The ionic conductivity of the KCl–KBr system was studied by Subba Rao and Hari Babu [7.48]. They found that the conductivity of mixed crystals was more than that of the pure crystals. They also evaluated the activation energy. As the experiments were made at moderate temperatures, they could evaluate only the activation energy of migration. The migration activation energy was found to be less for the mixed crystals than for the pure crystals. Subba Rao and Hari Babu [7.78] also studied the ionic conductivity of KBr–KI system. Here again the conductivity of mixed crystals was more than that of pure crystals. Further, they found that the activation energy in the intrinsic as well as extrinsic regions varies non-linearly with composition with negative deviations from linearity.

The dislocations in KCl–RbCl system were studied by Arends et al. [7.79] by the etching technique. They found that the density of dislocations is much more in the equimolar region than in the pure crystals. Qualitatively similar results were obtained by Veeresham et al. [7.80] for the KCl–KBr system.

7.2.9 Melting Temperature

Data are available for the melting temperatures of several alloy systems (Cs–K, K–Rb, Rb–Cs, W–Ta, Ta–V, V–W). These are quoted by Rabinovich et al. [7.81] and Corkill and Cohen [7.82]. In all these alloys, the composition dependence of the melting temperature is non-linear with negative deviations from linearity. Compared to the metal alloys, work on the melting temperatures of mixed crystals of compounds is limited. In the KCl–KBr system [7.11] and AgCl–AgBr system [7.83], it is again observed that the composition dependence is non-linear with negative deviations from linearity.

Lindemann's melting theory gives

$$\langle(\Delta r_{\mathrm{V}})^2\rangle/a^2 = x_{\mathrm{V}}^2 = (9\hbar^2 T_{\mathrm{m}})/(M k_{\mathrm{B}} \theta_{\mathrm{D}}^2 a^2), \tag{7.31}$$

where Δr_{V} is the deviation in the atomic position due to thermal vibration, a the lattice constant, $\hbar = (h/2\pi)$, M the mass, k_{B} the Boltzmann constant and θ_{D} the Debye temperature. (x_{V}^2) is the square of the reduced displacement. Rabinovich et al. [7.81] and Corkill and Cohen [7.82] extended Lindemann's equation to alloys with some modifications. They introduced the various parameters for the alloy in terms of the concentration of the alloy components as follows:

$$M = xM_1 + (1-x)M_2, \tag{7.32}$$

$$a = xa_1 + (1-x)a_2, \tag{7.33}$$

$$\theta_{\mathrm{D}}^{-2} = (\theta_{\mathrm{D}}^{-2})_1 + (\theta_{\mathrm{D}}^{-2})_2. \tag{7.34}$$

In addition they took into account the size disorder x_{s} given by

$$x_{\mathrm{s}}^2 = 4x(1-x)(r_1 - r_2)^2/a^2, \tag{7.35}$$

where r_1 and r_2 are the radii of the component atoms and a the lattice constant.

Further, they expressed the total displacement (Δr) and the reduced total displacement (x_{m}) as

$$\langle(\Delta r)^2\rangle/a^2 = x_{\mathrm{m}}^2 = x_{\mathrm{V}}^2 + x_{\mathrm{s}}^2. \tag{7.36}$$

With these modifications (7.31) becomes

$$T_{\mathrm{m}} = (M k_{\mathrm{B}} \theta_{\mathrm{D}}^2 a^2/9\hbar^2) x_{\mathrm{m}}^2 [1 - (x_{\mathrm{s}}^2/x_{\mathrm{m}}^2)]. \tag{7.37}$$

With this expression, Corkill and Cohen [7.82] were able to correctly explain the observed trends in the composition dependence of T_{m} of several alloy systems.

Sharma et al. [7.84] obtained an expression for the melting temperature (T_m) of an ionic mixed crystal using the simple Born potential:

$$\phi = (-z^2e^2/r) + (b/r^n). \tag{7.38}$$

The constants b and n can be obtained from the compressibility ψ. Subjecting the potential ϕ to the condition:

$$T_\mathrm{m} = (r_\mathrm{m}/2Ck_\mathrm{B})(\mathrm{d}\phi/\mathrm{d}r) \quad \text{at } r = r_\mathrm{m} \tag{7.39}$$

and introducing an arbitrary constant

$$C' = C + x(1-x). \tag{7.40}$$

Sharma et al. obtained

$$T_\mathrm{m} = (z^2\alpha e^2)[1-(r_0/r_\mathrm{m})^{n-1}]/2C'k_\mathrm{B}r_\mathrm{m}. \tag{7.41}$$

Here r and r_m are the values of the interionic distance at 0 K (or at room temperature) and at the melting point. In using (7.41) for mixed crystals, Sharma et al. used the following relations:

$$r_0 = x(r_0)_1 + (1-x)(r_0)_2, \tag{7.42}$$

$$\psi = x\psi_1 + (1-x)\psi_2. \tag{7.43}$$

Using (7.41), Sharma et al. [7.84] calculated the melting temperatures for a few alkali halide mixed crystal systems.

7.2.10 Pm3m ↔ Fm3m Transition in Mixed Crystals

A crystal assumes a structure for which the free energy is the least. In some cases the free energy will have very close values for two structures. In such cases the structure changes from one form to the other with a change in physical conditions. This is what happens in crystals with NaCl and CsCl structures. Several alkali halides with NaCl structure transform from the Fm3m NaCl structure to the Pm3m CsCl structure on the application of pressure. On the other hand CsCl, NH_4Cl and NH_4Br which have the Pm3m CsCl structure transform to the Fm3m structure at elevated temperatures.

The Pm3m ↔ Fm3m transition in CsCl takes place at 479°C [7.85]. The Pm3m ↔ Fm3m transition in CsBr cannot be observed as the transition temperature is above the melting point [7.86]. CsCl–CsBr mixed crystals do undergo the transition up to a CsBr concentration of 60% beyond which the transition temperatures are again above the melting points. The transition temperatures of CsCl–CsBr mixed crystals were reported by Weijma and Arends [7.86] and by Natarajan et al. [7.87]; their results are shown in Fig. 7.8.

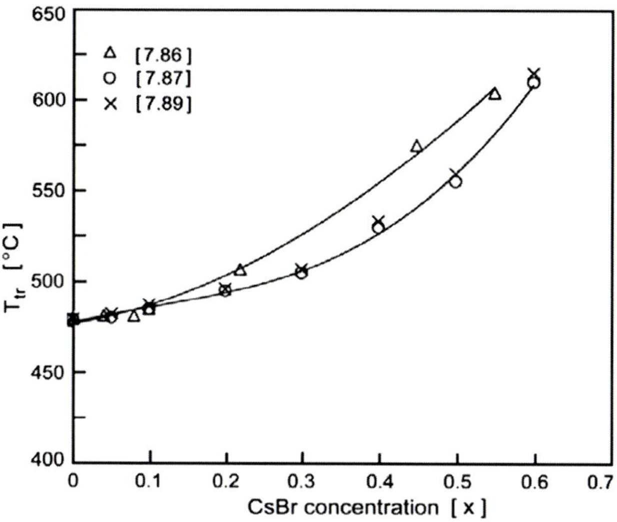

Fig. 7.8. Transition temperatures of $CsCl_{1-x}Br_x$

The transition temperature increases non-linearly as the CsBr content increases.

May [7.88] calculated the transition temperature of CsCl using the Born–Mayer theory according to which the lattice potential is expressed as

$$\phi = -(\alpha e^2/r) - (C/r^6) - (D/r^8) + be^{-r/\rho}. \tag{7.1}$$

He calculated the transition temperature T_{tr} from the relation

$$T_{tr} = (\phi_1 - \phi_2)/3k_B \log(A_1/A_2), \tag{7.44}$$

where $(-\phi_1)$ and $(-\phi_2)$ are the lattice energies for the two structures and A_1 and A_2 the force constants calculated from the short range terms. To obtain agreement with the experimental value of T_{tr}, May [7.88] multiplied the Van der Waal terms by a constant. Shanker et al. [7.89] extended May's model to calculate the transition temperatures of CsCl–CsBr mixed crystals. Their results are included in Fig. 7.8.

The composition dependence of the transition temperature in the $NH_4Cl_xBr_{1-x}$ system has been studied by Costich et al. [7.90] from cooling curves and by Murti and Prasad [7.91] from ionic conductivity measurements. Their results are shown in Fig. 7.9. There is a marked difference in the results. While Costich et al. observed a highly non-linear composition dependence with negative deviations from linearity, Murti and Prasad observed a linear composition dependence. There is no theoretical work on this system.

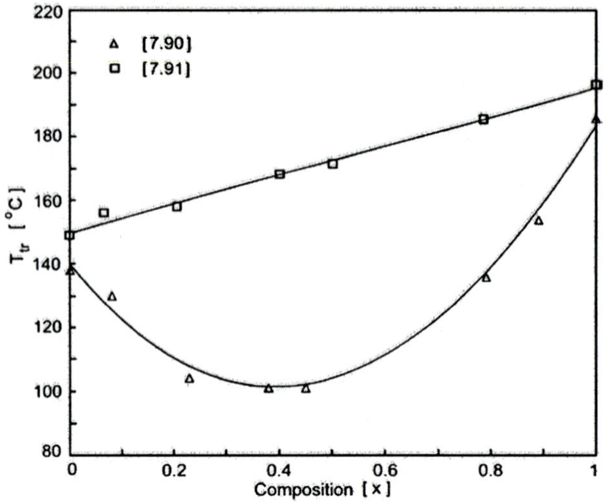

Fig. 7.9. Pm3m–Fm3m transition temperature Vs. composition (x) for $NH_4Cl_xBr_{1-x}$ mixed crystals

7.3 Some of our Results

Results of our studies of some solid state properties of pure crystals have been discussed in other chapters. In this section, results of our studies of mixed crystals are discussed.

7.3.1 Lattice Constants of Mixed Crystals

It has been mentioned in Sect. 7.2.1 that the composition dependence of lattice constants of mixed crystals is best described by either of the following equations:

$$\text{Retgers' law:} \quad a^3 = xa_1^3 + (1-x)a_2^3, \tag{7.4}$$

$$\text{Vegard's law:} \quad a = xa_1 + (1-x)a_2. \tag{7.5}$$

Srinivas et al. [7.92] determined the lattice constant of the $RbCl_xBr_{1-x}$ system from Debye–Scherrer photographs. The composition dependence is shown in Fig. 7.10; there are small deviations from linearity. The lattice constants were calculated from the above two equations and the deviations ($a_{\text{calc}} - a_{\text{exp}}$) are plotted against the composition parameter x (Fig. 7.11). It is seen that (7.5) gives a better fit.

Similarly, the lattice constants of the $RbBr_xI_{1-x}$ system were determined by Kumara Swamy et al. [7.93] with the help of a powder diffractometer. They evaluated the lattice constants from (7.4) and (7.5). The values of the lattice constant and the difference ($a_{\text{exp}} - a_{\text{calc}}$) are plotted in Fig. 7.12 . Here again, (7.5) is found to give a better fit.

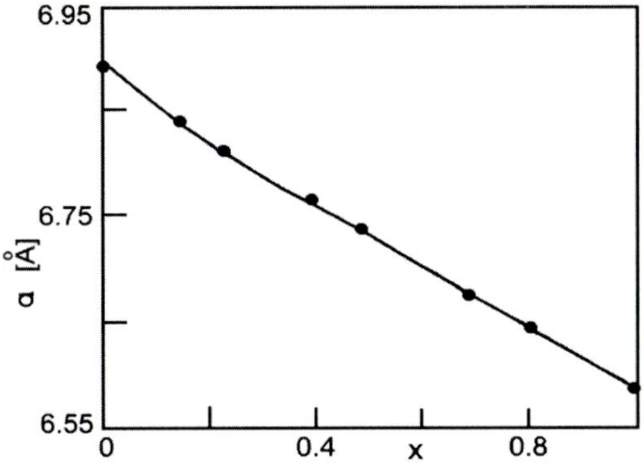

Fig. 7.10. Plot of lattice constant a vs. composition x for the $RbCl_xBr_{1-x}$ system

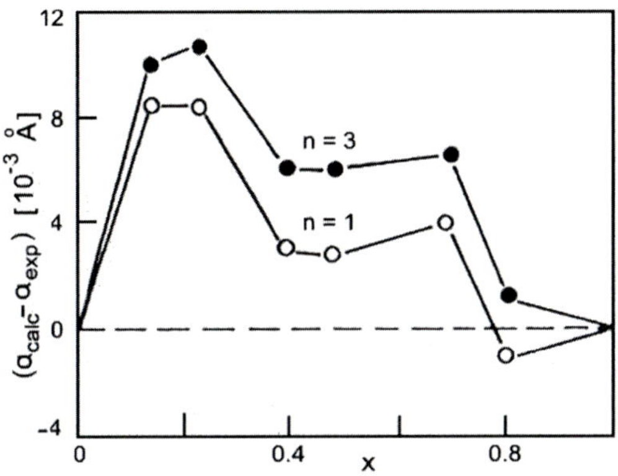

Fig. 7.11. Plot of $(a_{\text{calc}} - a_{\text{exp}})$ against composition x for the $RbCl_xBr_{1-x}$ system

It was mentioned in Sect. 7.2.1 that the lattice constants of some mixed crystal systems have shown strong deviations from Vegard's law as well as Retgers' law. Three of these systems have been reinvestigated in our laboratories with improved accuracy. These results are now discussed.

Suryanarayana and Sirdeshmukh [7.94] determined the lattice constants of $NH_4Cl_{1-x}Br_x$ mixed crystals using the Debye Scherrer method. The values of the lattice constant a_{exp} and the deviation $(a_{\text{exp}} - a_{\text{calc}})$ from (7.5) are given in Table 7.10. Negative deviations are observed in the Cl-rich regions. In this respect, these results agree with those reported by Anselmo and Smith [7.22]. The deviations observed in the Br-rich region, though positive, are also larger than the experimental error.

Subhadra and Hussain [7.95] determined the lattice constants of the $Na(ClO_3)_x(BrO_3)_{1-x}$ system with high accuracy using a symmetric focusing camera. They calculated the difference $(a_{\text{exp}} - a_{\text{calc}})$ between the experimental values and those calculated from Vegard's law. A plot of the $(a_{\text{exp}} - a_{\text{calc}})$ values from the results of Subhadra and Hussain [7.95] and those reported

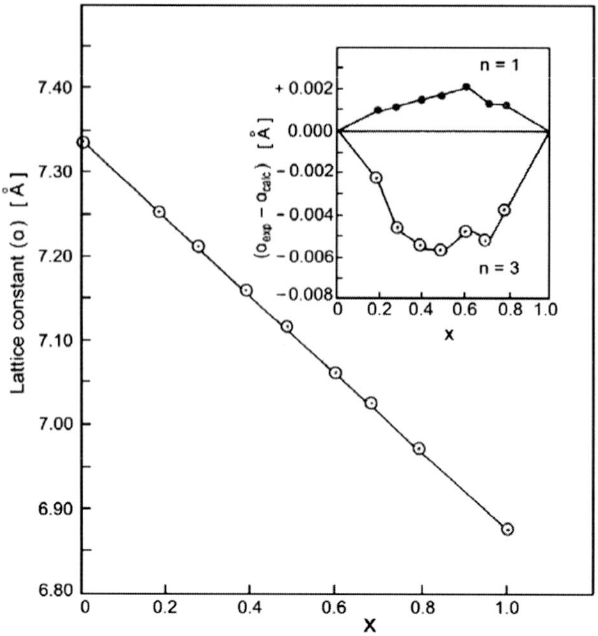

Fig. 7.12. Plots of lattice constant (a) as a function of composition x [inset, ($a_{\mathrm{exp}} - a_{\mathrm{calc}}$) vs. x] for $RbBr_xI_{1-x}$ system

Table 7.10. Lattice constants of the $NH_4Cl_{1-x}Br_x$ system

x	a_{exp} [Å]	$a_{\mathrm{exp}} - a_{\mathrm{calc}}$ [10^{-3} Å]	x	a_{exp} [Å]	$a_{\mathrm{exp}} - a_{\mathrm{calc}}$ [10^{-3} Å]
0.000	3.8759	0	0.158	3.8948	−10.2
0.010	3.8777	0	0.289	3.9285	−0.6
0.032	3.8811	−0.7	0.463	3.9623	+1.2
0.048	3.8837	−1.0	0.734	4.0190	+8.0
0.059	3.8855	−1.3	0.863	4.0378	+3.0
0.077	3.8896	−0.5	0.942	4.0504	+1.1
0.111	3.8927	−3.6	1.000	4.0600	0

by Mohanlal [7.1] is shown in Fig. 7.13 as a function of x. Mohanlal's data points lie on a double-peaked curve whereas those from the data of Subhadra and Hussain lie on a single-peaked curve. Further, the values of ($a_{\mathrm{exp}} - a_{\mathrm{calc}}$) in the data of Subhadra and Hussain are smaller in magnitude than those in Mohanlal's data. Mohanlal reported that he tried to fit his data to the equation

$$a^n = xa_1^n + (1-x)a_2^n \tag{7.7}$$

using values of 1, 2 and 3 for n but the fit was not satisfactory. Subhadra and Hussain's data also does not give a good fit for these values of n. As an exercise, n was allowed to take higher values up to $n = 30$. The best fit was

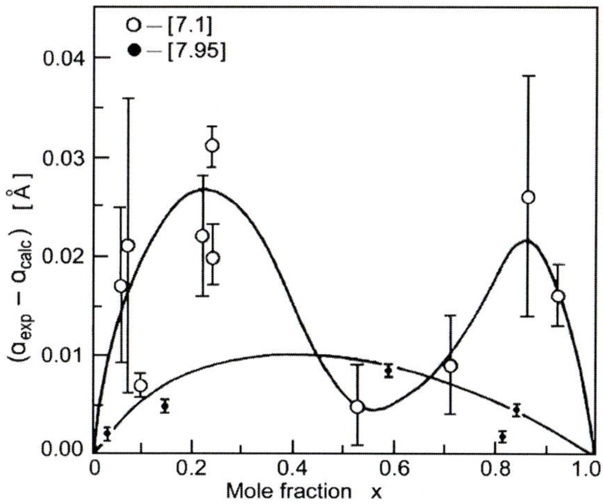

Fig. 7.13. Plot of $(a_{\text{exp}} - a_{\text{calc}})$ against composition x for the $Na(ClO_3)_x(BrO_3)_{1-x}$ system

obtained for $n = 25$ with an e.s.d. of 0.0022 Å for $(a_{\text{exp}} - a_{\text{calc}})$. Such a value is, of course, unacceptable.

A detailed study of the KBr_xI_{1-x} system was made by Hussain et al. [7.96]. It was mentioned in Sect. 7.1.1 that for mixed crystal formation, the difference in the lattice constants of the end members should not exceed 7–8%. In the KBr-KI system this difference is 7%; i.e. this system is a borderline case. Havighurst et al. [7.16], Tobolsky [7.97] and Nair and Walker [7.23] found that in this system mixed crystal formation was difficult and two phases appeared. Hussain et al. [7.96] also experienced difficulty in obtaining single phase samples. A powder X-ray diffractometer was employed for lattice constant determination. The half-widths of the lines were also measured. The composition dependence of the lattice constant is shown in Fig. 7.14. Clear deviations from linear (Vegard's law) behaviour were observed. These deviations were attributed to lattice strain caused by the instability due to the large difference in the lattice constants of KBr and KI. From the deviations, the lattice strain ε_l was calculated from the relation $\varepsilon_l = (\Delta a/a)$. The maximum value of ε_l was 4×10^{-3}. The lattice strain was independently determined from the half-widths using standard procedure [7.98]. A plot of ε_l thus obtained is also shown in Fig. 7.14 as a function of composition x. ε_l assumes a maximum value of 3.4×10^{-3} which is close to the value obtained from the excess lattice constant.

Thus the deviations from Vegard's law behaviour in this system can be interpreted as due to lattice strain caused by instability in the mixed crystals. It is possible that the deviations observed in the lattice constants of $Na(ClO_3)_x(BrO_3)_{1-x}$ are also due to lattice strains.

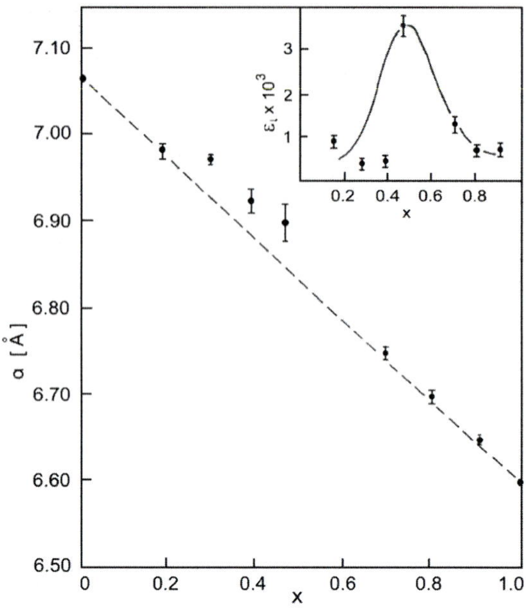

Fig. 7.14. Plot of lattice constant a vs. composition x for the KBr_xI_{1-x} system (*dashed line* represents additive behaviour); (*inset*) plot of strain ε_l vs. x (*continuous line*, guide to the eye)

7.3.2 Debye–Waller Factors

Various aspects of Debye–Waller factors of crystals have been discussed in Chap. 3. In Sect. 7.2.2, it has been pointed out that in mixed crystals, the static disorder (or size effect) contributes a static component B_{static} to the measured Debye–Waller factor; a simple method for estimating B_{static} has been discussed.

The Debye–Waller factors of a large number of mixed crystal systems have been studied as a function of composition. These are: KCl–KBr [7.99], KBr–RbBr [7.100], RbCl–RbBr [7.92], KCl–RbCl [7.101], RbBr–RbI [7.93], NaCl–NaBr and KI–RbI [7.102], AgCl–AgBr [7.103], CsCl–CsBr and NH_4Cl–NH_4Br [7.104]. Typical composition dependence of the Debye–Waller factors is shown in Fig. 7.15. These Debye–Waller factors are already corrected for the static disorder effect.

The Kopp–Neumann relation for the Debye temperatures of mixed crystals is:

$$\theta^{-3} = x\theta_1^{-3} + (1-x)\theta_2^{-3}. \tag{7.12}$$

At moderate temperatures, the Debye–Waller theory expression for B reduces to

$$B = 6h^2T/Mk_{\text{B}}\theta^2, \tag{7.45}$$

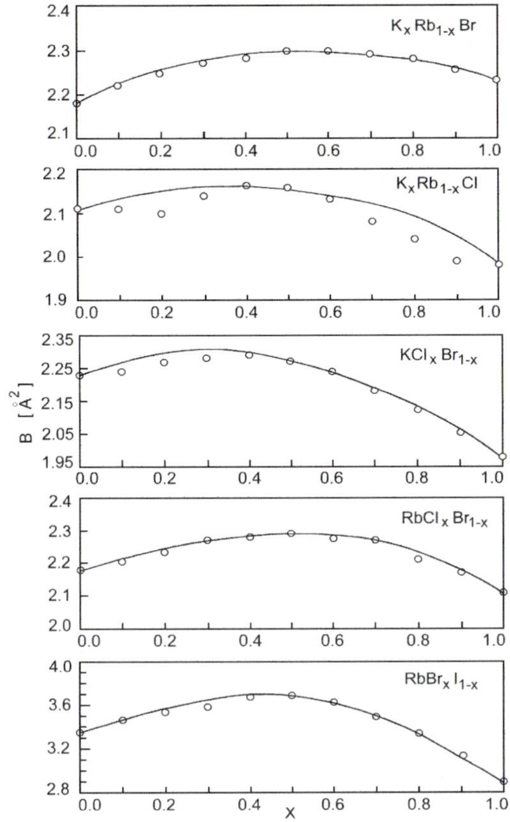

Fig. 7.15. Plots of Debye–Waller factor B_{thermal} vs. composition x for some alkali halide mixed crystal systems

where the various symbols have the usual meaning. Combining (7.12) and (7.45), Kumara Swamy [7.105] obtained

$$B = [x\{(M_1/M)B_1\}^{3/2} + (1-x)\{(M_2/M)B_2\}^{3/2}]^{2/3}, \tag{7.46}$$

where (B, M), (B_1, M_1) and (B_2, M_2) are the Debye–Waller factors and masses of the mixed crystal and the two end members.

The experimental data on Debye–Waller factors was examined in the light of (7.46). It was found that (7.46) predicts the same composition dependence as shown by the experimental data viz. non-linear variation with positive deviations from linearity. For a quantitative assessment, the values of $(B_{\text{exp}} - B_{\text{calc}})$ were calculated for each composition and its e.s.d. for each system was calculated. These e.s.d. values are given in Table 7.11. The e.s.d.s are of the same order as the error in B_{exp} thus indicating good agreement except in the case of RbBr–RbI.

Table 7.11. E.s.d. of ($B_{\text{thermal}} - B_{\text{calc}}$) for alkali halide mixed crystal systems

System	E.s.d. [Å^2]	System	E.s.d. [Å^2]
KCl–KBr	0.06	NaCl–NaBr	0.05
KCl–RbCl	0.03	KI–RbI	0.08
KBr–RbBr	0.06	CsCl–CsBr	0.02
RbCl–RbBr	0.09	NH_4Cl–NH_4Br	0.02
RbBr–RbI	0.35		

Table 7.12. Values of e.s.d. in ($\theta_E - \theta_{\text{calc}}$) for some mixed crystal systems

System	E.s.d. [K]		
	Equation (7.12)	Equation (7.47)	Equation (7.48)
KCl_xBr_{1-x}	2.7	6.2	0.85
$AgCl_xBr_{1-x}$	1.5	1.7	1.5
$Mg_xFe_{1-x}O$	74.6	86.7	8.4

Only in one system viz. KBr–KI, a different trend was observed. The data reported by Hussain et al. [7.96] show a non-linear composition dependence but with negative deviations. It may be mentioned that a similar composition dependence was observed by Ganesan and Girirajan [7.106] in the CsCl–CsBr system. But a redetermination by Balaiah [7.104] has led to results similar to the rest of the systems. Thus, the KBr–KI data seem to have been affected by some unidentified error.

7.3.3 Debye Temperatures of Mixed Crystals

Debye Temperatures from Elastic Constants

Nagaiah and Sirdeshmukh [7.107] calculated the Debye temperatures of three mixed crystal systems KCl_xBr_{1-x}, $AgCl_xBr_{1-x}$ and $Mg_xFe_{1-x}O$ from their elastic constants using the Voigt–Reuss–Hill averaging procedure discussed in Chap. 6. They compared these values (θ_E) with values (θ_{calc}) calculated from the following three equations:

$$\text{Kopp} - \text{Neumann} \quad \theta^{-3} = x\theta_1^{-3} + (1-x)\theta_2^{-3}, \tag{7.12}$$

$$\text{Empirical} \quad \theta^{2} = x\theta_1^{2} + (1-x)\theta_2^{2}, \tag{7.47}$$

$$\text{Empirical} \quad \theta^{-1} = x\theta_1^{-1} + (1-x)\theta_2^{-1}. \tag{7.48}$$

For each system the e.s.d. in ($\theta_E - \theta_{\text{calc}}$) was calculated. These e.s.d. values are given in Table 7.12. It is found that (7.48) has the lowest e.s.d. in all the three systems. Further, the e.s.d.s for the $Mg_xFe_{1-x}O$ system are too large.

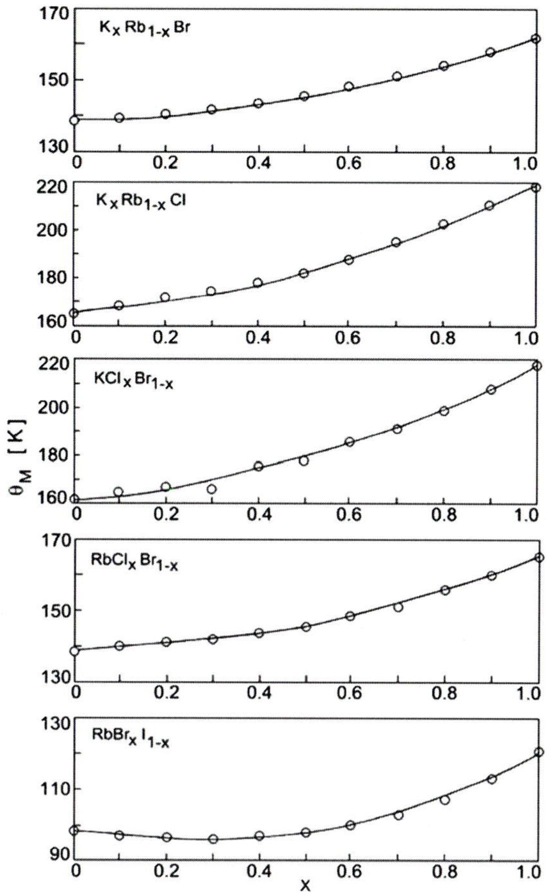

Fig. 7.16. Plot of Debye temperature θ_M vs. composition x for some alkali halide mixed crystal systems

Debye Temperatures from X-Ray Intensities

In Sect. 7.3.2 results on Debye–Waller factors of several mixed crystals were discussed. In Chap. 3, the procedure for evaluation of the X-ray Debye temperature (θ_M) has been discussed. Typical examples of the composition dependence of θ_M are shown in Fig. 7.16. It is seen that the composition dependence is slightly non-linear with negative deviations from linearity.

Geeta Krishna et al. [7.102] examined the data on θ_M for several mixed crystal systems vis-a-vis empirical relations. Apart from (7.12), (7.47) and (7.48), the following equations were also considered:

$$\theta^{-2} = x\theta_1^{-2} + (1-x)\theta_2^{-2}, \tag{7.37}$$

$$m\theta^2 = m_1 x\theta_1^2 + (1-x)m_2\theta_2^2. \tag{7.38}$$

To assess the relative merit of these equations, Geeta Krishna et al. calculated a parameter Ω defined by

$$\Omega = \left[\frac{\sum_n \{(\theta_{\text{calc}} - \theta_{\text{M}})/\theta_{\text{M}}\}^2}{n}\right]^{1/2} . \tag{7.49}$$

The values of Ω for the different equations for several alkali halide mixed crystal systems are given in Table 7.13. It is seen that for each system the least value is obtained for the Kopp–Neumann relation (7.12). Ω has the largest value for the RbBr–RbI system.

Table 7.13. Values of Ω

System	Source of θ_M (exp)	Ω				
		Equation (7.12)	Equation (7.47)	Equation (7.48)	Equation (7.37)	Equation (7.38)
KCl–KBr	[7.99]	0.016	0.023	0.032	0.058	0.030
KCl–RbBr	[7.100]	0.015	0.018	0.020	0.027	0.017
RbCl–RbBr	[7.92]	0.020	0.023	0.026	0.035	0.024
KCl–RbCl	[7.101]	0.008	0.014	0.020	0.044	0.017
KI–RbI	[7.102]	0.013	0.016	0.020	0.030	0.021
NaCl–NaBr	[7.102]	0.014	0.023	0.033	0.064	0.028
RbBr–RbI	[7.93]	0.079	0.084	0.089	0.104	0.092

Mixed crystals with single compositions in the systems TlBr–TlCl and TlBr–TlI were available for study. Their mean Debye–Waller factors and Debye temperatures were determined by Srinivas and Sirdeshmukh [7.108,7.109]. The θ_M values are given in Table 7.14 along with those for TlCl and TlBr [7.110]. It has been mentioned in Chap. 6 that the Debye temperatures of related crystals have a linear relationship with a^{-1} where a is the lattice constant. Such a plot is shown for the thallium halide crystals and their mixed crystals (Fig. 7.17). By extrapolation of this plot a value of 86 K is predicted for the θ_M of the CsCl phase of TlI.

Table 7.14. θ_M values of some Tl halides

Crystal	θ_M [K]	Reference
TlCl	101	[7.110]
$TlCl_{0.7}Br_{0.3}$ (KRS-6)	97	[7.108]
TlBr	95	[7.110]
$TlBr_{0.46}I_{0.54}$ (KRS-5)	90	[7.109]
TlI (CsCl)	86 (estimated)	

Debye Temperatures by Different Methods

Apart from the methods of determining the Debye temperatures from elastic constants and X-ray intensities, there are other methods also. Thus, the Debye temperature can be determined by fitting low-temperature specific heats to

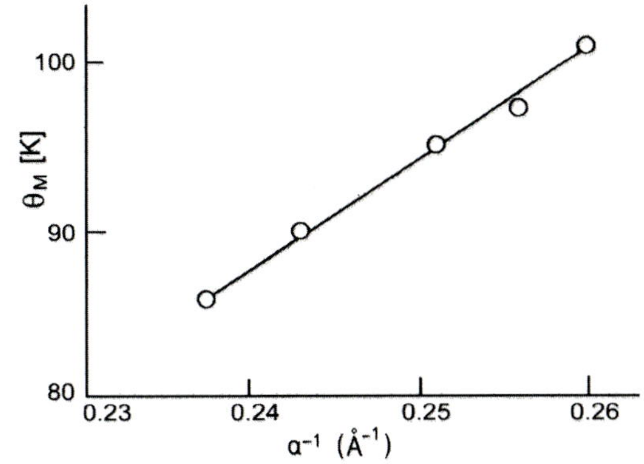

Fig. 7.17. Plot of Debye temperature (θ) against reciprocal of lattice constant (a^{-1}) for pure and mixed crystals of thallium halides

the Debye – T^3 equation. Further, the Debye temperature can be calculated from the melting point using the Lindemann melting point formula

$$\theta = C'(T_{\mathrm{m}}/MV^{2/3})^{1/2}, \tag{7.50}$$

where C' is a constant, T_{m} the melting point, M the molar mass and V the molar volume. Blackman [7.111] derived the following equation for the Debye temperature in terms of the compressibility:

$$\theta = (1/2\pi)(h/k_{\mathrm{B}})(5r/\mu\psi). \tag{7.51}$$

Here r is the interionic distance, μ the reduced mass and ψ the compressibility. The number 5 applies only to crystals with NaCl structure; it has a different value for other structures. Abrahams and Hsu [7.112] suggested the following empirical relation between the Debye temperature and hardness:

$$\theta = C''H_{\mathrm{V}}^{1/2}V^{1/6}M^{-1/2}. \tag{7.52}$$

Here C'' is a constant and H_{V} the Vickers hardness.

Srinivas and Sirdeshmukh [7.113] calculated the Debye temperatures of KCl_xBr_{1-x} mixed crystals by all these methods. Their results are given in Table 7.15. All the methods give comparable θ values and reveal the same trend in composition dependence. However, the θ values obtained from hardness data are systematically higher than the others. In Sect. 7.2.4 it has been pointed out that the static disorder (size effect) enhances the hardness of a mixed crystal. This leads to an enhancement in the value of the Debye temperature calculated from the measured hardness.

7.3.4 Hardness of Mixed Crystals

Various aspects of hardness of mixed crystals have been discussed in detail in Chap. 4. Results of some studies of hardness of mixed crystals have been discussed in Sect. 7.2.4.

Table 7.15. Debye temperature θ for the KCl_xBr_{1-x} system

x	θ [K]				
	Specific heats	Elastic constants	Melting points	Compressibility	Microhardness
0	172	170	170	184	170
0.20	180	179	178	190	270
0.40	190	190	188	197	315
0.62	202	201	200	207	334
0.83	218	216	215	220	320
1.00	232	230	230	233	230

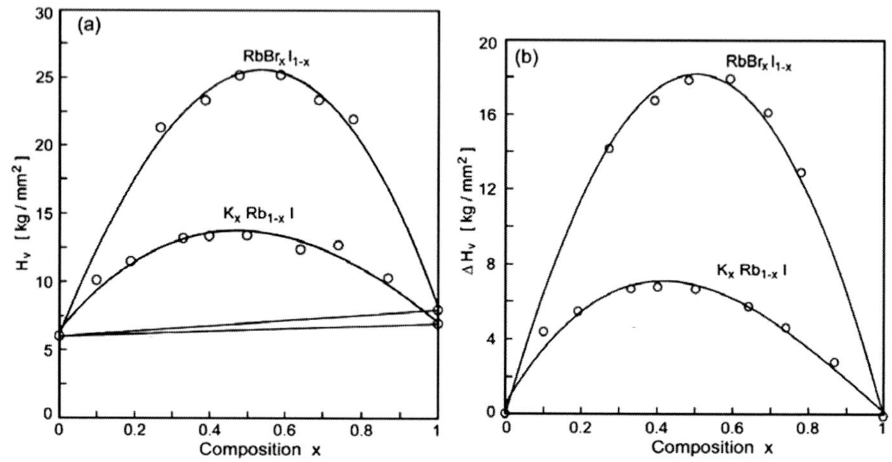

Fig. 7.18. Plot of (**a**) hardness, H_V and (**b**) ΔH_V against composition x for mixed crystals

Hardness of several mixed crystal groups have been studied in our laboratory. These are: $NaClO_3$–$NaBrO_3$ [7.114], RbCl–RbBr [7.115], RbBr–RbI and KI–RbI [7.116] and CsCl–CsBr and NH_4Cl–NH_4Br [7.117]. Typical plots of hardness as a function of composition are shown in Fig. 7.18. The common trend shown by all systems is that the hardness has a non-linear composition dependence with positive deviations from additivity. Plots of the difference between the measured hardness and that expected from linear behaviour (ΔH_V) are also shown in Fig. 7.18. Different models to explain this non-linear behaviour were discussed in Sect. 7.2.4. According to a simple model proposed by Subba Rao and Hari Babu [7.48], lattice forces cause a linear variation and defects and static disorder contribute a non-linear variation. The combined result is that

$$\Delta H = kx(1 - x), \tag{7.22}$$

where k is a constant. It can be seen that the maximum deviation of H from additivity is $0.25k$. Values of k for several mixed crystal systems are

Table 7.16. Values of constant k (7.22) for some mixed crystal systems

System	k [kg mm^{-2}]	Δr [Å]	Reference for H
NaCl–NaBr	96	0.15	[7.47]
KCl–KBr	53	0.15	[7.48]
KBr–KI	83	0.21	[7.49]
RbCl–RbBr	55	0.15	[7.115]
RbBr–RbI	76	0.21	[7.116]
KI–RbI	24	0.14	[7.116]
CsCl–CsBr	22	0.15	[7.117]
NH_4Cl–NH_4Br	120	0.15	[7.117]

given in Table 7.16. Values of Δr, the difference in the ionic radii of the mixing ions, are also given. Considering only the rubidium halide systems, Sirdeshmukh et al. [7.116] suggested that k depends on Δr, being less for the RbCl–RbBr system (with $\Delta r = 0.15$) and more for KBr–KI and RbBr–RbI systems ($\Delta r = 0.21$). But the large k values for NaCl–NaBr, CsCl–CsBr and NH_4Cl–NH_4Br systems (with $\Delta r = 0.15$) do not fit into this scheme.

7.3.5 Dielectric Properties

A detailed study of dielectric constant (ε) and dielectric loss (tan δ) was undertaken for two mixed crystal systems: KCl–KBr [7.118] and RbCl–RbBr [7.119]. The study included the determination of the composition dependence of ε, $\tan\delta$ and related parameters at room temperature and at elevated temperatures up to 350°C. The results are discussed in relation to earlier reports on mixed crystals mentioned in Sect. 7.2.5.

The static dielectric constant for $KCl_{1-x}Br_x$ is shown as a function of composition in Fig. 7.19 along with the values given by Fertel and Perry [7.57] and values obtained from (7.23). It can be seen that experimental values by Sirdeshmukh et al. [7.118] and those calculated by Kamiyoshi and Nigara [7.58] show a smooth variation with a maximum value at equimolar composition. On the other hand, the values by Fertel and Perry [7.57] differ from values in the other two reports [7.58, 7.118]. For $RbCl_{1-x}Br_x$ the experimental values from [7.119] are shown in Fig. 7.20 as a function of composition along with the values obtained from (7.23). Both the plots show a non-linear variation with a maximum around equimolar composition.

The variation of dielectric constant with composition at various temperatures is shown in Fig. 7.21 for $KCl_{1-x}Br_x$ and for $RbCl_{1-x}Br_x$ in Fig. 7.22. The variation is non-linear with a maximum value for equimolar composition. It is also seen that the difference between the values for the end members and the middle composition increases with temperature.

The temperature variation of dielectric constant can be expressed in the form of A, B, C terms on the basis of Havinga's theory which was discussed in

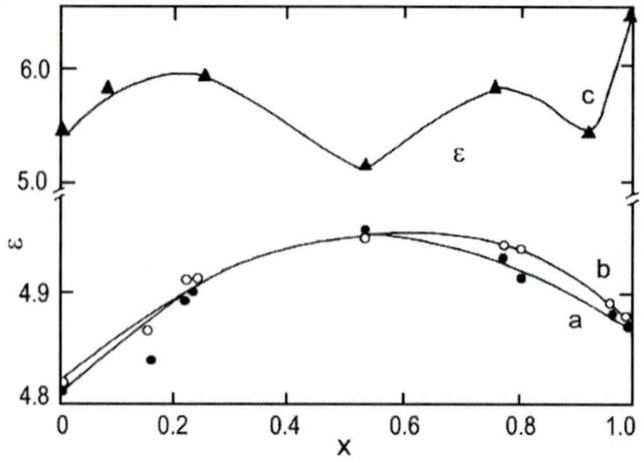

Fig. 7.19. Variation of ε with composition x of $KC_{1-x}Br_x$ at 100 KHz: (**a**) [7.118], (**b**) [7.58], (**c**) [7.57]

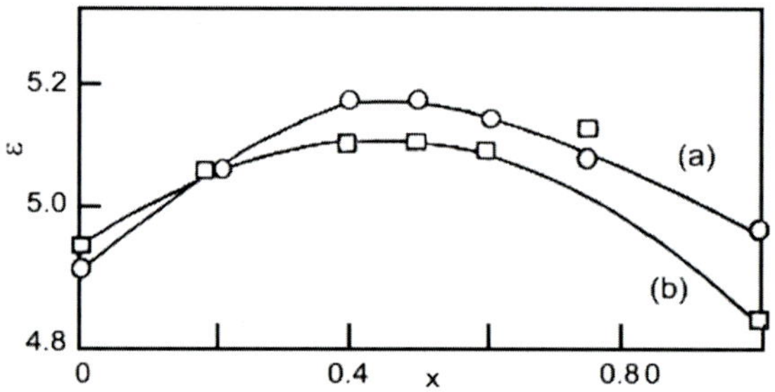

Fig. 7.20. Calculated and experimental values of dielectric constant (ε) with composition (x) for the $RbCl_{1-x}Br_x$ system: (**a**) [7.119], (**b**) [7.58]

Chap. 5. Further, following the treatment proposed by Samara [7.120], the anharmonic contribution to the dielectric constant G is related to the C term as

$$G = T(\varepsilon_s - 1)(\varepsilon_s + 2)C. \tag{7.53}$$

The input data and the calculated values of C and G are shown in Table 7.17. It is observed that the anharmonic contribution varies with the composition and it is maximum for the equimolar composition.

Samara [7.120] observed that for alkali halides at room temperature the anharmonic effects account for less than 10% of the lattice contribution to dielectric constant. From Table 7.17 it is seen that this is true for the mixed crystals also and it is maximum for the $KCl_{0.47}Br_{0.53}$ crystal. The non-linear variation with composition could be due to the enhanced anharmonic effects in the mixed crystals. Fertel and Perry [7.57] obtained the variation of damping constant with temperature and composition both experimentally and by theoretical calculation with virtual crystal model. The damping constant was found to show a non-linear variation with a maximum at the equimolar composition. The theoretical treatment [7.8] and the experimental work on the

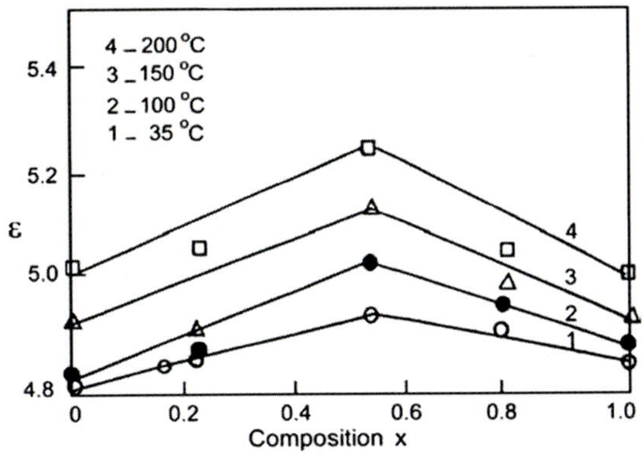

Fig. 7.21. Variation of ε of $KCl_{1-x}Br_x$ with composition (x) at 100 KHz at different temperatures

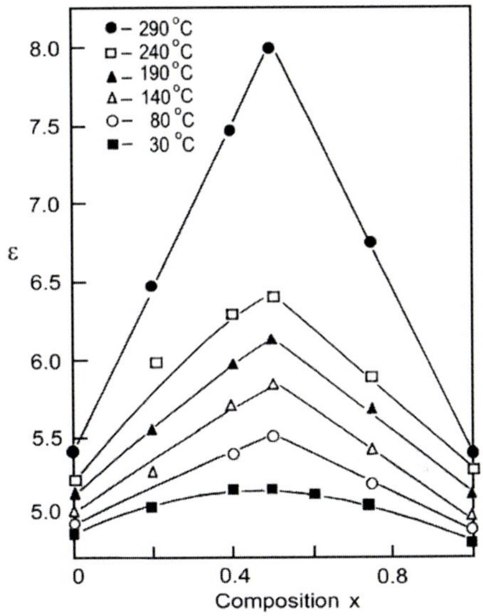

Fig. 7.22. Variation of ε of $RbCl_{1-x}Br_x$ with composition (x) at 100 KHz at different temperatures

infrared spectra of crystals [7.121] show that the resonance half-width or damping is a measure of anharmonic contributions to lattice vibrations. Subramaniam [7.122] evaluated the Gruneisen parameter (γ) for the KCl–KBr system and found that the average γ for the mixed crystal system is higher than the average γ for the pure alkali halides. The Gruneisen parameter being a measure of anharmonicity, it was concluded that atomic vibrations in mixed crystals are more anharmonic in nature than in the pure crystals. Plendl et al. [7.46] studied the vibrational spectra of the NiO–CoO and CaF_2–SrF_2

Table 7.17. Input data and values of A, B, C and G of $KCl_{1-x}Br_{-x}$ crystals; ε_s the static dielectric constant; β the volume thermal expansion coefficient and (r/ρ) repulsion index in the Born potential

x	$(A{+}B{+}C)^*$ [10^{-5}/°K]	$\beta/3$ [10^{-5}/°K]	(r/ρ)	$(A{+}B)$ [10^{-5}/°K]	C [10^{-5}/°K]	G
0.00	3.54	3.80	10.022	3.39	0.15	0.012
0.22	3.43	3.86	10.016	3.23	0.20	0.016
0.53	5.43	4.23	9.981	3.18	2.25	0.185
0.77	4.07	4.06	10.095	2.93	1.14	0.092
1.00	3.78	4.03	10.022	2.91	0.87	0.089

$^*(A+B+C) = \frac{10^5}{(\varepsilon_s-1)(\varepsilon_s+2)}\left(\frac{d\varepsilon_s}{dT} - \frac{\varepsilon_s\beta}{3}\right)$

mixed crystal systems. They obtained the spectral parameter and found that it shows a non-linear variation with composition. The spectral parameters for the intermediate compositions are higher than those for the pure crystals.

The anharmonic contribution G could not be evaluated for the RbCl–RbBr system for lack of required input data. But on the basis of above evidence and the results for KCl–KBr system, it is believed that the non-linear composition variation in both the systems is the result of anharmonic contributions to the dielectric constant.

7.3.6 Effective Ionic Charge

Employing the room temperature values of static dielectric constant, the effective ionic charge was evaluated for the KCl–KBr and RbCl–RbBr systems as a function of composition from the Szigeti relation (Sect. 7.2.6)

$$\varepsilon_s - \varepsilon_\infty = (\varepsilon_\infty + 2)^2(4\pi^2e^2)(q^*)^2/9\mu\omega_t^2V_i \tag{7.54}$$

with ($z = 1$). For the KCl–KBr system, the input data and the calculated values of q^* are shown in Table 7.18. The variation of q^* with composition evaluated using the frequency data of Fertel and Perry [7.57] is shown in Fig. 7.23 as plot 1. Plot 3 corresponds to values obtained from the frequency data of Ferraro et al. [7.123]. Non-linearity is more pronounced in plot 1 than in plot 3. Szigeti [7.124] gave a complete account of the lattice contribution to dielectric constant including the anharmonic effects (Chap. 5) and obtained the relation

$$\begin{aligned}\varepsilon_s - \varepsilon_\infty &= [(\varepsilon_\infty + 2)^2(4\pi^2e^2)(q^*)^2/9\mu\omega_t^2V_i] + G \\ &= \eta + G .\end{aligned} \tag{7.55}$$

Here, η is the lattice contribution in the absence of anharmonicity and G represents the contributions from higher order terms in the potential energy. Following Samara [7.120], the Szigeti charge q^* has been recalculated from

Table 7.18. Input data and values of q^* and q^*_C for $KCl_{1-x}Br_x$ crystals: volume V_i and ε_∞ from [7.58], ω_t from [7.123] and in parenthesis from [7.57], ε_s from [7.118], q^* and q^*_C calculations with data from [7.123] and in parenthesis from [7.57]

x	ε_s	V_i [Å^3]	μ	ε_∞	$\omega_t[10^{13}\,\mathrm{rad\,s^{-1}}]$	q^*	q^*_C
0.00	4.81	62.33	18.60	2.22	2.67 (2.66)	0.79 (0.79)	0.79 (0.78)
0.22	4.87	64.23	20.06	2.28	2.55 (2.68)	0.78 (0.82)	0.78 (0.82)
0.53	4.95	66.85	21.77	2.34	2.42 (2.58)	0.78 (0.83)	0.76 (0.80)
0.77	4.92	68.75	23.80	2.37	2.28 (2.42)	0.77 (0.81)	0.75 (0.80)
1.00	4.87	71.74	26.26	2.43	2.15 (2.14)	0.75 (0.75)	0.74 (0.73)

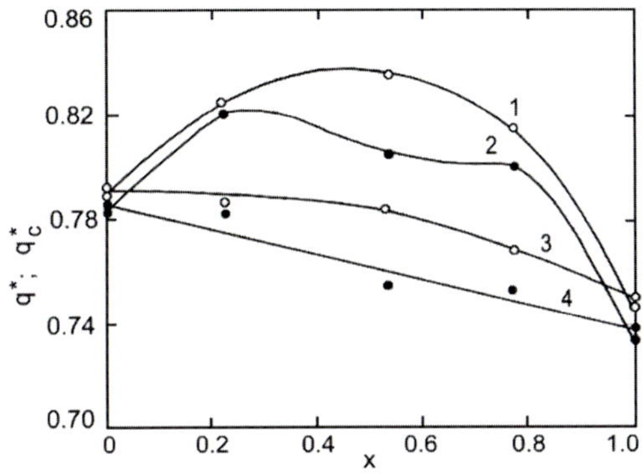

Fig. 7.23. Variation of q^* (*open circle*) and q^*_C (*filled circle*) with composition x for $KCl_{1-x}Br_x$: 1, 2 frequency data from [7.57] and 3, 4 from [7.123]

(7.55) using the values of G given in Table 7.17. The corrected values q^*_C are given in Table 7.18 and also shown in Fig. 7.23 as plots 2 and 4.

According to the criterion suggested by Chang and Mitra [7.8] $KCl_{1-x}Br_x$ is expected to show a normal one-mode behaviour in which each of the long wavelength optical mode frequencies varies continuously and approximately linearly with composition. The frequency data of Fertel and Perry [7.57] shows a non-linear variation while Ferraro et al. [7.123] have observed a linear variation with composition in agreement with the Chang and Mitra criterion. As shown in Fig. 7.23 the Szigeti charge evaluated by using the frequency data of the two reports differs considerably. Plendl et al. [7.46] observed a linear variation of the Szigeti charge with composition for NiO–CoO system. Ferraro et al. have made a generalized statement that the effective ionic charge varies linearly with composition for mixed crystals.

The corrected values q^*_C obtained using the frequency data of Ferraro et al. show a linear composition variation consistent with the above findings. On the other hand, q^* evaluated from the frequency data of Fertel and Perry show a highly non-linear variation even after applying the correction for anharmonicity. Hence, the frequency data of Ferraro et al. are physically more acceptable than those of Fertel and Perry.

Table 7.19. Input data and calculated values of q^* for $RbCl_{1-x}Br_{-x}$ crystals [7.125]

x	ε_s	V_i [$\mathring{A}^3$]	μ	ε_∞	$\omega_t[10^{13}\,\mathrm{rad\,s^{-1}}]$	q^*
0.00	4.91	71.26	25.06	2.19	2.22	0.84
0.20	5.06	64.23	27.20	2.22	2.11	0.86
0.40	5.17	66.85	29.74	2.25	1.99	0.86
0.50	5.18	68.75	31.19	2.27	1.94	0.86
0.60	5.13	71.74	32.80	2.28	1.88	0.85
0.75	5.08	79.12	35.54	2.30	1.80	0.84
1.00	4.88	81.74	41.30	2.33	1.65	0.81

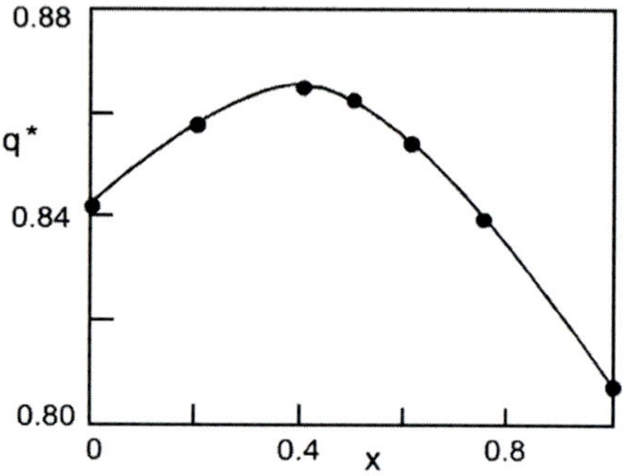

Fig. 7.24. Variation of q^* with composition (x) for the $RbCl_{1-x}Br_x$ system

For $RbCl_{1-x}Br_x$, q^* values have been evaluated assuming a linear variation of frequency with composition. The input data and the calculated values of q^* are given in Table 7.19. The values indicate a non-linear variation with composition as shown in Fig. 7.24.

7.3.7 Colour Centres in RbCl–RbBr Mixed Crystals

In Sect. 7.2.7 aspects of colour centres in mixed crystals were discussed. Srinivas et al. [7.92] studied the colour centres in $RbCl_xBr_{1-x}$ mixed crystals. The lattice constants were determined by X-ray diffraction. By colouring the crystals with γ-rays from a Co^{60} source, the F centres were recorded. The half-width of the F centre bands was also determined. The results are given in Table 7.20.

The composition dependence of the F centre wavelength (λ_F) is shown in Fig. 7.25. It is slightly non-linear. It has been mentioned (Sect. 7.2.7) that λ_F bears a relationship of the form

$$\lambda_F = Cr^s \tag{7.30}$$

with the interionic distance r; for NaCl structure $r = a/2$. A log–log plot of λ_F and r is shown in the same figure from which $s = 2.5$. Thus the

Table 7.20. Values of the F band peak position (λ_F), the F band half-width (W) and B_{stat} for the $RbCl_xBr_{1-x}$ system

x	λ_F [Å]	W [eV]	B_{stat} [Å^2]
0.0	7,060	0.344	0
0.25	6,900	0.369	0.0010
0.39	6,800	0.374	0.0014
0.50	6,720	0.377	0.0015
0.61	6,700	0.377	0.0013
0.80	6,540	0.372	0.0009
1.0	6,340	0.361	0

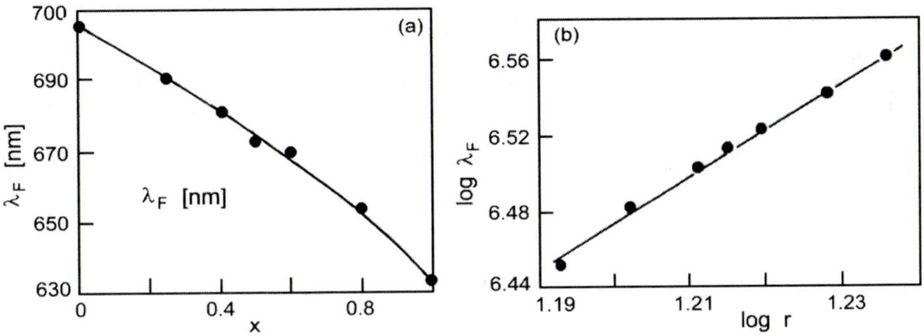

Fig. 7.25. Plot of (**a**) λ_F vs. x and (**b**) log λ_F vs. log r for $RbCl_xBr_{1-x}$

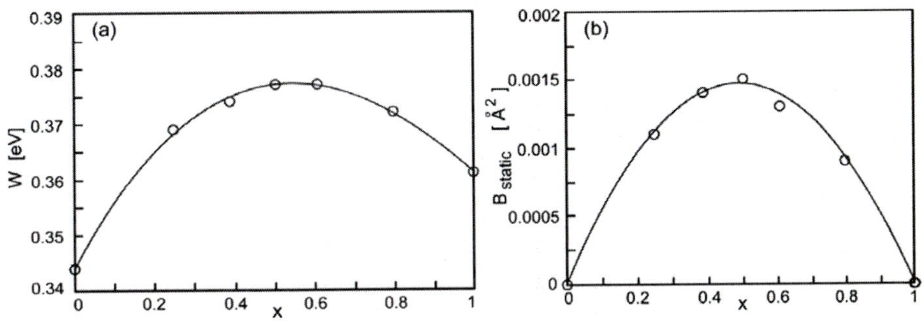

Fig. 7.26. Plot of (**a**) half-width W vs. x and (**b**) B_{stat} vs. x for $RbCl_xBr_{1-x}$

Mollwo–Ivey law holds in this system. Subramaniam and Bansigir [7.126] suggested that the 'size effect' affects the half-width of F centre bands. The size effect causes a static contribution B_{static} to the Debye–Waller factor. The values of B_{static} are given in Table 7.20 for the $RbCl_xBr_{1-x}$ system. Plots of half-width (W) and B_{static} against composition are shown in Fig. 7.26. They are similar, supporting the suggestion of Subramaniam and Bansigir [7.126].

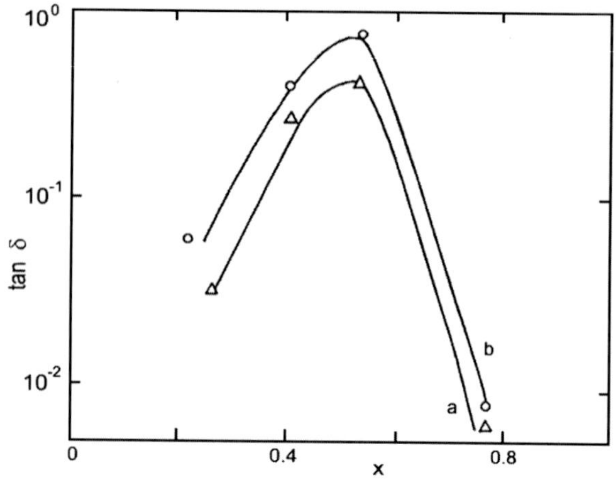

Fig. 7.27. Variation of tan δ at 100 kHz with composition (x) of $KCl_{1-x}Br_x$ for temperatures (**a**) 290, (**b**) 310°C

7.3.8 Defects in Mixed Crystals

Point Defects

The point defects in KCl_xBr_{1-x} and $RbCl_xBr_{1-x}$ mixed crystals were studied by Sirdeshmukh et al. [7.118] and Sathaiah and Sirdeshmukh [7.119], respectively, by making measurements of ionic conductivity and evaluation of activation energy. The A.C. conductivity σ was evaluated from the data on dielectric constant and loss using the relation $\sigma = \varepsilon\varepsilon_0\omega \tan\delta$.

The $\tan\delta$ values for $KCl_{1-x}Br_x$ and $RbCl_{1-x}Br_x$ are shown in Figs. 7.27 and 7.28, respectively. A non-linear variation with enhanced loss at equimolar composition is observed for both the systems. Conductivity too shows a non-linear composition dependence. The composition dependence of the activation energy for conduction (E) obtained from Arrhenius plots is shown in Figs. 7.29 and 7.30. It is to be noted that activation energy for motion of defects is less for intermediate compositions for both systems.

Hari Babu and Subba Rao [7.3] suggest that for the KCl–KBr system the non-linear variation in conductivity and the activation energy for conduction is the result of defects such as enhanced dislocations and grain boundaries. This could account for the variations observed in RbCl–RbBr system also.

Dislocations

Hussain et al. [7.114] studied the dislocations in the $NaClO_3-NaBrO_3$ system by the etching technique. The distribution of etch pits is shown in Fig. 7.31 and the dislocation density is shown in Fig. 7.32 as a function of composition. As in the alkali halide mixed crystals (Sect. 7.2.8) the density of dislocations is more in the mixed crystals than in the pure crystals.

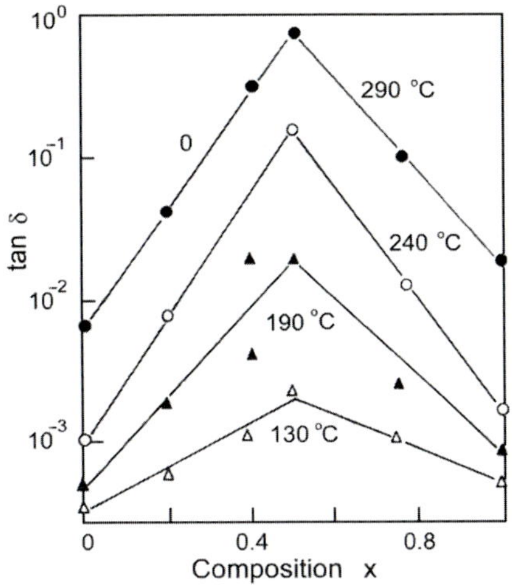

Fig. 7.28. Variation of tan δ at 10 KHz with composition (x) of $RbCl_xBr_{1-x}$ for different temperatures

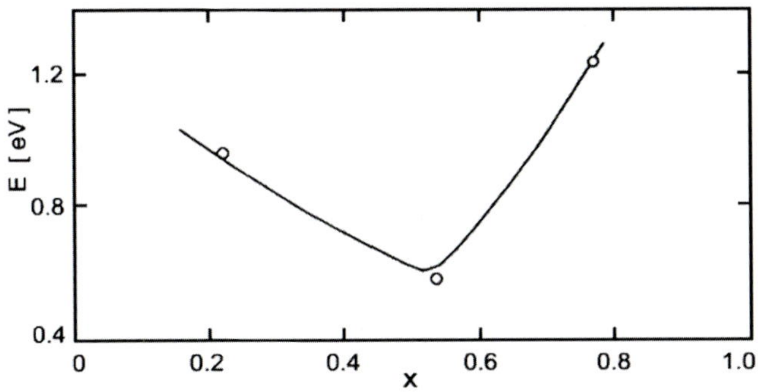

Fig. 7.29. Variation of activation energy E with composition x of $KCl_{1-x}Br_x$

7.3.9 Melting Temperatures of Mixed Crystals

It was pointed out in Sect. 7.2.9 that while there is considerable data on melting of alloys, there is very limited data on melting of compound mixed crystals. A systematic study of melting of compound mixed crystals was carried out in our laboratory. For this purpose, a hot stage microscope was set up by replacing the commonly used objective by a long distance objective and replacing the stage by a laboratory-built hot stage. The thermocouple tip was in contact with the sample. Thus the sample and thermocouple tip could always be seen together and when melting took place, the thermocouple recorded the actual sample temperature. By repeated measurements and from results on standard materials, the error in measured melting temperatures was estimated as ± 5 K.

The following systems were studied: $NaClO_3-NaBrO_3$ [7.127]; KCl–KBr, RbCl–RbBr, KBr–RbBr [7.128]; KCl–RbCl [7.129] and RbBr–RbI, KI–RbI

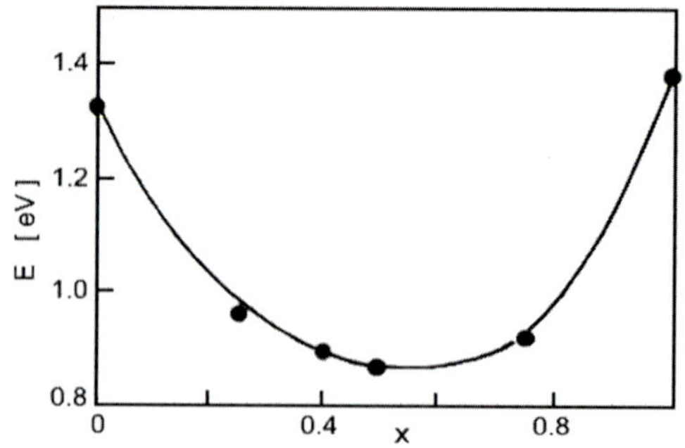

Fig. 7.30. Variation of activation energy E with composition x at 300°C for the $RbCl_{1-x}Br_x$ system

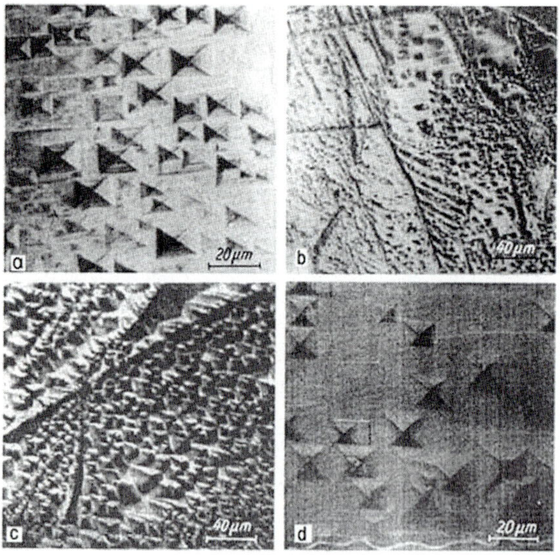

Fig. 7.31. Etch patterns on the (100) faces of $NaClO_3 - NaBrO_3$ mixed crystals (**a**) $NaBrO_3$, (**b**) $Na(ClO_3)_{0.93}$, (**c**) $Na(ClO_3)_{0.7}(BrO_3)_{0.3}$ and (**d**) $NaClO_3$

and NaCl–NaBr [7.130]. The composition dependence of melting point for some of these systems is shown in Fig. 7.33. The general trend is that the melting point is a slightly non-linear function of composition with negative deviations from additivity.

Sirdeshmukh et al. [7.130] compared these results with predictions from Lindemann's theory. Lindemann's melting point formula can be written in the form

$$T_m = CMa^2\theta^2, \tag{7.56}$$

where T_m is the melting temperature, M the mass, a the lattice constant and θ the Debye temperature. For a given mixed crystal system, C can be

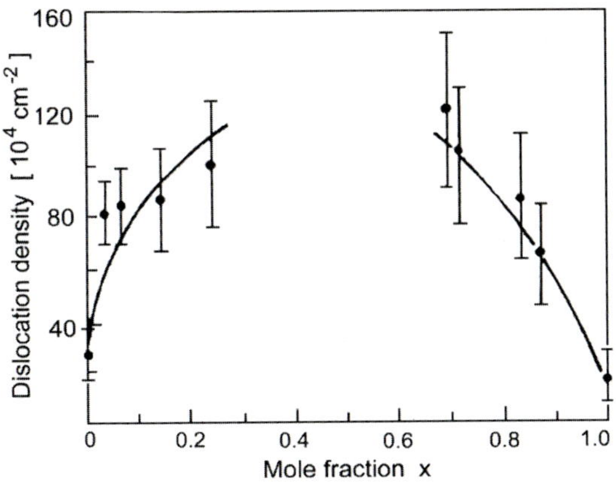

Fig. 7.32. Plot showing dislocation density as a function of composition x for $Na(ClO_3)_x(BrO_3)_{1-x}$ mixed crystals

calculated from the values of these parameters for pure end members. For mixed crystals, we have

$$M = xM_1 + (1 - x)M_2, \tag{7.57}$$

$$a = xa_1 + (1 - x)a_2, \tag{7.58}$$

$$\theta^{-3} = x\theta_1^{-3} + (1 - x)\theta_2^{-3}. \tag{7.59}$$

Thus with the help of (7.56)–(7.59), the melting points of mixed crystals can be calculated using only the parameters for the end members. These calculated values are also shown in Fig. 7.33.

For a quantitative assessment of these calculations, the e.s.d. in $[T_m(\exp)-T_m(\text{calc})]$ was calculated for each system. These e.s.d. values are given in Table 7.21. Since the experimental error was estimated at ± 5 K, the agreement between experimental results with values from (7.56) was considered good, fair or poor for e.s.d.s in the range 0–10, 10–20 K and more than 20 K. Thus, the agreement was found good in case of KI–RbI, KBr–KI and KCl–RbCl systems, fair in case of KBr–RbBr, KCl–KBr, NaCl–NaBr, CsCl–CsBr and RbCl–RbBr systems and poor in case of RbBr–RbI.

7.3.10 Pm3m → Fm3m Transition in NH_4Cl–NH_4Br System

It was pointed out in Sect. 7.2.10 that there is a marked difference in the composition dependence of the Pm3m–Fm3m transition temperatures of the NH_4Cl–NH_4Br system reported by Costich et al. [7.90] and Murti and Prasad [7.91]. A redetermination of the transition temperatures was carried out by Balaiah [7.104].

A Perkin-Elmer DSC 7 has been used to determine the transition temperatures. The operation of the DSC is based on the power compensation principle.

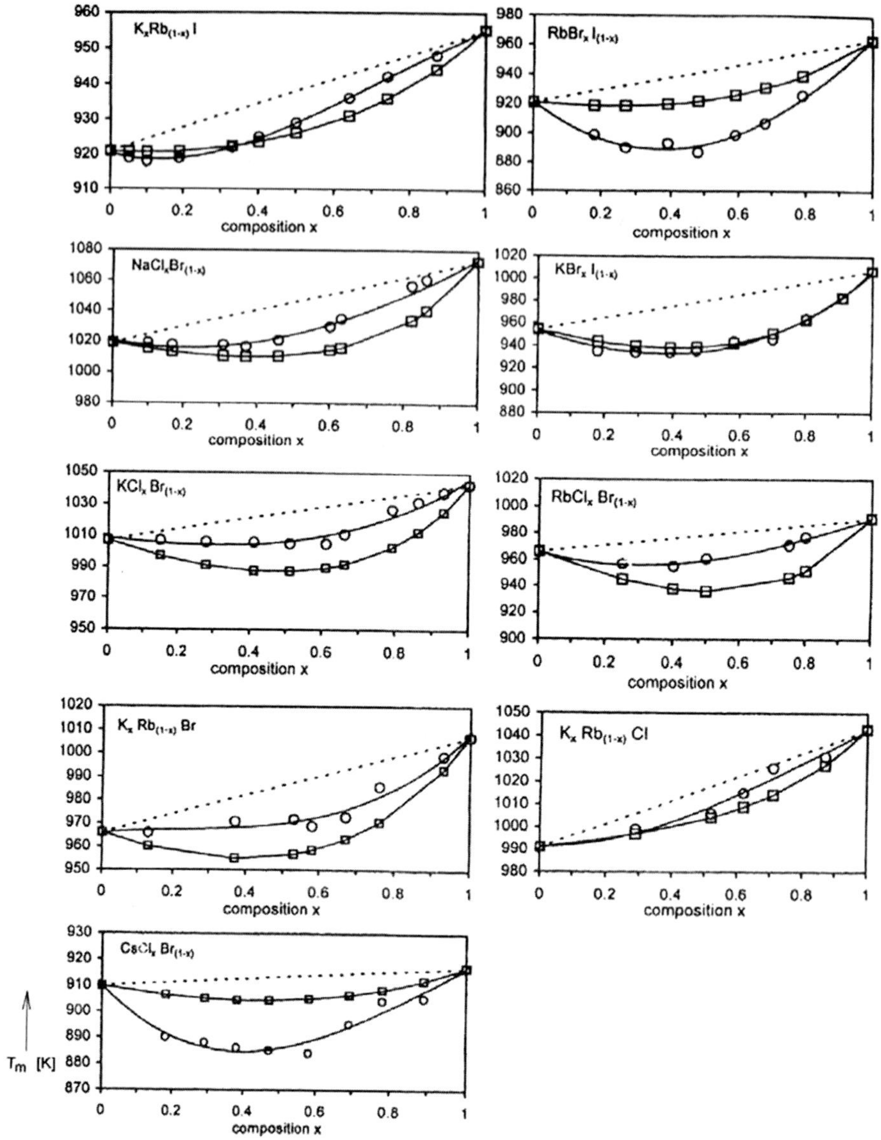

Fig. 7.33. Plots of T_m vs. composition x for alkali halide mixed crystal systems: experimental data for KCl_xBr_{1-x}, $RbCl_xBr_{1-x}$ and $K_xRb_{1-x}Br$ from [7.128], for KBr_xI_{1-x} and $K_xRb_{1-x}Cl$ from [7.129] and for others from [7.130]; symbols: *circle*, Expt.; *square*, calculated from (7.56)

Table 7.21. E.s.d.s in $[T_\mathrm{m}(\mathrm{exp}) - T_\mathrm{m}(\mathrm{calc})]$

System	E.s.d. [K]	System	E.s.d. [K]	System	E.s.d. [K]
KI–RbI	3	KBr–RbBr	10	CsCl–CsBr	14
KBr–KI	4	KCl–KBr	11	RbCl–RbBr	18
KCl–RbCl	5	NaCl–NaBr	13	RbBr–RbI	23

Table 7.22. Phase transition temperatures of $NH_4Cl_xBr_{1-x}$ mixed crystals

[7.90]		[7.91]		[7.104]	
x	T_tr[°C]	x	T_tr[°C]	x	T_tr[°C]
0	138	0	149	0	158
0.08	130	0.065	156	0.09	146
0.23	104	0.205	158	0.13	146
0.38	101	0.4	168	0.29	125
0.45	101	0.5	171	0.33	126
0.61	118	0.785	186	0.38	124
0.79	136	1	196	0.78	126
0.89	154			0.85	162
1	186			0.93	189
				0.96	193
				1	193

With power compensation DSC, the sample and reference material are placed in separate self-contained calorimeters. When the temperature rises or falls in the sample material, power is applied to or removed from the calorimeter to compensate for the sample energy. As a result, the system is maintained at a 'thermal null' at all times. The amount of power required to maintain the system equilibrium is directly proportional to the energy changes occurring in the sample. Hence the DSC based on the power compensation principle directly measures energy flow to and from the sample. The temperature sensor is a platinum resistance thermometer with an accuracy of ±0.1°C. The samples are encapsulated in aluminium holders. The phase transition temperatures are measured using the thermograms recorded at a scan rate of $10°\mathrm{C\,min}^{-1}$.

The results of the DSC measurements are given in Table 7.22. The transition temperatures reported by Costich et al. [7.90] and Murti and Prasad [7.91] are also included in the same table. The composition dependence of the transition temperatures obtained by Balaiah [7.104] is shown in Fig. 7.34. The composition dependence is clearly non-linear with negative deviations from linearity. The transition temperatures obtained by Costich et al. [7.90] and Murti and Prasad [7.91] are also plotted in Fig. 7.34 for the various compositions. The results of Balaiah [7.104] are qualitatively in agreement with the results of Costich et al. [7.90] which also show a non-linear composition dependence of the transition temperature with negative deviations from linearity. However, the results of Balaiah [7.104] deviate completely from the results of Murti and Prasad which indicate a linear composition

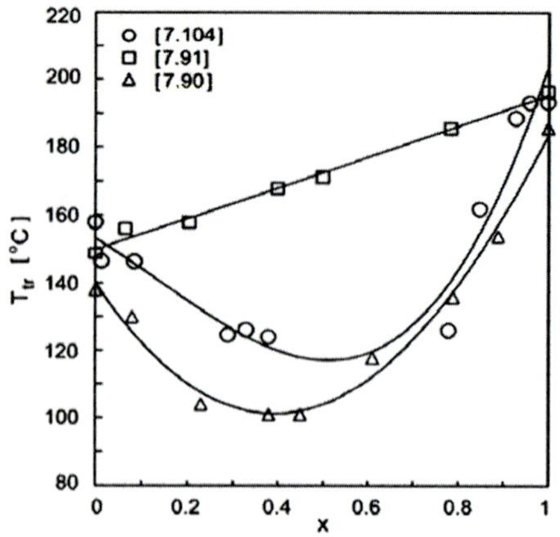

Fig. 7.34. *Pm3m–Fm3m* transition temperature (T_{tr}) vs. composition x for $NH_4Cl_xBr_{1-x}$ mixed crystals

dependence. The non-linear composition dependence of transition temperature in the NH_4Cl–NH_4Br system is also consistent with the non-linear composition dependence of transition temperature observed in the CsCl–CsBr system. Thus, the non-linear composition dependence of the Pm3m $\rightarrow$ Fm3m transition temperature of the NH_4Cl–NH_4Br mixed crystals is now clearly established.

8

Elastic Properties of Solids – A Critical Analysis

8.1 Introduction

The elastic properties of solids have a twofold importance. First, they are indicators of mechanical strength which is a matter of great practical significance. Second, on the scientific side, the elastic properties are inputs for determination of interatomic potential parameters and lattice dynamical calculations. When measured at high pressures, they provide information regarding the anharmonicity of the lattice. In view of this importance, a vast amount of information is now available on the elastic properties of solids. This information is scattered in various scientific journals. Exclusive compilations of data on elastic properties of solids have also been published [8.1–8.7].

Several experimental techniques are available for the determination of elastic properties. Some commonly employed techniques will be discussed in Sect. 8.2 along with a mention of their limitations and uncertainties in the results obtained from them.

8.2 Experimental Methods

Some of the commonly used experimental methods are briefly discussed in this section. For more details, reference may be made to the cited literature.

8.2.1 Piston Displacement Method

In this method, the sample is directly compressed by a piston or is enclosed in a medium which is compressed by a piston. The method originally introduced by Bridgman [8.8, 8.9] has undergone several modifications [8.10–8.12].

As a typical example, the arrangement used by Vaidya and Kennedy [8.12] is shown in Figs. 8.1 and 8.2. Only the important parts are described here. X is the sample held between the piston P and a fixed tungsten carbide element

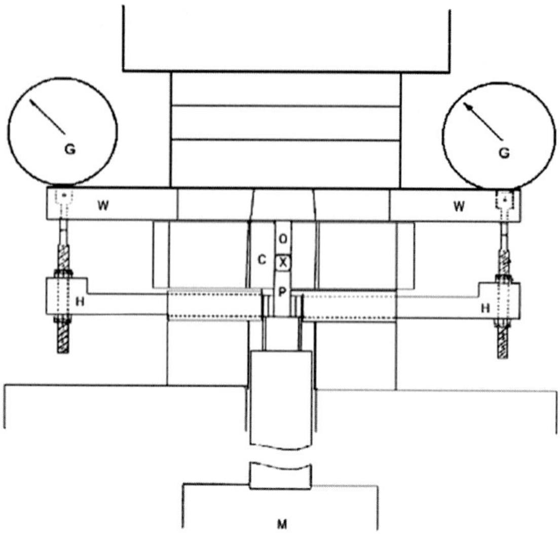

Fig. 8.1. Schematic diagram of the piston displacement set-up

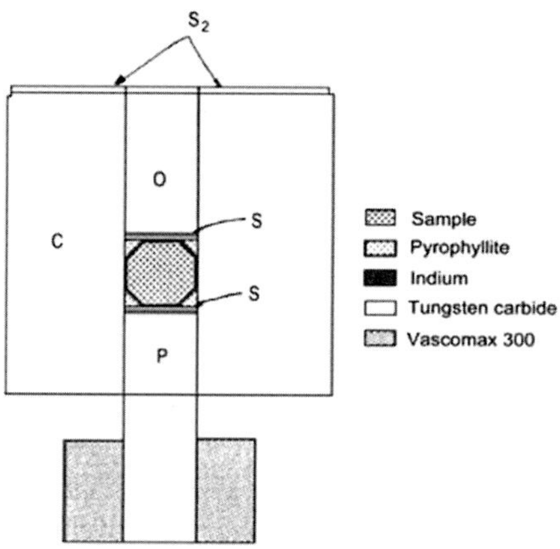

Fig. 8.2. Details of the sample assembly in the piston displacement set-up

O. The piston is driven by the hydraulic ram M. The pressure vessel C is made of tungsten carbide. The sample is covered by an indium sheath and the space between the sample and pressure vessel is filled with pyrophillite. The compression seen as a relative displacement between the lever arms W and H is measured by the dial gauges G.

The observed total compression has significant contributions from the compression of the press and the compression of the indium sheath. The ratio of these contributions and the sample compression have to be optimized. In Bridgman's experiments, the corrections were 75% of the total compression whereas the corrections were reduced by half in the set-up of Vaidya and Kennedy [8.12]. Care has to be taken to minimize the effects of (a) distortion of the press (b) horizontal fracture of the pressure vessel and (c) friction between the piston and the pressure vessel walls. This is a static method which gives isothermal values of the bulk moduli.

8.2.2 Shock Wave Method

In this method an explosive is detonated in an enclosed space. This results in the production of a shock wave. Either this shock wave is made to directly impinge on a sample or a projectile propelled by the shock wave is directed on to the sample. The shock wave velocity (u_S) and the particle velocity (u_p) are measured by recording the time of arrival of the waves at different designated points where electrical pin contactors are located. Whereas the shock wave velocity can be determined in a straight-forward way, the particle velocity has to be obtained from the free surface velocity (u_fs). A typical arrangement [8.13] is shown in Fig. 8.3, which is self-explanatory. The working equations are

$$\rho_0 u_\mathrm{S} = \rho_1 (u_\mathrm{S} - u_\mathrm{p}), \tag{8.1}$$

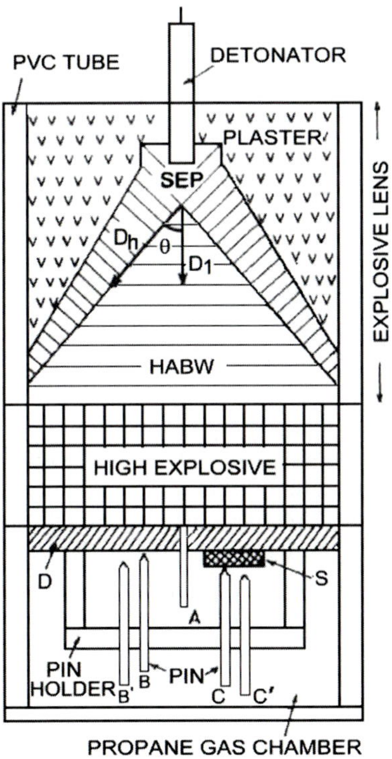

Fig. 8.3. Schematic diagram of a shock-wave set-up

$$P_1 - P_0 = \rho_0 \, u_S \, u_p, \tag{8.2}$$

where ρ is the density, P the pressure, u_S the shock velocity and u_p the particle velocity; $u_p = u_{fs}/2$ approximately. The subscripts 0 and 1 refer to the compression state ahead of the shock and the state immediately behind the shock front, respectively. The procedure for transforming the experimental velocities to pressure–compression points is discussed in detail by Rice et al. [8.14]. The P–V plot thus obtained is called the "Hugoniot"; from the Hugoniot, the bulk modulus can be calculated. Further details are given in [8.13–8.16].

This method is capable of, by far, the greatest pressure range, up to 2 Mbar [8.17]. It is a dynamic method. It is found that, in general, the compatibility of static (piston displacement) and shock wave results is good [8.12–8.14]. The method yields essentially isothermal values.

8.2.3 X-ray Diffraction Method

When the X-ray diffraction pattern is recorded at different pressures, the Bragg angles change. From this, the changes in lattice parameters can be calculated. This information, in turn, leads to the evaluation of the linear compressibilities and finally to the bulk modulus. The method yields isothermal values of elastic properties.

The pressure is applied through a pressure medium or, more commonly, by means of a diamond or tungsten carbide anvil. The sample may be a single crystal or in powder form. Both film and camera arrangements as well as diffractometer can be used replacing the common specimen holder by a pressure cell. While X-ray radiation is commonly employed, Bartholin et al. [8.18] used neutrons and Will et al. [8.19] used synchrotron radiation.

Considerable work on determination of compressional data by the X-ray diffraction method has been done [8.20–8.24]. One advantage of this method is that a small sample is required. Another unique advantage is that in a single experiment, linear compression can be studied in different directions. A common problem in high pressure work is that the pressure experienced by the sample is not necessarily the same as the applied pressure. In the X-ray diffraction method, this problem is overcome by using a 'marker' or an internal standard. On the other hand, the application of the method is limited by the use of Mo radiation which causes fluorescence in some materials. Also the short wavelength of the Mo radiation limits the accuracy with which the lattice parameters can be evaluated. Leger et al. [8.25] refer to the effect of uniaxial stresses on the results. Vaidya and Kennedy [8.12] consider the X-ray method to be of "extremely poor precision".

8.2.4 Optical Interferometric Method

Optical interferometry can be employed to determine the linear compressibilities and, from them, the bulk modulus. The optical interferometer, the

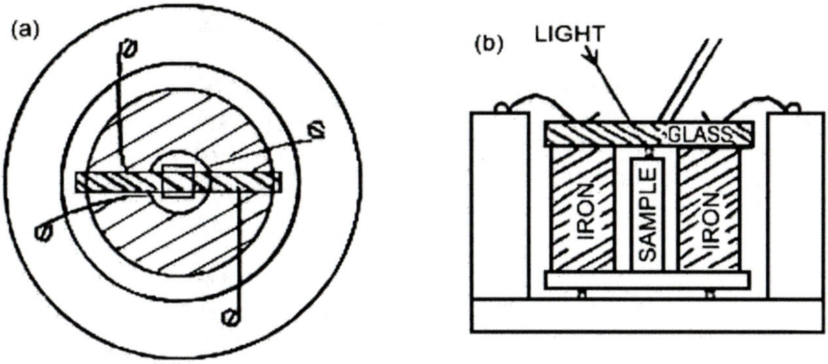

Fig. 8.4. Optical interferometer (**a**) view from above, (**b**) view from side

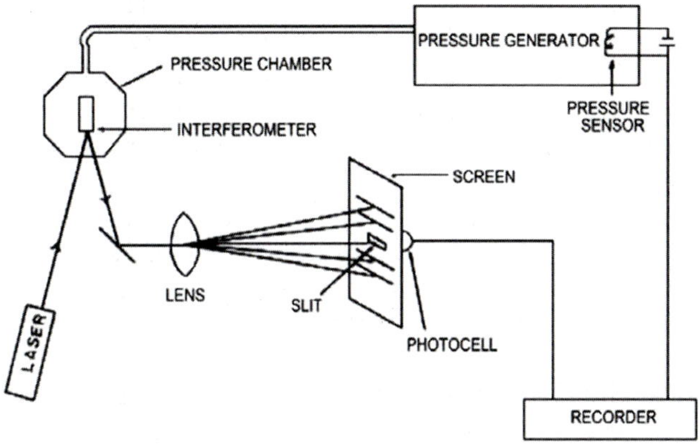

Fig. 8.5. Schematic diagram of the interferometric method

sample holder and the schematic diagram of the set-up used by Montalvo and Langer [8.26] are shown in Figs. 8.4 and 8.5. The interferometer consists of a glass plate resting over the surface of an iron cylinder. The sample with a flat face was held in the annular space of the iron cylinder. The incident light beam from an He–Ne laser was reflected from the lower surface of the glass plate and the top surface of the sample. The whole interferometer was placed in a pressure vessel containing isopentane as the pressure fluid. The interference fringes were sensed by a photocell coupled to a recorder. The experimental procedure was to count the fringes as the pressure changed from 1 to 10 kbar.

The optical interferometric measurements yield isothermal values of compressibilities and bulk modulus. Montalvo and Langer used this method to determine the linear compressibilities of several II–VI compound crystals and claimed an accuracy of 1%. However several of their values differ from accurate ultrasonic values by amounts larger than the isothermal-adiabatic conversion correction. Factors like the errors in the value of the compressibility of iron used in the calculations, the change in refractive index of the pressure fluid with pressure change and relative motion between glass and sample surfaces contribute to the final uncertainty in the results.

8.2.5 Ultrasonic Method

The measurement of velocity of ultrasonic waves is a popular method for determination of elastic constants of crystals. The sound velocity in a crystal is a function of the elastic constants. For example, for cubic crystals, we have

$$\rho\, \upsilon^2 = (C_{11} - C_{12})/2, \tag{8.3}$$

where ρ is the density, C_{11} and C_{12} are two of the three elastic constants and υ is the velocity of a sound wave propagating in the [110] direction with particle displacement in the $[1\bar{1}0]$ direction. There are similar equations for other combinations of elastic constants and also for the elastic constants of crystals of other symmetries. Relations between velocities and elastic constants for hexagonal and tetragonal crystals are given by Cline et al. [8.27] and Farley and Saunders [8.28], respectively.

Much of the earlier work on the determination of elastic constants was based on the composite oscillator method [8.29–8.31]. In more recent work, different versions of the pulse method [8.32–8.34] have been employed.

In a modification of the pulse comparison technique [8.35], two phase-coherent RF pulses are applied in quick succession to a transducer attached to the crystal surface. On transmitting through the sample, each pulse will generate its own train of echoes. When the separation of the second pulse is adjusted such that the echoes from the two coincide, destructive interference occurs successively at a number of frequencies. From a knowledge of these null frequencies and the thickness of the sample, the ultrasonic velocity can be estimated.

Other technical details of the method are given in [8.27, 8.36–8.38]. While the electronics is capable of measurement with high accuracy, various other factors like the characteristics of the bonding material, the accuracy of measurement of thickness and the accuracy of orientation of the sample introduce errors.

The ultrasonic method yields adiabatic values of elastic constants. It is by far the only method that gives elastic constants of single crystals. From the single crystal elastic constants, the isotropic bulk and shear moduli can be computed. Alternatively, the ultrasonic method can be used to find the sound velocities in polycrystalline samples from which the elastic moduli can be determined. In this method, very often a density correction has to be applied as the polycrystalline samples may not be packed to crystal density. The ultrasonic method is by far the most accurate. To cite an example, Bateman [8.39] has reported the elastic constants of ZnO with an accuracy of 0.1%.

8.2.6 Other Methods

1. *Ultrasonic-optic method.* In this method, a transparent crystal is excited into resonant vibration by a quartz plate. The result is that the crystal acts like a 3-d diffraction grating with respect to optical light. From the diffraction pattern, orientation of the crystal, crystal density and the resonant frequency, the elastic constants are evaluated. This is known as the Schaefer–Bergmann [8.40] method.
2. *Optical scattering method.* A method based on frequency shifts of the Brilluoin components appearing in the thermal scattering of light has been developed by Krishnan [8.41] to determine the elastic constants.
3. *X-ray diffuse scattering method.* Ramachandran and Wooster [8.42, 8.43] developed a method for determining the elastic constants of crystals from observations on the thermal diffuse scattering of X-rays. They used the method to determine the elastic constants of diamond, sodium chlorate and iron pyrites among other crystals.

8.2.7 Relative Merits and Limitations

The piston displacement method, the shock wave method and the X-ray diffraction methods are commonly employed for determination of bulk moduli whereas the ultrasonic method is used mostly to determine the single crystal elastic constants.

Bridgman was the pioneer in the development of the piston displacement method. He was continuously refining his equipment and procedure from the 1920s to 1940s. At a certain stage, Bridgman himself revised his earlier data by 10–20% [8.44]. Vaidya and Kennedy [8.12] call Bridgman's results 'remarkably discordant'. As mentioned in Sect. 8.2.1, several factors related to the high-pressure assembly need care and correction. Vaidya and Kennedy [8.12] claim that their own set-up and method are 3–10 times more precise than other methods.

Rice et al. [8.14] estimate an error of 0.7–3% in compression data obtained from shock-wave method. The shock-wave data has been found to be consistent with piston displacement data [8.12] and also with X-ray diffraction data [8.17]. However, Drickamer et al. [8.17] refer to the requirement of a large number of corrections to the shock-wave results.

The ultrasonic techniques are highly accurate but they have in some cases led to dubious results. There are also intrinsic analytical difficulties in retrieving elastic constant data from velocity data, particularly with regard to lower symmetry crystals.

8.3 Discrepancies in Elastic Properties

The methods discussed in Sect. 8.2 are capable of high accuracy. Some of the methods yield adiabatic values whereas others yield isothermal values. The difference between adiabatic and isothermal values ranges from 0.5 to 2%. These two values are mutually convertible by a thermodynamic correction. The uncertainties claimed in elastic properties range from 1 to 5%. With such high accuracy the values for a material by different methods or by different workers, generally agree within limits of errors.

However, there do exist cases where discrepancies much larger than the claimed error limits occur in values for the same material reported by different sources. In Table 8.1, values of bulk moduli at zero (or atmospheric) pressure are cited. It can be seen that in several cases the differences in bulk moduli are of the order of 20% or more and in some cases they are as much as a factor of 3–5. In Tables 8.2–8.4, data on single crystal elastic constants of some cubic, tetragonal, trigonal and hexagonal crystals are quoted. Here again the single crystal elastic constants differ by about 20% and in some cases by a factor of 3–15. In fact, there are cases where the elastic constants from different sources even differ in sign.

It may be mentioned that the list of crystals showing differences in reported elastic properties given in Tables 8.1–8.4 is not an exhaustive list. Further such conflicting data have been included in some compilations without any comments. Thus Simmons and Wang [8.7] call their compilation 'uncritical' and mention the inclusion of 'suspect elastic data' leaving it to the reader to 'choose the set which he believes better'.

These differences far exceed the admitted limits of uncertainty. When data differing so severely exist in literature, it becomes necessary to analyse them with a view to sifting the correct (or acceptable) data from the incorrect (or unacceptable). Several consistency checks for analysis of elastic properties of solids are reviewed in the following sections.

8.4 Consistency Checks for Bulk Moduli

Some consistency checks for bulk moduli are discussed in this section. These are categorized as (a) phenomenological relations, (b) theoretical consistency checks and (c) empirical relations as consistency checks.

8.4.1 Phenomenological Relations as Consistency Checks

Relations Between Elastic Moduli

The bulk modulus (B), Young's modulus (Y), shear modulus (G) and the Poisson's ratio (σ_P) are interrelated as follows:

$$Y = 2G\,(1 + \sigma_\mathrm{P}), \tag{8.4}$$

$$B = Y/[3\,(1 - 2\sigma_\mathrm{P})], \tag{8.5}$$

$$\sigma_\mathrm{P} = (1/2) - (Y/6B). \tag{8.6}$$

Table 8.1. Bulk moduli (B) of some solids

	Solid	Method	B [kbar]	Ref.
1.	C (Diamond)	(i) Ultrasonics	5800	[8.45]
		(ii) Doppler	4170	[8.46]
		(iii) scattering		
		(iv) Ultrasonics	4430	[8.38]
2.	Si	(i) Piston displacement	3120	[8.47]
		(ii) Ultrasonics	971	[8.48]
		(iii) Piston displacement	814	[8.49]
		(iv) Piston displacement	1007	[8.50]
3.	Iodine	(i) Piston displacement	87.03	[8.50]
		(ii) Shockwave	98.04	[8.51]
		(iii) Piston displacement	76.92	[8.52]
		(iv) X-ray diffraction	208.33	[8.53]
4.	Se (hex.)	(i) Piston displacement	92.7	[8.54] Bridgman's data
		(ii) Piston displacement	79	[8.50]
		(iii) Ultrasonics	174	[8.55, 8.56]
5.	Sb	(i) Piston displacement	369	[8.49]
		(ii) Piston displacement	404	[8.50]
		(iii) Shockwave	255	[8.51]
		(iv) Ultrasonics	380	[8.57]
6.	Zr	(i) Piston displacement	755	[8.49]
		(ii) Piston displacement	1028	[8.50]
		(iii) Shockwave	976	[8.51]
		(iv) Ultrasonics	945	[8.58]
7.	Y	(i) Piston displacement	337	[8.49]
		(ii) Piston displacement	449	[8.50]
		(iii) Shockwave	455	[8.51]
		(iv) Ultrasonics	491	[8.59]
8.	LiH	(i) Piston displacement	270	[8.60]
		(ii) Piston displacement	228	[8.61]
		(iii) Piston displacement	357	[8.62]

Table 8.1. Continued

	Solid	Method	B [kbar]	Ref.
9.	LiI	(i) Piston displacement	168.3	[8.63] Murnaghan equation.
		(ii) Piston displacement	135.4	[8.63] Modified Murnaghan equation.
		(iii) Shockwave	333	[8.51]
		(iv) Ultrasonics	171	[8.64]
10.	NaI	(i) Piston displacement	156	[8.49]
		(ii) Piston displacement	151	[8.63]
		(iii) Shockwave	200	[8.51]
		(iv) Ultrasonics	151	[8.64]
11.	CsBr	(i) Piston displacement	144	[8.63]
		(ii)Shockwave	221.8	[8.51]
		(iii) Ultrasonics	145.0	[8.65]
12.	AgI	(i) Piston displacement	20.1	[8.49]
		(ii) Piston displacement	28.5	[8.63]
13.	CaO	(i) Piston displacement	218	[8.66]
		(ii) Piston displacement	1120	[8.10]
		(iii) X-ray	1120	[8.67]
		(iv) Ultrasonics (S.C)	1090	[8.68]
14.	CaS	(i) Piston displacement	357	[8.69] Bridgman's data
		(ii) X-ray	455	[8.67]
		(iii) X-ray	699	Reanalysis of data in [8.67]
15.	SrO	(i) Piston displacement	1183	[8.10]
		(ii) Ultrasonics	880	[8.70]
16.	CeS	(i) Piston displacement	1300	[8.71]
		(ii) X-ray	820	[8.72]

Table 8.1. Continued

	Solid	Method	B [kbar]	Ref.
17.	SmS	(i) Piston displacement	151	[8.73]
		(ii) X-ray	476	[8.74]
		(iii) Ultrasonics	503	[8.75]
18.	ZnO	(i) Piston displacement	450	[8.76]
		(ii) Ultrasonics	1399	[8.77]
		(iii) Ultrasonics	1436	[8.39]
		(iv) Optical interferometry	1447	[8.26]
19.	BeO	(i) Piston displacement	3704	[8.10]
		(ii) Piston displacement	2433	[8.76]
		(iii) Ultrasonics	2198	[8.36]
		(iv) Ultrasonics	2140	[8.78]
20.	CdS	(i) Piston displacement	380	[8.76]
		(ii) Ultrasonics	615	[8.79]
21.	CdSe	(i) Piston displacement	281	[8.76]
		(ii) Ultrasonics	532	[8.79]
21.	CdSe	(i) Piston displacement	281	[8.76]
		(ii) Ultrasonics	532	[8.79]
22.	MnF_2	(i) Piston displacement	184	[8.80]
		(ii) Ultrasonics	883	[8.81]
23.	GeO_2 (Rutile)	(i) Ultrasonics	1950	[8.82]
		(ii) Ultrasonics	2576	[8.83]
24.	Stishovite (SiO_2, rutile)	(i) Piston displacement	3600	[8.84]
		(ii) Piston displacement	3000	[8.85]
25.	Magnetite	(i) Piston displacement	1830	[8.86]
		(ii) Ultrasonics	1590	[8.87]
		(iii) Ultrasonics	1390	[8.88]
		(iv) X-ray	1830	[8.89]
26.	Benzil	(i) Piston displacement	30.7	[8.90]
		(ii) Piston displacement	66.5	[8.91]

Table 8.2. Elastic constants C_{ij} [kbar], bulk modulus B [kbar] and Poisson's ratio σ_P of some cubic crystals

	Crystal		C_{11}	C_{12}	C_{44}	B	σ_P	Ref.
1.	FeS_2 (Pyrite)	(i)	3585	−529	1035	842	−0.17	[8.2]
		(ii)	3818	310	1094	1479	0.07	[8.92]
2.	Sodium chlorate	(i)	618.2	−208.3	119.6	67	−0.50	[8.1]
		(ii)	509	155	118	273	0.23	[8.6]
3.	PbTe	(i)	1040	−44	130	317	−0.04	[8.93]
		(ii)	1080	80	130	413	0.07	[8.94]
4.	PbSe	(i)	4040	3190	162	3473	0.44	[8.95]
		(ii)	1130	150	130	477	0.11	[8.96]
		(iii)	1240	190	160	540	0.13	[8.97]

Table 8.3. Elastic constants C_{ij} [kbar] and bulk modulus B [kbar] for some tetragonal crystals

Crystal		C_{11}	C_{12}	C_{13}	C_{33}	C_{44}	C_{66}	C_{16}	B	Ref.
1. Zircon ($ZrSiO_4$)	(i)	570	250	50	460	140	320		190	[8.98]
	(ii)	3300	1075	1540	3805	733	397		2040	[8.99]
	(iii)	4230	703	1490	4900	1136	485		2030	[8.100]
2. ADP	(i)	676	59	199	336	87	66		2820	[8.101]
	(ii)	758	−243	133	296	87	61		2060	[8.5]
3. KDP	(i)	714	−49	129	562	127	63		2680	[8.5]
	(ii)	785	320	387	763	123	61		5020	[8.5]
4. $SrMoO_4$	(i)	1275	886	501	1034	347	213	40		[8.102]
		213	−2400				1275	−40		
		665	−490				823	527		
		823	−989				665	−527		
	(ii)	1154	599	444	1042	350	475	121		[8.103]
	(iii)	1190	620	480	1040	349	420	−120		[8.104]

The following features may be noted:

(a) If any two parameters are known, the other two can be calculated and checked with experimental values.
(b) The elastic moduli are all positive.
(c) The Poisson's ratio is positive. Further, from (8.6), it follows that $0 < \sigma_P < 0.5$. Empirically, σ_P is found to be close to 0.3 for most solids.

As an example of the use of phenomenological relations as consistency checks, we shall consider the bulk modulus of selenium. Gschneidner [8.54] quoted literature values of 591 and 92.7 kbar for the Young's modulus and the bulk modulus; the bulk modulus was from Bridgman's work. Substituting these values in (8.6), one gets a value of −0.562 for σ_P which is physically unacceptable in view of (c) and indicates an error in the value of Y/B. Since the value of Y is reasonable from various considerations, Gschneidner concluded that there is an error in Bridgman's value for B for Se, which should have been much larger. Subsequently, Mort [8.56] reported single crystal elastic

Table 8.4. Elastic constants C_{ij} [kbar], S_{ij} [kbar^{-1}], bulk modulus B [kbar] and Debye temperature θ[K] for some hexagonal and trigonal crystals

Crystal		S_{11}	$-S_{12}$	$-S_{13}$	$2S_{14}$	S_{33}	$4S_{44}$	B	θ	Source
1. Calcite (trigonal)	(i)	0.00103	0.00035	0.00080	0.0012	0.00175	0.0038	−6300		[8.105]
	(ii)	0.001	0.00038	0.00048	0.0009	0.00175	0.0041	700		[8.106]
Crystal		C_{11}	C_{12}	C_{13}	C_{33}	C_{44}	C_{66}	B	θ	**Source**
2. BeO (hex.)	(i)	4700	1680	1190	4940	1530	1520	2495	1260	[8.36]
	(ii)	4606	1265	885	4916	1477	1670	2244	1280	[8.27]
3. $LaCl_3$ (hex.)	(i)	3065	1238	1653	3611	1677	913	2050	600	[8.107]
	(ii)	2585	1670	1517	3419	1731	458	1979	500	[8.107]
	(iii)	2952	1564	1607	3648	1742	694	2100	507	[8.107]
	(iv)	190	80	100	230	100	60	128	~150	[8.108]

constants of Se from which Sirdeshmukh and Subhadra [8.55] calculated the isotropic elastic moduli and obtained the values $B = 174$ kbar, $Y = 234$ kbar and $\sigma_P = 0.27$. This value of B is much larger than the Bridgman value. Further, σ_P is positive and close to 0.3. Thus these values are phenomenologically consistent. The much lower value of 92.7 kbar quoted from Bridgman's work for the bulk modulus of Se (hex.) has to be ignored and the ultrasonic value of 174 kbar is recommended.

It may be mentioned that values for B for Se larger than that reported by Bridgman were predicted by Gschneidner [8.54] from an empirical relation between B and the cohesive energy and by Sirdeshmukh [8.109] by a theoretical method; these will be discussed later in this section. It may be also mentioned that a value of 79 kbar reported later by Vaidya and Kennedy [8.50] is much lower than Bridgman's value and open to the same criticism as Bridgman's value.

It is to be noted that the condition that σ_P cannot be negative is useful only in finding whether the B value is physically valid or not. At best, it helps in estimating a lower bound for B but it cannot estimate, by itself, the correct value of B.

Relations Between Elastic Constants

The quadratic stress (or strain) energy is positive definite. This condition results in interrelationship between the single crystal elastic constants [8.110–8.114]. These relationships will be used in analyzing data on some crystals in Sects. 8.4.2 and 8.5.

8.4.2 Theoretical Consistency Checks

The bulk modulus may be calculated by theoretical methods and the calculated value may be compared with the experimental values. Alternatively, the experimental bulk modulus may be used to calculate other physical properties theoretically; these calculated properties may then be compared with their respective experimental values. Some approaches of this type are now considered.

Knopoff's Universal Relation

By an interpolation of the equations of state obtained from the finite strain theory and the Thomas–Fermi model, Knopoff [8.115] derived an expression for the bulk modulus B of a solid in terms of the atomic volume (V_a), atomic number (Z), the Bohr radius (a_0) and the electron charge (e). The relation is:

$$\left[BZ^{-10/3}\right]\left[ZV_\mathrm{a}\right]^{7/3} = \left[\pi^{10/3}6^{2/3}/15\right] a_0^3\, e^2 = 34.17 \quad \mathrm{Mbar}\,\mathrm{\AA}^7, \tag{8.7}$$

For compounds, V_a is the mean atomic volume and Z is given by

$$Z^{2/3} = \sum_i n_i Z_i^{5/3} \Big/ \sum_i n_i Z_i \tag{8.8}$$

where n_i is the number of times an atom i with atomic number Z_i occurs in the chemical formula.

The various reported values of B for selenium are given in Table 8.1. Since Gschneidner [8.54] had expressed doubts regarding the soundness of the Bridgman value on the basis of the "negative σ_P criterion", Sirdeshmukh [8.109] applied Knopoff's method to estimate the bulk moduli of some elemental solids including selenium. These calculated values are given in Table 8.5 along with experimental values quoted by Gschneidner. It is seen that agreement between experimental and calculated values is good in the case of Zr and W but otherwise there are differences of as much as 30–50% in other cases. The largest difference occurs in the case of selenium. The calculated value of 526 kbar is much larger than Bridgman's value of 92 kbar. Further, the calculated value combined with the Young's modulus value quoted by Gschneidner yields a value of 0.3 for the Poisson's ratio.

Table 8.5. Values of B for some elemental solids

Element	B [kbar]	
	Calc. from (8.7)	Expt.
Al	645	735
Ti	709	1075
Mg	270	361
Zr	901	850
W	4,000	3,333
Li	80	118
Se	526	92.7

However, as mentioned in Sect. 8.4.1, later determination of elastic constants of Se have led to a consistent set of elastic moduli which yield a value of 174 kbar for the bulk modulus and a Poisson's ratio value of 0.27. In view of the high accuracy of ultrasonic results, the value obtained from Knopoff's relation appears to be an overestimate.

Knopoff's method is now used to analyse the data on MnF_2 and Fe_3O_4. Values of 90 kbar and 2,800 kbar, respectively, are obtained for the bulk modulus of these two compound crystals. This value for MnF_2 is lower than both the values given in Table 8.1 but between the two values, it is closer to the piston-displacement value. In the case of magnetite, the Knopoff-value is much higher than all the values quoted in Table 8.1 but is closer to the piston-displacement and X-ray values.

Knopoff's method is simple and universal; it is independent of bonding and structure. Yet, in view of the cases discussed above, we may conclude that Knopoff's method may be used only to provide corroborative rather than conclusive evidence.

Bulk Modulus from Molecular Data

From simple considerations, it can be shown that there is a direct relation between the bulk modulus (B) and the interatomic force constant (k). Waser and Pauling [8.116] showed that

$$B = (N r^2 / 9 V_c) k, \tag{8.9}$$

where N is the number of equivalent bonds and V_c the unit cell volume. Yean and Riter [8.117] proposed an equivalent relation:

$$B = (2\rho N_A r^2 / 9 A) k, \tag{8.10}$$

where N_A is the Avogadro number, ρ the density and A the mean atomic weight.

While Waser and Pauling used (8.9) to estimate the force constant of several elemental and compound crystals from known values of their bulk modulus, Yean and Riter proceeded in the reverse direction and estimated the bulk modulus from (8.10) using known values of the force constant. In doing so, Yean and Riter made the important assumption that the interatomic force constant between two atoms in the solid state is the same as the stretching force constant for the same two atoms in the molecular state. The stretching force constant for molecules can be evaluated from IR and Raman spectroscopic data. The force constants and bulk moduli estimated by Yean and Riter for some systems are given in Table 8.6.

The value of 960 kbar estimated for the bulk modulus of Ge compares well with Bridgman's value of 787 kbar quoted in [8.54] and the value of 778 kbar quoted in [8.7]. Further, Yean and Riter predicted the value of 2,240 kbar for the bulk modulus of SiC without the knowledge of any experimental value.

Table 8.6. Stretching force constants (k) and estimated bulk moduli (B) for some crystals

Molecule	k [millidyne cm^{-1}]	Crystal	B from (8.10) [kbar]
C_2H_6	4.36	C (diamond)	4,070
Si_2H_6	1.73	Si	1,060
Ge_2H_6	1.62	Ge	960
Sn_2H_6	1.40	Sn (α)	720
CH_3SiH_3	2.19	SiC (β, cubic)	2,240

However, the bulk modulus of SiC was earlier determined by Einspruch and Clairborne [8.118] who obtained a value of 2,140 kbar from ultrasonic measurements on a polycrystalline sample. Thus, the method seems to work well. Comparing the value of the bulk modulus of Si calculated from the spectroscopic data with the values given in Table 8.1, Yean and Riter recommended rejection of the abnormally large literature value of 3,120 kbar. It is clear that this method is suitable only for molecular and crystalline systems having a common bond and which are amenable to spectroscopic measurements.

Bulk Modulus from Szigeti's Theory of Dielectrics

Considering the dielectric polarization in ionic crystals, Szigeti [8.119] derived the relation

$$B = (1/\psi) = \left[r^2 \mu \omega_t^2 (\varepsilon_s + 2)\right] / \left[3V_i(\varepsilon_\infty + 2)\right], \tag{8.11}$$

where ψ is the compressibility, r the interionic distance, μ the reduced mass, ω_t the transverse optical frequency obtained from IR reflectivity curve, ε_s the static dielectric constant, V_i the volume per ion pair and ε_∞ the optical dielectric constant. Using appropriate values for the various quantities in (8.11), Szigeti calculated the bulk modulus. The calculated and experimental values of the bulk modulus ($B_{calc.}$ and B_{exp}, respectively) and the ratio $B_{exp.}/B_{calc.}$ for some crystals are given in Table 8.7.

The values of $B_{exp.}/B_{calc.}$ for the alkali halides are all close to unity showing that (8.11) works well, particularly for perfectly ionic crystals like alkali halides. For MgO, Szigeti got a low value for $B_{exp.}/B_{calc.}$. Anderson and Glynn [8.120] felt that the value of ω_t used by Szigeti was inaccurate. They redetermined ω_t from the reflectivity curve for MgO and recalculated B. The revised value for $B_{exp.}/B_{calc.}$ is, again, close to unity.

This method is now applied to analyze the bulk modulus data for CaO and SrO given in Table 8.1. For CaO, the value of $B_{exp.}/B_{calc.}$ obtained by Szigeti using Bridgman's value for B is low. The alkaline earth oxides are not as ionic as the alkali halides and some deviations $B_{exp.}/B_{calc.}$ from unity can be expected. But the value 0.13 obtained by Szigeti is too low. The values for this ratio have been recalculated using the other experimental values of the bulk modulus for CaO given in Table 8.1. It is seen that with the new values

Table 8.7. Calculated and experimental values of bulk moduli for some crystals

Crystal	B_{calc} (8.11) [kbar]	B_{exp} [kbar]	$B_{\mathrm{exp}}/B_{\mathrm{calc}}$
LiF	670	670	1.00
NaCl	242	239	0.99
KBr	160	152	0.95
MgO	3570	1,690	0.47
	1,640[a]		1.04[a]
CaO	1,660	218	0.13
		1,120[a]	0.67[a]
		1,090[a]	0.65[a]
SrO	680	880[a]	1.29[a]
		1183[a]	1.73[a]

[a]See discussion in text; rest of the data are from [8.119, 8.120]

of B, the ratio $B_{\mathrm{exp.}}/B_{\mathrm{calc.}}$ has a value ~0.66 which is much better than the Szigeti value of 0.13. Clearly, (8.11) is able to show that Bridgman's value of bulk modulus of CaO is in error and the more recent values are reasonable.

For SrO again, Szigeti calculated the value of B from (8.11); data on $B_{\mathrm{exp.}}$ was not available to him to make a comparison. Now that the bulk modulus of SrO has been determined, the values of $B_{\mathrm{exp.}}/B_{\mathrm{calc.}}$ have been calculated using data in Table 8.1 leading to values of 1.29 and 1.73; the lower value of 880 kbar [8.70] appears more reasonable than the larger value of 1,183 kbar [8.10] for the bulk modulus of SrO.

Theoretical Calculation of Physical Properties Using the Bulk Modulus as an Input Cohesive Energy

For an ionic crystal, the interaction energy u_{ij} between two ions i and j may be represented by

$$u_{ij} = -z_i z_j e^2 r_{ij}^{-1} - c_{ij}(6) r_{ij}{}^{-6} - d_{ij}(8) r_{ij}{}^{-8} + b_i b_j f_{ij} \exp(-r_{ij}/\rho), \quad (8.12)$$

where the terms represent the Coulomb, dipole–dipole, dipole–quadrupole and repulsion interactions, respectively [8.121]. For a crystal as a whole, the cohesive energy is obtained by summation over all ions. The cohesive energy may be represented by

$$U = U_e + U_d(6) + U_d(8) + U_r, \quad (8.13)$$

where U is the cohesive energy and the four terms, again, represent the contributions of the four interactions. The parameters in the expression, particularly those in the repulsion term are calculated from the conditions:

$$\mathrm{d}U/\mathrm{d}r = 0, \quad (8.14)$$

$$r^2(\mathrm{d}^2U/\mathrm{d}r^2) = 9VB, \quad (8.15)$$

where r is the interatomic distance, V the molar volume and B the bulk modulus. Once these parameters are known, one may go back to (8.13) to get the cohesive energy. In this way, the bulk modulus comes into the calculation of the cohesive energy. Comparison of the calculated value and the 'experimental' value of cohesive energy helps to judge the reliability of the value of the bulk modulus used as input.

Benson et al. [8.122] employed this approach in the case of ThO_2 and UO_2. They assumed various values of B to calculate U from (8.13) and constructed a B–U plot (Fig. 8.6) from which they read off the values of B corresponding to the experimental values of U ($-2,413$ and $-2,461\,\mathrm{kcal\,mol^{-1}}$ for ThO_2 and UO_2, respectively, [8.122]). These values are given in Table 8.8 along with experimental values which were available from ultrasonic measurements on sintered samples. There is a difference of 10–30% between these two sets of values. Benson et al. [8.122] suggested that this difference could be because the experimental B values were based on experiments on sintered samples. However, later accurate ultrasonic measurements on single crystals yielded values which show the same (or even larger) difference from the B–U plot values. Thus, the difference between experimental values and those from the B–U plot persists.

It may be cautiously concluded that the procedure of calculating the bulk modulus by comparing the calculated value of the cohesive energy with the 'experimental' value leads only to an 'order-of-magnitude' check on the B-values.

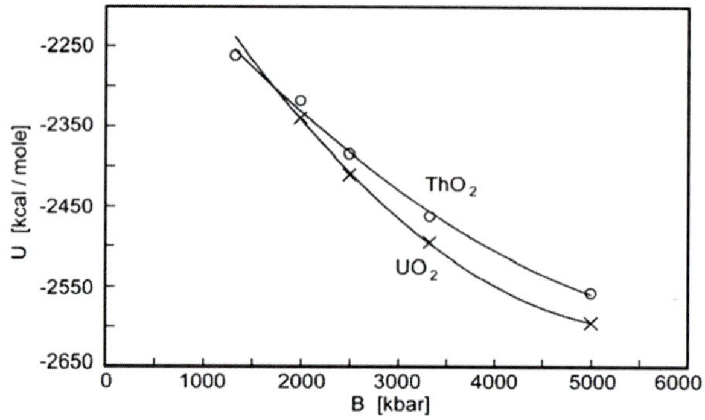

Fig. 8.6. Bulk modulus (B) vs. cohesive energy (U) plot for ThO_2 and UO_2

Table 8.8. Bulk moduli of ThO_2 and UO_2

Method	B [kbar]	
	ThO_2	UO_2
Expt. (polycrystalline samples)	2403	2127
Predicted from B–U plot	2770	2958
Expt. (single crystal)	1930	2127
Minimum predicted from B–σ plot	1500	2000

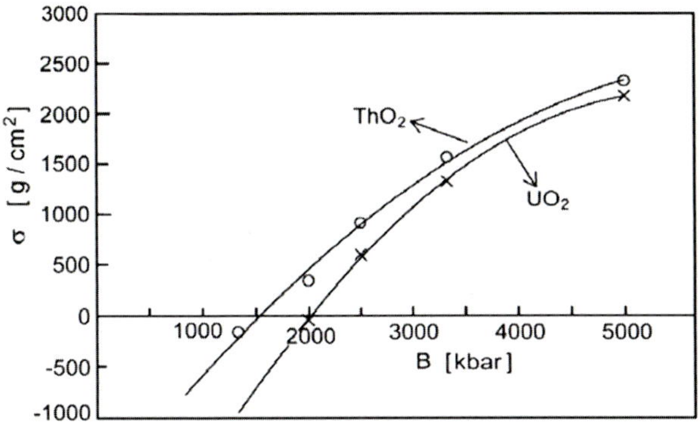

Fig. 8.7. Bulk modulus (B) vs. surface energy (σ) plot for ThO_2 and UO_2

Surface Energy

Starting with (8.12), one can proceed to obtain an expression for the surface energy. The expression depends on the structure of the crystal and also the plane for which the surface energy is to be calculated. As in the case of the cohesive energy, the surface energy (σ) will also have four terms:

$$\sigma = \sigma_e + \sigma_d(6) + \sigma_d(8) + \sigma_r. \tag{8.16}$$

Benson et al. [8.122] calculated σ for the (111) plane of ThO_2 and UO_2 assuming different values of the bulk modulus to evaluate the repulsion parameters. The exact expressions for (8.16) for the (111) plane of the fluorite structure have been worked out [8.123].

The variation of σ with the assumed value of B is shown in the B–σ plot (Fig. 8.7). If the surface energy is known from experiment, one may read off the value of B from the plot corresponding to the experimental value of σ. Experimental values of σ are, however, not known for ThO_2 and UO_2. Some information can still be extracted from the B–σ plot. It can be seen from Fig. 8.7 that σ depends on B in a sensitive manner. In fact, σ assumes positive and negative values for different ranges of B. Physically, σ cannot assume negative values. Thus, we can read off the B value for which σ changes from positive to negative values. This is the minimum value for B. The correct value for B will be larger than this value. These minimum values estimated from Fig. 8.7 are given in Table 8.8. The two sets of experimental values of B as well as those estimated from the B–U plot are all larger than the minimum value from Fig. 8.7. Thus, again, surface energy values serve only as an order-of-magnitude check on bulk modulus values.

Single Crystal Elastic Constants

The Krishnan–Roy theory [8.124] provides a method for evaluation of elastic constants of crystals. Krishnan and Roy assume the following simple expression for the cohesive energy of ionic crystals like the alkali halides:

$$U = N_{\mathrm{A}}\,[-\alpha_{\mathrm{M}}\, z^2\, e^2/r] + b \exp(-r/\rho), \tag{8.17}$$

where α_{M} is the Madelung constant, z the valence and b and (r/ρ) are constants in the repulsion term. Application of (8.14), (8.15) and (8.17) leads to

$$B = \left[\frac{z^2\alpha_{\mathrm{M}}e^2}{18r^4}\right]\,[(r/\rho) - 2], \tag{8.18}$$

(r/ρ) can be determined from the bulk modulus by the use of (8.18). Krishnan and Roy obtained the following relations for the elastic constants:

$$C_{11} = \{2\alpha_{\mathrm{M}}[1 + (r/\rho)] - 6X\}\,(z^2e^2/r^4), \tag{8.19}$$

$$C_{12} = C_{44} = 3(X - \alpha_{\mathrm{M}})\,(z^2e^2/r^4), \tag{8.20}$$

X is a lattice sum which depends on the crystal structure. For the NaCl structure, $X = 3.14$. Thus, with B, as input, (r/ρ) can be evaluated and with (r/ρ) as input, the elastic constants C_{11} and C_{12} can be evaluated. The equality of C_{12} and C_{44} is a consequence of the assumption of central forces. Krishnan and Roy calculated the elastic constants of several alkali halides and found good agreement with experimental values. Mathur et al. [8.69] calculated the elastic constants of alkaline earth chalcogenides from the Krishnan and Roy equations. Sirdeshmukh and Subhadra [8.125] applied the Krishnan–Roy theory to some transition metal oxides with NaCl structure.

Subhadra and Sirdeshmukh [8.126] used the method of Krishnan and Roy [8.124] to analyse the several differing values of the bulk modulus of lithium hydride which has the NaCl structure and is considered highly ionic. Apart from the values 270, 228 and 357 kbar quoted in Table 8.1, Pretzel et al. [8.127] quoted values of 400, 312 and 208 (kbar) for the bulk modulus of LiH by other methods. Subhadra and Sirdeshmukh [8.126] calculated the values of the elastic constants of LiH from (8.19) and (8.20) using values of bulk moduli in the range 200–400 kbar as input. Some of their results are given in Table 8.9. They plotted the values of C_{11} and C_{11}/C_{12} against the values of B used as input. The plots (Figs. 8.8 and 8.9) are smooth lines. Subhadra and Sirdeshmukh [8.126] would have used experimental values of C_{11} and C_{12} to read off the correct value of B. However, such data was not available to them.

An interesting feature of the C_{11} vs. B and (C_{11}/C_{12}) vs. B plots is that both C_{11} and C_{11}/C_{12} are very sensitive to the value of B assumed for the

Table 8.9. Values of elastic constants of LiH from (8.19), (8.20)

B [kbar]	C_{11} [kbar]	$C_{12} = C_{44}$ [kbar]	C_{11}/C_{12}
500	580	460	1.25
357	140	460	0.29
270	−110	460	−0.24
228	−240	460	−0.52

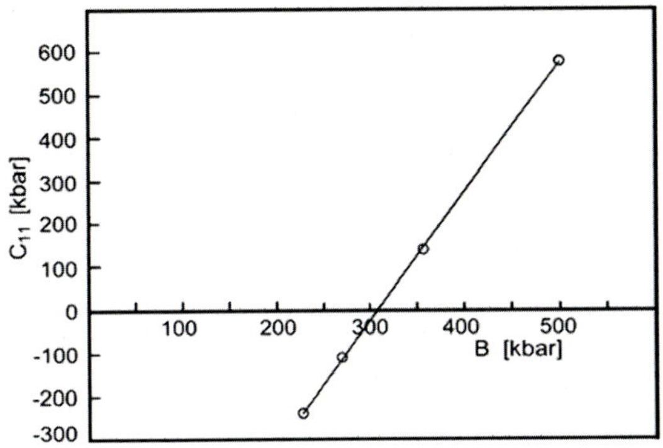

Fig. 8.8. B vs. C_{11} plot for LiH

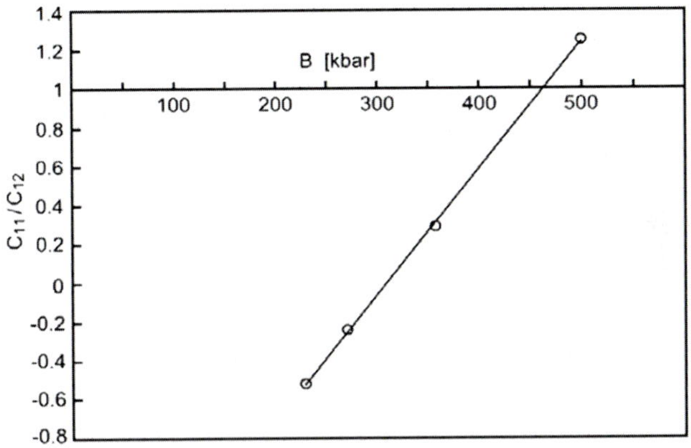

Fig. 8.9. B vs. C_{11}/C_{12} plot for LiH

calculations. In fact C_{11} and C_{11}/C_{12} assume–negative values for a range of B values. Subhadra and Sirdeshmukh [8.126] invoked the stability conditions proposed by Born and Huang [8.111] and Alers and Neighbours [8.113] according to which, in cubic crystals

$$C_{11} > 0 \tag{8.21}$$

and

$$(C_{11}/C_{12}) > 1. \tag{8.22}$$

From Figs. 8.8 and 8.9, it is seen that for both the stability conditions to be satisfied B should be greater than 465 kbar. Thus, 465 kbar is the lower bound for the bulk modulus of LiH. It may be noted that the Krishnan–Roy theory assumes central forces, neglects the Van der Waal interactions and treats the crystal as completely ionic. Subsequent to the analysis by Subhadra and Sirdeshmukh [8.126], Haussuhl and Skorezyk [8.128] measured the single crystal elastic constants of LiH. They observed a large difference in the values of C_{12} and C_{44} indicating considerable departure from ionicity. In view of these limitations, it may not be proper to fix a sharp limit for the lower bound of B. Thus the value of 357 kbar reported by Stephen and Lilley [8.62] may be treated as the best among the several B values.

8.4.3 Empirical Relations as Consistency Checks

There are several empirical relations between the bulk modulus and other physical properties. Although these correlations do not have a rigorous theoretical support, they are otherwise well-established and can be used to check data on bulk moduli. Some of these empirical relations are discussed in this section.

Relation Between Bulk Modulus (B) and Volume (V)

Keyes [8.129] showed from dimensional analysis that the bulk modulus varies inversely as a^4 where a is the lattice constant. He verified this relationship for crystals with zinc blende structure. Mitra and Marshall [8.130] found that for alkali halides the bulk modulus is inversely proportional to a^3. Anderson and Nafe [8.131] drew log B vs. log V plots for several systems and proposed a relation:

$$B\,V = \text{constant}. \tag{8.23}$$

Subsequently linear log B vs. log V plots have been reported for several systems (Table 8.10).

It was shown [8.131–8.133] that a relationship close to (8.23) can be derived on the basis of interatomic forces. Neumann [8.134] suggested that a better fit with experimental data is obtained by using a modified equation:

$$B = b\,a^{-m}\,(1 - g\,f_i), \tag{8.24}$$

where 'a' is the lattice constant, f_{i} the Phillips ionicity and b, m and g are constants for a family of crystals.

However, the simple log B vs. log V plots have proved quite useful. Jayaraman et al. [8.133] pointed out that these plots are useful in scaling

Table 8.10. Systems for which linear log B vs. log V plots have been reported

System(s)	Ref.
Oxides, alkaline earth fluorides	[8.131]
Wurtzite type crystals	[8.27]
Alkali halides, divalent chalcogenides and trivalent rare earth chalcogenides	[8.133]
Divalent oxides and lead chalcogenides	[8.135]
Uranium Pnictides	[8.25]
Rare earth chalcogenides	[8.72]

bulk modulus data, in estimating ionic charge and in predicting bulk moduli where they are not available. Sirdeshmukh and Subhadra [8.135] pointed out another application of the log B–log V plots viz. in sifting the correct value of bulk modulus when several differing values are reported. Thus, they were able to show that among the three values for PbSe given in Table 8.2, the value 3,470 kbar deviates severely from the log B vs. log V plot (or the B vs. V^{-1} plot) and is to be disregarded. Similarly, Vedel et al. [8.72] showed that between the two values for CeS given in Table 8.1, the value 820 kbar is close to the log B vs. log V plot for the rare earth chalcogenides, in contrast to the other value of 1,300 kbar which shows a strong deviation.

The log B vs. log V plots have now been redrawn for the alkali halides with NaCl structure (Fig. 8.10), alkali halides with cesium halide structure (Fig. 8.11), NaCl-type oxides (Fig. 8.12), chalcogenides with NaCl structure (Fig. 8.13) and wurtzite type chalcogenides (Fig. 8.14) including several new

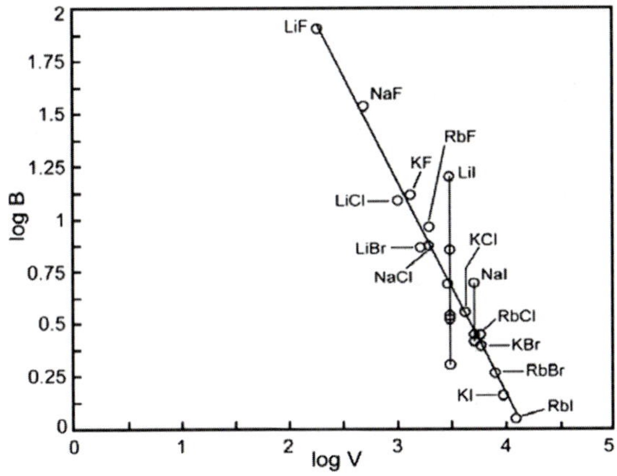

Fig. 8.10. Plot of log B vs. log V for NaCl type alkali halides

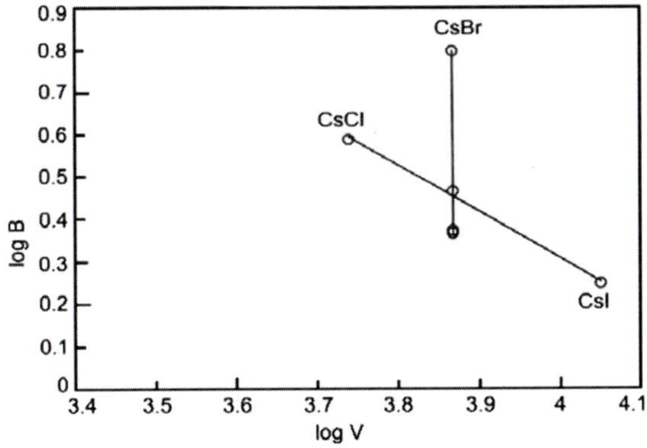

Fig. 8.11. Plot of log B vs. log V for CsCl type alkali halides

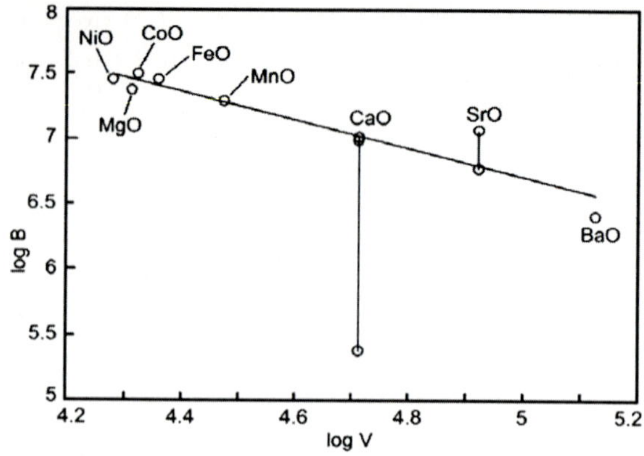

Fig. 8.12. Plot of log B vs. log V for oxides with NaCl structure

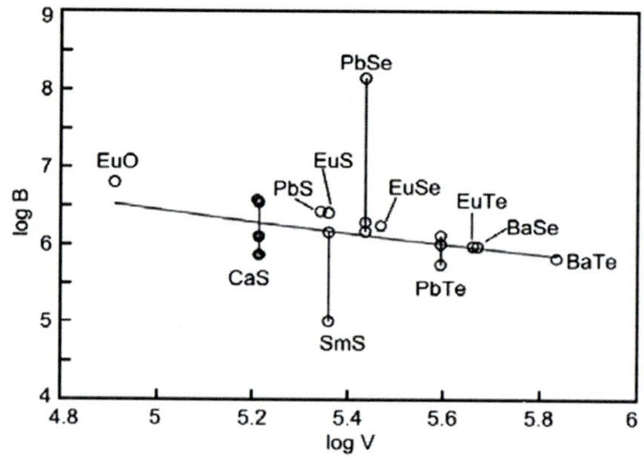

Fig. 8.13. Plot of log B vs. log V for divalent chalcogenides

B-values. It is seen that in Fig. 8.10, the plot is able to differentiate between the several largely differing values of B for LiI and also in the case of NaI where the difference is about 25%. Again in Fig. 8.11, the plot is able to show that in the case of CsBr, between the three values differing by about 60%, the largest value deviates very much from the plot and the two lower values are preferable. Figure 8.12 clearly indicates that the Bridgman value of 218 kbar for CaO is to be discarded in comparison with the other values which are larger by a factor of 5. However, in the case of SrO, the deviation of the data point for the larger value for SrO is only marginal. In Fig. 8.13, the plot clearly shows that the largest among the B values for PbSe and the lowest among the B values for SmS deviate too much from the plot and deserve to be ignored. However, the plot does not differentiate between the different data for CaS and PbTe with the same clarity. Lastly, in Fig. 8.14, the lesser value for ZnO

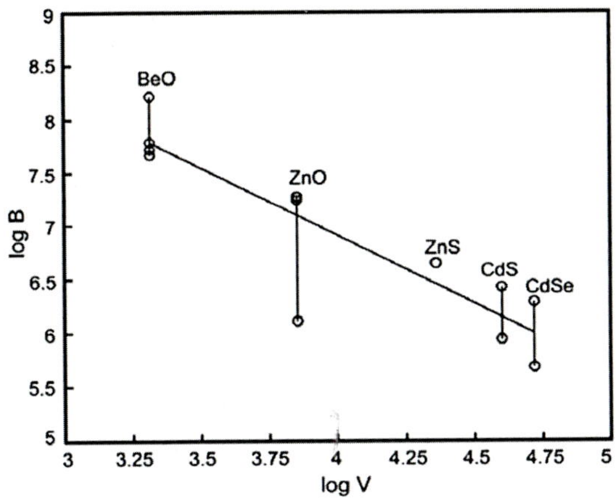

Fig. 8.14. Plot of log B vs. log V for wurtzite type chalcogenides

is clearly off the plot but the same cannot be said about the data points for BeO, CdS and CdSe.

Thus, the log B vs. log V plots can be effectively used to distinguish different Bvalues particularly if the differences are large.

Relation Between Bulk Modulus and Cohesive Energy of Elements

Gschneidner [8.54] pointed out that an empirical linear relation exists between the cohesive energy U and the bulk modulus of elements. The relationship is of the type

$$U = mB + b, \tag{8.25}$$

where m and b are constants for a family of related crystals.

Gschneidner plotted the values of U against B for the elements S, Se, Te and Po. He found that the data point for Se (based on Bridgman's results) deviates severely from the straight line joining the data points for the other three elements. The value of B read off from the straight line suggested a much larger value for B of Se than the Bridgman value. It is shown in other sections that other consistency checks also supported this suggestion and a later ultrasonic measurement indeed resulted in a much larger value.

The relation between $U$and B is all too empirical. In fact, Gschneidner [8.54] pointed out that the slope m of the straight line plot has positive as well as negative values for different systems. Thus the plot for any family must be clearly established before it can be put to use to analyze B values for a given solid.

Relation Between Bulk Modulus and Thermal Expansion

Sirdeshmukh [8.136] proposed a linear relation between the reciprocal of bulk modulus ($1/B$) and the product of the coefficient of expansion (α) and volume

(V). He obtained linear plots of $\alpha\, V$ vs. $(1/B)$ for several families of metals and the alkali halides.

Relation Between Bulk Modulus and Debye Temperature

The Debye temperature is empirically given by:

$$\theta = c\, B^{1/2} \rho^{-1/6} m^{-1/3}, \tag{8.26}$$

where θ is the Debye temperature, ρ the density, m the mass and c a constant. This relation is known in literature as the Madelung formula. Deus and Schneider [8.137] modified this relation as follows:

$$\theta = a_1\, x + a_2, \tag{8.27}$$

where x is $(B^{1/2} \rho^{-1/6} m^{-1/3})$ and a_1 and a_2 are constants for a family of crystals.

Relation Between Bulk Modulus and Hardness

Reddy et al. [8.138] proposed an empirical relation between the bulk modulus, hardness (H), refractive index (n) and Phillips ionicity (f_{i}) as follows:

$$B = c\, H\, n^{0.13} f_{\mathrm{i}}, \tag{8.28}$$

where c is a constant. The relation was found to work well in ternary chalcopyrites of the $A^{I}B^{III}C^{VI}$ and $A^{II}B^{IV}C^{V}$ types with appropriate values for the constant c.

8.5 Consistency Checks for Single Crystal Elastic Constants

Single crystal elastic constants $C_{ij}$are almost invariably determined by ultrasonic methods which are highly accurate (Sect. 8.2). While the velocity determination, per se, is accurate, inaccuracies enter by way of factors like (a) sample history, wall reflections and bonding problems and (b) analytical difficulty in converting velocities into C_{ij}'s. Several examples of strong differences in reported values of C_{ij}'s are cited in Tables 8.2–8.4. In this section, some consistency checks are considered which facilitate sifting of correct data from several reported values of elastic constants. Again, the checks are based on (a) phenomenological constraints and (b) use of C_{ij}'s to calculate other physical properties.

In analyzing data on single crystal elastic constants use is made of interrelations between elastic constants resulting from stability conditions. Some of these relations and constraints are as follows:

(a) Cubic [8.111, 8.113]:
$C_{11}, C_{12}, C_{44} > 0$; $C_{11} - C_{12} > 0$
(b) Hexagonal [8.112]:
$C_{44} > 0$; $C_{11} > |C_{12}|$; $(C_{11} + C_{12})C_{33} > 2C_{13}^2$
(c) Tetragonal [8.114]:
$C_{11}, C_{33}, C_{44}, C_{66} > 0$; $C_{11} > |C_{12}|$; $C_{11}C_{66} > C_{16}^2$; $C_{11}C_{33} > C_{13}^2$

Alers and Neighbours [8.113] pointed out that for all crystal classes the diagonal elements of the elastic constant matrix are all positive. They also discussed the stability conditions for the trigonal class. Sundara Rao [8.110] discussed the inter-relationships between the elastic constants for orthorhombic crystals.

8.5.1 Cubic Crystals

In the first three crystals for which data are quoted in Table 8.2, the value of C_{12} differs in sign. Since the Poisson's ratio equals $C_{12}/(C_{11}+C_{12})$ a negative value for C_{12} results in a negative value for the Poisson's ratio. Phenomenologically, σ_P cannot be negative. Poisson's ratio values for different sets of elastic constants are given in Table 8.2. The sets of elastic constants for FeS_2, $NaClO_3$ and PbTe with negative values of C_{12} can be straight away disregarded.

In the last case viz., PbSe, the elastic constants reported by Chudinov [8.95] are larger than the values from the other two sources by a factor of 3 in the case of C_{11} and a factor of 20 in the case C_{12}. This results in the bulk moduli differing by a factor of about 7. The log B vs. log Vplot (Fig. 8.12) clearly shows that the larger value of B does not fit into the correlation and hence the corresponding set of elastic constants [8.95] has to be ignored.

8.5.2 Tetragonal Crystals

Zircon

In Table 8.3, three sets of C_{ij}'s are quoted for zircon. Sets (ii) and (iii) broadly agree with each other, although there are differences between the values for C_{11}, C_{12} and C_{44}. However, both sets differ from set (a) by factors ranging from 1.5 (C_{66}) to 30 (C_{13}).

Sirdeshmukh and Subhadra [8.139] analysed these differing results using several approaches. Their results are given in Table 8.11. Sirdeshmukh and Subhadra [8.139] calculated the Voigt–Reuss–Hill bulk modulus from the three sets of C_{ij}'s and compared the values with those obtained from other independent methods. It is seen that the value of the bulk modulus from set (i) is very low. It is lower than the values from sets (ii) and (iii) by a factor of 8–9. It is also lower than typical values for several mineral crystals quoted by Anderson et al. [8.142]. The bulk modulus of zircon has been estimated by other independent methods. These are given in Table 8.11. It may be noted that the value from set (i) is in disagreement with all these values whereas the

Table 8.11. Bulk modulus (B), Debye temperature (θ) and Gruneisen constant (γ) of zircon

Set	B[kbar]		θ [K]		γ
	From C_{ij}	From other methods	From C_{ij}	From sp. heats	
i	190		300		0.11
ii	2,040		601		0.89
iii	2,030		725	600–900[c]	0.99
		3731[a] (X-ray)			
		3875[a] (Neutron)			
		2702[b]			

[a] See [8.140]
[b] See [8.115, 8.139]
[c] See [8.141]

values from sets (ii) and (iii) show at least an order-of-magnitude agreement with these independently estimated values. This comparison of bulk moduli clearly indicates that the elastic constants in set (i) are unacceptable.

Sirdeshmukh and Subhadra [8.139] further used the bulk moduli to calculate the thermal Gruneisen constant γ given by

$$\gamma = 3\,\alpha V B / C_V, \tag{8.29}$$

where α is the linear coefficient of thermal expansion, V the volume and C_V the specific heat. Using the values of the bulk moduli from the three sets together with values for the other quantities taken from literature, Sirdeshmukh and Subhadra [8.139] obtained the values given in Table 8.11 for γ. In analyzing the results on γ, they used the empirical fact that for most solids, the value of γ lies in the range 1–3. On this basis again, the value of $\gamma = 0.11$ originating from set (i) is far too low. On the other hand, the other two values of γ originating from sets (ii) and (iii) are close to ~ 1 and indicate the superiority of C_{ij}'s in sets (ii) and (iii).

Finally, Sirdeshmukh and Subhadra [8.139] calculated the Debye temperature of zircon from the elastic constants using Anderson's method [8.143]. These values are included in Table 8.11 along with values from specific heats. The value of Debye temperature from set (i) is about half the values from sets (ii) and (iii). Also, the values from sets (ii) and (iii) are within the range of values quoted from specific heat data unlike the value from set (i). This again indicates the superiority of sets (ii) and (iii) vis-à-vis set (i).

Thus, it has been clearly established that the elastic constants in set (i) are physically unacceptable. This analysis is an example of a comprehensive check on elastic constant data.

KDP Type Crystals

Two sets of elastic constants each for potassium dihydrogen phosphate (KDP) and ammonium dihydrogen phosphate (ADP) are given in Table 8.3. In both cases there are strong differences in the values of C_{13}. More seriously, there is a difference in the sign for C_{12}. As mentioned earlier, for tetragonal crystals, one stability condition is $C_{11} > |C_{12}|$. Thus, there is no constraint on the sign of C_{12}. Both sets satisfy the requirement $C_{11} > |C_{12}|$.

The bulk moduli calculated from the elastic constants also show a difference of 40% in the case of ADP and a difference of 80% in the case of KDP. However, the bulk moduli are not known from any other independent method. Being an important family of crystals, the elastic constant data deserves further analysis.

Strontium Molybdate

Strontium molybdate ($SrMoO_4$), along with $CaWO_4$ and $CaMoO_4$, belongs to T II Laue symmetry. There are several intrinsic difficulties in the determination of elastic constants of these crystals. First, there are seven (instead of the usual 6) independent elastic constants; the existence of the seventh elastic constant C_{16} complicates measurements. Second, the axes of acoustic symmetry in the (001) plane do not coincide with the <100> and <110> directions which makes the determination of C_{16} ambiguous. Third, the equations for the elastic constants C_{11}, C_{66}, C_{12} and C_{16} have four different solutions.

In Table 8.3, three sets of elastic constants are quoted for $SrMoO_4$. The first set is reported by Chung and Li [8.102]. Chung and Li [8.102] gave in their paper the four solutions for C_{11}, C_{66}, C_{12} and C_{16} referred earlier. These are reproduced in Table 8.3. It can be seen that the values of C_{ij}'s in these solutions not only differ by several factors but also in sign. Chung and Li [8.102] chose the first of these solutions as the most reasonable, on the basis of the stability conditions discussed by Alton and Barlow [8.114].

Comparing the values of C_{ij}'s reported by Chung and Li [8.102] with C_{ij}'s of the related crystals $CaWO_4$ and $CaMoO_4$, James [8.103] noticed a similarity in the values of all C_{ij}'s except C_{16} and C_{66}. For crystals with this symmetry, the following relation must hold:

$$\sum_{i=1}^{3} v_i^2 = S_1/\rho = (C_{11} + C_{44} + C_{66})/\rho \,, \tag{8.30}$$

where ρ is the density, S_1 a constant and v_i are velocities of propagation of sound waves in the x, y plane. James [8.103] found that this condition is satisfied in $CaWO_4$ and $CaMoO_4$ but not by the C_{ij}'s of $SrMoO_4$ reported by Chung and Li [8.102]. James [8.103] carried out a reanalysis of the velocity data of Chung and Li [8.102] by calculating the parameter |SUMSQ| defined by

$$|\mathrm{SUMSQ}| = \sum_{n=1}^{n} \left[\frac{{v_i}^2(\mathrm{calc})}{{v_i}^2(\mathrm{meas})} - 1 \right]^2 . \tag{8.31}$$

The input C_{ij}'s for calculating v_i were varied till |SUMSQ| was minimum. The values thus obtained by James [8.103] are given in Table 8.3 as set (ii);

the new C_{16} and C_{66} values are different from those given by Chung and Li [8.102] but are comparable with those for $CaWO_4$ and $CaMoO_4$.

The C_{ij}'s of $SrMoO_4$were subsequently redetermined by Farley et al. [8.104] taking care regarding the choice of axes which resulted in set (iii). Farley et al. [8.144] showed that the sign of C_{16} is negative for all crystals with scheelite structure.

8.5.3 Trigonal and Hexagonal Crystals

Calcite

Calcite is trigonal with six independent elastic constants. There are two reports on the low temperature elastic constants of calcite [8.105, 8.106]. The values from these two reports are given in Table 8.4. These values were read off from a diagram given by Dandekar and Ruoff [8.106] and pertain to a temperature of 160K. This is the only crystal for which the elastic constants have been given in the S_{ij} notation in this chapter. This has been done as the data are given in this form in the original papers and conversion into C_{ij}'s may introduce errors. There are differences of 30%, 60% and 10% in the values of S_{14}, S_{13} and S_{44}. Dandekar and Ruoff [8.106] calculated the bulk modulus for the two sets and found values of −6300 and 700 kbar. The negative value of bulk modulus is phenomenologically unacceptable and hence the set of elastic constants reported by Ramamurthy and Reddy [8.105] is to be rejected.

BeO (Hex.)

For hexagonal crystals, there are only five independent elastic constants. The sixth constant C_{66} is some times given as it is generally measured independently; it is equal to $(C_{11} - C_{12})/2$.

The two sets of C_{ij}'s for BeO given in Table 8.4 yield B values which agree with each other and are close to the log Bvs. log V line (Fig. 8.14). The Debye temperatures calculated from these two sets also are close to each other and with the specific heat value. However, there is a difference of 32% and 35% in the values of C_{12} and C_{13} which is much larger than the estimated errors of 5–10% mentioned by Bentle [8.36] and Cline et al. [8.27]. Bentle obtained the values 1,520 and 1,630 kbar for C_{66} but retained the lower value. As mentioned, C_{12} is obtained from the difference between C_{11} and 2 C_{66}. If the value 1,630 kbar is used for C_{66}, the difference between the values of C_{12} in the two reports will reduce to 12% which is reasonable. The difference in C_{13}, however, remains. The solution of the velocity equation gives positive and negative values for C_{13}. But only the positive value is retained, though there is no physical reason for doing so [8.27].

$LaCl_3$ (Hex.)

In Table 8.4, values of C_{ij}'s for $LaCl_3$ are given. Sets (i)–(iii) are theoretical values reported by Stedman and Newman [8.107] from three models. There are differences in the C_{ij} values for the three models ranging from 10% (C_{13}) to 100% (C_{66}). These differences may be of significance in relation to the models. The bulk moduli calculated from these three sets are not very different. So it is not possible to differentiate between these sets on the basis of bulk modulus calculations.

Carlson et al. [8.108] calculated the Debye temperatures from these three sets of elastic constants and obtained values 600, 500 and 507 K, respectively. The marginal difference in these values is consistent with marginal differences in the C_{ij}'s from which they originate. However, all these Debye temperature values are larger by a factor of 3–4 than the value of 145 K obtained from specific heat data. Carlson et al. pointed out that the θ from elastic constants can be reconciled with the θ from specific heats if all the elastic constants are reduced uniformly by a factor of 16. These reduced C_{ij}'s are shown as set (iv). They yield a Debye temperature of $\sim$150 K which is close to the specific heat value. The bulk moduli pertaining to sets (i)–(iii) have values in the range 1,980–2,100 kbar. On the other hand, set (iv) yields a bulk modulus value of 128 kbar. Experimental determination of the bulk modulus and the elastic constants will help to sort out the differences in these theoretical values.

8.6 Conclusions

Several methods are now available for the determination of elastic moduli and elastic constants with a high degree of accuracy. In spite of the accuracy of the methods, there are numerous cases where the values of bulk moduli and single crystal elastic constants reported by different workers differ by amounts far exceeding the limits of errors. When such differences exist, it is necessary to subject the reported data to consistency checks. Several consistency checks are reviewed to analyse data on bulk modulus. These may be broadly classified as (a) phenomenological constraints, (b) theoretical estimates of bulk moduli, (c) theoretical estimates of other physical properties using the bulk modulus as input and (d) empirical checks. Differing results on the single crystal elastic constants are, again, analysed by using phenomenological constraints and by comparison of properties calculated from the elastic constants with independently determined values.

A common feature of most of the consistency checks discussed here is that they are able to distinguish between values (or data sets) which differ severely and cannot be forcefully applied when the differences are small. In the analysis of single crystal elastic constants, the consistency check of calculating other physical properties like the bulk modulus and Debye temperature has the limitation that these calculations involve several or all elastic constants. Thus, it is possible to comment on an entire set of elastic constants and not on each elastic constant. Another feature is that some of the checks may help to single out an unacceptable value but cannot decide which among the acceptable values is best. Finally, a single check may not be enough to analyse data; application of several criteria may yield better information; the case of zircon is an example.

References

Chapter 1

1.1 H.S. Peiser, H.P. Rooksby, A.J.C. Wilson, *X-ray Diffraction by Polycrystalline Materials*, Institute of Physics, London (1955)

1.2 B.D. Cullity, *Elements of X-ray Diffraction*, Addison-Wesley, Reading, MA (1956)

1.3 H.P. Klug, L.E. Alexander, *X-ray Diffraction Procedures*, Wiley, New York (1967)

1.4 E.F. Kaelble, *Handbook of X-rays*, McGraw Hill, New York (1967)

1.5 M. Ateequddin, K.A. Hussain, K.G. Subhadra, D.B. Sirdeshmukh, Pramana, **43**, 101 (1994)

1.6 D.B. Sirdeshmukh, V.T. Deshpande, Proc. Indian Natl Sci. Acad., **A38**, 167 (1972)

1.7 A.J.C. Wilson, Proc. Phys. Soc., **54**, 487 (1942)

1.8 W.L. Bond, Acta Crystallogr., **13**, 814 (1960)

1.9 M.V. Cohen, Rev. Sci. Instrum., **6**, 68 (1935)

1.10 R.W.G. Wyckoff, *Crystal Structures*, Vols. 1–3, Interscience, New York, (1963)

1.11 M. Wilkins, quoted by W. Parrish [1.12]

1.12 W. Parrish, Acta Crystallogr., **13**, 838 (1960)

1.13 A.F. Ievins, quoted by W. Parrish [1.12]

1.14 M. Hart, J. Cryst. Growth, **55**, 409 (1981)

1.15 R. Donnay, C. Nowacki, Memoir No. 60, Geological Society of America (1954)

1.16 W.B. Pearson, *A Handbook of Lattice Spacings and Structure of Metals and Alloys*, Pergamon, Oxford (1958)

1.17 R.L. Barns, Mater. Res. Bull., **2**, 273 (1967)

1.18 V. Lisiovan, R. Dikovskaya, Instrum. Exp. Technol., **4**, 992 (1969)

1.19 A. Segmuller, Adv. X-ray Anal., **13**, 455 (1970)

1.20 C.R. Hubbard, H.E. Swanson, F.A. Mauer, J. Appl. Crystallogr., **8**, 45 (1975)

1.21 H.W. King, C.W. Russel, Adv. X-ray Anal., **8**, 1 (1968)
1.22 H.W. King, C.M. Preece, Adv. X-ray Anal., **10**, 354 (1970)
1.23 N.D. Tap, M. Schenk, P. Schafer, Cryst. Res. Technol., **18**, K33 (1983)
1.24 P. Gille, M. Schenk, Phys. Status Solidi A, **84**, K121 (1984)
1.25 J.P. Stott, B.A. Bellamy, L.W. Hobbs, L.V. Hobbs, W. Hayes, A.E. Hughes, J. Phys. C, **9**, L223 (1976)
1.26 W.B. Zimmerman, Phys. Rev. B, **5**, 4704 (1972)
1.27 C. Belouet, M. Monnier, R. Crouzier, J. Cryst. Growth, **30**, 151 (1975)
1.28 F.H. Spedding, B.J. Beaudry, J. Less Common Metals, **25**, 61 (1971)
1.29 A. Van Den Bosch, J. Dresselaers, J. Vansummeren, M. Hovi, Phys. Status Solidi A, **11**, 479 (1972)
1.30 D.B. Sirdeshmukh, K. Srinivas, J. Mater. Sci., **21**, 4117 (1986)
1.31 E. Kaldis, B. Fritzler, W. Peteler, Z. Naturforsch., **34**, 55 (1979)
1.32 E.V. Sampathkumaran, R. Vijayaraghavan, Phys. Rev. Lett., **56**, 2861 (1986)
1.33 Y. Kida, K. Tajima, Y. Shinoda, K. Hayashi, H. Ohsumi, J. Phys. Soc. Jpn, **68**, 650 (1999)
1.34 J.A. Stapien, J. Auleytner, K. Lukaszewicz, Phys. Status Solidi A, **10**, 631 (1972)
1.35 H.D. Holland, D. Gottfried, Acta Crystallogr., **8**, 291 (1955)
1.36 F.W.C. Boswell, Proc. Phys. Soc., **64**, 465 (1951)
1.37 G.C. Benson, K.S. Yun, J. Chem. Phys., **42**, 3085 (1965)
1.38 W.T. Barrett, W.E. Wallace, J. Am. Chem. Soc., **76**, 366 (1954)
1.39 M.E. Straumanis, Acta Crystallogr., **2**, 82 (1949)
1.40 S.A. Semiletov, Sov. Phys. Crystallogr., **12**, 277 (1967)
1.41 G.A. Rozgonyi, P.M. Petroff, M.B. Panish, J. Cryst. Growth, **27**, 106 (1974)
1.42 D.B. Sirdeshmukh, Ph.D. Thesis, Osmania University, Hyderabad, India (1963)
1.43 M. Ateequddin, Ph.D. Thesis, Kakatiya University, Warangal, India (1991)
1.44 K.G. Subhadra, Ph.D. Thesis, Osmania University, Hyderabad, India (1976)
1.45 K.A. Hussain, D.B. Sirdeshmukh, Cryst. Res. Technol., **28**, 1147 (1993)
1.46 M.J.M. Rao, Ph.D. Thesis, Kakatiya University, Warangal, India (1977)
1.47 D.B. Sirdeshmukh, K.G. Subhadra, K.A. Hussain, N. Gopi Krishna, B. Raghavendra Rao, Cryst. Res. Technol., **28**, 15 (1993)
1.48 V.T. Deshpande, D.B. Sirdeshmukh, V.M. Mudholker, Acta Crystallogr., **12**, 257 (1959)
1.49 K.A. Hussain, K.G. Subhadra, M. Ateequddin, D.B. Sirdeshmukh, Cryst. Res. Technol., **26**, K15 (1991)
1.50 L. Vegard, Z. Phys., **9**, 395 (1922)
1.51 N.A. Razik, J. Mater. Sci. Lett., **7**, 569 (1988)
1.52 R.W. Keyes, J. Appl. Phys., **33**, 3371 (1962)
1.53 S.S. Mitra, R. Marshall, J. Chem. Phys., **41**, 3158 (1964)

1.54 B. Nagaiah, M. Ram Babu, D.B. Sirdeshmukh, Indian J. Pure Appl. Phys., **17**, 838 (1979)
1.55 N. Gopi Krishna, K.G. Subhadra, D.B. Sirdeshmukh, Indian J. Pure Appl. Phys., **22**, 618 (1984)
1.56 T. Thirmal Rao, D.B. Sirdeshmukh, Cryst. Res. Technol., **26**, K53 (1991)
1.57 D.B. Sirdeshmukh, K.G. Subhadra, K. Kishan Rao, T. Thirmal Rao, Cryst. Res. Technol., **30**, 861 (1995)
1.58 D.B. Sirdeshmukh, L. Sirdeshmukh, K.G. Subhadra, K. Kishan Rao, S. Bal Laxman, Bull. Mater. Sci., **24**, 469 (2001)
1.59 D.B. Sirdeshmukh, P. Geeta Krishna, K.G. Subhadra, Bull. Mater. Sci., **25**, 203 (2002)
1.60 W. Hussain, D.B. Sirdeshmukh, Pramana, **29**, 583 (1987)
1.61 T. Kumara Swamy, K.G. Subhadra, D.B. Sirdeshmukh, Pramana, **43**, 33 (1994)

Chapter 2

2.1 W.J. Campbell, *Thermal Expansion of MgO*, Report-6115, US Bureau of Mines (1962)
2.2 G. Kommichau, H. Newmann, E. Schmitz, B. Schumann, Cryst. Res. Technol., **21**, 1583 (1986)
2.3 H. Fullam, Mater. Res. Bull., **7**, 289 (1972)
2.4 C. Kittel, *Introduction to Solid State Physics*, Wiley, New York (1966)
2.5 R.A. Levy, *Principles of Solid State Physics*, Academic, New York (1968)
2.6 R.S. Krishnan, *Progress in Crystal Physics*, S. Viswanathan, Madras (1958)
2.7 H.D. Megaw, Ph.D. Thesis, Cambridge University (1935)
2.8 R.R. Reeber, J. Appl. Phys., **41**, 5063 (1970)
2.9 K.V.K. Rao, S.V. Nagender Naidu, L. Iyengar, J. Am. Ceram. Soc., **53**, 1970 (1970)
2.10 K.V.K. Rao, S.V. Nagender Naidu, L. Iyengar, J. Am. Ceram. Soc., **51**, 461 (1968)
2.11 K.V.K. Rao, S.V. Nagender Naidu, L. Iyengar, Acta Crystallogr., **6**, 136 (1973)
2.12 W.A Wooster, *A Text Book on Crystal Physics*, Cambridge University Press, Cambridge (1949)
2.13 V.T. Deshpande, A.A. Khan, Acta Crystallogr., **16**, 936 (1963)
2.14 V.T. Deshpande, R.R. Pawar, S.V. Suryanarayana, Curr. Sci., **36**, 513 (1967)
2.15 H.D. Megaw, Mater. Res. Bull., **6**, 1007 (1971)
2.16 E. Gruneisen, Ann. Phys., **12**, 257 (1912)

2.17 J.F. Vetelino, K.V. Namjoshi, S.S. Mitra, J. Appl. Phys., **41**, 5141 (1970)
2.18 J.F. Vetelino, K.V. Namjoshi, S.S. Mitra, Phys. Rev. B, **2**, 2167 (1970)
2.19 B. Yates, *Thermal Expansion*, Plenum, New York (1972)
2.20 R.S. Krishnan, R. Srinivasan, S. Devanarayanan, *Thermal Expansion of Crystals*, Thomson, Faridabad (1979)
2.21 H. Fizeau, Ann. Chim. Phys., **4**, 146 (1864); for further references see [2.20]
2.22 H. Fizeau, Ann. Chim. Phys., **8**, 535 (1866)
2.23 F.D. Enck, D.G. Engle, K.I. Marks, J. Appl. Phys., **36**, 389 (1965)
2.24 A.C. Bailey, B. Yates, Proc. Phys. Soc., **91**, 390 (1967)
2.25 A.J. Kirkham, B. Yates, J. Phys. C: Solid State Phys., **1**, 1162 (1968)
2.26 N. Waterhouse, B. Yates, Cryogenics, **8**, 267 (1968)
2.27 A.F. Pojur, B. Yates, J. Phys. E, **6**, 63 (1973)
2.28 D.B. Frazer, A.C.H. Hollis-Hallett, Can. J. Phys., **43**, 193 (1965)
2.29 P.P.M. Meincke, G.M. Graham, Can. J. Phys., **43**, 1853 (1965)
2.30 R.K. Kirby, J. Res. Natl Bur. Stand., **A71**, 363 (1967)
2.31 T.S. Aurora, S.M. Day, V. King, D.O. Pederson, Rev. Sci. Instrum., **55**, 2 (1984)
2.32 D. Bijl, H. Pullan, Physica, **21**, 285 (1955)
2.33 P.W. Dheer, S.L. Surange, Philos. Mag., **8**, 665 (1958)
2.34 N. Madaiah, G.M. Graham, Can. J. Phys., **42**, 221 (1964)
2.35 R.H. Carr, R.D. McCammon, G.K. White, Proc. R. Soc. London, Ser A, **280**, 72 (1964)
2.36 G.K. White, Proc. R. Soc. London, Ser A, **286**, 204 (1965)
2.37 R.V. Jones, J.C.S. Richards, J. Phys. E, **6**, 589 (1973)
2.38 K. Ema, K. Hamano, I. Hatta, J. Phys. Soc. Jpn, **39**, 726 (1975)
2.39 H.P. Klug, L.E. Alexander, *X-ray Diffraction Procedures*, Wiley, New York (1967)
2.40 H.S. Peiser, H.P. Rooksby, A.J.C. Wilson, *X-ray Diffraction by Polycrystalline Materials*, Institute of Physics, London (1955)
2.41 E.F. Kaelble, *Handbook of X-rays*, McGraw-Hill, New York (1967)
2.42 B. Janson, C.A. Sjoblom, Z. Naturforsch., **B25**, 1115 (1970)
2.43 B.K. Rao, Ph.D. Thesis, Kakatiya University, Warangal, India (1980)
2.44 N.H. Cook, E. Rabinowicz, *Physical Measurement and Analysis*, Addison-Wesley, Reading, MA (1963)
2.45 R.K. Kirby, T.A. Hahn, *NBS Certificate of Analysis for Standard Material*, 736 (1969)
2.46 J.E. Rapp, H.D. Merchant, J. Appl. Phys., **44**, 3919 (1973)
2.47 K.K. Srivastava, H.D. Merchant, J. Phys. Chem. Solids, **34**, 2069 (1973)
2.48 R. Srinivasan, J. Indian Inst. Sci., **A37**, 232 (1955)
2.49 P.D. Pathak, N.G. Vasavada, Acta Crystallogr., **A26**, 655 (1970)
2.50 T.H. Nielsen, M. Leipold, J. Am. Ceram. Soc., **46**, 381 (1963)
2.51 S, Jaakkola, J. Poyhonen, K. Simola, Ann. Acad. Sci. Fenn., **AVI**, 15 (1968)

2.52 J.D. Foster, I. Finnie, Rev. Sci. Instrum., **36**, 654 (1968)
2.53 R.S. Shrivastava, D.S. Joshi, Am. Inst. Phys. Conf. Proc., **3**, 44 (1972)
2.54 E. Mantysalo, Phys. Lett., **16**, 17 (1965)
2.55 E. Mantysalo, Ann. Acad. Sci. Fenn., **AVI**, 168 (1965)
2.56 L.O. Heflinger, H. Spetzler, R.F. Wuerker, Rev. Sci. Instrum., **44**, 629 (1973)
2.57 W.D. Drotning, Rev. Sci. Instrum., **50**, 1567 (1979)
2.58 V.T. Deshpande, V.M. Mudholker, Indian J. Phys., **35**, 434 (1961)
2.59 Y.S. Touloukian, R.K. Kirby, R.E. Taylor, T.Y.R. Lee, *Thermophysical Properties of Matter Vol.* 13, *Thermal Expansion of Nonmetallic Solids*, Plenum, New York (1977)
2.60 G.K. White, Philos. Mag., **6**, 1425 (1961)
2.61 S.S. Sharma (see references in [2.20])
2.62 A.J. Falzone, F.D. Stacey, Phys. Chem. Miner., **8**, 212 (1982)
2.63 D. Taylor, Br. Ceram. Trans. J., **83**, 32 (1984)
2.64 K.V.K. Rao, V.T. Deshpande, *Thermal Expansion*, Plenum, New York (1978)
2.65 M. Somi Reddy, S.V. Suryanarayana, J. Mater. Sci. Lett., **2**, 171 (1983)
2.66 M. Somi Reddy, S.V. Suryanarayana, J. Mater. Sci. Lett., **3**, 325 (1984)
2.67 C.E. Guillaume, *Nobel Lectures*, Elsevier, Amsterdam (1920)
2.68 T.H.K. Barron, Philos. Mag., **46**, 720 (1955)
2.69 O.G. Peterson, D.N. Batchelder, R.O. Simmons, Phys. Rev., **150**, 703 (1966)
2.70 D.N. Batchelder, D.L. Losee, R.O. Simmons, Phys. Rev., **162**, 767 (1967)
2.71 D.L. Losee, R.O. Simmons, Phys. Rev., **172**, 944 (1968)
2.72 C.R. Tilford, C.A. Swenson, Phys. Rev. B, **5**, 719 (1972)
2.73 L.G. Van Uitert, H.M. O'Bryan, M.E. Lines, H.J. Guggenheim, G. Zydzik, Mater. Res. Bull., **12**, 261 (1977)
2.74 R.E. Hannemann, H.C. Gatos, J. Appl. Phys., **36**, 1794 (1965)
2.75 J. Askil, Phys. Status Solidi, **11**, k49 (1965)
2.76 R.O. Simmons, R.W. Balluffi, Phys. Rev., **117**, 52 (1960)
2.77 R.O. Simmons, R.W. Balluffi, Phys. Rev., **119**, 600 (1960)
2.78 R.O. Simmons, R.W. Balluffi, Phys. Rev., **125**, 862 (1962)
2.79 R.O. Simmons, R.W. Balluffi, Phys. Rev., **129**, 1533 (1963)
2.80 A.S. Nowick, R. Feder, Phys. Rev. B, **5**, 1238 (1972)
2.81 R. Feder, A.S. Nowick, Phys. Rev. B, **5**, 1244 (1972)
2.82 C. Janot, T. Mallejac, B. George, Phys. Rev. B, **2**, 3088 (1970)
2.83 S.S. Sharma, Proc. Indian Acad. Sci., **A31**, 83 (1950)
2.84 V.T. Deshpande, V.M. Mudholker, Acta Crystallogr., **13**, 483 (1960)
2.85 A.L. Sutton, V.C. Howard, J. Nucl. Mater., **7**, 58 (1962)
2.86 O.D. Slagle, J. Appl. Phys., **38**, 1993 (1967)
2.87 R.J. Price, J.C. Bohras, J. Appl. Phys., **36**, 1965 (1965)
2.88 M.F. Merriam, D.A. Wiegand, R. Smoluchowski, Phys. Rev., **125**, 52 (1962)

2.89 J.A. Stapien-Damm, D. Kucharczyk, J. Urbanowicz, K. Lukaszewicz, Bull. Acad. Polon. Sci., **23**, 985 (1975)
2.90 B.A. Nesterenko, A.D. Borodkin, O.V. Snitko, Sov. Phys. Solid State, **15**, 1731 (1974)
2.91 H.U. Finzel, H. Frank, H. Hoinkes, M. Luschka, H. Nahr, H. Wilsch, U. Wonka, Surf. Sci., **49**, 577 (1975)
2.92 Y. Watanabe, A. Sakai, T. Sakurai, J. Phys. Soc. Jpn, **66**, 649 (1997)
2.93 H.K. Swanson, E. Tatge, *Standard X-ray Diffraction Patterns*, NBS Report, Vol. I (1953)
2.94 V.E. Kenner, R.E. Allen, Phys. Rev. B, **8**, 2916 (1973)
2.95 R. Boehler, G.C. Kennedy, J. Phys. Chem. Solids, **41**, 517 (1980)
2.96 R. Boehler, G.C. Kennedy, J. Phys. Chem. Solids, **41**, 1019 (1980)
2.97 T. Yagi, J. Phys. Chem. Solids, **39**, 563 (1978)
2.98 M. Kumar, Indian J. Phys., **69A**, 517 (1995)
2.99 E. Gruneisen, Handbuch Phys., **10**, 1 (1926) [English translation, NASA Publication, Washington]
2.100 J.K. Roberts, *Heat and Thermodynamics*, Blackie, Glasgow (1960)
2.101 O.L. Anderson, *Thermal Expansion Symposium*, Gaithersburg (1968)
2.102 W. Hume-Rothery, Proc. Phys. Soc., **57**, 209 (1945)
2.103 H.F. Fitschmeister, Acta Crystallogr., **9**, 416 (1956)
2.104 D.N. Talwar, M. Vandevyver, K. Kune, M. Zigone, Phys. Rev. B, **24**, 741 (1981)
2.105 D.B. Sirdeshmukh, L. Sirdeshmukh, K.G. Subhadra, *Alkali Halides – A Handbook of Physical Properties*, Springer, Berlin Heidelberg New York (2001)
2.106 R.W. Munn, J. Phys. C, **5**, 535 (1972)
2.107 M. Blackman, Philos. Mag., **3**, 831 (1958)
2.108 D. Gerlich, J. Phys. Chem. Solids, **35**, 1026 (1974)
2.109 E. Gruneisen, E. Goens, Z. Phys., **29**, 141 (1924)
2.110 R.D. McCammon, G.K. White, Philos. Mag., **11**, 1125 (1965)
2.111 D.P. Riley, J. Phys. Soc., **A57**, 486 (1945)
2.112 D.J. Pastine, Phys. Rev., **148**, 748 (1966)
2.113 V.T. Deshpande, D.B. Sirdeshmukh, Acta Crystallogr., **14**, 353 (1961)
2.114 D.B. Sirdeshmukh, V.T. Deshpande, Curr. Sci., **33**, 428 (1964)
2.115 V.T. Deshpande, D.B. Sirdeshmukh, V.M. Mudholker, RI 6115, US Bureau of Mines (1962)
2.116 D.B. Sirdeshmukh, V.T. Deshpande, Indian J. Pure Appl. Phys., **11**, 366 (1973)
2.117 K.A. Hussain, D.B. Sirdeshmukh, Cryst. Res. Technol., **28**, 1147 (1993)
2.118 D.B. Sirdeshmukh, V.T. Deshpande, Acta Crystallogr., **22**, 438 (1967)
2.119 W. Hussain, Ph.D. Thesis, Kakatiya University, Warangal, India (1989)
2.120 W. Hussain, D.B. Sirdeshmukh, Pramana, **29**, 583 (1987)
2.121 D.B. Sirdeshmukh, B.K. Rao, *Thermal Expansion*, American Institute of Physics, New York (1973)
2.122 B.K. Rao, D.B. Sirdeshmukh, Indian J. Pure Appl. Phys., **18**, 56 (1980)

2.123 B.K. Rao, K.G. Subhadra, D.B. Sirdeshmukh, Indian J. Pure Appl. Phys., **19**, 87 (1981)
2.124 V.T. Deshpande, D.B. Sirdeshmukh, Acta Crystallogr., **14**, 355 (1961)
2.125 D.B. Sirdeshmukh, V.T. Deshpande, Acta Crystallogr., **A26**, 295 (1970)
2.126 D.B. Sirdeshmukh, V.T. Deshpande, Curr. Sci., **36**, 630 (1967)
2.127 D.B. Sirdeshmukh, V.T. Deshpande, Curr. Sci., **41**, 210 (1972)
2.128 D.B. Sirdeshmukh, B.K. Rao, J. Appl. Phys., **44**, 894 (1973)
2.129 D.B. Sirdeshmukh, J. Appl. Phys., **39**, 349 (1968)
2.130 D.B. Sirdeshmukh, Curr. Sci., **37**, 162 (1968)
2.131 D.B. Sirdeshmukh, K.G. Subhadra, J. Appl. Phys., **46**, 3681 (1975)
2.132 R. Srinivasan, Proc. Ind. Acad. Sci., **A41**, 49 (1955)
2.133 B. Krukowska-Fulde, T. Neimisky, J. Cryst. Growth, **1**, 183 (1967)
2.134 L.A. Acuna, M.F. Ortiz, J. Phys. Chem. Solids, **46**, 401 (1985)
2.135 T.S. Aurora, (1984) quoted by M.O. Manasreh, D.O. Pederson, Phys. Rev. B, **30**, 3482 (1985)
2.136 S.S. Sharma, Proc. Indian Acad. Sci., **A31**, 261 (1950)
2.137 V.G. Kuzmin, G.V. Ivanova, O.U. Mazova, K.B.A. Sarachen, Izmerit Tekh., **8**, 46 (1979)
2.138 B. Schumann, H. Neumann, Cryst. Res. Technol., **21**, 1583 (1984)
2.139 D.N. Batchelder, R.O. Simmons, J. Chem. Phys., **41**, 2324 (1964)
2.140 R.B. Foster, J.A. Brewer, S.R. Montgomery, D.O.C. Pederson, J. Acoust. Soc. Am., **73**, 352 (1983)
2.141 H. Neumann, G. Kommichau, W. Schmitz, B. Schumann, J. Mater. Sci. Lett., **5**, 1131 (1986)
2.142 S.S. Sharma, Proc. Indian Acad. Sci., **A34**, 72 (1951)
2.143 R.S.B. Chrystal, Trans. Faraday Soc., **512**, 1811 (1965)
2.144 V.T. Deshpande, Ph.D. Thesis, Osmania University, Hyderabad, India (1955)
2.145 V.T. Deshpande, D.B. Sirdeshmukh, Acta Crystallogr., **15**, 294 (1962)
2.146 V.T. Deshpande, D.B. Sirdeshmukh, Proc. Symp. Solid State Physics, IISc, Bangalore (1960)
2.147 D.B. Sirdeshmukh, Ph.D. Thesis, Osmania University, Hyderabad, India (1963)
2.148 B. Szigeti, Trans. Faraday Soc., **45**, 155 (1949)
2.149 D.B. Sirdeshmukh, J. Chem. Phys., **45**, 2333 (1966)
2.150 D.B. Sirdeshmukh, J. Appl. Phys., **38**, 4083 (1967)

Chapter 3

3.1 P. Debye, Ann. Phys., **43**, 49 (1914)
3.2 I. Waller, Zeit. Phys., **17**, 398 (1923)
3.3 H. Faxen, Ann. Phys., **54**, 615 (1918)
3.4 H. Faxen, Zeit. Phys., **17**, 266 (1923)

3.5 J. Laval, CR Acad. Sci. Paris, **207**, 169 (1938)
3.6 J. Laval, CR Acad. Sci. Paris, **208**, 1512 (1939)
3.7 J. Laval, Bull. Soc. Franc. Miner., **62**, 137 (1939)
3.8 R.W. James, *Optical Principles of the Diffraction of X-rays*, G. Bell, London (1967)
3.9 B.E. Warren, *X-ray Diffraction*, Addison, New York, USA (1969)
3.10 F.H. Herbstein, Philos. Mag. Supp., **10**, 313 (1961)
3.11 L.S. Birks, Naval Research Laboratory, Washington, Report No. H-2517 (1945)
3.12 N.N. Sirota, Acta Crystallogr., **A25**, 223 (1969)
3.13 G.B. Mitra, T. Chattopadhyay, Indian J. Phys., **49**, 218 (1975)
3.14 M. Inagaki, T. Furuhashi, T. Ozeki, B. Mugishima, S. Naka, J. Mater. Sci., **6**, 1520 (1971)
3.15 P. Suortti, J. Appl. Crystallogr., **5**, 325 (1972)
3.16 R.J. Blattner, L.K. Walford, T.O. Baldwin, J. Appl. Phys., **43**, 935 (1972)
3.17 S. Togawa, J. Phys. Soc. Jpn, **20**, 742 (1965)
3.18 D.R. Chipman, A. Paskin, J. Appl. Phys., **30**, 1998 (1959)
3.19 D.R. Chipman, Acta Crystallogr., **A25**, 209 (1969)
3.20 D.T. Cromer, J.T. Waber, Acta Crystallogr., **18**, 104 (1965)
3.21 D.T. Cromer, D. Liberman, J. Chem. Phys., **53**, 1891 (1970)
3.22 T.O. Baldwin, C.W. Tompson, J. Chem. Phys., **41**, 1420 (1964)
3.23 L.K. Walford, Mat. Res. Bull., **4**, 137 (1969)
3.24 S.V.N. Naidu, C.R. Houska, J. Appl. Phys., **42**, 4971 (1971)
3.25 G.C. Benson, E.K. Gill, *Table of Integral Functions Related to Debye–Waller Factor*, National Research Council of Canada, Ottawa (1966)
3.26 B.S. Borie, Acta Crystallogr., **9**, 617 (1956)
3.27 F.H. Herbstein, B.L. Averbach, Acta Crystallogr., **8**, 843 (1955)
3.28 A. Paskin, Acta Crystallogr., **12**, 290 (1959)
3.29 R.W. James, Philos. Mag., **49**, 585 (1925)
3.30 R.W. James, E.M. Furth, Proc. R. Soc. London, Ser A, **117**, 62 (1927)
3.31 I. Waller, R.W. James, Proc. R. Soc. London, Ser A, **117**, 214 (1927)
3.32 R.W. James, G.W. Brindley, Proc. R. Soc. London, Ser A, **121**, 155 (1928)
3.33 R.W. James, G.W. Brindley, R.G. Wood, Proc. R. Soc. London, Ser A, **125**, 401 (1929)
3.34 G.W. Brindley, Philos. Mag., **9**, 193 (1930)
3.35 J.J. Shonka, Phys. Rev., **43**, 947 (1933)
3.36 G.W. Brindley, P. Ridley, Philos. Mag., **23**, 96 (1938)
3.37 H.S. Ribner, E.O. Wollan, Phys. Rev., **53**, 972 (1938)
3.38 G.W. Brindley, P. Ridley, Philos. Mag., **24**, 69 (1939)
3.39 *International Tables for X-ray Crystallography*, Vol. III, Kynoch Press, Birmingham, UK (1968)
3.40 N.M. Butt, J. Bashir, B.T.M. Willis, G. Hegger, Acta Crystallogr., **A44**, 396 (1988)

3.41 N.M. Butt, J. Bashir, M.N. Khan, Acta Crystallogr., **A49**, 171 (1993)
3.42 N. Gopi Krishna, D.B. Sirdeshmukh, Acta Crystallogr., **A54**, 513 (1998)
3.43 D.B. Sirdeshmukh, L. Sirdeshmukh, K.G. Subhadra, *Alkali Halides*, Springer, Berlin Heidelberg New York (2001)
3.44 M. Renninger, Acta Crystallogr., **5**, 711 (1952)
3.45 H. Witte, E. Wolfel, Rev. Mod. Phys., **30**, 51 (1958)
3.46 S. Vihinen, Ann. Acad. Sci. Fenn., **AVI**, 51 (1960)
3.47 H.A. Levy, P. Agron, W.R. Busing, Paper E-7, ACA Meeting, Cambridge, MA (1963)
3.48 S.C. Abrahams, J.L. Bernstein, Acta Crystallogr., **18**, 926 (1965)
3.49 M. Merisalo, T. Paakkari Acta Crystallogr., **23**, 1107 (1967)
3.50 S. Gottlicher, Acta Crystallogr., **B24**, 122 (1968)
3.51 M.V. Linkoaho, Acta Crystallogr., **A25**, 450 (1969)
3.52 P.P.M. Groenewegen, C. Huiszoon, Acta Crystallogr., **A28**, 164 (1972)
3.53 J.L. Lawrence, Acta Crystallogr., **A29**, 94 (1973)
3.54 S. Hosoya, T. Yamagishi, M. Tokonami, J. Phys. Soc. Jpn, **24**, 363 (1968)
3.55 R. Stewart, Acta Crystallogr., **A29**, 602 (1973)
3.56 B. Dawson, C.J. Howard, Abstracts of International Crystal Conference, Melbourne, Australia (1974)
3.57 R. Parthasarathi, Acta Crystallogr., **13**, 802 (1960)
3.58 B.W. Batterman, Phys. Rev., **126**, 1461 (1962)
3.59 B.W. Batterman, Phys. Rev., **127**, 686 (1963)
3.60 S. Annaka, J. Phys. Soc. Jpn, **18**, 1342 (1963)
3.61 B.W. Batterman, H. Strock, Acta Crystallogr., **A25**, 109 (1969)
3.62 G.B. Carpenter, J. Chem. Phys., **32**, 525 (1960)
3.63 B. Post, Acta Crystallogr., **13**, 1103 (1960)
3.64 T.H.K. Barron, M.L. Klein, A.J. Leadbetter, J.A. Morrison, L.S. Salter, Proceedings of the Eighth International Conference Low Temperature Physics, Butterworth, London (1963)
3.65 F.R.L. Schoening, L.A. Vermuelen, Solid State Commun., **7**, 15 (1969)
3.66 G.B. Mitra, A.K. Chaudhari, Indian J. Pure Appl. Phys., **10**, 119 (1972)
3.67 J.M. Blakeley, *Introduction to Properties of Crystal Surfaces*, Pergamon, New York (1973)
3.68 K. Ohshima, J. Harada, J. Phys. C, **17**, 1607 (1984)
3.69 P.A. Flinn, G.M. McManus, J.A. Rayne, Phys. Rev., **123**, 809 (1961)
3.70 K. Alexopoulos, J. Boskovits, S. Morikis, M. Boilos, Acta Crystallogr., **19**, 349 (1965)
3.71 M. Inagaki, T. Furuhashi, T. Ozeki, S. Naka, J. Mater. Sci., **8**, 312 (1973)
3.72 R.E. Gilbert, K. Lonsdale, Acta Crystallogr., **9**, 697 (1956)
3.73 D.B. McWhan, M. Marezio, J.P. Remeiko, P.D. Dernier, Phys. Rev. B, **10**, 490 (1974)

3.74 A. Matsumoro, M. Kobayashi, T. Kikegawa, M. Senoo, J. Appl. Phys., **68**, 2719 (1990)
3.75 T.J. Bastow, S.L. Mair, S.W. Wilkins, J. Appl. Phys., **48**, 494 (1977)
3.76 M.A. Viswamitra, A.R. Jayalaxmi, Acta Crystallogr., **A28**, S189 (1972)
3.77 M.J. Cooper, K.D. Rouse, B.T.M. Willis, Acta Crystallogr., **A24**, 484 (1968)
3.78 S.L. Mair, Z. Barnea, M.J. Cooper, K.D. Rouse, Acta Crystallogr., **A30**, 806 (1974)
3.79 M.H. Dickens, W. Hayes, M.T. Hutchings, C. Smith, J. Phys. C, **12**, L97 (1979)
3.80 E.A. Owens, R.W. Williams, Proc. R. Soc. London, Ser A, **188**, 509 (1947)
3.81 R.H. Wilson, E.F. Skelton, J.L. Katz, Acta Crystallogr., **21**, 635 (1966)
3.82 M.A. Krivoglaz,E.A. Tekhonova, Sov. Phys. Crystallogr., **6**, 399 (1961)
3.83 H. Hahn, W. Ludwig, Z. Phys., **161**, 401 (1961)
3.84 A.A. Maraduddin, P.A. Flinn, Phys. Rev., **129**, 2529 (1963)
3.85 Y. Kashiwase, J. Phys. Soc. Jpn, **20**, 320 (1965)
3.86 B.T.M. Willis, Acta Crystallogr., **A25**, 277 (1969)
3.87 Jyoti Prakash, Sureshchandra, M.P. Hemkar, Indian J. Pure Appl. Phys., **11**, 305 (1973)
3.88 Sureshchandra, Jyoti Prakash, M.P. Hemkar, Indian J. Phys., **46**, 579 (1972)
3.89 Sneh, Indian J. Pure Appl. Phys., **12**, 89 (1974)
3.90 C.K. Shepard, J.G. Mullen, G. Schupp, Phys. Rev. B, **57**, 889 (1998)
3.91 R.K. Gupta, T.M. Haridasan, Acta Crystallogr., **A29**, 579 (1973)
3.92 R.K. Singh, S.P. Sanyal, Phys. Status Solidi B, **114**, 637 (1982)
3.93 A.W. Pryor, Acta Crystallogr., **20**, 138 (1966)
3.94 R.K. Sen, Physica, **30**, 1655 (1964)
3.95 C.J. Howard, R.G. Khadake, Acta Crystallogr., **A30**, 296 (1974)
3.96 A.M. Ratner, Sov. Phys. Crystallogr., **3**, 744 (1959–1960)
3.97 W.J.L. Buyers, T.J. Smith, Phys. Chem. Solids, **25**, 483 (1964)
3.98 J.S. Reid, T. Smith, J. Phys. Chem. Solids, **31**, 2689 (1970)
3.99 H.Y. Gao, L.M. Peng, J.M. Zuo, Acta Crystallogr., **A55**, 1014 (1999)
3.100 J.S. Reid, Acta Crystallogr., **A39**, 1 (1983)
3.101 F. Lindemann, Z. Physik, **11**, 609 (1910)
3.102 M.S. Kushwaha, Nuovo Cimento, **65B**, 285 (1981)
3.103 J.F. Vetelino, S.P. Gaur, S.S. Mitra, Phys. Rev. B, **5**, 2360 (1972)
3.104 H. Brooks, S.C. Yu, Unpublished (1964)
3.105 Y.F. Tsay, B. Gong, S.S. Mitra, J.F. Vetelino, Phys. Rev. B, **6**, 2330 (1972)
3.106 C. Keffer, T.M. Hayes, A. Bienenstock, Phys. Rev. Lett., **21**, 1676 (1968)
3.107 C. Keffer, T.M. Hayes, A. Bienenstock, Phys. Rev. B, **2**, 1966 (1970)
3.108 V. Prakash, Ph.D. Thesis, Harvard University (1967)
3.109 C.N. Koumelis, Acta Crystallogr., **A27**, 246 (1971)

3.110 P.J. Florias, C.N. Koumelis, Acta Crystallogr., **A30**, 506 (1974)
3.111 Y.J. Yosida, Appl. Phys., **78**, 3036 (1995)
3.112 L.W. Barr, A.B. Lidiard, Physical Chemistry, Vol. X, Academic, New York (1970)
3.113 K. Mukherji, Philos. Mag., **12**, 915 (1965)
3.114 N.H. March, Phys. Lett., **20**, 231 (1966)
3.115 H.R. Glyde, J. Phys. Chem. Solids, **28**, 2061 (1967)
3.116 V.K. Tewary, J. Phys., **F3**, 704 (1973)
3.117 P.V. Sastry, B.G. Mulimani, Philos. Mag., **20**, 659 (1969)
3.118 P.D. Pathak, J.M. Trivedi, Proceedings of the Nuclear Physics and Solid State Physics Symposium, Rourkee, India (1969)
3.119 O. Inkinen, M. Jarvinen, Phys. Kond. Mater., **7**, 372 (1968)
3.120 L.K. Patomaki, M.V. Linkoaho, Acta Crystallogr., **A25**, 304 (1969)
3.121 M. Jarvinen, O. Inkinen, Phys. Status Solidi, **21**, 127 (1967)
3.122 M.J. Cooper, K.D. Rouse, Acta Crystallogr., **A27**, 622 (1971)
3.123 J.A. Wasastjerna, Soc. Sci. Fenn. Commun. Phys. Maths., **13**, 1 (1946)
3.124 M.M. Beg, S. Mahmood, N. Ahmad, J. Aslam, Q.H. Khan, N.M. Butt, Phys. Status Solidi B, **106**, K43 (1981)
3.125 P.D. Dernier, W. Weber, L.D. Longinotti, Phys. Rev. B, **14**, 3635 (1976)
3.126 E.N. Baker, T.L. Blundell, J.F. Cutfield, S.M. Cutfield, E.J. Dodson, G.G. Dodson, D.M.C. Hodgkin, R.E. Hubbard, N.W. Isaacs, C.D. Reynolds, K. Sakabe, N. Sakabe, N.M. Vijayan, Philos. Trans. R. Soc. London, Ser B, **319**, 369 (1988)
3.127 V.M.S. Lenin, S. Parthasarathy, M.R.N. Murthy, Curr. Sci., **78**, 1098 (2000)
3.128 N. Gopi Krishna, D.B. Sirdeshmukh, Indian J. Pure Appl. Phys., **25**, 503 (1987)
3.129 N. Gopi Krishna, D.B. Sirdeshmukh, B. Rama Rao, B.J. Beaudry, K.A. Gschneidner Jr., Phys. Status Solidi A, **89**, k37 (1985)
3.130 N. Gopi Krishna, Ph.D. Thesis, Kakatiya University, Warangal, India (1985)
3.131 N. Gopi Krishna, D.B. Sirdeshmukh, B. Rama Rao, B.J. Beaudry, K.A. Gschneidner Jr., Indian J. Pure Appl. Phys., **24**, 324 (1986)
3.132 N. Gopi Krishna, D.B. Sirdeshmukh, K.A. Gschneidner Jr., Indian J. Pure Appl. Phys., **26**, 724 (1988)
3.133 M. Shankar Narayana, N. Gopi Krishna, D.B. Sirdeshmukh, Acta Crystallogr., **A57**, 218 (2000)
3.134 P. Geeta Krishna, Ph.D. Thesis, Kakatiya University, Warangal, India (1997)
3.135 K. Srinivas, D.B. Sirdeshmukh, Pramana-J. Phys., **31**, 221 (1988)
3.136 K.Srinivas, D.B. Sirdeshmukh, Pramana-J. Phys., **25**, 71 (1985)
3.137 P. Geeta Krishna, K.G. Subhadra, T. Kumara Swamy, D.B. Sirdeshmukh, Pramana-J. Phys., **28**, 81 (1999)
3.138 K. Srinivas, M. Ateequddin, D.B. Sirdeshmukh, Pramana-J. Phys., **28**, 81 (1987)

3.139 T. Kumara Swamy, K.G. Subhadra, D.B. Sirdeshmukh, Pramana-J. Phys., **43**, 33 (1994)
3.140 K. Srinivas, D.B. Sirdeshmukh, Pramana, **23**, 595 (1984)
3.141 K.G. Subhadra, D.B. Sirdeshmukh, Pramana, **10**, 357 (1978)
3.142 K.G. Subhadra, D.B. Sirdeshmukh, Pramana, **9**, 223 (1977)
3.143 K.G. Subhadra, B. Raghavendra Rao, D.B. Sirdeshmukh, Pramana-J. Phys., **38**, 681 (1992)
3.144 K.G. Subhadra, D.B. Sirdeshmukh, Natl Acad. Sci. Lett., **3**, 126 (1980)
3.145 B. Nagaiah, Ph.D. Thesis, Kakatiya University, Warangal, India (1983)
3.146 M.J.M. Rao, Ph.D. Thesis, Kakatiya University, Warangal, India (1977)
3.147 E. Balaiah, Ph.D. Thesis, Kakatiya University, Warangal, India (2003)
3.148 K.G. Subhadra, D.B. Sirdeshmukh, Pramana, **10**, 597 (1978)
3.149 K. Srinivas, D.B. Sirdeshmukh, Cryst. Res. Technol., **21**, 157 (1986)
3.150 K. Srinivas, D.B.Sirdeshmukh, Cryst. Res. Technol., **21**, 165 (1986)
3.151 K.G. Subhadra, D.B. Sirdeshmukh, Curr. Sci. **46**, 845 (1977)
3.152 K.G. Subhadra, N. Satyanarayana, E. Balaiah, B. Raghavendra Rao, D.B. Sirdeshmukh, Proc. AP Akad. Sci., **5**, 173 (2001)
3.153 W. Menz, Z. Phys., **211**, 183 (1968)
3.154 C. Huiszoon, P.P.M. Groenewegen, Acta Crystallogr., **A28**, 170 (1972)
3.155 U. Korhonen, M. Linkoaho, Ann. Acad. Sci. Fenn., **AVI**, No. 195 (1966)
3.156 L.K. Walford, J.A. Schoeffel, Philos. Mag., **21**, 375 (1970)
3.157 W.C. Hollenberg, B.W. Batterman, Phys. Rev. B, **10**, 2148 (1974)
3.158 R.G. Kulkarni, G.K. Bichile, Acta Crystallogr., **A31**, 619 (1975)
3.159 S. Mahmood, N.M. Butt, N. Ahmed, M.M. Beg, Q.H. Khan, J. Aslam, Acta Crystallogr., **A36**, 147 (1980)
3.160 J.L. Feldman, Acta Crystallogr., **A30**, 301 (1974)
3.161 C. Scheringer, Acta Crystallogr., **A29**, 82 (1973)
3.162 P.L. Sanger, Acta Crystallogr., **A25**, 694 (1969)
3.163 Y. Noda, S. Ohba, S. Sato, Y. Saito, Acta Crystallogr., **B39**, 312 (1983)
3.164 V. Meisalo, O. Inkinen, Ann. Acad. Sci. Fenn., **A VI**, 243 (1967)
3.165 B. Nagaiah, D.B. Sirdeshmukh, Curr. Sci., **53**, 26 (1984)
3.166 M. Gnanasoundari, K. Ramachandran, Curr. Sci., **55**, 459 (1986)
3.167 C. Vidal-Valat, J.P. Vidal, K. Kurki-Suonio, Acta Crystallogr., **A34**, 584 (1978)
3.168 M. Linkoaho, E. Rantavouri, M. Suvanen, Phys. Fenn., **9**, 167 (1974)
3.169 R.K. Gupta, Phys. Rev. B, **12**, 4452 (1975)
3.170 J. Govindarajan, Acta Crystallogr., **A29**, 576 (1973)
3.171 B.S. Agrawal, J.P. Beaver, J.W. Weymouth, J.R. Hardy, Acta Crystallogr., **A31**, 249
3.172 J.L. Feldman, Phys. Rev. B, **12**, 813 (1975)
3.173 W.A. Wooster, *A Text-Book of Crystal Physics*, Cambridge University Press, Cambridge (1938)
3.174 W. Boas, J.K. Mackenzie, Prog. Met. Phys., **2**, 90 (1950)
3.175 R.W.G. Wyckoff, *Crystal Structures*, Vol. 1, Interscience, New York, (1960)

3.176 Y. Watanabe, H. Iwasaki, S. Ogawa, Jap. J. Appl. Phys., **10**, 786 (1971)
3.177 N. Gopi Krishna, D.B. Sirdeshmukh, Indian J. Pure Appl. Phys., **31**, 198 (1993)
3.178 D.B. Sirdeshmukh, K.G. Subhadra, K.A. Hussain, N. Gopi Krishna, B. Raghavendra Rao, Cryst. Res. Technol., **28**, 15 (1993)
3.179 N.N. Sirota, V.D. Yanovich, *Chemical Bonds in Semiconductors and Solids*, Consultants Bureau, New York (1967)
3.180 D. De Noble, Phillips Tech. Rep., **14**, 489 (1959)
3.181 J.C. Phillips, Rev. Mod. Phys., **42**, 317 (1970)
3.182 D.B. Sirdeshmukh, B.S. Shah, J. Res. M.S. University, Baroda, India, **14**, 49 (1965)
3.183 E. Burstein, P.H. Egli, Adv. Electron. Electron. Phys., **7**, 56 (1955)
3.184 A.F. Wells, *Structural Inorganic Chemistry*, Oxford University Press, London (1950)
3.185 C.R. Houska, Technical Report No. C-17, Union Carbide Research Institute (1963)
3.186 M. Blackman, *Handbuch der Physik*, VII/I, Springer, Berlin Heidelberg New York (1955)
3.187 K.A. Gschneidner Jr., Solid State Phys., **16**, 276 (1964)
3.188 L.S. Salter, Adv. Phys., **14**, 1 (1965)
3.189 C. Zener, S. Bilinsky, Phys. Rev., **50**, 101 (1936)
3.190 T.H.K. Barron, A.J. Leadbetter, J.A. Morrison, L.S. Salter, Acta Crystallogr., **20**, 125 (1966)
3.191 R.D. Horning, J.L. Staudenmann, Acta Crystallogr., **A44**, 136 (1988)
3.192 Z. Barnea, B. Post, Acta Crystallogr., **21**, 181 (1966)
3.193 C.S.N. Murthy, Y.V.G.S. Murti, Proceedings of Nuclear Physics and Solid State Physics Symposium, Madurai, India (1970)
3.194 F.A. Kroger, J. Chem. Phys., **51**, 4025 (1969)
3.195 A.D. Franklin, J. Phys. Chem. Solids, **29**, 823 (1968)
3.196 R.W. Ure, J. Chem. Phys., **26**, 1363 (1957)
3.197 C.R.A. Catlow, M.J. Norgett, J. Phys. C, **6**, 1325 (1973)
3.198 S. Mizuta, K. Shirasawa, H. Yanagida, Denki-Kagaku, **41**(12), 913 (1973)
3.199 B.K. Rao, Ph.D. Thesis, Kakatiya University, Warangal, India (1980)
3.200 F. Bailly, Phys. Status Solidi, **25**, 317 (1968)
3.201 M.P. Dariel, *Handbook on the Physics and Chemistry of Rare Earths*, Vol. 1, Editors K.A. Gschneidner Jr., L. Eyring, North Holland, Amsterdam (1978)

Chapter 4

4.1 D. Tabor, *Hardness of Metals*, Clarendon, Oxford (1951)
4.2 N.A. Ashby, NZ Eng., **6**, 33 (1951)

4.3 J.H. Westbrook, H. Conrad, in *The Science of Hardness Testing and its Research Applications*, American Society of Metals, Ohio (1973)
4.4 J.J. Gilman, *The Science of Hardness Testing and its Research Applications*, American Society of Metals, Ohio (1973)
4.5 B.W. Mott, *Microhardness Indentation Testing*, Butterworths, London (1956)
4.6 M. Braunovic, *The Science of Hardness Testing and its Research Applications*, American Society of Metals, Ohio (1973)
4.7 M.C. Shaw, *The Science of Hardness Testing and its Research Applications*, American Society of Metals, Ohio (1973)
4.8 W.C. Oliver, G.M. Pharr, J. Mater. Res., **7**, 1564 (1992)
4.9 K. Kishan Rao, D.B. Sirdeshmukh, Pramana, **23**, 251 (1984)
4.10 J.R. Hopkins, J.A. Miller, J.J. Martin, Phys. Status Solidi A, **19**, 591(1973)
4.11 K. Inabe, K. Emoto, K. Sakamaki, N. Tokenchi, Jpn. J. Appl. Phys., **11**, 1743 (1972)
4.12 J.J. Gilman, J. Appl. Phys., **46**, 5110 (1975)
4.13 S.M. Hu, J. Appl. Phys., **46**, 1470 (1975)
4.14 I.G. Berzina, I.B. Berman, P.A. Savintsev, Sov. Phys. Crystallogr., **9**, 483 (1965)
4.15 U.V. Subba Rao, V. Hari Babu, Pramana, **11**, 149 (1978)
4.16 G.Y. Chin, L.G. Van Uitert, M.L. Green, G.J. Zydzik, Scr. Met., **6**, 475 (1972)
4.17 J.N. Plendl, P.J. Gielisse, Phys. Rev., **125**, 828 (1962)
4.18 D.B. Sirdeshmukh, K.G. Subhadra, K. Kishan Rao, T. Thirmal Rao, Cryst. Res. Technol., **30**, 861 (1995)
4.19 D.B. Sirdeshmukh, L. Sirdeshmukh, K.G. Subhadra, *Alkali Halides*, Springer, Berlin Heidelberg New York (2001)
4.20 R.F. King, D Tabor, Proc. R. Soc. London, Ser A, **223**, 225 (1954)
4.21 M.J. Joshi, B.S. Shah, Cryst. Res. Technol., **19**, 1107 (1984)
4.22 G. Constantinidis, R.D. Tomlinson, Philos. Mag. Lett., **57**, 91 (1988)
4.23 C. Ascheron, C. Haase, G. Kuhn, H. Neumann, Cryst. Res. Technol., **24**, K33 (1989)
4.24 C.C. Desai, V. John, Kristall Technik, **14**, 439 (1979)
4.25 M.J. Joshi, A.V. Antony, P. Mohan Rao, Kristall Technik, **15**, 743 (1980)
4.26 B.S. Shah, Cryst. Res. Technol., **17**, K27 (1982)
4.27 R. K. Marwaha, B.S. Shah, Cryst. Res. Technol., **23**, K63 (1988)
4.28 R. Bajpai, S.C. Datt, Indian J. Pure Appl. Phys., **24**, 254 (1986)
4.29 R. Bajpai, J.M. Kelkar, S.C. Datt, Proceedings of Symposium, IIT, Kanpur (1986)
4.30 P. Parashar, R., Bajpai, J.M. Kelkar, S.C. Datt, Makromol. Chem. Symp. (1988)
4.31 W. Rostoker, J. Inst. Met., **77**, 175 (1950)
4.32 D.B. Sirdeshmukh, B.S. Shah, J. M.S. Univ., Baroda, **14**, 3 (1965)

4.33 B. Vengatesan, N. Kannaiah, P. Ramasami, III National Seminar on Crystal Growth, Madras (1987)
4.34 K.J. Pratap, V. Hari Babu, Bull. Mater. Sci., **2**, 43 (1980)
4.35 P.N. Kotru, II National Seminar on Crystal Growth, Madras (1984)
4.36 C. Hayes, E.G. Kendall, Metallography, **6**, 275 (1973)
4.37 U.V. Subba Rao, V. Hari Babu, Indian J. Phys., **54A**, 147 (1980)
4.38 U.C. Shrivastava, J. Appl. Phys., **51**, 1510 (1980)
4.39 J.N. Plendl, Appl. Opt., **10**, 1134 (1971)
4.40 J.S. Dryden, S. Morimoto, J.S. Cook, Philos. Mag., **10**, 379 (1965)
4.41 G.Y. Chin, L.G. Van Uitert, M.L. Green, G.J. Zydzik, T.Y. Kometani, J. Am. Ceram. Soc., **56**, 369 (1973)
4.42 J.J. Gilman, J. Appl. Phys., **45**, 508 (1974)
4.43 C.C. Chen, A.A. Hendrickson, *The Science of Hardness Testing and its Research Applications*, American Society of Metals, Ohio (1973)
4.44 E. Aerts, S. Amelinckx, W. Deckeyser, Acta Metall., **7**, 29 (1959)
4.45 J.S. Nadeau, J. Appl. Phys., **33**, 3480 (1962)
4.46 J.S. Nadeau, J. Appl. Phys., **35**, 1248 (1964)
4.47 Y.S. Boyarskaya, R.P. Zhitaru, Phys. Status Solidi, **42**, k29 (1970)
4.48 P. Veeresham, U.V. Subba Rao, V. Hari Babu, Indian J. Pure Appl. Phys., **17**, 488 (1979)
4.49 U.V. Subba Rao, P. Veeresham, V. Hari Babu, Cryst. Res. Technol., **18**, 1423 (1983)
4.50 A.P. Gerk, J. Mater. Sci., **12**, 735 (1977)
4.51 R. Krishnamoorthy, T.S. Murthy, Indian J. Pure Appl. Phys., **19**, 1021 (1981)
4.52 M.N. Ansary, R. Pandey, P. Parashar, S.C. Datt, Indian J. Pure Appl. Phys., **23**, 103 (1985)
4.53 R.C. Rau, K. Lacefield, Phys. Status Solidi, **21**, k69 (1967)
4.54 A.R. Patel, C.C. Desai, Indian J. Pure Appl. Phys., **7**, 562 (1969)
4.55 A.R. Patel, K.S. Raju, J. Phys. Chem. Solids, **31**, 331 (1970)
4.56 C. Ascheron, H. Neumann, G. Dlubek, R. Krause, J. Mater. Sci. Lett., **5**, 891 (1986)
4.57 S. Mukherjee, Curr. Sci., **83**, 263 (2002)
4.58 L.I. Evanov, A.I. Dedurin, I.V. Borovitskoya, O.N. Krokhni, V.Y. Nikolin, S.N. Polukhin, A.A. Tikhonirov, A.S. Fedorov, Pramana, **61**, 1179 (2003)
4.59 A.S. Powarjonnych, Geologie, 93 (1959)
4.60 A. Julg, Phys. Chem. Minerals, **3**, 45 (1978)
4.61 K.S. Pillay, Indian J. Pure Appl. Phys., **20**, 46 (1982)
4.62 J.J. Gilman, J. Appl. Phys., **44**, 982 (1973)
4.63 G.Y. Chin, *Deformation of Ceramic Materials*, Plenum, New York (1975)
4.64 V.I. Barbashov, Y.B. Tkachenko, Sov. Phys. Solid State, **33**, 1564 (1991)
4.65 A.G. Atkins, D. Tabor, Proc. R. Soc. London, Ser A, **292**, 441 (1966)

4.66 H.D. Merchant, G.S. Murty, S.N. Bahadur, L.T. Dwivedi, J. Mater. Sci., **8**, 437 (1973)
4.67 G.C. Jain, A.R. Patel, Indian J. Phys., **56A**, 76 (1982)
4.68 Y.S. Boyarskaya, D.Z. Grabko, D.S. Pishkova, Phys. Status Solidi A, **87**, 175 (1985)
4.69 W.A. Wooster, Rep. Progr. Phys., **16**, 62 (1953)
4.70 C.A. Brookes, R.P. Burnand, *The Science of Hardness Testing and its Research Applications*, American Society of Metals, Ohio (1973)
4.71 D.I. Goland, G. Meyer, *The Science of Hardness Testing and its Research Applications*, American Society of Metals, Ohio (1973)
4.72 C.A. Brookes, J.B. O'Neill, B.A.W. Redfern, Proc. R. Soc. London, Ser A, **322**, 73 (1971)
4.73 R.H.J. Hannink, D.L. Kohlstadt, M.J. Murray, Proc. R. Soc. London, Ser A, **326**, 409 (1972)
4.74 G.Y. Chin, L.G. Van Uitert, M.L. Green, G. Zydzik, Scr. Met., **6**, 503 (1972)
4.75 G.Y. Chin, M.L. Green, L.G. Van Uitert, W.A. Hargreaves, J. Mater. Sci., **8**, 1421 (1973)
4.76 E.L. Offenbacher, I.C. Roselman, Nature (Phys. Sci.), **234**, 112 (1971)
4.77 H.G. Gallaghar, A. Littlejohn, D.B. Sheen, J.N. Sherwood, Proceedings of the V Europhysical Conference on Defects, Madrid, Spain (1986)
4.78 Z. Morlin, Phys. Status Solidi, **8**, k77 (1965)
4.79 G.P. Upit, S.A. Varchenya, Phys. Status Solidi, **17**, 831 (1966)
4.80 N. Gane, Proc. R. Soc. London, Ser A, **317**, 367 (1970)
4.81 G.P. Upit, S.A. Varchenya, *The Science of Hardness Testing and its Research Applications*, American Society of Metals, Ohio (1973)
4.82 K.B. Yoder, D.S. Stone, R.A. Hoffman, J.C. Lin, J. Mater. Res., **13**, 3214 (1998)
4.83 K.S. Rajam, H.C. Barshilia, Private communication (2005)
4.84 H.C. Barshilia, K.S. Rajam, Bull. Mater. Sci., **27**, 35 (2004)
4.85 A.E. Smirnov, A.A. Urusovskaya, Sov. Phys. Solid State, **29**, 485 (1987)
4.86 V.U. Golovin, R.B. Moryunov, D.Y. Lopatin, A.A. Baskorov, Phys. Status Solidi A, **160**, R3 (1997)
4.87 A. Sasaki, M. Iwata, Phys. Status Solidi, **85**, k105 (1984)
4.88 A. Kailer, Y.G. Gogotsi, K.G. Nickel, Mater. Res. Soc. Symp. Proc., **499** (1998)
4.89 A. Kailer, Y.G. Gogotsi, K.G. Nickel, Proc. Fifth Eur. Conf. Adv. Mater. Process Appl., **4**, 171 (1997)
4.90 T. Thirmal Rao, Ph.D. Thesis, Kakatiya University, Warangal, India (1991)
4.91 E.H. Nickel, A.H. Webster, L.G. Ripley, Can. Miner., **10**, 773 (1971)
4.92 S.N. Demisroy, Jb. Min. Mh., 323 (1969)
4.93 T. Thirmal Rao, D.B. Sirdeshmukh, Cryst. Res. Technol., **29**, k112 (1994)
4.94 R.F. Flint, B.J. Skinner, *Physical Geology*, Wiley, New York (1974)

4.95 W.J. Phillips, N. Phillips, *An Introduction to Mineralogy for Geologists*, Wiley, New York (1980)
4.96 B.H.T. Chai, *CRC Handbook of Laser Science and Technology*, CRC, Boca Raton (1995)
4.97 H. Brown, A.A. Ballman, G.Y. Chin, J. Mater. Sci., **10**, 1157 (1975)
4.98 G. Dhanraj, H.L. Bhat, P.S. Narayanan, Ferroelectrics, **157**, 7 (1994)
4.99 K.G. Subhadra, K. Kishan Rao, D.B. Sirdeshmukh, Bull. Mater. Sci., **23**, 147 (2000)
4.100 T. Thirmal Rao, D.B. Sirdeshmukh, Cryst. Res. Technol., **26**, k53 (1991)
4.101 J.J. Gilman, Aust. J. Phys., **13**, 327 (1960)
4.102 D.B Sirdeshmukh, J. Phys. Chem. Solids, **27**, 1557 (1966)
4.103 K. Kishan Rao, D.B. Sirdeshmukh, Bull. Mater. Sci., **5**, 449 (1983)
4.104 R. Srinivasan, Proc. Indian Acad. Sci., **A41**, 49 (1955)
4.105 M.F.C. Ladd, W.H. Lee, Prog. Solid State Chem., **1**, 37 (1964)
4.106 B.K. Rao, Ph.D. Thesis, Kakatiya University, Warangal, India (1980)
4.107 D.B. Sirdeshmukh, L. Sirdeshmukh, K.G. Subhadra, K. Kishan Rao, S. Bal Laxman, Bull. Mater. Sci., **24**, 469 (2001)
4.108 A.M. Hofmeister, K.R. Campbell, J. Appl. Phys., **72**, 638 (1992)
4.109 R. Brout, Phys. Rev., **113**, 43 (1959)
4.110 G.Y. Chin, Trans. Am. Cryst. Assoc., **11**, 1 (1975)
4.111 F. Euler, J.A. Bruce, Acta Crystallogr., **19**, 971 (1965)
4.112 K. Kishan Rao, D.B. Sirdeshmukh, Indian J. Pure Appl. Phys., **16**, 860 (1978)
4.113 S. Haussuhl, Z. Kristallogr., **120**, 401 (1964)
4.114 A.A. Khan, W.H. Baur, Acta Crystallogr., **B29**, 272 (1973)
4.115 R.S. Adhav, J. Appl. Phys., **46**, 2828 (1975)
4.116 I.D. Brown, R.D. Shannon. Acta Crystallogr., **B29**, 266 (1973)
4.117 P. Hartman, Acta Crystallogr., **9**, 721 (1956)
4.118 A.J. Dekker, *Solid State Physics*, Macmillan, London (1981)
4.119 S. Radhakrishna, A.M. Karguppiker, J. Phys. Soc. Jpn, **35**, 578 (1973)
4.120 C.Ramasastry, Y.V.G.S. Murty, Indian J. Pure Appl. Phys., **2**, 35 (1964)
4.121 C. Ramasastry, Y.V.G.S. Murty, Bull. Natl Inst. Sci. India, **30**, 36 (1965)
4.122 K. Kishan Rao, D.B. Sirdeshmukh, Phys. Status Solidi A, **80**, k117 (1983)
4.123 M. Thomas, H. Von Seggeren, A.Winnacker, J. Appl. Phys., **76**, 1800 (1994)
4.124 D.B. Sirdeshmukh, T. Kumara Swamy, P.G. Krishna, K.G. Subhadra, Bull. Mater. Sci., **26**, 261 (2003)
4.125 D.B. Sirdeshmukh, P.G. Krishna, K.G. Subhadra, J. Mater. Sci., **38**, 2001 (2003)
4.126 N. Gopi Krishna, Ph.D. Thesis, Kakatiya University, Warangal, India (1984)

4.127 T.E. Scott, *Handbook of the Physics and Chemistry of Rare Earths*, Vol. I, North Holland, Amsterdam (1978)
4.128 K. Kishan Rao, D.B. Sirdeshmukh, Pramana, **24**, 887 (1985)
4.129 K. Kishan Rao, D.B. Sirdeshmukh, Pramana, **34**, 151 (1990)
4.130 K. Ito, Science Papers, Tohoku University, Sendai, **12**, 137 (1923)
4.131 V.P. Shishokin, Z. Anorg. Chem., **189**, 263 (1930)
4.132 D.B. Sirdeshmukh, K.G. Subhadra, R. Bauri, Proc. AP Akad. Sci., **7**, 203 (2003)
4.133 R.W.G. Wyckoff, *Crystal Structures*, Vol. 1, Interscience, New York, (1965)
4.134 J.D. Axe, Phys. Rev. A, **139**, 1215 (1969)
4.135 S.S. Mitra, R. Marshall, J. Chem. Phys., **41**, 3158 (1964)
4.136 G.C. Benson, J. Chem. Phys., **35**, 2113 (1961)
4.137 G.C. Benson, E. Dempsey, Proc. R. Soc. London, Ser A, **266**, 344 (1962)
4.138 T. Thirmal Rao, K. Kishan Rao, D.B. Sirdeshmukh, Cryst. Res. Technol., **26**, k189 (1991)
4.139 K. Kishan Rao, D.B. Sirdeshmukh, Cryst. Res. Technol., **18**, 1125 (1983)
4.140 D.B. Sirdeshmukh, K. Kishan Rao, J. Mater. Sci. Lett., **7**, 567 (1988)
4.141 Y.S., Boyarskaya, D.Z. Grabko, E.I. Purich, J. Mater. Sci., **14**, 737 (1979)

Chapter 5

5.1 C.P. Smyth, *Dielectric Behaviour and Structure*, McGraw-Hill, New York (1955)
5.2 A.R. Von Hippel, *Dielectric Materials and Applications*, Wiley, New York (1961)
5.3 V.V. Daniel, *Dielectric Relaxation*, Academic, London (1967)
5.4 N.E. Hill, W.E. Vaugan, A.H. Price, M. Davies, *Dielectric Properties and Molecular Behaviour*, Von Nostrand and Reinhold, London (1969)
5.5 F.C. Brown, *The Physics of Solids*, Benjamin, New York (1967)
5.6 K.V. Rao, A. Smakula, J. Appl. Phys., **37**, 319 (1966)
5.7 A.J. Dekker, *Solid State Physics*, Prentice-Hall, New York (1957)
5.8 A.B. Lidiard, *Handbuch der Physik*, Vol. 20, Springer, Berlin Heidelberg New York (1957)
5.9 J.C. Anderson, *Dielectrics*, Reinhold, New York (1964)
5.10 D.A.G. Bruggeman, Ann. Phys. Lpz., **24**, 636 (1935)
5.11 C.J.F. Bottcher, *Theory of Electric Polarization*, Elsevier, New York (1952)
5.12 H. Looyenga, Physica, **31**, 401 (1965)
5.13 D.C. Dube, R. Prasad, J. Phys. D, **3**, 677 (1970)
5.14 S. Chandra, Can. J. Phys., **47**, 969 (1969)

5.15 A. Kahan, J.W. Goodram, R.S. Singh, S.S. Mitra, J. Appl. Phys., **42**, 4444 (1971)
5.16 J.N. Plendl, L.C. Mansur, S.S. Mitra, I.F. Chang, Solid State Commun., **7**, 109 (1969)
5.17 H. Finkenrath, N. Uhle, Solid State Commun., **5**, 875 (1967)
5.18 M.H. Brodsky, G. Lucovsky, M.F. Chen, T.S. Plaskett, Phys. Rev. B, **2**, 3303 (1970)
5.19 P. Denham, G.R. Field, P.L.R. Morse, G.R. Wilkinson, Proc. R. Soc. London, Ser A, **317**, 55 (1970)
5.20 K.S. Cole, R.H. Cole, J. Chem. Phys., **9**, 341 (1941)
5.21 K. Hojendahl, Z. Phys. Chem. B, **20**, 54 (1933)
5.22 S. Haussuhl, Z. Naturforsch., **12A**, 445 (1961)
5.23 A.J. Bosman, E.E. Havinga, Phys. Rev., **129**, 1593 (1963)
5.24 B.W. Jones, Philos. Mag., **16**, 1085 (1967)
5.25 K. Kamiyoshi, Y. Nigara, Phys. Status Solidi A, **3**, 735 (1970)
5.26 J. Fontanella, C. Andeen, D.E. Schuele, Phys. Rev. B, **6**, 582 (1972)
5.27 J.H. Fertel, C.H. Perry, Phys. Rev., **184**, 874 (1969)
5.28 C. Kittel, *Introduction to Solid State Physics*, Wiley-Eastern, New Delhi (1990)
5.29 B. Tareev, *Physics of Dielectric Materials*, Mir, Moscow (1979)
5.30 I.S. Zheludev, *Physics of Crystalline Dielectrics*, Vol. 2, Plenum, New York (1971)
5.31 R.S. Krishnan, *Progress in Crystal Physics*, S. Viswanathan, Madras (1958)
5.32 K.F. Young, H.P.R. Frederikse, Phys. Chem. Ref. Data, **2**, 313 (1973)
5.33 J.R. Stefan, *Digest of Literature on Dielectrics*, National Academy of Sciences, Washington (1979)
5.34 N.A.G. Ahmed, M. Norman, C.W. Smith, Collect. Phenom., **3**, 67 (1980)
5.35 R. Pethig, Annu. Rev. Phys. Chem., **43**, 177 (1992)
5.36 R.P. Lowndes, D.H. Martin, Proc. R. Soc. London, Ser A, **308**, 473 (1969)
5.37 R.P. Lowndes, D.H. Martin, Proc. R. Soc. London, Ser A, **316**, 351 (1970)
5.38 K.V. Rao, Phys. Status Solidi, **30**, 391 (1975)
5.39 S. Chandra, J.Prakash, Can. J. Phys., **50**, 1053 (1972)
5.40 G.A. Samara, Phys. Rev., **165**, 959 (1967)
5.41 G.A. Samara, Proc. Nato Adv. Study Inst. (1970)
5.42 D.B. Sirdeshmukh, L. Sirdeshmukh, K.G. Subhadra, *Alkali Halides – A Handbook of Physical Properties*, Springer, Berlin Heidelberg New York (2001)
5.43 G.A. Samara, Phys. Rev., **165**, 959 (1968)
5.44 K.V. Rao, A. Smakula, J. Appl. Phys., **36**, 2034 (1965)
5.45 K.V. Rao, K.S. Rao, Philos. Mag., **23**, 1053 (1971)
5.46 C. Andeen, J. Fontanella, D.E. Schuele, Phys. Rev. B, **2**, 5068 (1970)

5.47 C. Andeen, J. Fontanella, D.E. Schuele, J. Appl. Phys., **42**, 2216 (1971)
5.48 C. Andeen, D.E. Schuele, J. Fontanella, Phys. Rev. B, **6**, 591 (1972)
5.49 K.F. Young, H.P.R. Frederikse, J. Appl. Phys., **40**, 3115 (1969)
5.50 G.A. Samara, Phys. Rev. B, **13**, 4529 (1976)
5.51 W. Hayes, *Crystals with Fluorite Structures*, Clarendon, Oxford (1974)
5.52 S. Haussuhl, Z. Kristallogr., **116**, 371 (1961)
5.53 A.H. Rama Rao, M.R. Srinivasan, H.L. Bhat, P.S. Narayanan, Ferroelectrics, **21**, 433 (1978)
5.54 Y.A. Badr, R. Kamel, J. Phys. Chem. Solids, **41**, 1127 (1980)
5.55 Y.A. Badr, R. Kamel, Phys. Status Solidi A, **53**, k161 (1979)
5.56 F. El-Kabbany, G. Said, S. Mahrous, N.H. Taher, Phys. Status Solidi A, **99**, 105 (1987)
5.57 Kishan Lal, H.K. Jahns, J. Phys. C, **10**, 1315 (1977)
5.58 H.B. Lal, B.K. Verma, V. Ram Yadav, J. Mater. Sci., **17**, 3317 (1982)
5.59 Kh.S. Bagdasarov, A.I. Munchaev, M.V. Remizov, Sov. Phys. Crystallogr., **36**, 263 (1991)
5.60 S.R. Rotman, H.L. Tuller, C. Warde, J. Appl. Phys., **71**, 1209 (1992)
5.61 R.D. Shannon, M.A. Subramanian, G.R. Rossman, J. Appl. Phys., **67**, 3798 (1990)
5.62 R. Shelby, J. Fontanella, C. Andeen, J. Phys. Chem. Solids, **41**, 69 (1980)
5.63 G.A. Samara, P.S. Peercy, Phys. Rev. B, **7**, 1131 (1973)
5.64 P.H. Fang, W.S. Brower Jr., Phys. Rev., **129**, 1561 (1963)
5.65 W.S. Brower Jr., P.H. Fang, Phys. Rev., **149**, 646 (1966)
5.66 W.S. Brower Jr., P.H. Fang, J. Appl. Phys., **40**, 4988 (1969)
5.67 H.B. Lal, N. Dar, Ashok Kumar, J. Phys. C, **7**, 4335 (1974)
5.68 H.B. Lal, B.K. Verma, N. Dar, Indian J. Cryogen., **1**, 119 (1976)
5.69 A.E. Dugan, H.K. Henisch, J. Phys. Chem. Solids, **28**, 971 (1967)
5.70 A.M. Fernandez, O.N. Srivastava, Solid State Commun., **34**, 137 (1980)
5.71 A. Mansingh, S.S. Bawa, J. Phys. Soc. Jpn, **35**, 1136 (1973)
5.72 K.S. Rao, A. Bhanumati, Curr. Sci., **47**, 207 (1978)
5.73 A. K. Goswami, J. Appl. Phys., **40**, 619 (1969)
5.74 B. Sing, C.W. Smith, R. Hughes, Med. Biol. Eng. Comput., **17**, 45 (1979)
5.75 R. Agarwal, V.R. Singh, Indian J. Pure Appl. Phys., **29**, 524 (1991)
5.76 V. Saraswati, Bull. Mater. Sci., **18**, 831 (1995)
5.77 R. Pethig, *Dielectric and Related Molecular Processes*, Vol. 3, The Chemical Society, London (1977)
5.78 J. Yamashita, Prog. Theor. Phys., **8**, 280 (1952)
5.79 J. Yamashita, J. Prog. Theor. Phys., **12**, 454 (1954)
5.80 P.T. Narasimhan, Proc. Natl Inst. Sci., **21**, 133 (1955)
5.81 J.C. Owens, Phys. Rev., **181**, 1228 (1969)
5.82 P.A. Varotsos, J. Phys. Lett., **39**, L-79 (1978)
5.83 E.E. Havinga, J. Phys. Chem. Solids, **18**, 253 (1961)
5.84 E.E. Havinga, A.J. Bosman, Phys. Rev. A, **140**, 292 (1965)

5.85 B. Szigeti, Trans. Faraday Soc., **45**, 155 (1949)
5.86 B. Szigeti, Proc. R. Soc. London, Ser A, **204**, 51 (1950)
5.87 M.R. Srinivasan, P.S. Narayanan, Pramana, **19**, 117 (1982)
5.88 M. Born, G. Mayer, *Handbuch der Physik*, Vol. XXIV/2, Springer, Berlin Heidelberg New York (1933)
5.89 H. Born, K. Huang, *Dynamical Theory of Crystal Lattices*, Oxford University Press, New York (1954)
5.90 B. Donovan, J.F. Angress, *Lattice Vibrations*, Chapman & Hall, London (1971)
5.91 P.J. Gielisse, S.S. Mitra, J.N. Plendl, R.D. Griffis, L.C. Mansur, R. Marshall, E.A. Pascoe, Phys. Rev., **115**, 1039 (1967)
5.92 K.G. Bansigir, Nature, **216**, 256 (1967)
5.93 S.S. Mitra, R. Marshall, J. Chem. Phys., **41**, 3158 (1964)
5.94 R.A. Bartels, P.A. Smith, Phys. Rev. B, **7**, 3885 (1973)
5.95 K.V. Rao, A. Smakula, Mater. Res. Bull., **6**, 1047 (1971)
5.96 B. Szigeti, Proc. R. Soc. London, Ser A, **252**, 217 (1959)
5.97 B. Szigeti, Proc. R. Soc. London, Ser A, **261**, 274 (1961)
5.98 R.A. Cowley, Adv. Phys., **12**, 421 (1963)
5.99 E.R. Cowley, J. Phys. C., **5**, 1345 (1972)
5.100 E.R. Cowley, R.A. Cowley, Proc. R. Soc. London, Ser A, **287**, 259 (1965)
5.101 A.A. Maraduddin, A.E. Fein, Phys. Rev., **128**, 2589 (1962)
5.102 D.H. Martin, Adv. Phys., **12**, 421 (1965)
5.103 I.F. Chang, S.S. Mitra, Phys. Rev. B, **5**, 4094 (1972)
5.104 R. Sano, J. Phys., **C9**, 849 (1976)
5.105 R.H. Lyddane, R.G. Sachs, E. Teller, Phys. Rev., **59**, 673 (1941)
5.106 W. Cochran, Adv. Phys., **18**, 157 (1969)
5.107 S. Chandra, J. Rolfe, Can. J. Phys., **48**, 412 (1970)
5.108 K.S. Rao, K.V. Rao, Z. Phys., **216**, 300 (1968)
5.109 J. Appel, Solid State Phys., **21**, 193 (1968)
5.110 A.J. Bosman, H. Van Daal, Adv. Phys., **19**, 1 (1970)
5.111 T. Holstein, Ann. Phys., **8**, 343 (1959)
5.112 I.G. Austin, N.F. Mott, Adv. Phys., **18**, 41 (1969)
5.113 N.F. Mott, E.A. Davis, *Phonons and Polarons in Electronic Processing in Non-Crystalline Materials*, Clarendon, Oxford (1971)
5.114 H. Frohlich, Adv. Phys., **3**, 325 (1954)
5.115 G.R. Allcock, Adv. Phys., **5**, 412 (1956)
5.116 V.N. Bogomolov, E.K. Kudinov, N. Mirlin, Yu.A. Firsov, Sov. Phys. Solid State, **9**, 1630 (1968)
5.117 D. Adler, J. Feinleib, Phys. Rev. B, **2**, 3112 (1970)
5.118 H.B., Lal, B.K. Verma, N. Dar, Indian J. Cryogen., **1**, 119 (1976)
5.119 S. Roberts, Phys. Rev., **76**, 1215 (1949)
5.120 T.R. Tessman, A.H. Kahn, W. Shockley, Phys. Rev., **92**, 890 (1953)
5.121 R.D. Shannon, NISTY Spec. Publ., **804**, 457 (1991)
5.122 R.D. Shannon, J. Appl. Phys., **73**, 348 (1993)

5.123 D. Rosen, Trans. Faraday Soc., **59**, 2178 (1963)
5.124 I.R. Kuntz, W. Kauzmann, Adv. Protein Chem., **28**, 239 (1974)
5.125 S. Takashima, H.P. Schwan, J. Phys. Chem., **69**, 4176 (1965)
5.126 B.A. Lawton, Z.H. Zu, R. Pethig, Y. Wei, J. Mol. Liq., **42**, 83 (1989)
5.127 S.C. Harvey, P. Hoekstra, J. Phys. Chem., **76**, 2987 (1972)
5.128 S. Bone, R. Pethig, J. Mol. Biol., **181**, 323 (1985)
5.129 R. Pethig, A.S. Gyorgyi, Biophysics, Proc. Natl Acad. Sci., USA, **74**, 226 (1977)
5.130 S. Bone, Biophysics, Proc. Natl Acad. Sci., USA, **75**, 315 (1978)
5.131 K.V. Rao, J. Phys. Chem. Solids, **20**, 193 (1961)
5.132 A. Subrahmanyam, K.V.S. Badarinath, Phys. Status Solidi A, **84**, k93 (1984)
5.133 A. Subrahmanyam, Solid State Commun., **55**, 1127 (1985)
5.134 K.V. Rao, Z. Phys., **172**, 258 (1963)
5.135 S. Govinda, K.V. Rao, Indian J. Pure Appl. Phys., **13**, 175 (1975)
5.136 J. Fontanella, R.L. Johnston, G.H. Sigel Jr., C. Andeen, J. Non-Cryst. Solids, **31**, 401 (1978)
5.137 L. Sirdeshmukh, Y.R. Reddy, Indian J. Phys., **52A**, 564 (1978)
5.138 L. Sirdeshmukh, Curr. Sci., **43**, 712 (1974)
5.139 L. Sirdeshmukh, Curr. Sci., **44**, 690 (1975)
5.140 Prameela Devi, L. Sirdeshmukh, Bull. Mater. Sci., **4**, 17 (1982)
5.141 Y.R. Reddy, L. Sirdeshmukh, Solid State Commun., **40**, 353 (1981)
5.142 Y.R. Reddy, Ph.D. Thesis, Kakatiya University, Warangal, India (1980)
5.143 L. Sirdeshmukh, Y.R. Reddy, *Vistas in Molecular, Solid State and Biophysics*, A Commemmorative Research Volume in honour of P.G. Puranik (1977)
5.144 G. Sathaiah, L. Sirdeshmukh, Proceedings of Third National Seminar on Ferroelectrics and Dielectrics, IISc, Bangalore (1984)
5.145 G. Sathaiah, L. Sirdeshmukh, Indian J. Pure Appl. Phys., **23**, 257 (1985)
5.146 G. Sathaiah, Ph.D. Thesis, Kakatiya University, Warangal, India (1988)
5.147 L. Sirdeshmukh, Y.R. Reddy, Pramana, **22**, 559 (1984)
5.148 Y.R. Reddy, L. Sirdeshmukh, Solid State Commun., **51**, 407 (1984)
5.149 Prameela Devi, Ph.D. Thesis, Kakatiya University, Warangal, India (1983)
5.150 L. Sirdeshmukh, Y.R. Reddy, Ferroelectrics, **64**, 247 (1985)
5.151 L. Sirdeshmukh, Y.R. Reddy, Bull. Mater. Sci., **2**, 61 (1980)
5.152 Y.R. Reddy, L. Sirdeshmukh, Phys. Status Solidi A, **103**, k157 (1987)
5.153 Prameela Devi, L. Sirdeshmukh, Proceedings of the Third National Seminar on Ferroelectrics and Dielectrics, IISc, Bangalore (1984)
5.154 Prameela Devi, L. Sirdeshmukh, Proceedings of the Second National Seminar on Ferroelectrics and Dielectrics, Ravishankar University, Raipur (1982)
5.155 M. Srinivas, L. Sirdeshmukh, Cryst. Res. Technol., **23**, 209 (1988)
5.156 M. Srinivas, L. Sirdeshmukh, Cryst. Res. Technol., **23**, 1121 (1988)

5.157 M. Srinivas, L. Sirdeshmukh, *Solid State Ionics Materials and Applications*, World Scientific, Singapore (1992)
5.158 K. Krishna Kumar, V. Balaram, L. Sirdeshmukh, Bull. Mater. Sci., **15**, 279 (1992)
5.159 K. Krishna Kumar, Ph.D. Thesis, Kakatiya University, Warangal, India (1991)
5.160 K. Krishna Kumar, G. Sathaiah, L. Sirdeshmukh, Ferroelectrics, **155**, 213 (1994)
5.161 K. Krishna Kumar, A. Ramakrishna, G. Sathaiah, L. Sirdeshmukh, J. Kor. Phys. Soc., **32**, S321 (1998)
5.162 L. Sirdeshmukh, K. Krishna Kumar, S. Bal Laxman, A. Ramakrishna, G. Sathaiah, Bull. Mater. Sci., **21**, 219 (1998)
5.163 A. Mallikarjuna Reddy, A. Ramakrishna, G. Sathaiah, L.Sirdeshmukh, *Condensed Matter Physics*, Narosa, New Delhi (1999)
5.164 S. Bal Laxman, Ph.D. Thesis, Kakatiya University, Warangal, India (1993)
5.165 K. Krishna Kumar, L. Sirdeshmukh, Indian J. Pure Appl. Phys., **34**, 559 (1996)
5.166 K. Krishna Murthy, K. Krishna Kumar, L. Sirdeshmukh, Indian J. Pure Appl. Phys., **36**, 769 (1998)
5.167 B. Jagannadh, L. Sirdeshmukh, Bull. Mater. Sci., **3**, 423 (1981)
5.168 K. Krishna Murthy, Ph.D. Thesis, Kakatiya University, Warangal, India (1997)
5.169 T. Ramakrishna Rao, P. Raghunath Rao, P. Lingaiah, L. Sirdeshmukh, J. Indian Chem. Soc., **67**, 862 (1990)
5.170 T. Ramakrishna Rao, P. Raghunath Rao, P. Lingaiah, L. Sirdeshmukh, Die Angew. Makromol. Chem., **191**, 177 (1991)
5.171 T. Ramakrishna Rao, Ph.D. Thesis, Kakatiya University, Warangal, India (1990)
5.172 J. Shankar, V.C. Jain, A.K.G. Lashkari, Indian J. Phys., **53A**, 420 (1979)
5.173 M. Srinivas, Ph.D. Thesis, Kakatiya University, Warangal, India (1989)
5.174 L. Sirdeshmukh, D.B. Sirdeshmukh, Curr. Sci., **41**, 558 (1972)
5.175 R. Srinivasan, Proc. Indian Acad. Sci., **A41**, 49 (1955)
5.176 R.P. Lowndes, A. Rastogi, Phys. Rev. B, **14**, 3598 (1976)
5.177 C.K. Kim, A. Feldman, D. Horowitz, R.M. Waxler, Solid State Commun., **25**, 397 (1978)
5.178 K.K. Srivastava, H.D. Merchant, J. Phys. Chem. Solids, **34**, 2069 (1973)
5.179 K.A. Gschneidner Jr., L. Eyring, *Handbook of Physics and Chemistry of Rare Earths*, Vol. 3, North-Holland, New York (1979)
5.180 S. Lawrence, K. Milan, V.J. Flynn, J. Solid State Chem., **2**, 137 (1970)
5.181 A.M. Hofmeister, K.R. Campbell, J. Appl. Phys., **72**, 638 (1992)
5.182 A.N. Petrov, C.V. Denisov, V.M. Zhukovskii, Inorg. Mater. (USA), **22**, 579 (1986)
5.183 W.P. Mason, Phys. Rev., **70**, 529 (1946)

5.184 A.D. Prasad Rao, P.D.R. Andrade, S.P.S. Porto, Phys. Rev. B, **9**, 1077 (1974)
5.185 P. Dawson, Phys. Status Solidi B, **50**, 571 (1972)
5.186 L.W. Barr, A.B. Lidiard, *Physical Chemistry – An Advanced Treatise, Defects in Ionic Crystals*, Vol. 10, Academic, New York (1968)
5.187 M.D. Agrawal, Physica, **72**, 397 (1974)
5.188 G. Sathaiah, L. Sirdeshmukh, Proceedings Sixth National Seminar on Ferroelectrics and Dielectrics, Warangal (1990)
5.189 T. Thirmal Rao, Ph.D. Thesis, Kakatiya University, Warangal, India (1991)
5.190 R. Subrahmanyam, Solid State Commun., **85**, 1127 (1985)
5.191 K. Tagaya, Jpn. J. Appl. Phys., **23**, 168 (1984)
5.192 K. Tagaya, Jpn. J. Appl. Phys., **24**, 75 (1985)
5.193 D. Bhatta, J.K. Mishra, S.R. Mohanty, Indian J. Chem., **19A**, 466 (1984)

Chapter 6

6.1 K.S. Krishnan, S.K. Roy, Proc. R. Soc. London, Ser A, **207**, 447 (1951)
6.2 K.S. Krishnan, S.K. Roy, Proc. R. Soc. London, Ser A, **210**, 487 (1952)
6.3 R.S. Krishnan, *Progress in Crystal Physics*, S. Viswanathan, Madras (1958)
6.4 T.B. Kinney, M. O'Keefe, Solid State Commun., **32**, 2634 (1969)
6.5 D.B. Sirdeshmukh, K.G. Subhadra Indian J. Pure Appl. Phys., **11**, 938 (1973)
6.6 N. Uchide, S. Saito, J. Acoust. Soc. Am., **51**, 1602 (1972)
6.7 K.G. Subhadra, D.B. Sirdeshmukh, J. Appl. Phys., **40**, 2357 (1969)
6.8 H.T. Smyth, J. Am. Ceram. Soc., **38**, 140 (1955)
6.9 M.N. Sharma, M.P. Madan, Indian J. Phys., **38**, 231 (1964)
6.10 C.M. Kachhava, S.C. Saxena, Indian J. Phys., **39**, 145 (1965)
6.11 M.P. Tosi, Solid State Phys., **16**, 1, (1964)
6.12 G.C. Benson, E. Dempsey, Proc. R. Soc. London, Ser A, **266**, 344 (1962)
6.13 D.B. Sirdeshmukh, Indian J. Pure Appl. Phys., **4**, 323 (1966)
6.14 S.K. Joshi, S.S. Mitra, J. Chem. Phys., **34**, 1462 (1961)
6.15 D.B. Sirdeshmukh, J. Appl. Phys., **37**, 3910 (1966)
6.16 D.B. Sirdeshmukh, Curr. Sci., **36**, 372 (1967)
6.17 K.A Gschneidner Jr., Solid State Phys., **16**, 276 (1964)
6.18 G.A. Alers, *Physical Acoustics*, Vol. 3B, p. 1, Editor W.P. Mason, Academic, New York (1965)
6.19 O.L. Anderson, *Physical Acoustics*, Vol. 3B, p. 43, Editor W.P. Mason, Academic, New York (1965)
6.20 J. de Launay, J. Chem. Phys., **21**, 1975 (1953)
6.21 J. de Launay, J. Chem. Phys., **22**, 1676 (1954)

6.22 P.N. Marcus, Quoted from *Physical Acoustics*, Vol. 3B, p. 10, Editor W.P. Mason, Academic, New York (1965)
6.23 L. Hopf, G. Lechner, Verhandl. Deut. Physik Ges., **16**, 643 (1914)
6.24 S.L. Quimby, P.M. Sutton, Phys. Rev., **91**, 1122 (1953)
6.25 P.M. Sutton, Phys. Rev., **99**, 1926 (1955)
6.26 G. Leibfried, *Handbuch der Physik*, Vol. VII, Editor Flugge, Springer, Berlin Heidelberg New York (1955)
6.27 V.V. Houston, Rev. Mod. Phys., **20**, 161 (1948)
6.28 A.B. Bhatia, G.E. Tauber, Philos. Mag., **46**, 108 (1955)
6.29 D.D. Betts, A.B. Bhatia, M. Wyman, Phys. Rev., **104**, 43 (1956)
6.30 D.D. Betts, Can. J. Phys., **39**, 233 (1961)
6.31 A. Konti, T.P. Varshni, Can. J. Phys., **47**, 2021 (1969)
6.32 A. Konti, T.P. Varshni, Can. J. Phys., **49**, 3115 (1971)
6.33 D.D. Betts, A.B. Bhatia, M. Wyman, Phys. Rev., **104**, 37 (1956)
6.34 F.I. Fedorov, Sov. Phys. Crystallogr., **8**, 159, 312 (1963)
6.35 F.I. Fedorov, Sov. Phys. Crystallogr., **10**, 125 (1965)
6.36 F.I. Fedorov, Sov. Phys. Dokl., **10**, 896 (1965)
6.37 F.I. Fedorov, *Theory of Elastic Waves in Crystals*, Plenum, New York (1968)
6.38 F.I. Fedorov, T.G. Bystrova, Sov. Phys. Crystallogr., **11**, 333 (1966)
6.39 W. Boas, *Introduction to Physics of Metals and Alloys*, Wiley, New York (1947)
6.40 L. Thomson, J. Geophys. Res., **77**, 315 (1972)
6.41 W. Voigt, Ann. Phys. Lpz., **38**, 573 (1928)
6.42 A. Reuss, Z. Angew Math. Mech., **9**, 49 (1929)
6.43 R.W. Hill, Proc. Phys. Soc., **A65**, 349 (1952)
6.44 A.V. Hershey, J. Appl. Mech., **21**, 236 (1954)
6.45 E. Kroner, Z. Phys., **151**, 504 (1958)
6.46 E. Hashin, S. Shtrikman, J. Mech. Phys. Solids, **10**, 343 (1962)
6.47 K.S. Aleksandrov, Sov. Phys. Dokl., **10**, 893 (1966)
6.48 V.P. Binnie, Phys. Rev., **103**, 579 (1956)
6.49 J.J. Gilvarry, Phys. Rev., **103**, 1701 (1956)
6.50 O.L. Anderson, J. Phys. Chem. Solids, **24**, 909 (1963)
6.51 J.K.D. Verma, M.D. Aggarwal, J. Appl. Phys., **46**, 2841 (1975)
6.52 J.K.D. Verma, M.D. Aggarwal, J. Appl. Phys., **47**, 2778 (1976)
6.53 J.K.D. Verma, D. Basu, M.D. Aggarwal, Acta Phys. Slov., **30**, 209 (1980)
6.54 D. Basu, J.K.D. Verma, Acta Phys. Slov., **31**, 221 (1981)
6.55 P.J. Reddy, Physica, **29**, 63 (1963)
6.56 D.I. Bolef, M. Menes, J. Appl. Phys., **31**, 1010 (1960)
6.57 S.K. Joshi, S.S. Mitra, Proc. Phys. Soc., **76**, 295 (1960)
6.58 S.K. Joshi, S.S. Mitra, Indian J. Phys., **34**, 532 (1960)
6.59 R.A. Robie, J.L. Edwards, J.Appl. Phys., **37**, 2659 (1966)
6.60 J.T. Lewis, A. Lehoczky, C.V. Briscoe, Phys. Rev., **161**, 877 (1967)
6.61 F. Michard, P. Plioque, A. Zarembowitch, J. Appl. Phys., **44**, 572 (1973)

6.62 T. Seshagiri Rao, B. Revathi, M. Purnanandam, Indian J. Pure Appl. Phys., **9**, 797 (1971)
6.63 M.D. Aggarwal, J.K.D. Verma, Indian J. Pure Appl. Phys., **10**, 838 (1972)
6.64 M.D. Aggarwal, V. Raju, J.K.D. Verma, Indian J. Pure Appl. Phys., **11**, 216 (1973)
6.65 D. Basu, J.K.D. Verma, Curr. Sci., **49**, 141 (1980)
6.66 H.N. Ledbetter, J. Appl. Phys., **44**, 1451 (1973)
6.67 N.T. Padial, L.M. Brescansin, M.M. Shukla, Acta Phys. Pol., **A57**, 129 (1980)
6.68 K.G. Subhadra, D.B. Sirdeshmukh, Indian J. Pure Appl. Phys., **11**, 938 (1978)
6.69 N. Gopi Krishna, K.G. Subhadra, D.B. Sirdeshmukh, Indian J. Pure Appl. Phys., **22**, 618 (1984)
6.70 T.O. Baldwin, C.W. Tompson, J. Chem. Phys., **41**, 1420 (1964)
6.71 M.E. Mullen, B. Luthi, P.S. Wang, Phys. Rev. B, **10**, 186 (1974)
6.72 Y. Shapira, T.B. Reed, AIP Conf. Proc., **15**, 837 (1971)
6.73 B. Nagaiah, M. Rambabu, D.B. Sirdeshmukh, Indian J. Pure Appl. Phys., **17**, 838 (1979)
6.74 S.S. Haussuhl, D. Mateika, Z. Naturforsch., **27A** 1522 (1972)
6.75 S.S. Haussuhl, D. Mateika, W. Tolksdorf, Z. Naturforsch., **31A**, 390 (1976)
6.76 B.K. Rao, K.G. Subhadra, D.B. Sirdeshmukh, Indian J. Pure Appl. Phys., **19**, 87 (1981)
6.77 H. Schweppe, IEEE Trans., **SU-16**, 219 (1969)
6.78 I.M. Nedyukha, V.G. Chernyi, Fiz. Met. and Metalloved, **23**, 1117 (1967)
6.79 D.B. Sirdeshmukh, M.J.M. Rao, Phys. Status Solidi B, **44**, k105 (1971)
6.80 S.S. Haussuhl, Phys. Status Solidi, **28**, 127 (1968)
6.81 D.B. Sirdeshmukh, K.G. Subhadra, J. Appl. Phys., **46**, 3682 (1975)
6.82 J. Bhimasenachar, G. Venkataratnam, J. Acoust. Soc. Am., **27**, 922 (1955)
6.83 T.V. Ryzhova, K.S. Aleksandrov, V.M. Korobhova Izv. Earth Phys., **2**, 63 (1966)
6.84 H. Ozkan, L. Cartz, J.C. Jamieson, J. Appl. Phys., **45**, 556 (1974)
6.85 D.B. Sirdeshmukh, M.J.M. Rao, Indian J. Phys., **49**, 241 (1975)
6.86 F.A. Lindemannn, Z. Physik, **11**, 609 (1910)
6.87 J.M. Farley, G.A. Saunders, D.Y. Chung, J. Phys. C: Solid State Phys., **6**, 2010 (1973)
6.88 S.V. Suryanarayana, Indian J. Pure Appl. Phys., **3**, 553 (1975)
6.89 D. Basu, J.K.D. Verma, J. Phys. Chem. Solids, **41**, 809 (1980)
6.90 J.C. Slater, *Introduction to Chemical Physics*, McGraw-Hill, New York (1939)
6.91 C.D. Das, H.V. Keer, R.V.G. Rao, Z. Phys. Chem., **224**, 377 (1963)
6.92 C.M. Kachhava, S.C. Saxena, J. Appl. Phys., **39**, 2973 (1968)

6.93 D.B. Sirdeshmukh, B.K. Rao, J. Phys. Chem. Solids, **36**, 355 (1975)
6.94 D.B. Sirdeshmukh, B.K. Rao, J. Chem. Phys., **57**, 577 (1972)
6.95 D.B. Sirdeshmukh, B.K. Rao, AIP Conf. Proc., **17**, 56 (1973)
6.96 D.B. Sirdeshmukh, B.K. Rao, J. Appl. Phys., **44**, 894 (1973)
6.97 D.B. Sirdeshmukh, K.G. Subhadra, Phys. Status Solidi B, **150**, K11 (1988)
6.98 D.B. Sirdeshmukh, L. Sirdeshmukh, K.G. Subhadra, *Alkali Halides – A Handbook of Physical Properties*, Springer, Berlin Heidelberg New York (2001)
6.99 W. Hussain, Ph.D. Thesis, Kakatiya University, Warangal, India (1989)
6.100 L. Knopoff, J.N. Shapiro, J. Geophys. Res., **74**, 1439 (1969)
6.101 B.K. Rao, Ph.D. Thesis, Kakatiya University, Warangal, India (1980)
6.102 D.H. Chung, W.R. Buessem, J. Appl. Phys., **38**, 3010 (1967)
6.103 R.W. Roberts, C.S. Smith, J. Phys. Chem. Solids, **31**, 619 (1970)
6.104 R.W. Roberts, C.S. Smith, J. Phys. Chem. Solids, **31**, 2397 (1970)
6.105 K.O. McLean, C.S. Smith, J. Phys. Chem. Solids, **33**, 279 (1972)
6.106 J.R. Ferraro, S.S. Mitra, A. Quattrochi, J. Appl. Phys., **42**, 3677 (1971)
6.107 R.P. Lowndes, J. Phys. C, **4**, 3083 (1971)
6.108 R. Ruppin, J. Phys. Chem. Solids, **33**, 85 (1972)
6.109 J.R. Kessler, E. Monberg, M. Nicol, J. Chem. Phys., **60**, 5057 (1974)
6.110 B. Szigeti, Trans. Faraday Soc., **45**, 155 (1949)
6.111 B. Szigeti, Proc. R. Soc. London, **204**, 51 (1950)
6.112 B.W. Jones, Philos. Mag., **26**, 1085 (1967)
6.113 C. Andeen, D. Schuele, A. Quattrochi, Phys. Rev. B, **6**, 591 (1972)
6.114 J.D. Axe, Phys. Rev. A, **139**, 1215 (1965)
6.115 Prameela Devi, Ph.D. Thesis, Kakatiya University, Warangal, India (1983)
6.116 T.H.K. Barron, A. Batana, Philos. Mag., **20**, 619 (1969)

Chapter 7

7.1 S.K. Mohanlal, *Mixed Crystals*, Madurai University, Madurai (1971)
7.2 A.I. Kitaigorodski, *Mixed Crystals*, Springer, Berlin Heidelberg New York (1984)
7.3 V. Hari Babu, U.V. Subba Rao, Prog. Cryst. Growth Charact., **8**, 189 (1984)
7.4 D.B. Sirdeshmukh, K. Srinivas, J. Mater. Sci., **21**, 4117 (1986)
7.5 W.E. Wallace, J. Chem. Phys., **17**, 1095 (1949)
7.6 C.V.K. Krishnamurthy, Y.V.G.S. Murti, Phys. Rev. B, **43**, 14206 (1991)
7.7 C.V.K. Krishnamurthy, Y.V.G.S. Murti, Phys. Rev. B, **43**, 14219 (1991)
7.8 I.F. Chang, S.S. Mitra, Phys. Rev., **172**, 924 (1968)
7.9 Y.S. Chen, W. Shockley, G.L. Pearson, Phys. Rev., **151**, 648 (1966)
7.10 J. Retgers, Z. Phys. Chem., **3**, 497 (1889)
7.11 O.D. Slagle, H.A. Mckinstry, J. Appl. Phys., **38**, 446 (1967)

7.12 J.S. Wollam, W.E. Wallace, J. Phys. Chem., **60**, 1654 (1956)
7.13 W.T. Barrett, W.E. Wallace, J. Am. Chem. Soc., **76**, 366 (1954)
7.14 L. Vegard, Z. Phys., **5**, 17 (1921)
7.15 H.G. Grimm, K.F. Herzfeld, Z. Phys., **16**, 77 (1923)
7.16 R.J. Havighurst, E. Mack, F.C. Blake, J. Am. Chem. Soc., **47**, 29 (1925)
7.17 O.D. Slagle, H.A. Mckinstry, Acta Crystallogr., **21**, 1013 (1966)
7.18 M. Ahtee, Ann. Acad. Sci. Fenn., **AVI**, 3 (1969)
7.19 P.J. Gielisse, J.N. Plendl, L.C. Mansoor, R. Marshall, S.S. Mitra, R. Mykolajewycz, A. Smakula, J. Appl. Phys., **36**, 2446 (1965)
7.20 W.B. Zimmerman, Phys. Rev. B, **54**, 704 (1972)
7.21 K.A. Gschneidner Jr., G.H. Vineyard, J. Appl. Phys., **33**, 3444 (1962)
7.22 V.C. Anselmo, N.O. Smith, J. Phys. Chem., **63**, 152 (1970)
7.23 I.R. Nair, C.T. Walker, Phys. Rev. B, **7**, 2740 (1973)
7.24 D. Sridevi, K.V. Reddy, Mater. Res. Bull., **20**, 929 (1985)
7.25 J.A. Wasastjerna, Soc. Sci. Fenn. Comm., **XIII**, 1 (1946)
7.26 M. Ahtee, A. Pesonen, P. Salmo, O. Inkinen, Z. Naturforsch., **25**, 1761 (1970)
7.27 S.K. Mohanlal, K.S. Chandrasekaran, C. Sanjeeviraja, J. Phys. C: Solid State Phys., **15**, 4235 (1982)
7.28 J. Heitala, Ann. Acad. Sci. Fenn., **AVI**, 122 (1963)
7.29 B. Borie, Acta Crystallogr., **10**, 89 (1957)
7.30 R.J. Weiss, *X-ray Determination of Electron Distribution*, North Holland, Amsterdam (1966)
7.31 P.D. Dernier, W. Weber, L.D. Longinotti, Phys. Rev. B, **14**, 3635 (1976)
7.32 P. Debye, Ann. Phys., **39**, 789 (1912)
7.33 M. Blackman, *Handbuch der Physik*, Vol. VII (Part I), Springer, Berlin Heidelberg New York (1955)
7.34 F.H. Herbstein, Philos. Mag. Suppl., **10**, 313 (1961)
7.35 S.S. Mitra, J. Sci. Indian Res., **21A**, 76 (1962)
7.36 G.A. Alers, *Physical Acoustics*, Vol. 3B, Academic, New York (1965)
7.37 A.V. Karlsson, Phys. Rev. B, **2**, 3332 (1970)
7.38 A.K. Giri, G.B. Mitra, Indian J. Phys., **59A**, 318 (1985)
7.39 A.V. Sharko, A.A. Botaki, Sov. J. Phys., **14**, 1710 (1971)
7.40 K.S. Cholokov, E.N. Novikov, V.A. Grishukov, A.A. Botaki, Izv. Vyss Uch. Zav. Fiz., **12**, 157 (1970)
7.41 A.A. Botaki, J.N. Gyrbu, V.L. Ul'yanov, A.V. Sharko, Izv. Vyss Uch. Zav. Fiz., **12**, 147 (1970)
7.42 B.D. Nathan, L.F. Lou, R.H. Tait, Solid State Commun., **19**, 615 (1976)
7.43 A.A. Botaki, J.N. Gyrbu, A.V. Sharko, Sov. J. Phys., **15**, 917 (1972)
7.44 V.E. Avericheva, A.A. Botaki, G.A. Dvornikov, A.V. Sharko, Izv. Vyss Uch. Zav. Fiz., **16**, 148 (1973)
7.45 M.M. Beg, J. Aslam, N. Ahmed, Q.H. Khan, N.M. Butt, Phys. Status Solidi B, **94**, K45 (1979)
7.46 J.N. Plendl, P.J. Gielisse, L.C. Mansur, S.S. Mitra, A. Smakula, P.C. Tarte, Appl. Opt., **10**, 1129 (1971)

7.47 T. Bhimasankaram, Ph.D. Thesis, Osmania University, Hyderabad, India (1974)
7.48 U.V. Subba Rao, V. Hari Babu, Pramana, **11**, 149 (1978)
7.49 U.V. Subba Rao, V. Hari Babu, Indian J. Phys., **54A**, 147 (1980)
7.50 M.V.S. Sarma, S.V. Suryanarayana, J. Mater. Sci. Lett., **5**, 1277 (1986)
7.51 I.V. Bodnar, B.V. Korzun, A.P. Chernyakova, Phys. Status Solidi A, **101**, 409 (1987)
7.52 J.N. Plendl, P.J. Gielisse, Z. Kristallogr., **118**, 404 (1963)
7.53 U.C. Shrivastava, J. Appl. Phys., **51**, 1510 (1980)
7.54 T. Kataoka, T. Yamada, Jpn. J. Appl. Phys., **16**, 1119 (1977)
7.55 D. Tabor, *The Hardness of Metals*, Clarendon, Oxford (1956)
7.56 Y.S. Boyarskaya, D.Z. Gabko, D.S. Pishkova, Phys. Status Solidi A, **87**, 175 (1985)
7.57 J.H. Fertel, C.H. Perry, Phys. Rev., **184**, 874 (1969)
7.58 K. Kamiyoshi, V. Nigara, Phys. Status Solidi A, **6**, 223 (1971)
7.59 P. Varotsos, Phys. Status Solidi B, **100**, K133 (1980)
7.60 P.K. Sinha, J. Shanker, Indian J. Pure Appl. Phys., **32**, 675 (1994)
7.61 B. Szigeti, Trans. Faraday Soc., **45**, 155 (1949)
7.62 B. Szigeti, Proc. R. Soc. London, Ser A, 204, **51** (1950)
7.63 J. Shanker, G.D. Jain, Phys. Status Solidi B, **119**, 185 (1983)
7.64 J.F. Scott, Phys. Rev. B, **4**, 1360 (1971)
7.65 F. Gervais, Solid State Commun., **18**, 191 (1976)
7.66 K. Wakamura, T. Arai, J. Appl. Phys., **62**, 1750 (1987)
7.67 J.J. Markham, *F-Centres in Alkali Halides*, Academic, New York (1966)
7.68 W.B. Fowler, *Physics of Colour Centres*, Academic, New York (1968)
7.69 F. Seitz, Rev. Mod. Phys., **18**, 384 (1946)
7.70 W.D. Compton, H. Rabin, Solid State Phys., **16** (1964)
7.71 D.B. Sirdeshmukh, L. Sirdeshmukh, K.G. Subhadra, *Alkali Halides – A Handbook of Physical Properties*, Springer, Berlin Heidelberg New York (2001)
7.72 A. Smakula, N.C. Maynard, A. Repucci, Phys. Rev., **130**, 113 (1963)
7.73 R. Thyagarajan, J. Phys. Chem. Solids, **27**, 218 (1966)
7.74 R.A. Reitz, W.A. Butler, J.R. Brandenberger, J. Chem. Phys., **37**, 1893 (1962)
7.75 M. Hovi, Cryst. Lattice Defects, **1**, 261 (1970)
7.76 Y.V.G.S. Murti, K.L.N. Prasad, Solid State Commun., **41**, 691 (1982)
7.77 W.E. Wallace, R.A. Flinn, Nature, **172**, 681 (1953)
7.78 U.V. Subba Rao, V. Hari Babu, Cryst. Lattice Defects, **8**, 21 (1978)
7.79 J. Arends, H.W.D. Hartog, A.J. Dekker, Phys. Status Solidi, **10**, 105 (1965)
7.80 P. Veeresham, U.V. Subba Rao, V. Hari Babu, Cryst. Growth, **52**, 943 (1981)
7.81 S. Rabinovich, A. Voronel, L. Peretzman, J. Phys. C: Solid State Phys., **21**, 5943 (1988)
7.82 J.L. Corkill, M.L. Cohen, J. Phys. Chem. Solids, **54**, 51 (1993)

7.83 I. Singh, Indian J. Technol., **8**, 52 (1970)
7.84 O.P. Sharma, G.S. Sharma, S. Prakash, Pramana, **39**, 323 (1992)
7.85 L.J. Wood, W. Secunda, C.H. McBride, J. Am. Chem. Soc., **80**, 307 (1958)
7.86 H. Weijma, J. Arends, Phys. Status Solidi, **35**, 205 (1969)
7.87 M. Natarajan, K.J. Rao, C.N.R. Rao, Trans. Faraday Soc., **66**, 2497 (1970)
7.88 A. May, Phys. Rev., **52**, 539 (1937)
7.89 J. Shanker, S.C. Sharma, M. Kumar, Phys. Status Solidi B, **141**, 409 (1987)
7.90 P.S. Costich, G.J. Maass, N.O. Smith, J. Chem. Eng. Data, **8**, 26 (1963)
7.91 Y.V.G.S. Murti, P.S. Prasad, Physica B, **79**, 243 (1975)
7.92 K. Srinivas, M. Ateequddin, D.B. Sirdeshmukh, Pramana, **28**, 81 (1987)
7.93 T. Kumara Swamy, K.G. Subhadra, D.B. Sirdeshmukh, Pramana, **43**, 33 (1994)
7.94 K. Suryanarayana, D.B. Sirdeshmukh, Proceedings of National Seminar on Crystallography, Calcutta (1981)
7.95 K.G. Subhadra, K.A. Hussain, Cryst. Res. Technol., **22**, 1165 (1987)
7.96 K.A. Hussain, K.G. Subhadra, D.B. Sirdeshmukh, Cryst. Res. Technol., **27**, 211 (1992)
7.97 A.V. Tobolsky, J. Chem. Phys., **10**, 187 (1942)
7.98 E.F. Kaelble, *Handbook of X-Rays*, McGraw-Hill, New York (1967)
7.99 K. Srinivas, M.Phil. Thesis, Kakatiya University, Warangal, India (1981)
7.100 K. Srinivas, D.B. Sirdeshmukh, Pramana, **25**, 71 (1985)
7.101 K. Srinivas, D.B. Sirdeshmukh, Pramana, **31**, 221 (1988)
7.102 P. Geeta Krishna, K.G. Subhadra, D.B. Sirdeshmukh, Pramana J. Phys., **52**, 503 (1999)
7.103 K. Srinivas, D.B. Sirdeshmukh, Curr. Sci., **54**, 740 (1985)
7.104 E. Balaiah, Ph.D. Thesis, Kakatiya University, Warangal, India (2003)
7.105 T. Kumara Swamy, Ph.D. Thesis, Kakatiya University, Warangal, India (1996)
7.106 V. Ganesan, K.S. Girirajan, Pramana J. Phys., **30**, 407 (1988)
7.107 B. Nagaiah, D.B. Sirdeshmukh, Indian J. Pure Appl. Phys., **18**, 903 (1980)
7.108 K. Srinivas, D.B. Sirdeshmukh, Cryst. Res. Technol., **21**, K165 (1986)
7.109 K. Srinivas, D.B. Sirdeshmukh, Cryst. Res. Technol., **21**, 157 (1986)
7.110 K.G. Subhadra, D.B. Sirdeshmukh, Pramana, **10**, 597 (1978)
7.111 M. Blackman, Proc. R. Soc. London, Ser A, **181**, 58 (1942)
7.112 S.C. Abrahams, F.S.L. Hsu, J. Chem. Phys., **63**, 1162 (1975)
7.113 K. Srinivas, D.B. Sirdeshmukh, Indian J. Pure Appl. Phys., **24**, 95 (1986)
7.114 K.A. Hussain, K.G. Subhadra, K. Kishan Rao, Cryst. Res. Technol., **23**, 171 (1988)

7.115 T. Thirmal Rao, D.B. Sirdeshmukh, Cryst. Res. Technol. **26**, K53 (1991)
7.116 D.B. Sirdeshmukh, T. Kumara Swamy, P. Geeta Krishna, K.G. Subhadra, Bull. Mater. Sci., **26**, 261 (2003)
7.117 K.G. Subhadra, E. Balaiah, D.B. Sirdeshmukh, Bull. Mater. Sci., **25**, 31 (2002)
7.118 L. Sirdeshmukh, G. Sathaiah, P. Devi, Phys. Status Solidi A, **99**, 631 (1987)
7.119 G. Sathaiah, L. Sirdeshmukh, Pramana, **29**, 303 (1987)
7.120 G.A. Samara, Phys. Rev. B, **13**, 4529 (1976)
7.121 J.R. Jasperse, A.H. Kahn, J.N. Plendl, S.S. Mitra, Phys. Rev., **146**, 526 (1966)
7.122 B. Subramaniam, Indian J. Pure Appl. Phys., **18**, 57 (1980)
7.123 J.R. Ferraro, C. Postmus, S.S. Mitra, C.J. Hoskins, Appl. Opt., **9**, 5 (1970)
7.124 B. Szigeti, Proc. R. Soc. London, Ser A, **261**, 274 (1961)
7.125 G. Sathaiah, Ph.D. Thesis, Kakatiya University, Warangal, India (1988)
7.126 B. Subramaniam, K.G. Bansigir (1981) Pramana J. Phys., **17**, 445 (1981)
7.127 K.A. Hussain, K.G. Subhadra, Cryst. Res. Technol., **22**, 827 (1987)
7.128 D.B. Sirdeshmukh, K.G. Subhadra, K.A. Hussain, Pramana, **34**, 235 (1990)
7.129 K.A. Hussain, D. Thirupathi Reddy, K.G. Subhadra, Cryst. Res. Technol., **29**, K37 (1994)
7.130 D.B. Sirdeshmukh, K.G. Subhadra, K. Srinivas, T. Kumara Swamy, P. Geeta Krishna, E. Balaiah, Proc. AP Akad. Sci. **8**, 119 (2004)

Chapter 8

8.1 R.F.S. Hearmon, Rev. Mod. Phys., **18**, 409 (1946)
8.2 R.F.S. Hearmon, Adv. Phys., **5**, 323 (1956)
8.3 R.F.S. Hearmon, *Landolt Bornstein Tables*, Vol. I, Group III, Springer, Berlin Heidelberg New York (1966)
8.4 R.F.S. Hearmon, *Landolt Bornstein Tables*, Vol. II, Group III, Springer, Berlin Heidelberg New York (1969)
8.5 H.B. Huntington, Solid State Phys., **7**, 213 (1958)
8.6 K.S. Alexandrov, T.V. Ryzhova, Sov. Phys. Crystallogr., **6**, 228 (1961)
8.7 G. Simmons, H. Wang, *Single Crystal Elastic Constants: A Handbook*, MIT, London (1971)
8.8 P.W. Bridgman, Proc. Am. Acad. Arts Sci., **58**, 165 (1923)
8.9 P.W. Bridgman, *The Physics of High Pressures*, Bell, London (1931)
8.10 C.E. Weir, J. Res. Natl Bureau Stand., **56**, 187 (1956)
8.11 D.R. Stephens, J. Phys. Chem. Solids, **25**, 423 (1964)

8.12 S.N. Vaidya, G.C. Kennedy, J. Phys. Chem. Solids, **31**, 2329 (1970)
8.13 Y. Syono, T. Goto, J. Nakai, Y. Nakagawa, H. Iwasaki, J. Phys. Soc. Jpn, **37**, 442 (1974)
8.14 M.H Rice, R.G. McQueen, J.M. Walsh, Solid State Phys., **6**, 1 (1958)
8.15 R.Y. Goranson, D. Bancroft, B.L. Burton, T. Blechar, E.E. Houston, E.F. Gittings, S.A. Landeen, J. Appl. Phys., **26**, 1472 (1955)
8.16 H.D. Mallory, J. Appl. Phys., **26**, 553 (1955)
8.17 H.G. Drickamer, R.W. Lynch, R.L. Clendenen, E.A. Perez-Albuerne, Solid State Phys., **19**, 135 (1966)
8.18 H. Bartholin, D. Florence, G. Parisot, J. Paureau, O. Vogt, Phys. Lett., **60A**, 47 (1977)
8.19 G. Will, E. Hinze, W. Nuding, Phys. Chem. Miner., **6**, 157 (1984)
8.20 E.A. Perez-Albuerne, K.F. Forsgren, H.G. Drickamer, Rev. Sci. Instrum., **35**, 29 (1964)
8.21 J.C. Jamieson, Z. Kristallogr., **107**, 65 (1956)
8.22 D.B. McWhan, A. Jayaraman, Appl. Phys. Lett., **3**, 129 (1963)
8.23 J.D. Barnett, H.J. Hall, Rev. Sci. Instrum., **35**, 175 (1964)
8.24 T. Takahashi, W.A. Bassette, Science, **145**, 3631 (1964)
8.25 J.M. Leger, K. Oki, A.M. Redon, I. Vedel, J. Rosset-Mignaud, O. Vogt, Phys. Rev. B, **33**, 1986 (1986)
8.26 R.A. Montalvo, D.W. Langer, J. Appl. Phys., **41**, 4101 (1970)
8.27 C.P. Cline, H.L. Dunegan, G.W. Henderson, J. Appl. Phys., **38**, 1944 (1967)
8.28 J.M. Farley, G.A. Saunders, Solid State Commun., **9**, 965 (1971)
8.29 L. Balamuth, Phys. Rev., **45**, 715 (1934)
8.30 F.C. Rose, Phys. Rev., **49**, 50 (1936)
8.31 S. Bhagavantam, J. Bhimasenachar, Proc. Indian Acad. Sci., **A20**, 298 (1944)
8.32 H.B. Huntington, Phys. Rev., **72**, 321 (1947)
8.33 J. Williams, J. Lamb, J. Acoust. Soc. Am., **30**, 308 (1958)
8.34 H.J. McSkimin, J. Acoust. Soc. Am., **33**, 12 (1961)
8.35 V.K. Raju, P.J. Reddy, J. Phys. E, **12**, 230 (1974)
8.36 G.G. Bentle, J. Am. Ceram. Soc., **49**, 125 (1966)
8.37 D.H. Chung, D.J. Silversmith, B.B. Chick, Rev. Sci. Instrum., **40**, 718 (1969)
8.38 H.J. McSkimin, P. Andreatch Jr., P. Glynn, J. Appl. Phys., **43**, 985 (1972)
8.39 J.B. Bateman, J. Appl. Phys., **33**, 3309 (1962)
8.40 C. Schaefer, L. Bergmann, Naturwissenchaften, **22**, 685 (1934)
8.41 R.S. Krishnan, Proc. Indian Acad. Sci., **A41**, 91 (1955)
8.42 G.N. Ramachandran, W.A. Wooster, Nature, **164**, 839 (1949)
8.43 G.N. Ramachandran, W.A. Wooster, Acta Crystallogr., **4**, 335, 431 (1951)
8.44 R. Grover, R.N. Keeler, F.J. Rogers, G.C. Kennedy, J. Phys. Chem. Solids, **30**, 2091 (1969)

8.45 S. Bhagavantam, J. Bhimasenachar, Proc. R. Soc. London, Ser A, **187**, 381 (1946)
8.46 R.S. Krishnan, V. Chandrasekharan, E.S. Rajagopal, Nature, **182**, 518 (1958)
8.47 *International Critical Tables*, Vol. 3, McGraw-Hill, New York (1927)
8.48 D.J. Bolef, J. Appl. Phys., **32**, 100 (1961)
8.49 G.C. Kennedy, quoted in [8.12]
8.50 S.N. Vaidya, G.C. Kennedy, J. Phys. Chem. Solids, **33**, 1377 (1972)
8.51 R.N Keeler, quoted in [8.12]
8.52 E. Gruneisen, *Handbuch der Physik*, Vol. 10, Springer, Berlin Heidelberg New York (1926)
8.53 R.W. Lynch, H.G. Drickamer, J. Chem. Phys., **45**, 1020 (1966)
8.54 K.A. Gschneidner Jr., Solid State Phys., **16**, 275 (1964)
8.55 D.B. Sirdeshmukh, K.G. Subhadra, J. Appl. Phys., **40**, 5404 (1970)
8.56 J. Mort, J. Appl. Phys., **38**, 3415 (1967)
8.57 S. Epstein, A.P. Brettervelle, Phys. Rev., **138**, 771 (1965)
8.58 A. Myers, Philos. Mag., **5**, 927 (1960)
8.59 J.F. Smith, Trans. AIME, **209**, 1212 (1957)
8.60 R. Well, A.W. Lawson, J. Chem. Phys., **37**, 2730 (1962)
8.61 F.F. Voronov, V.A.Gancherva, O.V. Stalgorova, T.A. Agapova, Sov. Phys. Solid State, **8**, 1313 (1966)
8.62 D.R. Stephens, E.M. Lilley, J. Appl. Phys., **39**, 177 (1968)
8.63 S.N. Vaidya, G.C. Kennedy, J. Phys. Chem. Solids, **32**, 951 (1971)
8.64 D.B. Sirdeshmukh, L. Sirdeshmukh, K.G. Subhadra, *Alkali Halides*, Springer, Berlin Heidelberg New York (2001)
8.65 R. Barsch, Z.P. Chang, Phys. Status Solidi, **19**, 139 (1967)
8.66 P.W. Bridgman, Proc. Am. Acad. Arts Sci., **67**, 345 (1932)
8.67 E.A. Perez-Albuerne, H.G. Drickamer, J. Chem. Phys., **43**, 1381 (1965)
8.68 H.E. Hite, K.J. Kearney, J. Appl. Phys., **38**, 5424 (1967)
8.69 S. Mathur, C.M. Kacchava, S.C. Saxena, Proc. Natl Inst. Sci. (India), **33**, 289 (1967)
8.70 R. Son, R.A. Bartels, J. Phys. Chem. Solids, **33**, 819 (1972)
8.71 M. Croft, A. Jayaraman, Solid State Commun., **29**, 9 (1980)
8.72 I. Vedel, A.M. Redon, J.M. Leger, J. Rossat-Mignot, B. Vogt, J. Phys. C, **19**, 6297 (1986)
8.73 A. Chatterjee, A.K. Singh, A. Jayaraman, Phys. Rev. B, **6**, 2258 (1972)
8.74 E. Kaldis, P. Wachter, Solid State Commun., **11**, 907 (1972)
8.75 T. Hailing, G.A. Saunders, H. Bach, Phys. Rev. B, **29**, 1848 (1984)
8.76 C.F. Cline, D.R. Stephens, J. Appl. Phys., **36**, 2869 (1965)
8.77 N. Soga, O.L. Anderson, J. Appl. Phys., **38**, 2985 (1967)
8.78 R.E. Fryxell, B.A. Chandler, J. Am. Ceram. Soc., **47**, 283 (1964)
8.79 D. Berlincourt, H. Jaffe, L.R. Shiozawe, Phys. Rev., **129**, 1009 (1963)
8.80 R. Stevenson, Can. J. Phys., **44**, 281 (1966)
8.81 S. Haussuhl, Phys. Status Solidi, **28**, 127 (1968)
8.82 N. Soga, J. Geophys. Res., **76**, 3983 (1971)

8.83 H. Wang, G. Simmons, J. Geophys. Res., **78**, 1262 (1973)

8.84 T.J. Arhens, D.L Anderson, A.E. Ringwood, Rev. of Geophys. Space Res., **7**, 667 (1969)

8.85 T.J. Arhens, T. Takahashi, G. Davies, J. Geophys. Res., **75**, 310 (1970)

8.86 P.W. Bridgman, Am. J. Sci., **10**, 483 (1925)

8.87 M.S. Doraiswamy, Proc. Indian Acad. Sci., **A25**, 413 (1947)

8.88 G. Simmons, A.W. England, Phys. Earth Planet. Sci., **2**, 69 (1969)

8.89 H. Mao, T. Takahashi, W.A. Basset, G.L. Kinsland, L. Merril, J. Geophys. Res., **79**, 1165 (1974)

8.90 P.W. Bridgman, Proc. Am. Acad. Arts Sci., **76**, 9 (1945)

8.91 S.N. Vaidya, G.C. Kennedy, J. Chem. Phys., **55**, 987 (1971)

8.92 G. Simmons, F. Birch, J. Appl. Phys., **34**, 2736 (1963)

8.93 A.A. Chudinov, Sov. Phys. Solid State, **4**, 553 (1962)

8.94 B. Houston, R.E. Strakna, H.S. Belson, J. Appl. Phys., **39**, 3913 (1968)

8.95 A.A. Chudinov, Sov. Phys. Crystallogr., **8**, 374 (1963)

8.96 S. Rabi, Phys. Rev., **167**, 801 (1968)

8.97 G. Lippmann, P. Kastener, W. Wanninger, Phys. Status Solidi A, **6**, k159 (1971)

8.98 J. Bhimasenachar, G. Venkataratnam, J. Acoust. Soc. Am., **27**, 922 (1955)

8.99 T.V. Ryzhova, K.S. Alexandrov, V.M. Korobhova, Izv. Earth Phys., **2**, 63 (1966)

8.100 H. Ozkan, L. Cartz, J.C. Jamieson, J. Appl. Phys., **45**, 556 (1974)

8.101 R.S. Adhav, J. Acoust. Soc. Am., **43**, 839 (1968)

8.102 D.H. Chung, Y. Li, Phys. Status Solidi A, **5**, 669 (1971)

8.103 B.W. James, Phys. Status Solidi A, **13**, 89 (1972)

8.104 J.M. Farley, G.A. Saunders, D.Y. Chung, J. Phys. C: Solid State, **6**, 2010 (1973)

8.105 L. Ramamurthy, P.J. Reddy, J. Phys. Chem. Solids, **28**, 2131 (1967)

8.106 D.P. Dandekar, A.L. Ruoff, J. Appl. Phys., **39**, 6004 (1968)

8.107 G.E. Stedman, D.J. Newman, J. Chem. Phys., **55**, 152 (1971)

8.108 E.H. Carlson, D.H. Current, C.L. Foils, J. Chem. Phys., **55**, 5831 (1971)

8.109 D.B. Sirdeshmukh, Acta Crystallogr., **A24**, 318 (1968)

8.110 R.V.G. Sundara Rao, Proc. Indian Acad. Sci., **A32**, 365 (1950)

8.111 M. Born, K. Huang, *Dynamical Theory of Crystal Lattices*, Clarendon, Oxford (1954)

8.112 J.F. Nye, *Physical Properties of Crystals*, Clarendon, Oxford (1957)

8.113 G.A. Alers, J.R. Neighbours, J. Appl. Phys., **28**, 1514 (1957)

8.114 W.J. Alton, A.J. Barlow, J. Appl. Phys., **38**, 3817 (1967)

8.115 L. Knopoff, Phys. Rev. A, **5**, 1445 (1965)

8.116 J. Waser, L. Pauling, J. Chem. Phys., **18**, 747 (1950)

8.117 D.H. Yean, J.R. Riter, J. Phys. Chem. Solids, **32**, 653 (1971)

8.118 N.G. Einspruch, L.T. Clairborne, J. Acoust. Soc. Am., **35**, 925 (1963)

8.119 B. Szigeti, Proc. R. Soc. London, **204**, 51 (1950)

8.120 O.L. Anderson, P. Glynn, J. Phys. Chem. Solids, **26**, 1961 (1965)

8.121 M.P. Tosi, Solid State Phys., **16**, 1 (1964)
8.122 G.C. Benson, P.I. Freeman, E. Dempsey, J. Am. Ceram. Soc., **46**, 43 (1963)
8.123 G.C. Benson, E. Dempsey, Proc. R. Soc. London, Ser A, **266**, 344 (1962)
8.124 K.S. Krishnan, S.K. Roy, Proc. R. Soc. London, **210**, 487 (1952)
8.125 D.B. Sirdeshmukh, K.G. Subhadra, Indian J. Pure Appl. Phys., **11**, 938 (1973)
8.126 K.G. Subhadra, D.B. Sirdeshmukh, J. Appl. Phys., **40**, 2357 (1969)
8.127 F.E. Pretzel, G.N. Rupert, C.L. Mader, E.K. Storms, G.V. Gritton, C.C. Rushing, J. Phys. Chem. Solids, **16**, 10 (1960)
8.128 S. Haussuhl, W. Skorezyk, Z. Kristallogr., **130**, 340 (1969)
8.129 R.W. Keyes, J. Appl. Phys., **33**, 3371 (1962)
8.130 S.S. Mitra, R. Marshall, J. Chem. Phys., **41**, 3158 (1964)
8.131 O.L. Anderson, J.E. Nafe, J. Geophys. Res., **70**, 3951 (1965)
8.132 D.L. Anderson, O.L. Anderson, J. Geophys. Res., **75**, 3494 (1970)
8.133 A. Jayaraman, B. Battlogg, R.G. Maines, H. Bach, Phys. Rev. B, **26**, 3347 (1982)
8.134 H. Neumann, Cryst. Res. Technol., **22**, 99 (1987)
8.135 D.B. Sirdeshmukh, K.G. Subhadra, J. Appl. Phys., **59**, 276 (1986)
8.136 D.B.Sirdeshmukh, J. Appl. Phys., **38**, 4083 (1967)
8.137 P. Deus, H.A. Schneider, Cryst. Res. Technol., **20**, 867 (1985)
8.138 R.R. Reddy, Y.N. Ahammed, K. Ramagopal, D.V. Raghuram, Indian J. Pure Appl. Phys., **37**, 25 (1999)
8.139 D.B. Sirdeshmukh, K.G. Subhadra, J. Appl. Phys., **46**, 3681 (1975)
8.140 T.G. Worlton, L. Cartz, A. Niravath, H. Ozkan, High Temp. High Press., **4**, 463 (1972)
8.141 H. Ozkan, L. Cartz, AIP Conf. Proc., **17**, 21 (1974)
8.142 O.L. Anderson, E. Schreiber, R.C. Leibermann, Rev. Geophys., **6**, 491 (1968)
8.143 O.L. Anderson, J. Phys. Chem. Solids, **24**, 909 (1963)
8.144 J.M. Farley, G.A. Saunders, D.Y. Chung, J. Phys. C, **8**, 780 (1975)

Index

Springer Series in

MATERIALS SCIENCE

Editors: R. Hull R. M. Osgood, Jr. J. Parisi H. Warlimont

20 **Microcluster Physics**
By S. Sugano and H. Koizumi
2nd Edition

21 **The Metal-Hydrogen System**
By Y. Fukai 2nd Edition

22 **Ion Implantation in Diamond, Graphite and Related Materials**
By M. S. Dresselhaus and R. Kalish

23 **The Real Structure of High-T_c Superconductors**
Editor: V. Sh. Shekhtman

24 **Metal Impurities in Silicon-Device Fabrication**
By K. Graff 2nd Edition

25 **Optical Properties of Metal Clusters**
By U. Kreibig and M. Vollmer

26 **Gas Source Molecular Beam Epitaxy**
Growth and Properties of Phosphorus Containing III–V Heterostructures
By M. B. Panish and H. Temkin

27 **Physics of New Materials**
Editor: F. E. Fujita 2nd Edition

28 **Laser Ablation**
Principles and Applications
Editor: J. C. Miller

29 **Elements of Rapid Solidification**
Fundamentals and Applications
Editor: M. A. Otooni

30 **Process Technology for Semiconductor Lasers**
Crystal Growth and Microprocesses
By K. Iga and S. Kinoshita

31 **Nanostructures and Quantum Effects**
By H. Sakaki and H. Noge

32 **Nitride Semiconductors and Devices**
By H. Morkoç

33 **Supercarbon**
Synthesis, Properties and Applications
Editors: S. Yoshimura and R. P. H. Chang

34 **Computational Materials Design**
Editor: T. Saito

35 **Macromolecular Science and Engineering**
New Aspects
Editor: Y. Tanabe

36 **Ceramics**
Mechanical Properties, Failure Behaviour, Materials Selection
By D. Munz and T. Fett

37 **Technology and Applications of Amorphous Silicon**
Editor: R. A. Street

38 **Fullerene Polymers and Fullerene Polymer Composites**
Editors: P. C. Eklund and A. M. Rao

39 **Semiconducting Silicides**
Editor: V. E. Borisenko

40 **Reference Materials in Analytical Chemistry**
A Guide for Selection and Use
Editor: A. Zschunke

41 **Organic Electronic Materials**
Conjugated Polymers and Low Molecular Weight Organic Solids
Editors: R. Farchioni and G. Grosso

42 **Raman Scattering in Materials Science**
Editors: W. H. Weber and R. Merlin

43 **The Atomistic Nature of Crystal Growth**
By B. Mutaftschiev

44 **Thermodynamic Basis of Crystal Growth**
P–T–X Phase Equilibrium and Non-Stoichiometry
By J. Greenberg

45 **Thermoelectrics**
Basic Principles and New Materials Developments
By G. S. Nolas, J. Sharp, and H. J. Goldsmid

46 **Fundamental Aspects of Silicon Oxidation**
Editor: Y. J. Chabal

Springer Series in
MATERIALS SCIENCE

Editors: R. Hull R. M. Osgood, Jr. J. Parisi H. Warlimont

49 **Alkali Halides**
A Handbook of Physical Properties
By D. B. Sirdeshmukh, L. Sirdeshmukh, and K. G. Subhadra

50 **High-Resolution Imaging and Spectrometry of Materials**
Editors: F. Ernst and M. Rühle

51 **Point Defects in Semiconductors and Insulators**
Determination of Atomic and Electronic Structure from Paramagnetic Hyperfine Interactions
By J.-M. Spaeth and H. Overhof

52 **Polymer Films with Embedded Metal Nanoparticles**
By A. Heilmann

53 **Nanocrystalline Ceramics**
Synthesis and Structure
By M. Winterer

54 **Electronic Structure and Magnetism of Complex Materials**
Editors: D.J. Singh and D. A. Papaconstantopoulos

55 **Quasicrystals**
An Introduction to Structure, Physical Properties and Applications
Editors: J.-B. Suck, M. Schreiber, and P. Häussler

56 **SiO_2 in Si Microdevices**
By M. Itsumi

57 **Radiation Effects in Advanced Semiconductor Materials and Devices**
By C. Claeys and E. Simoen

58 **Functional Thin Films and Functional Materials**
New Concepts and Technologies
Editor: D. Shi

59 **Dielectric Properties of Porous Media**
By S.O. Gladkov

60 **Organic Photovoltaics**
Concepts and Realization
Editors: C. Brabec, V. Dyakonov, J. Parisi and N. Sariciftci

61 **Fatigue in Ferroelectric Ceramics and Related Issues**
By D.C. Lupascu

62 **Epitaxy**
Physical Principles and Technical Implementation
By M.A. Herman, W. Richter, and H. Sitter

63 **Fundamentals of Ion-Irradiated Polymers**
By D. Fink

64 **Morphology Control of Materials and Nanoparticles**
Advanced Materials Processing and Characterization
Editors: Y. Waseda and A. Muramatsu

65 **Transport Processes in Ion-Irradiated Polymers**
By D. Fink

66 **Multiphased Ceramic Materials**
Processing and Potential
Editors: W.-H. Tuan and J.-K. Guo

67 **Nondestructive Materials Characterization**
With Applications to Aerospace Materials
Editors: N.G.H. Meyendorf, P.B. Nagy, and S.I. Rokhlin

68 **Diffraction Analysis of the Microstructure of Materials**
Editors: E.J. Mittemeijer and P. Scardi

69 **Chemical–Mechanical Planarization of Semiconductor Materials**
Editor: M.R. Oliver

70 **Applications of the Isotopic Effect in Solids**
By V.G. Plekhanov